Electromagnetic Modeling and Simulation

IEEE Press
445 Hoes Lane
Piscataway, NJ 08854

Electromagnetic Modeling and Simulation

Levent Sevgi

IEEE Antennas and Propagation Society, *Sponsor*

The IEEE Press Series on Electromagnetic Wave Theory
Andreas C. Cangellaris, *Series Editor*

IEEE PRESS

WILEY

Library of Congress Cataloging-in-Publication Data:

Sevgi, Levent.
Electromagnetic modeling and simulation / Levent Sevgi.
pages cm
Includes bibliographical references and index.
ISBN 978-1-118-71618-2 (hardback)
1. Electromagnetism–Computer simulation. I. Title.
QC760.54.S48 2014
621.380285'53–dc23

2013047870

Printed in the United States of America

10 9 8 7 6 5 4 3 2 1

Yes, You Can!

In Memory of

Leopold B. Felsen

(May 7, 1924–September 24, 2005)

Contents

Preface

Today, addressing the technical challenges posed by system complexities requires a broad range of innovative, multidisciplinary, physics-based, problem-matched analytical and computational skills that are not adequately covered in conventional electrical-electronics (EE) engineering curricula. A great many higher educational institutions are now actively engaged in efforts to define "what makes a modern engineer" and to design curricula for teaching the necessary skills to a computer-weaned generation of students, with access to the internet and consequent globalization of information. The explosive growth of computer capabilities has revolutionized communication and the analysis of complex systems, and has made interdisciplinary exposure necessary in modern EE engineering. Physics-based modeling, observation-based parameterization, computer-based simulations, and code calibration against canonical problems (i.e., problems that have mathematical exactness and numerically computable forms) are the key issues of these challenges.

As phrased by Einstein, *"in the matter of physics, first lessons should contain nothing but what is experimental and interesting to see—experimentation and hands-on training are the key issues in engineering education at least at undergraduate level."* On the other hand, with the development of new computer technologies, interactive multimedia programming languages, and the Internet, it is now possible to simulate engineering and science laboratory projects of all sorts on a computer all around the world. Experiment-oriented problems can be offered without the overhead incurred when maintaining a full laboratory. At this point the question arises: should an intelligent balance be established between real and virtual experimentations and how? Another similar problem is the balance to be maintained between teaching essentials (theory) and cranking the gear (blind computer applications). It is a general observation that the motto "*I did it, it works*" is widespread among the youth of today, without really grasping the general principles and boundaries of validity of the underlying phenomena and what is even worse with a false sense of satisfaction.

Engineering, as given in the American Society for Engineering Education web site (www.asee.org), is "*the art of applying scientific and mathematical principles, experience, judgment, and common sense to make things that benefit people*." That is, it is the process of producing a technical product or system to meet a specific need in a society. Engineering education is a university education, where knowledge of *mathematics* and *natural sciences* are gained, followed up by lifetime self-education where *experience* is piled up with *practice*. Therefore, the four keywords mathematics, physics, experience, and practice are the untouchables of engineering education.

Many applications in science and technology rely increasingly on field theory and circuit theory computations in either man-made or natural complex structures. Wireless communication systems, for example, pose challenging problems with respect to field propagation prediction, microwave hardware design, compatibility issues, biological hazards, and so on. Nanotechnologies, on the other hand, have challenges of locating multimillion circuit elements and subsystems on a few square centimeter chips, with very low emissions and immune to environmental interference. Moreover, need to and use of these theories are not limited to EE applications only; they are exploited in a very wide spectrum ranging from biomedical to geophysical applications. Since different problems have their own combination of geometrical features and scales, frequency ranges, material properties, and so on, no single method or approach is best suited for handling all possible cases; instead, a combination of methods, "hybridization," is needed to attain the greatest flexibility and efficiency in engineering. Relations between field theory and network (circuit) theory play an important role in this respect.

The necessity for hybrid methods has already been recognized in the past: for example, in scattering and antenna problems, techniques have been devised that combine the method of moments (MoM) and the geometrical theory of diffraction (GTD) or physical theory of diffraction (PTD). Similarly, numerical methods such as finite elements (FE) or finite differences (FD) have been considered in conjunction with MoM, with integral equations, with boundary integrals, with modal techniques, with multipole methods, and so on. Combinations of other methods, for example, boundary contour and mode matching or hybrid electric field integral equations (EFIE) and magnetic field integral equations (MFIE) denoted as HEM, have also been proposed.

Physics-based modeling and observable-based parameterization are very important in EE engineering education. The models that are established via well-known Maxwell equations (field theory) and transmission line equations (circuit theory) in both time and frequency domains parameterize a complex physical problem well defined that guarantees existence, uniqueness, and convergence. Field and circuit theories are dual; that is, any field problem (e.g., antenna radiation) can be transformed into a circuit theory problem and solved there (or vice versa). Starting after World War II, circuit formulations of field problems have also been employed extensively in the design of microwave, optical, and other closed and open *waveguiding* and *radiating* systems.

Another very important occurrence of hybridization is the increase in integrated circuit (IC) performance being exponential in time at rates of more than 100/decade, with the critical device dimensions shrinking and the interconnects between devices becoming smaller and more closely spaced, *interconnect delays (ID)* started to dominate over *gate delays (GD)*; the ratio GD/ID of the order of 7–8 in favor of interconnects in early nineties is expected to be of the order of nearly 1/20 in favor of gates within a couple of years. As the count of active devices exceeds several tens of millions and the number of interconnects among these devices grows superlinearly with this count, efficient evaluation of time delays and signal integrity becomes more difficult and important. Devices with operating frequencies exceeding a hundred gigahertz have already appeared and today's circuits contain millions of transistors per unit area as opposed to 1970s SPICE targeted for circuits with a few hundred transistors. Hence the need arises for a new generation of simulators with improved numerical methods using, if possible, analytic solution techniques to handle very large circuits.

Engineering as defined above is based on *practice*. The minima of this practice should be given during the EE education. This has become more and more comprehensive and expensive parallel to high-technology devices developed and presented to societies: computers and other microprocessor-based devices make EE engineering education not only very complex but also interdisciplinary as well. The cost of building undergraduate labs in EE may vary from 1 unit to 10^5 units;

for example, a spectrum or a network analyzer may cost few 10^4 units, whereas a simple software of 1 unit with or without the addition of specific cards costing 10^2 units may turn a regular personal computer (PC) into a virtual lab. The key question therefore is to establish a balance between virtual and real labs, so as to optimize cost problems, while graduating sophisticated engineers with enough practice.

Doing numerical simulations in EE engineering has become as easy (as well as difficult) as doing measurements. It is easy because one can purchase commercial codes that do almost everything, such as supplying computer-controlled devices for measurements. The simulation packages are user friendly, have self-checking routines for control, and all can be calibrated, like most of high-tech measurement devices. On the other hand all the efforts of simulation can be in vain if one does not know how to interpret the resulting numbers. In addition, they are capable of doing only what has already been planned and included by the developer. Moreover, important concepts such as *accuracy*, *precision*, and *resolution*, in short the *underlying theory*, should be well understood by engineers.

This book aims to introduce simple, easy-to-use, but effective short codes as well as virtual tools that can be used in broad range of EE engineering lectures. The book itself may serve as a textbook for several lectures such as electromagnetic modeling and simulation, computational electromagnetics, transmission line theory, guided wave theory, diffraction theory, and others. Almost all of the virtual tools are coded in MATLAB; therefore, the reader is strongly advised to get used to working with MATLAB. The book contains 16 chapters. Roughly speaking, the first five chapters are introductory, the next five chapters are for analytical modeling, and the last six chapters are for numerical modeling and simulation.

People had to simplify problems as much as possible a century ago in order to get the *feeling* on the results. This is why we have had excellent canonical problems with simple analytical models. Today, we very often revisit these problems for (i) teaching electromagnetics and (ii) validation, verification, and calibration of numerical models. That is why we included many canonical problems in this book (although not very often, we also use numerical models in validation and verification of analytical models).

The first chapter is a kind of introduction which contains some fundamental modeling and simulation concepts. Validation, verification, and calibration are briefly reviewed. Core models are given and model selection criteria are discussed. A novel electromagnetic modeling and simulation course outline, content, and flow are proposed. Then, two-level electromagnetic guided waves lecture is discussed and a highly attractive virtual tools that can be used in these lectures are reviewed. Finally, virtual tools for some advanced topics are outlined.

The second chapter presents fundamental concepts of engineering: measurement and calculation. Engineers speak with numbers therefore the true meaning of the numbers we use shall clearly be understood. Terms such as accuracy, precision, resolution, error, and uncertainty are reviewed in this chapter. Error analysis and propagation of error are given. The use of statistics in engineering, basic statistical evaluations, confidence level, and hypothesis testing (decision making) are outlined.

Some numerical analysis issues are revisited in Chapter 3. Root search of a nonlinear function, numerical integration, and differentiation, and solving systems of linear equations are reviewed and simple MATLAB scripts are introduced. Taylor's expansion of a function is also discussed.

Chapter 4 is reserved for Fourier transformation and Fourier series representation of a function. Fourier transform has been widely used in circuit analysis and synthesis, from filter design to signal processing, image reconstruction, and so on. Fourier transform is also used in electromagnetics from antenna analysis to imaging and nondestructive measurements, even in propagation problems. In order to numerically compute the Fourier transform on a computer, discretization and numerical integration are required. This is an approximation of the mathematically defined Fourier transform in a synthetic (digital) environment; therefore, the reader should be aware of the problems related to this approximation. The true meaning of the Fourier transform and Fourier series representations, their mathematical definitions, and computer implementations are discussed in detail. Effects of discretization and finite nature of discrete Fourier transformation (DFT) and fast Fourier transformation (FFT) are shown with several examples prepared with simple MATLAB scripts.

Ramdomness in electromagnetics is briefly discussed in Chapter 5 together with some statistical discussions. Stochastic modeling of communication and/or radar signals is outlined. Computer generation of useful signal, noise, clutter, and interference is presented.

Chapter 6 is reserved for some basic electromagnetic terms, concept, identities, and definitions. Coordinate systems mostly used in electromagnetic, point/line source concepts, transverse electric and magnetic models in guided wave representations, and electromagnetic plane waves–transmission line analogy are presented.

Sturm–Liouville equation represents both eigenvalue (source-free) and Green's function (source-driven) electromagnetic problems. This is briefly discussed in Chapter 7. 1D eigenvalue and Green's function representations for both finite and infinite regions are given. The connection between these two problems established via the completeness and orthonormality relations is outlined.

A canonical two-dimensional (2D) wave propagation problem is discussed in Chapter 8. Many natural or man-made guiding environments are characterized by physical parameters that render the wave equation nonseparable in any of the standard coordinate systems. Propagation inside 2D parallel perfectly electrical conductor (PEC) plates is used to review ray optical and modal solutions which represent source-driven and source-free wave problems. The 2D Helmholtz wave equation is reviewed. 1D and 2D spectral representations in terms of a Green's function problem and eigenfunction (normal mode) summation are summarized accordingly. The classical ray–mode interchange approach and their hybridization to home in best features of both rays and modes are also discussed. The MATLAB-based virtual tools RAYMODE and HYBRID included in this chapter can be used to visualize individual contributions of rays and modes which are very important for the reader in order to get a physical insight to this problem.

In the absence of transverse–longitudinal separability, it is not possible to define discrete or continuous normal modes that individually satisfy the transverse boundary conditions and propagate longitudinally without coupling to other modes. When transverse–longitudinal separability is only weakly perturbed, one may define local (adiabatic) modes which adapt smoothly, without intermode coupling, to the slowly changing conditions. Adiabatic modes fail in cutoff regions and can be uniformized there by intrinsic modes, which are synthesized by a spectral continuum of adiabatic modes. These concepts can be elucidated and validated by investigating the wave dynamics in a simple test environment: A wedge waveguide with nonpenetrable boundaries. This canonical problem is discussed in Chapter 9.

High frequency asymptotics (HFA) are presented in Chapter 10. HFA approaches have long been used in electromagnetics. HFA methods include geometrical optics (GO), geometrical theory of diffraction (GTD), its uniform extension uniform theory of diffraction (UTD), physical optics (PO), physical theory of diffraction (PTD), elementary edge waves (EEWs), and parabolic equation (PE) methods. The GO, GTD, and UTD are *ray-based* diffraction methods; PO and PTD are *induced-source-based* methods. HFA techniques are reviewed through a classical, canonical problem: EM wave scattering from a wedge shaped object with PEC boundaries. A MATLAB-based virtual diffraction tool WedgeGUI, using analytical exact as well as HFA techniques, is also introduced. Both line source and plane wave illuminations are analyzed. Effects of various parameters on the reflection, refraction, and diffraction are investigated. Comparisons among GO, GTD/UTD, PO/PTD, and PE models through many scenarios are presented.

Chapter 11 deals with antenna arrays. Arrays of isotropic radiators are taken into account and boresight/end-fire beam forming, beam steering are discussed on linear, circular, and planar arrays of isotropic radiators. A simple MATLAB-based virtual tool ARRAY is prepared and introduced for the students and lecturers who think that 2D and 3D visualizations of array patterns could be very effective in learning/teaching antenna arrays. The ground effects on the vertical radiation patterns are also investigated via a simple MATLAB script.

Chapter 12 presents another interesting MATLAB virtual tool which explores guided wave propagation modeling using the *Snell law*. Electromagnetic guiding occurs along regions which are physically bounded transversely and/or where waves are confined due to the medium parameter variations. For example, ground waves propagate through surface and/or elevated ducts formed by the refractivity variations of the atmosphere. The MATLAB-based ray shooting virtual tool SNELL may be used to explore transverse ray confinement due to linear, bilinear, and trilinear refractivity variations. It is amazing to see how a simple equation—Snell law—can numerically be used to visualize all the ray behaviors under different refractivity conditions. The formation of various surface and or elevated ducts may be observed only by changing the refractivity profiles.

Chapter 13 introduces method of moments (MoM): one of the earliest numerical models. The model is based on segmentation of the object/environment under investigation. Surface currents caused by any specified electromagnetic source are obtained by using Green's function solution of the problem. Then, the scattered field caused by each segment is calculated and total scattered field is obtained by the application of the superposition principle over all segments. The method is first explained through a simple electrostatic application: parallel plate capacitor modeling. Then, propagation over nonpenetrable flat earth is discussed and MoM results are compared with analytical reference data. Scattering from infinitely long PEC cylinders with arbitrary cross section is also modeled with MoM. MoM modeling of a few complex radiation and scattering problems are also discussed. Finally, wedge diffraction and propagation inside a wedge waveguide are modeled using MoM.

Chapter 14 presents finite-difference time-domain (FDTD) method. FDTD method is first used to represent plane waves and discretization effects, stability, numerical dispersion, source-injection, boundary implementation are all reviewed. Then, 1D FDTD is used to model transmission lines. Starting with the fundamental TL theory, basic concepts such as characteristic impedance, voltage and current reflections under different terminations, and standing wave formation are reviewed. The generic time-domain TL equations are discussed using the analogy between plane waves and transmission lines. An effective MATLAB-based virtual tool TDRMETER is introduced to numerically test time-domain reflectometers. Time-domain reflectometers have long been used in EE engineering; initially by the power and telephone engineers to find out SC or OC problems along hundreds of kilometers long transmission lines without doing physical search. Fault location and identification using Fourier and Laplace transformations are also included. The FDTD problem is also discussed through second-order decoupled 1D wave equations. Then, 2D FDTD modeling is discussed through three different virtual tools. First, MGL-2D is presented where broad range of 2D scenarios can be created by the user and electromagnetic waves can be visualized. It includes modeling of TE and TM problems. Any penetrable and/or nonpenetrable rectangular, circular, or triangular material can easily be located inside the 2D computation space and electromagnetic scattering effects can be visualized. Perfectly reflecting or absorbing termination can be specified at the four edges of the space. Both narrowband and broadband (pulsed) wave propagation can be taken into account. Finally, the WedgeFDTD virtual tool is given for the observation of wedge-diffracted waves and FDTD versus HFA comparisons.

Chapter 15 presents parabolic equation (PE) modeling. The PE model has long been used in propagation modeling and can handle boundary irregularities as well as medium variations. Both split-step parabolic equation (SSPE) and finite-element method (FEM) implementation are introduced. Simple MATLAB scripts for SSPE and FEMPE solutions of both narrow- and wide-angle models are given. These models are validated against analytical model for a 2D surface duct problem where ducting is controlled by a vertically decreasing linear refractivity profile. This problem has an analytical exact mode summation solution in terms of Airy functions. Then, a 2D groundwave propagator GrSSPE is presented. GrSSPE developed in a way to let the user draw his/her own longitudinal terrain profile by just clicking the computer mouse and supply various input parameters. The rest is done automatically by the virtual tool. The output is a 3D color plot representing field strength versus range/height. The formation of various surface and or elevated ducts may be observed by changing the parameters and terrain profiles. Chapter 14 also discusses analytical and numerical modeling of the 2D dielectric slab waveguide (i.e., optical film). The exact wave solutions can be derived from the 2D wave equation in terms of mode summation. Mathematical representations of the modes can be extracted directly or through a simplification of even- and odd-symmetry applications. An interesting MATLAB-based virtual tool DiSLAB is developed which compares mathematical solutions against the split-step parabolic equation results. The reader may also enjoy extraction of eigenvalues from propagation characteristics inside the slab waveguide by using the longitudinal correlation function and its Fourier transform.

Chapter 16 revisits the 2D parallel plate nonpenetrable waveguide and analytical and numerical models are compared. Tables of stand alone codes for mode summation, ray summation, image method, SSPE, MoM, and FDTD are given.

Appendix A presents introductory material for MATLAB use. Appendix B lists some useful books for both electromagnetics and computational electromagnetics. Most of these books have become classical, e.g., the book by Felsen and Marcuvitz, by Don Dudley, R. E. Collin, J. D. Kraus, R. F. Harrington, and by W. L. Stutzman, G. A. Thiele should be among the books the reader have in their library.

Finally, Appendix C lists tutorials for both fundamental issues and electromagnetic virtual tools that are discussed in this book. The reader is referred to these papers for the complete references of each virtual tool.

All the virtual tools are gathered and uploaded in the publisher's website. Sample scenarios and canonical comparisons are included. Moreover, some nice video clips recorded during characteristic simulations are also presented.

Note that MATLAB scripts are mostly prepared with the classical coding approach in order to make them compact and easily traceable. On the other hand, MATLAB-based virtual tools use all advanced level coding approaches and commands. Note also that this book primarily is neither about fundamental EM theory nor about numerical models included (FDTD, MoM, and SSPE). There are excellent books which has become classic and are listed in Appendix B. This book is primarily about EM modeling and simulation with the emphasis on the modeling and simulation side. It is a "how to" or "let's do it!" type book. It touches upon wide range of problems and discusses several canonical scenarios. Because of these, some of the chapters, on purpose, are reviewed briefly with one or two short MATLAB scripts, while others are discussed in detail using several scripts as well as virtual tools. Each chapter is pretty much stand alone.

LEVENT SEVGI

About the Author

Levent Sevgi

Born in Akhisar, Turkey, in 1958, Dr. Sevgi graduated from the Electronics and Communications Engineering Department of Istanbul Technical University (ITU) and completed his PhD studies with Professor Leopold B. Felsen at the Weber Research Institute/Polytechnic University in New York.

He has been with several institutions for the last two decades: Istanbul Technical University (1991–1998), the Scientific and Technological Council of Turkey—Marmara Research Institute (1999–2000), as the chair of Electronics Systems Department, with Weber Research Institute/Polytechnic University (1988–1990), the Scientific Research Group of Raytheon Systems, Canada (1998–1999), during Canadian East Coast Integrated Surveillance System trials, the Center for Defense Studies, ITUV-SAM (1993–1998) during large scale system designs for the Turkish Navy and (2000–2002) during Maritime Vessel Traffic Monitoring System design for Turkish Straits, and the Department of Electrical and Computer Engineering, University of Massachusetts, Lowell (2012–2013), for his sabbatical term. Since February 2001, he has been with Doğuş University, Istanbul.

He has been involved with complex electromagnetic problems and systems for more than two decades. His research study has focused on analytical and numerical methods in electromagnetics, high frequency asymptotic (HFA) approaches, FDTD, TLM, FEM, SSPE, and MoM techniques and their applications, propagation in complex environments, diffraction modeling, microwave circuit design, EMC/EMI modeling and measurement, sensors and integrated surveillance systems, surface wave HF radars, RCS modeling, and bioelectromagnetics. He is also interested in novel approaches in engineering education, teaching electromagnetics via virtual tools, and popular science topics such as science, technology and society, and public understanding of science.

He is a fellow of the IEEE, associate editor of the IEEE *Antennas and Propagation* magazine, the writer/editor of the "Testing Ourselves" Column, and member of the IEEE Antennas and Propagation Society AdCom (2013–2015) and Education Committee.

Acknowledgments

Most of the materials used in this book have come or derived from studies carried out under my direction and/or supervision, and some in collaboration with Professor Leopold B. Felsen, who passed away in 2005 (right after receiving an honorary doctorate degree from Doğuş University of Istanbul), and with Professor Pyotr Ya Ufimtsev. We had the chance and honor of hosting them at Doğuş on different occasions. Almost all the virtual tools presented in this book were first presented to the scientific society and were published in prestigious journals and magazines as either education papers or as feature articles.

I express my gratitude to my students who had worked in preparation of these virtual tool codes during their undergraduate, graduate, and PhD theses.

Among others, I would like to thank Funda Akleman, Gonca Çakır, Çağatay Uluışık, Mustafa Çakır, and Feray Hacıvelioglu for their valuable contributions in the preparation of some of these virtual tools. I would like to thank Gokhan Apaydın and M. Alper Uslu, in particular, not only for their contributions in the preparation of some of these tools but also for their direct contributions to this book.

L.S.

CHAPTER 1

Introduction to MODSIM

Today's technical challenges posed by system complexities require a range of multidisciplinary, physics-based, problem-matched analytical and computational skills [1]. In electrical engineering (EE), real-life systems (from nanoscale to kilometer-wide) are among the most complex ones and almost totally involved in electromagnetics (EMs). EM theory is well established with Maxwell's equations but teaching/lecturing is always a challenge. Experimentation and hands-on measurement are the fundamentals of EM engineering education; however, strong theoretical background and numerical simulations are also essential. An intelligent approach is to use physics-based modeling, hands-on training, numerical-based modeling, and computer simulations all together.

Computer simulations, either in-house prepared or commercially available, have been effectively used in EMs. Key issues are model validation, code verification, and calibration (VV&C) and physical interpretation of the numbers obtained. The four critical words are *mathematics*, *physics*, *experience*, and *practice* [1]. A good EM modeling and simulation (EM-MODSIM) course should cover them all. Note that despite a correctly presented physical model, numerical simulation of the model contains errors caused by the numerical method itself, simplification of the physical structure, assumptions made there,

---

Electromagnetic Modeling and Simulation, First Edition. Levent Sevgi.

machine computation limitations, and so forth. It is a challenge to establish a confidence in the results of numerical simulations.

Numerical models can be viewed differently from the developer's and the user's perspectives [2]. Developers, in addition to their concern about accuracy, cost/competition, and user friendly graphical user interface (GUI) design, deal primarily with the conceptual suitability and implementational steps of the code (verification), while the users are more involved in computation and application. Both are concerned with validation, although users are often tempted to apply code in a manner that it is not designed for, thus making validation an especially a sensitive topic. From the developer's perspective, the process of developing a numerical model involves conceptualization, formulation, numerical implementation, computation, and validation. On the other hand, the user's perspective involves the problem, that is, choosing a problem-matched approach and the application steps. A developer needs to know how the code works, but the user needs to know what it can do.

The fair metrics in characterization and comparison of numerical simulators are accuracy, reliability, efficiency, and, finally, applicability. Someone who deals with numerical simulation is in a quite similar position to that of an experimenter. They both need to understand requirements for their particular problem in terms of basic EM phenomena, but both also need to depend on complex tools that they did not design in order to accomplish their particular goals.

1.1 MODELS AND MODELING

When not measuring or soldering, engineers continuously deal with models when analyzing, designing, and implementing [1–3]. The two phases in modeling are *utilization* and *creation*. The utilization is common practice and does not necessitate further comment other than "make the right choice and use the model carefully." The creation phase requires delicate measurements, extreme perception, and/or excellent imagination.

Maxwell's well-known equations establish the physics of EE, well define the interaction of electromagnetic waves with matter, and form the basis for a real understanding of EE problems and their solutions. Moreover, circuit theory equations are also derived from Maxwell's

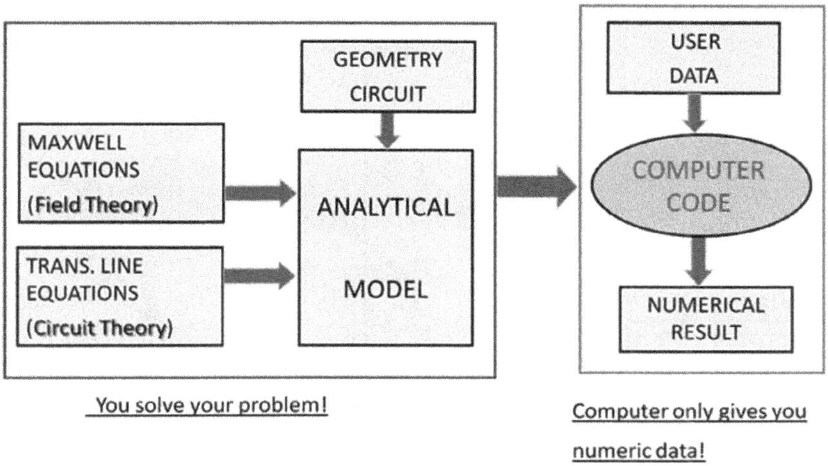

Figure 1.1. Analytical-based modeling.

equations. There are two different solution approaches: analytical formulations and direct numerical simulation methods. Analytical and numerical model-based approaches are schematized in Fig. 1.1 and Fig. 1.2, respectively. The key difference between these two is the placement of problem geometry (i.e., boundary conditions [BCs]). Analytical-based approaches are problem (geometry) dependent. Once geometry changes, the problem has to be resolved. On the other hand, numerical-based approaches are problem (geometry) independent.

The model is derived from Maxwell's equations under a given problem geometry (i.e., for a given boundary conditions and medium parameters) for the analytical model-based approach. These models express solutions for independent variables, such as electric and magnetic field components or input–output voltages and currents, in terms of analytic functions (such as sine/cosine functions, Bessel/Hankel series, etc.). A computer program is required only to calculate an output value for a given input supplied by the user.

On the other hand, the principal algorithm models the intrinsic behavior of fields/circuits without reference to specific boundary and material configurations. Some well-known and widely used numerical approaches are also listed in the figure. The generic *numerical model*

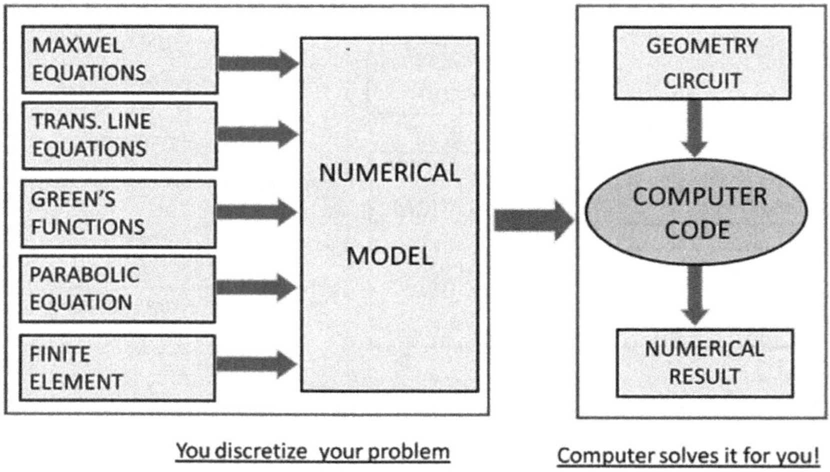

Figure 1.2. Numerical-based modeling.

is applied from the *very beginning* and is augmented by boundary simulators and/or other peripheral units, such as near-field far-field transformations. Different problems (with respect to geometry and medium parameters) can be accommodated using such models.

Whether analytical or numerical, models need to be coded for calculations on a computer. While the model used in analytical solutions is constructed according to the geometry of the problem (i.e., boundary conditions and medium parameters), the numerical model is general and the geometry of the problem (together with the input parameters) is supplied after the model is built. That is, the boundary and/or initial conditions are supplied externally to the numerical model together with the medium parameters, operating frequency, signal bandwidth, and so forth. Once they are specified, simulations are run and sets of observable-based output parameters are computed for a given set of input parameters.

So the challenging question for an engineer becomes "which model to use, when?" No easy answer exist other than "experience." One may wonder, "why not use the most sophisticated one?" The simple reply is that, to use a more complex model than needed may consume huge amount of computation time and, more importantly, obscure the insight to the problem.

1.2 VALIDATION, VERIFICATION, AND CALIBRATION

Engineers, when analyzing, designing, calculating, simulating, and so on, always use models. The crucial point here is to choose the right model for the physical problem under consideration. The same physical problem may be assigned different models, with increasing order of complexity, depending on the conditions.

EM-MODSIM requires high performance computing, applied mathematics and physics, intelligent systems, and information technologies. It is a relatively novel and cost-effective way of solving complex EM problems beyond the reach of analytical methods, and the outcome is powerful software packages and virtual tools for the students, lecturers, researchers, and scientists.

EM-MODSIM requires basic understanding of fundamental concepts such as modeling, analytical solution, numerical solution, analytical- and numerical-based modeling, simulation, model validation, code verification through canonical tests/comparisons, accreditation, and so on (see Fig. 1.3 for the fundamental building blocks and interrelations [4]).

EM MODSIM is extremely valuable in engineering if based on physics-based modeling and observable-based parameterization. It starts with the definition of a real-life problem. Its conceptual model is the basic theory behind the real-world problem. Maxwell's equations establish the mathematical model for field theory, define the interaction of EM waves with matter, and form the foundation for a real understanding of electrical problems and their solutions. All of the frequency and time-domain methods use either differential or integral form of Maxwell's equations.

A generic chart of EM-MODSIM is pictured in Fig. 1.4. A real-world engineering problem is first represented by a conceptual (mathematical or not) model and then is implemented in the discrete world with a computer code. The code may be augmented some numerical analysis methods, such as root search, numerical differentiation or

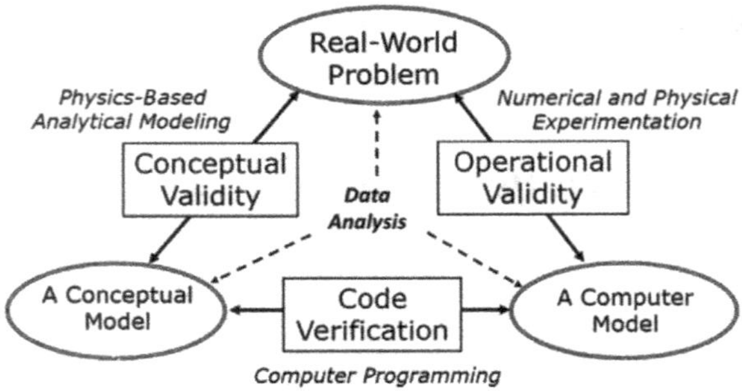

Figure 1.3. Fundamental blocks of numerical MODSIM.

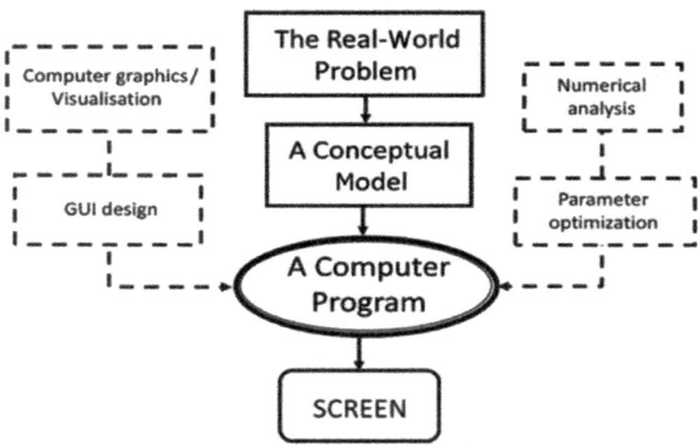

Figure 1.4. A generic chart of EM-MODSIM.

integration, solving systems of equations with a huge number of unknowns, etc., and some optimization algorithms. Visualization capabilities may also be inserted and a GUI can be designed.

EM scattering problems may be categorized under three groups: (1) antenna and radiation problems, (2) scattering and radar cross-section (RCS) problems, and (3) guided wave problems. All these are represented by Maxwell's equations with proper BCs. The solution then exists and is unique.

Antenna and radiation problems can better be handled via auxiliary vector potential functions since these functions perfectly matched with the excitation. In scattering and RCS problems, one needs to separate total EM fields into incident and scattered components. The word scattering includes reflections, refractions, and diffractions. The first two are primary effects and diffraction usually can be neglected in the presence of former two. In shadow regions, only diffracted fields contribute. Therefore, edges and/or tips necessitate special treatment. Finally, guided wave problems should be decomposed into transverse and longitudinal components prior to searching a solution. In this context, Sturm–Liouville equation and interrelations plotted in fig. 7.3 of Chapter 7 must be well understood.

Model calibration is performed through a VV&C procedure. This may be achieved via the following:

- tests and comparisons against analytical reference data
- tests and comparisons between numerical solution
- real-world measurements.

all of which are extremely challenging. Analytical exact solution does not necessarily mean analytic reference data; it necessitates accurate numerical computations. Comparisons among different numerical models may yield almost perfect agreement, but they may all be incorrect if used beyond their range of validity.

1.3 AVAILABLE CORE MODELS

Well-known and widely used EM-MODSIM models are FEM, MoM, PE, FDTD, and TLM. An EM-MODSIM course should touch upon all.

Finite element model (FEM) [5] is used for finding approximate numerical solutions mostly to partial differential equations (PDEs) and rarely integral equations. FEM discretizes the solution region of a problem into a finite number of subregions or elements, derive the governing equations for a typical element, assemble all elements in the solution region, and solve the system of equations obtained for a given BC. Its been widely used in electromagnetic, fluid dynamics, structural mechanics, civil engineering, and so on. FEM's capability in handling complicated geometries (and boundaries) is very high. The main difficulty encountered in FEM is element generation; therefore, special mesh generators have been developed for this purpose.

Method of moments (MoM) [6] is a numerical technique used to approximately solve linear operator equations such as differential equations or integral equations. The equation solved by MoM generally has the form of an *electric field integral equation* (EFIE) or a *magnetic field integral equation* (MFIE). The unknown function is approximated by a finite series of known expansion functions with unknown expansion coefficients. The approximate function is substituted into the original operator equation, and the resulting approximate equation is tested so that the weighted residual is zero. This results into a number of simultaneous algebraic equations for the unknown coefficients. These equations are then solved using matrix calculus. MoM can also be applied in time domain, but majority of MoM studies are in frequency domain. MoM has been used to solve vast number of electromagnetic problems during the last five decades.

Parabolic equation (PE) model, also known as the beam propagation method (BPM) in optics (see Refs. 7 and 8 and references therein), widely used in propagation scenarios, implements the solution of one-way parabolic-type wave equation with fast Fourier transform (FFT) or FEM. They are called split-step PE (SSPE) and FEMPE [8], respectively. The PE model neglects backscatter effects and is valid in regions close to near-axial propagation (two-way PE models have also been introduced lately [9–11]). It is an initial value problem and transverse and/or longitudinal characteristics (e.g., irregular terrain profiles as well as atmospheric refractivity variations) can be included, but boundary conditions must be satisfied artificially.

Finite-difference time-domain (FDTD) method [13] discretizes Maxwell's equations by replacing derivatives with their finite-difference (FD) approximations directly in the time domain. It is simple, easy to code, but has the open-form (iterative) solution, therefore, conditionally stable; one needs to satisfy stability condition. The FDTD volume is finite; therefore, it may model only closed regions. Free-space simulation is an important task in FDTD and various effective boundary terminations have been developed for the last two decades. Broadband (pulse) excitation is possible in FDTD but inherits numerical dispersion problem. Finally, only near fields can be simulated around the object under investigation; far fields can be extrapolated using equivalence principle (e.g., Stratton–Chu equations) [12].

Finally, transmission line matrix (TLM) method [14] uses Huygen's principle (i.e., is based on circuit theory) and three-dimensional (3D) fields are discretized in an array of 3D lumped elements. It is based on the analogy between the EM fields and a mesh of transmission lines (TLs), therefore discretizes the computational domain (volume in 3D, surface in 2D) in terms of a mesh of TLs, interconnected at nodes. TLM technique involves two basic steps: (1) replacing the field problem by the equivalent network and establishing the connection between the field and network quantities; (2) solving the equivalent network by iterative methods. TLM also inherits almost all FDTD problems, such as stability, numerical dispersion, incapability of far fields, and others.

1.4 MODEL SELECTION CRITERIA

Which of these models is going to be used, when, where, and why? It is also one of the aims of this book to answer these questions. Start with the user requirements: What kind of a problem are you interested in? Are you looking for narrowband or broadband response? How do you classify your problem; as a circuit model or an EM model? What kind of problem geometry do you have, is it simple or too complex in shape? Do you need to model specific details? What are the largest and smallest dimensions in terms of wavelength in your geometry? Which coordinate frame fits in better, rectangular, cylindrical, or spherical? Are you looking for something fast? How much computer memory do you need? What accuracy do you expect?

For example, time-domain simulations (e.g., with FDTD and/or TLM) are preferred because of their ability to handle complex EM environments and broadband responses obtained from a single simulation. Parameter selection, in this case, should be as follows (see Chapter 14 for the details of FDTD method):

- Specify frequency response (minimum, maximum frequencies $f_{\min}$, $f_{\max}$, and the frequency resolution Δf). For example, the radiation characteristics or input impedance of an antenna, RCS behavior of a chosen target, transmission and/or reflection characteristics of a microstrip network, propagation characteristics of a waveguide, resonance frequencies of a closed enclosure, shielding effectiveness of an aperture, and so on are among important problems.

- Choose the source waveform (pulse width) that contains maximum frequency of interest.
- In FFT, maximum frequency determines minimum time step, that is, $\Delta t_{\mathrm{FFT}} = 1/(2 \times f_{\max})$. This is the hard limit for frequency analysis.
- The frequency resolution Δf determines the observation time $T_{\max} = 1/\Delta f$. Remember, time-domain simulations model transient effects; therefore, simulations should continue until all transients are over. If the structure has resonances (ringing in time domain), much longer observation time may be required.
- Watch out Courant stability criteria [11] and numerical dispersion. Spatial mesh sizes Δx, Δy, Δz are chosen according to numerical dispersion requirements. It is nothing but satisfying Nyquist sampling criteria in spatial domain. The minimum wavelength $\lambda_{\min}$ component must be sampled with at least two samples, that is, max of $\{\Delta x, \Delta y, \Delta z\} \leq \lambda_{\min}/2$. In practice, at least $\lambda_{\min}/10$ is required for acceptable results ($\lambda_{\min}/20$ would be enough in many cases, but as much as $\lambda_{\min}/100 - \lambda_{\min}/120$ may be required in some problems).
- The time step Δt_{FDTD} may directly be chosen from the Courant stability criteria (choose $\Delta t = 1/\sqrt{3}c$ in 3D simulations, where c is the speed of light). Usually, Δt_{FDTD} is much less than Δt_{FFT}, therefore take Δt_{FDTD} into account.
- At the end, apply offline DFT/FFT on the FDTD data and get the frequency response.

Both FDTD and TLM suffer from long distances around the object under investigation since the whole computation volume needs to be discretized. Alternatively, near-to-near or near-to-far field transformations may be used for this purpose, but they are extremely time consuming. MoM, on the other hand, is very handy if the observer is quite far from the object under investigation, because the Green's function takes care of that.

Realistic problems have neither analytical exact nor approximate solutions; therefore, comparisons must be done against measurement or some other numerical model. It might be smart and fruitful to use two different numerical models and compare them with each other. It would be easy if both models are in time or frequency domains. If not, one needs to transform time data to frequency data or vice versa. For frequency-domain comparisons, time-domain data need to be converted via DFT/FFT procedure.

For example, in order to obtain input impedance of a microstrip circuit or an antenna versus frequency by using FDTD and MoM, one needs a single FDTD simulation and then an FFT procedure, but MoM simulations have to be repeated for a number of desired frequencies. If the field versus time inside a resonator is of interest, one needs only a single FDTD simulation. On the other hand, a three-step procedure is essential for MoM simulations. First, MoM simulations are repeated for a number of desired frequencies. These responses are normalized with the frequency response of the FDTD excitation source and then inverse DFT/FFT is applied. Field versus time of the MoM simulations is then obtained.

1.5 GRADUATE LEVEL EM MODSIM COURSE

Students who are willing to take this course are encouraged to have References 4, 12, and 14–18 as reference and to revisit fundamental EM theory. Although topics to be covered in this course are self-contained, an intermediate level EM theory knowledge is essential.

1.5.1 Course Description and Plan

EM-MODSIM course should be about giving fundamentals of the numerical models mentioned earlier and training on programming with canonical applications. It must establish a balance among EM theory (physics), analytical modeling (math), and computer implementation (numerical simulation). First, PDEs; elliptic, hyperbolic, and parabolic forms; Maxwell's equations and BCs; and one-dimensional (1D), two-dimensional (2D), and 3D decomposition should be discussed. Radiation, diffraction, and guided wave problems, coupled/uncoupled equations, and wave equation should also be given. Source-free (eigenvalue) and source-driven (Green's function) problem should be presented via the 1D Sturm–Liouville problem.

Propagation modeling inside a nonpenetrable, infinitely long parallel plate waveguide is an excellent structure in terms of simplicity and gaining physical insight. The Green's function representation, and analytical modeling in terms of exact mode and ray summations, should be reviewed before using the numerical models mentioned earlier. The steps of producing numerical reference data from analytical exact solutions should be discussed in this content. This necessitates time- and frequency-domain comparisons as well as VV&C procedure. A typical course plan may be as given in Table 1.1.

1.5.2 Available Virtual EM Tools

Several, simple, user-friendly, free EM virtual tools have been presented in this Magazine for nearly a decade. Some of them aimed only education, but many can also be used in research. Even source codes of many are available for further user improvements. These virtual tools that can be used in teaching EM-MODSIM with the content discussed earlier are listed in Table 1.2, Table 1.3, and Table 1.4.

1.6 EM-MODSIM LECTURE FLOW

Make a short introduction and talk about real world problems. Explain (1) why and when we need simulations, why analytical representations are not sufficient; (2) what mathematical models we are trying to solve with 1D Sturm–Liouville equation, what the Green's function and Eigenvalue problems are; (3) numerical simulation techniques FD, FDTD, MoM, PEM, TLM, FEM; (4) which problem domains are suitable for which technique, (e.g., TLM for patch antennas, MoM for radiation problems, etc.). Discuss current status of and future trends in parallel computing, cloud computing, and so on.

Start with fundamental issues in numerical analysis. Quickly review root search algorithms, N by N system of equations, numerical integration. Discuss Taylor expansion of an analytic function and use it in FD approximation of partial derivatives. Present simple MATLAB scripts and also use built-in MATLAB commands such as ***roots()***, ***diff***(), ***gradient***(), ***trapz()***, ***inv()***, and so on. Write a short MATLAB code which solves Poisson and Laplace equations in 1D and 2D with FD method. Then, write down coupled equations of a plane wave propagating along z-direction directly in time domain and introduce the FDTD in its simplest form (give students [25] and individual MATLAB projects).

Table 1.1.
A Typical EM MODSIM Course Plan

Week	Topics—Fundamentals
1	Engineers speak with numbers [18,47]
	Numerical analysis, Fourier transform, DFT, FFT, and Fourier series [19]
	Deterministic/stochastic modeling [20], hypothesis testing [21]
2	EM theory and Maxwell's equations [4]
	Elliptic, hyperbolic, and parabolic equations
	Radiation, diffraction, and guided wave problems
	Coupled/uncoupled and wave equation in 1D, 2D, and 3D
	2D TE and TM problems [22]
3	Source-free and source-driven problems
	1D Sturm–Liouville equation [23]
	Guided waves and eigenvalue extraction [24]
4	Laplace, Poisson, and wave equations [25]
	FD discretization
5	MODSIM [4]: Analytical and numerical modeling
	Validation, verification, and calibration (VV&C)
	Time-domain and frequency-domain comparisons
	Reference numerical data generation
6	Two-dimensional (2D) parallel plate waveguide [26,27]
	Ray and mode summation models
7	FDTD method [13], Yee cell, discretization, stability, numerical dispersion, open boundaries (PML), near- to far-field transformations
	1D plane wave propagator and time-domain reflectometer [29]
	MGL-2D package and cylindrical wave scattering [28]
	MTM-FDTD package [29]
8	General review and midterm exam
9	Method of moments (MoM) [6,31]
	MoM in resonating structures [32]
10	Antenna arrays and beam forming [4,33]
11	EM scattering (reflections, refractions, and diffractions)
	High frequency asymptotics (GO,PO,GTD,PTD,UTD) [34]
	High frequency models and FDTD [35]
12	PE and propagation modeling [36–40]
	Flat earth, knife-edge problem and terrain modeling
	Propagation through atmosphere
	MoM versus PE (GrMoMPE package) [39]
13	3D EM virtual tools
	FDTD-based microstrip design and analysis tool [41]
	FDTD-based RCS prediction tool [42]
14	Ground-wave propagation and mixed-path problem [43–45]

Table 1.2.
Basic Time-Domain EM Virtual Tools

VTool	Explanation
1DFDTD	A MATLAB-based FDTD simulation of plane wave propagation in time domain through single-, double-, or three-layer media. EM parameters are supplied by the user [25].
TDRMeter	A virtual time-domain reflectometer virtual tool. It is used to locate and identify faults in all types of metallic paired cable. Fourier and Laplace analyses are also possible [30].
MGL2D	A general purpose 2D FDTD package for both TE- and TM-type problems. Any 2D scenario may be created by the user by just using the mouse [28].
MTM-FDTD	Modified version of MGL-2D to simulate cylindrical wave propagation through metamaterials (MTMs) [29].
WedgeFDTD	A 2D MATLAB-based simulator for the modeling of EM diffraction from a semi-infinite nonpenetrable wedge using high frequency asymptotics and FDTD [35].

Table 1.3.
Basic Frequency-Domain EM Virtual Tools

VTool	Explanation
RAYMODE	Ray/mode representations inside a parallel plate nonpenetrable waveguide. RAY serves as a tool to compute and display eigenray trajectories between specified source/observer locations and to analyze their contributions to wave fields individually [27].
ARRAY	A MATLAB package designed to investigate wave propagation through a 2D dielectric waveguide. Both analytical formulations and the SSPE propagator are used for comparisons.
SNELL	A simple MATLAB package for the visualization of ray contributions between a source/receiver pair above a 2D ground using the ray shooting technique. A number of user-specified rays are shot through a propagation medium characterized by various linear vertical refractivity profiles [48].

Table 1.3.
(Continued)

VTool	Explanation
WedgeGUI	A 2D MATLAB-based simulator for the modeling of EM diffraction from a semi-infinite nonpenetrable wedge using GO, GTD/UTD, PO, and PTD [34].
WedgeGUIDE	A 2D MATLAB-based simulator for the modeling of Green's function in terms of NM, AM, and IM representations [46].

Table 1.4.
Advanced-Level EM Virtual Tools

VTool	Explanation
GrMoMPE	A MATLAB package which modifies MoM by the application of forward–backward spectral acceleration (FBSA) technique and integrate it with the SSPE method. SSPE versus MoM comparisons are possible [39].
MSTRIP	An FDTD-based EM simulator for the broadband investigation of microstrip circuits. The user only needs to picture the microstrip circuit via computer mouse on a rectangular grid to specify basic dimensions and operational needs such as the frequency band, simulation length [41].
MGL-RCS	An FDTD-based EM simulator for RCS prediction. The user only needs to locate a 3D image file of the target in 3DS graphics format, specify dimensions, and supply other user parameters. The simulator predicts RCS versus angle and/or RCS versus frequency [42].
DrMIX	A MATLAB-based Millington package prepared for the mixed-path, path loss predictions. The effects of the number of multimixed paths, path lengths, electrical parameters of each propagation section, and the frequency can be investigated [43].

Review fundamental engineering terms and discuss the meanings of the numbers obtained either by computation or by measurement. Go over the meanings of accuracy, precision, sensitivity, resolution, error, and uncertainty. Show how error propagates. Review stochastic modeling, randomness, statistics, and confidence level. In this context, discuss a stand-alone radar simulation problem. Then, use MATLAB and simulate complex radar signal environment including target, noise, clutter, and interference. Discuss radar/communication receiver unit and explain (as radar/communication engineers use) *decision making* or (as mathematicians use) *hypothesis testing* (give students [19,20] and individual MATLAB projects).

Spend extra time on Fourier transform, its implementation in a discrete environment (i.e., computer), and Fourier series expansion of a function. The former is important in obtaining frequency responses of the TD MODSIM models. The latter is especially important in teaching MoM. Discuss nonphysical effects introduced by discretization when using discrete/fast Fourier transformations (DFT/FFT) such as spectral leakage, aliasing, periodicity, and so on. Write together short MATLAB code; one for DFT and/or FFT which gives frequency spectrum, the other for Fourier series representation of any given function. Let students exercise and solve the problems alone.

Take 1D Sturm–Liouville equation into account and discuss it in details. Review source-free (eigenvalue) and source-driven (Green's function) EM wave problems and their relations via completeness and orthonormality properties. List 1D eigenvalue and Green's functions in finite and infinite domains and discuss alternative spectral representations.

Then, take propagation inside a 2D nonpenetrable parallel plate waveguide problem and discuss exact solutions in terms of mode summation and ray summation models. Mention important steps in generating numerical reference data from analytical exact solutions. Show the procedure of eigenvalue extraction from propagation characteristics of the problems. Optionally, discuss the wedge waveguide and review normal modes (NMs), adiabatic modes (AMs), and intrinsic modes (IMs) there. This concludes the first part.

Let us start the second part by discussing fundamental MODSIM terms and concepts such as model, analytical model, numerical model, simulation, validation, verification, calibration, and so on. Mention the importance of physics-based modeling and observable-based parameterization. Discuss open (iterative)- and closed (matrix)-form discrete representations of models and their advantages and disadvantages. Discuss stability and numerical dispersion. Pay attention to excitation and discuss point source, line source, beam, aperture antenna, and other excitations. Accurate source modeling is essential in VV&C. Give some examples from presented virtual tools starting from simple to extremely complex ones. Students now realize broad range of EM problems that can be handled with MODSIM.

Discuss major numeric models: FDTD, MoM, PE, TLM, FEM. State problems where, when, and how they are applicable or not. Give examples of EM problems which are solved by at least two of these methods and present some comparisons. Discuss VV&C. Discuss TD and FD comparisons. Show cases where these methods agree very well but still incorrect. Present some virtual tools. This concludes the second part.

Let us reserve the last part for a simple canonical problem where all analytical and numerical models are applicable. Revisit propagation inside a 2D parallel plate waveguide. Discuss mode and ray summation models, image method, PE, MoM, FDTD, and FEM altogether. Present short MATLAB code for each and then do tests and comparisons. Let the students choose a topic among microstrip circuit design, groundwave propagation, and RCS problems, and test with one of the sophisticated EM virtual tools (e.g., MSTRIP, DrMIX, MGL-RCS).

This will finalize the course. The progress and how far to go depend on students' individual interests and willingness.

1.7 TWO LEVEL EM GUIDED WAVE LECTURE

As will be discussed in detail throughout the book, guided wave problems are important in teaching EM. In most of these problems such as TLs, 2D parallel plate waveguides, and 3D rectangular and/or circular cross-section waveguides, analytical exact solutions are known. This is specifically important in (1) understanding the problem and gaining physical insight and (2) generating reference data for measurements and numerical simulations. A two-level guided wave theory can be taught as follows.

Level I (Undergraduate)

- Decompose Maxwell's equations into transverse and longitudinal components.
- Discuss Sturm–Liouville equation in 1D and establish characteristic relations between source-free (homogeneous) and source-driven (inhomogeneous) representations.
- Discuss *orthogonality* and *completeness*.

- Discuss Fourier series expansion of a function in a given interval at this point (use supplied F***Series.m*** script for various examples).
- Give mathematical details of 2D parallel plate waveguide with nonpenetrable boundaries (use ***PPlate.m*** script for various examples).
- Give mathematical details of 2D surface duct formation above flat Earth and discuss trapping effects of refractivity variations (use ***SDuct.m*** script for various examples).
- Briefly review 2D parabolic wave equation and give details of numerical SSPE propagator in terms of FFT.
- Discuss the formation of longitudinal correlation function and give steps of eigenvalue extraction procedure (use ***PPCorel.m*** script for various examples).
- Review 2D dielectric film waveguide and derive eigensolutions (use ***DiSlab.Fig*** and ***DiSlab.m*** scripts for various examples).

Level II (Graduate)

- Start with the Ray-mode representations inside the 2D parallel plate waveguide. Give details of alternative integral representations and derivations by contour deformations, residue series, and others on various complex planes.
- Discuss hybridization in terms of rays and modes partially.
- Use MATLAB-based virtual tools and visualize effects of rays, modes individually, and in hybrid form on various user-specified scenarios.
- Take propagation inside the 2D wedge waveguide into account and give mathematical details of NM, AM, and IM.
- Establish (exact) NMs in cylindrical coordinates.
- Establish (approximate) AMs in Cartesian coordinates.
- Give IM integral representation and its reduction to AM away from critical, for example, cutoff, transitions.
- Use MATLAB-based virtual tools and visualize the effects of different modal expansions on various user-specified scenarios.

1.8 CONCLUSIONS

A generic electromagnetic modeling and simulation (EM-MODSIM) course outline is introduced and a teaching flow is proposed. The course description/outline, a weekly course plan, and some virtual tools are presented. Wave propagation inside an infinitely long parallel plate waveguide with nonpenetrable boundaries is chosen as a canonical structure and fundamental analytical (ray-mode summation) and numerical models (FDTD, PE, MoM, TLM, and FEMs) are reviewed. The process of VV&C is also presented. Note that there are excellent books which describe different analytical and numerical models in details which serve as reference in EM-MODSIM studies; some of them are listed in Appendix B.

REFERENCES

1. L. Sevgi, C. Goknar, "An Intelligent Balance between Numerical and Physical Experimentation," *IEEE Potentials Magazine*, **23** (4), 2004, pp. 40–44.
2. L. B. Felsen, L. Sevgi, "Electromagnetic Engineering in the 21st Century: Challenges and Perspectives," *ELEKTRIK, Turkish Journal of Electrical Engineering and Computer Sciences*, **10** (2), 2002, pp. 131–145.
3. C. Goknar, "Models and Modeling: Be Careful and Use Your Imagination," *IEEE Antennas and Propagation Magazine*, **50** (5), 2008, pp. 215–221.
4. C. Balanis, *Advanced Engineering Electromagnetics*, 2nd ed., Wiley Press, New York, 2012.
5. J.-M. Jin, *The Finite Element Method in Electromagnetics*, John Wiley Press, New York, 2002.
6. R. F. Harrington, *Field Computation by Moment Method*, 1st ed., 1968, IEEE Press, New York, 1993.
7. V. A. Fock, *Electromagnetic Diffraction and Propagation Problems*, Pergamon, Oxford, 1965.
8. M. Levy, *Parabolic Equation Methods for Electromagnetic Wave Propagation*, IEE, Institution of Electrical Engineers, London, 2000.
9. M. D. Collins, R. B. Evans, "A Two-Way Parabolic Equation for Acoustic Backscattering in the Ocean," *The Journal of the Acoustical Society of America*, **91** (3), 1992, pp. 1357–1368.
10. O. Ozgun, "Recursive Two-Way Parabolic Equation Approach for Modeling Terrain Effects in Tropospheric Propagation," *IEEE Transactions on Antennas and Propagation*, **57** (9), 2009, pp. 2706–2714.
11. G. Apaydın, Ö. Özgün, M. Kuzuoğlu, L. Sevgi, "A Novel Two-Way Finite-Element Parabolic Equation Groundwave Propagation Tool: Tests With

Canonical Structures and Calibration," *IEEE Transactions on Geoscience and Remote Sensing*, **49** (8), 2011, pp. 2887–2899.

12. R. Garg, *Analytical and Computational Methods in Electromagnetics*, Artech House, Boston, 2008.
13. K. S. Yee, "Numerical Solution of Initial Boundary Value Problems Involving Maxwell's Equations in Isotropic Media," *IEEE Transactions on Antennas and Propagation*, **14**, 1966, pp. 302–307.
14. P. B. Johns, R. L. Beurle, "Numerical Solution of 2-dimensional Scattering Problems Using a Transmission Line Matrix," *Proceedings of the IEEE*, **119** (9), 1971, pp. 1203–1208.
15. L. Sevgi, "A New Electromagnetic Engineering Program and Teaching via Virtual Tools," *PIER, Progress in Electromagnetics Research B (EWS 2006 Special Issue)*, **6**, 2008, pp. 205–224.
16. M. N. O. Sadiku, *Numerical Techniques in Electromagnetics with MATLAB*, 3rd ed., CRC Press, Boca Raton, FL, 2009.
17. J. M. Jin, *Theory and Computation of Electromagnetic Fields*, John Wiley, Hoboken, NJ, 2011.
18. L. Sevgi, *Complex Electromagnetic Problems and Numerical Simulation Approaches*, IEEE Press/Wiley, Piscataway, NJ, 2003.
19. L. Sevgi, "Numerical Fourier Transforms DFT and FFT," *IEEE Antennas and Propagation Magazine*, **49** (3), 2007, pp. 238–243.
20. L. Sevgi, "Synthetic Radar Signal Environment: Computer Generation of Signal, Noise, and Clutter," *IEEE Antennas and Propagation Magazine*, **49** (5), 2007, pp. 192–198.
21. L. Sevgi, "Hypothesis Testing and Decision Making: Constant-False-Alarm Rate," *IEEE Antennas and Propagation Magazine*, **51** (3), 2009, pp. 218–224.
22. L. Sevgi, "Guided Waves and Transverse Fields: Transverse to What?," *IEEE Antennas and Propagation Magazine*, **50** (6), 2008, pp. 221–225.
23. L. Sevgi, "Sturm-Liouville Equation: The Bridge between Eigenvalue and Green's Function Problems," *ELEKTRIK, Turkish Journal of Electrical Engineering and Computer Sciences*, **14** (2), 2006, pp. 293–311.
24. L. Sevgi, "Guided Waves and Eigenvalue Extraction from Propagation Characteristics," *IEEE Antennas and Propagation Magazine*, **50** (2), 2008, pp. 222–234.
25. L. Sevgi, "Review of Discrete Solutions of Poisson, Laplace, and Wave Equations," *IEEE Antennas and Propagation Magazine*, **50** (1), 2008, pp. 246–254.
26. L. B. Felsen, A. H. Kamel, "Hybrid Ray-Mode Formulation of Parallel Plate Waveguide Green's Functions," *IEEE Transactions on Antennas and Propagation*, **29** (4), 1981, pp. 637–649.

27. L. B. Felsen, F. Akleman, L. Sevgi, "Wave Propagation inside a Two-Dimensional Perfectly Conducting Parallel Plate Waveguide: Hybrid Ray-Mode Techniques and Their Visualizations," *IEEE Antennas and Propagation Magazine*, **46** (6), 2004, pp. 69–89.

28. G. Çakır, M. Çakır, L. Sevgi, "A Multipurpose FDTD-Based Two Dimensional Electromagnetic Virtual Tool," *IEEE Antennas and Propagation Magazine*, **48** (4), 2006, pp. 142–151.

29. M. Çakır, G. Çakır, L. Sevgi, "A Two-Dimensional FDTD-Based Virtual Metamaterial—Wave Interaction Visualization Tool," *IEEE Antennas and Propagation Magazine*, **50** (3), 2008, pp. 166–175.

30. L. Sevgi, Ç. Uluışık, "A MATLAB-Based Transmission Line Virtual Tool: Finite-Difference Time-Domain Reflectometer," *IEEE Antennas and Propagation Magazine*, **48** (1), 2006, pp. 141–145.

31. E. Arvas, L. Sevgi, "A Tutorial on the Method of Moments," *IEEE Antennas and Propagation Magazine*, **54** (3), 2012, pp. 260–275.

32. G. Apaydin, L. Sevgi, "Method of Moment (MoM) Modeling for Resonating Structures: Propagation inside a Parallel Plate Waveguide," *ACES, International Journal on Applied Computational Electromagnetics*, **27** (10), 2012, pp. 842–849.

33. L. Sevgi, Ç. Uluışık, "A MATLAB-Based Visualization Package for Planar Arrays of Isotropic Radiators," *IEEE Antennas and Propagation Magazine*, **47** (1), 2005, pp. 156–163.

34. F. Hacıvelioğlu, M. A. Uslu, L. Sevgi, "A MATLAB-Based Simulator for the Electromagnetic Wave Scattering from a Wedge with Perfectly Reflecting Boundaries," *IEEE Antennas and Propagation Magazine*, **53** (6), 2011, pp. 234–243.

35. M. A. Uslu, L. Sevgi, "MATLAB-Based Virtual Wedge Scattering Tool for the Comparison of High Frequency Asymptotics and FDTD Method," *ACES, International Journal on Applied Computational Electromagnetics*, **27** (9), 2012, pp. 697–705.

36. G. Apaydin, L. Sevgi, "A Canonical Test Problem for Computational Electromagnetics (CEM): Propagation in a Parallel Plate Waveguide," *IEEE Antennas and Propagation Magazine*, **54** (4), 2012, pp. 290–315.

37. G. Apaydin, L. Sevgi, "Validation, Verification and Calibration in Applied Computational Electromagnetics," *International Journal on Applied Computational Electromagnetics*, **25** (12), 2010, pp. 1026–1035.

38. G. Apaydın, L. Sevgi, "The Split Step Fourier and Finite Element Based Parabolic Equation Propagation Prediction Tools: Canonical Tests, Systematic Comparisons, and Calibration," *IEEE Antennas and Propagation Magazine*, **52** (3), 2010, pp. 66–79.

39. F. Akleman, L. Sevgi, "A Novel MoM- and SSPE-Based Groundwave Propagation Field Strength Prediction Simulator," *IEEE Antennas and Propagation Magazine*, **49** (5), 2007, pp. 69–82.

40. O. Ozgun, G. Apaydin, M. Kuzuoglu, L. Sevgi, "PETOOL: MATLAB-Based One-Way and Two-Way Split Step Parabolic Equation Tool for Radiowave Propagation over Variable Terrain," *Computer Physics Communications (CPC)*, Elsevier, **182**, 2011, pp. 2638–2654.
41. G. Çakır, M. Çakır, L. Sevgi, "A Novel Virtual FDTD-Based Microstrip Circuit Design and Analysis Tool," *IEEE Antennas and Propagation Magazine*, **48** (6), 2006, pp. 161–173.
42. G. Çakır, M. Çakır, L. Sevgi, "Radar Cross section (RCS) Modeling and Simulation: Part II—A Novel FDTD-Based RCS Prediction Virtual Tool," *IEEE Antennas and Propagation Magazine*, **50** (2), 2008, pp. 81–94.
43. L. Sevgi, "A Mixed-Path Groundwave Field Strength Prediction Virtual Tool for Digital Radio Broadcast Systems in Medium and Short Wave Bands," *IEEE Antennas and Propagation Magazine*, **48** (4), 2006, pp. 19–27.
44. G. Apaydin, L. Sevgi, "Numerical Investigations of and Path Loss Predictions for Surface Wave Propagation over Sea Paths Including Hilly Island Transitions," *IEEE Transactions on Antennas and Propagation*, **58** (4), 2010, pp. 1302–1314. Apr.
45. L. Sevgi, "Modeling and Simulation Strategies for Electromagnetic Wave Propagation in Complex Environments: Groundwave Path Loss Prediction Virtual Tools," *IEEE Transactions on Antennas and Propagation*, **55** (6), 2007, pp. 1591–1598.
46. L. Sevgi, F. Akleman, L. B. Felsen, "Visualizations of Wave Dynamics in a Wedge-Waveguide with Non-Penetrable Boundaries: Normal, Adiabatic, and Intrinsic Mode Representations," *IEEE Antennas and Propagation Magazine*, **49** (4), 2007, pp. 76–94.
47. L. Sevgi, "Innumeracy: The Meaning of the Numbers We Use," *IEEE Antennas and Propagation Magazine*, **49** (2), 2007, pp. 195–190.
48. L. Sevgi, "A Ray Shooting Visualization MATLAB Package for 2D Groundwave Propagation Simulations," *IEEE Antennas and Propagation Magazine*, **46** (4), 2004, pp. 140–145.

CHAPTER 2

Engineers Speak with Numbers

2.1 INTRODUCTION

Every day, we hear on the TV or read in the newspaper phrases such as "The conducted measurements yield that the highest peak of the Earth, peak Everest, is 8872 m above sea level," "The National Institute of Statistics (NIS) declares that the unemployment rate, which was 9.7% last year, reduced to 9.5% this year," or "Scientists measured the speed of light as 299.793 km/s." Have you ever thought about the meaning of giving numbers, or speaking with numbers? For example, what conclusions can be drawn from the NIS's declaration? Is it obvious that the unemployment rate has been falling since last year? What can you say if somebody claims that the unemployment is escalated? Think about another phrase you listen to on the radio, such as "The polls show that with a 95% confidence level, the vote of Democrats was $(32 \pm 2)\%$." What does it mean with 95% confidence level? How confident is the confidence level itself? What if the confidence level is 99%?

Engineers speak with numbers. Speaking with numbers recalls a measurement or a numerical calculation [1]; a measurement or a numerical calculation means using a model. A model means a mathematical representation after a series of observations, experiments,

Electromagnetic Modeling and Simulation, First Edition. Levent Sevgi.

assumptions, approximations, and simplifications. It necessitates a good understanding of the fundamental concepts of a measurement and/or a numerical calculation. Every assumption and approximation means neglecting something, and introducing extra error and/or uncertainty. It is essential to specify the error and uncertainty bounds of a numerical result if it is *scientific*. Concepts, such as error, uncertainty, accuracy, precision, sensitivity, and resolution, should be well understood. You may meet someone who presents his/her results with, for example, 12-digit accuracy, while the numerical error limits it to eight digits (for example), and while only two digits are meaningful because of the approximations made. What do all of these mean?

Here are a few examples of expressing numbers: they belong to people from academic society and presented in various scientific conferences I attended in the past; a path loss of a radiolink at 10 GHz at 20 km was presented as 121.7812 dB. Is it ***meaningful***, why? A surface impedance of an inverse problem was given as 25.128765341296 Ω. Is this ***acceptable***, why? A ratio of 189 patients out of 550 was expressed as 34.363636%. Is there *anything* **wrong** with this, **why?**

2.2 MEASUREMENT, CALCULATION, AND ERROR ANALYSIS

A number is given as a result of a measurement or a computation. What quantities can be measured or calculated? Basic quantities, such as length, weight, time, voltage, current, stress, temperature, electric and magnetic fields, and so on, can be measured; any other quantity that can be specified in terms of these can be calculated. For example, the value of a resistor can be calculated via $R = V/I$ if the voltage across its terminals and the current flowing through the resistor are measured. The results of multiple measurements using different devices with different precisions—even measurements with the same device—are observed to have distributions. What is the meaning of device precision, parameter sensitivity, or numerical accuracy?

Before giving the answers to all these questions, let us agree on these basic concepts and on their definitions [2,3].

Accuracy is the closeness of the measured or calculated quantity to its exact value. Measurement accuracy is the ability of a device to measure a quantity within a stated error bounds. Numerical accuracy is the closeness of a result obtained via a numerical model that represents a physical system to the exact value of the quantity. The accuracy is expressed in terms of error.

Error is the difference between a measured or calculated value of a quantity and its exact value. The error may be systematic or random.

Systematic error is error that plagues experiments or calculations caused by negative factors. For example, a dc voltage component that is unintentionally present because of a failure of the blockage capacitor during an ac voltage measurement is a systematic error. Another example would be an ammeter that only displays 85% of the true voltage because of calibration problems. Systematic errors can be complex, but can be removed once understood or discovered via careful controls and calibration.

Random error is error that is always present, but varies unpredictably in size and direction. Random errors are related to the scatter in the data obtained under fixed conditions that determine the repeatability (precision) of the measurement. Random errors (fortunately) (usually) follow well-behaved statistical rules. Their effects can be reduced by repeating the measurement as often as possible. Error can be expressed in one of two ways as follows.

Absolute error is an error that is expressed in physical units. It is the absolute value of the difference between the measured value and the true value of a quantity (or the average value, if the true value is not known).

Relative error is an error expressed as a fraction relating the absolute error to the true (or average) value of a quantity. It is always given as a percentage.

Uncertainty is a range that is likely to contain the true value of a quantity being measured or calculated. Uncertainty can be expressed in absolute or relative terms.

The terms uncertainty and error have different meanings in modeling and simulation. Modeling uncertainty is defined as the potential deficiency due to a lack of information. On the other hand, modeling error is the recognizable deficiency, not due to a lack of information, but due to the approximations and simplifications made there. *Measurement error* is the difference between the measured and true values, while *measurement uncertainty* is an estimate of the error in a measurement. Modeling and simulation uncertainties occur during the phases of conceptual modeling of the physical system, mathematical modeling of the conceptual model, discretization and computer modeling of the mathematical model, and numerical computations. Numerical uncertainties occur during computations due to the discretization, round-off, nonconvergence, artificial dissipations, and so on.

Measurement error is the difference between the measured and true value of a quantity; measurement uncertainty is the amount of predicted error.

Confidence level is the probability that the true value of the measurement or calculation falls within a given range of uncertainty caused by the inherent random nature of the measurement or calculation. Confidence levels can be defined through a good understanding of the nature (probabilistic distributions) of the errors.

Precision is a measure of the closeness of the value obtained via multiple measurements to the true value. It is the total amount of random error present. A very precise measurement means a small random error. Precision is given as the percentage of the ratio of the value region to the true (or average) value of the quantity being measured. The value region is the difference between the maximum and minimum values in multiple measurements. Precision, which is proportional to the sensitivity of the measurement device, does not necessarily mean accurate results: *High precision implies high sensitivity*!

Sensitivity is the smallest change in a physical quantity that a measurement device can detect or sense. Statistical test sensitivity is the smallest probabilistic change.

Ideally, a measurement device is expected to have high precision and high accuracy. High precision alone does not guarantee high accuracy. For example, a device may have high precision; all measurements may fall in a narrow value region, but the results may still be incorrect because of a systematic error present in the device.

Resolution is the ability to resolve, to discriminate. It may also be defined as the smallest physically indicated division that an instrument displays or has marked.

Range resolution, *azimuth resolution*, or *velocity resolution* in a radar are the minimum distance, azimuth angle, or velocity difference when two targets are no longer resolved in range, azimuth, or velocity, respectively.

Numerical picture resolution, screen resolution, and camera resolution are used for digital pictures, and are defined as the distance between two nearby pixels. These are often given as the number of pixels.

Tolerance is the amount of uncertainty in a material's physical characteristics. For example, resistor tolerance, inductor tolerance, and so on.

2.3 SIGNIFICANT DIGITS, TRUNCATION, AND ROUND-OFF ERRORS

Numbers are expressed using figures (numerals). The first nonzero numeral at the left of a number (which contributes the largest amount to the number) is called *the most-significant digit*. The first nonzero numeral at the right of a number (which contributes the least amount) is *the least-significant digit*. If a number has fractional terms, then all the digits are significant. For example, each of these numbers has five significant digits:

45,302, 45,257, 4.8010, 10,001, 3920.0, 1000.0, 3.0000, 45,032,000, 5,000,100

A number (in a computer) is represented with a finite (fixed) number of digits called the *word length*. The precision of a measurement or the accuracy of a computation bounds the number of significant digits. All nonzero digits beyond the number of significant digits at the right of a number are removed in one of two ways: truncation or round off. Leaving out the digits beyond the least-significant digit is truncation, and the error introduced via this process is called *truncation error*. Rounding the least-significant digit according to the most-significant digit of the left-out part is called round off, and the error introduced via this process is called *round-off error*. If the most-significant digit of the truncated part is greater than 5, then the least-significant digit of the number is increased by 1, otherwise it is kept unchanged. For example, the number 53.0534 has six significant digits. If the number is going to be represented only by four significant digits, both the truncation and round-off processes yield 53.05. On the other hand, if the number of significant digits is to be three, then the truncation and round-off processes yield 53.0 and 53.1, respectively.

Truncation error is also defined for mathematical modeling. For example, Taylor's expansion or Fourier-series representation is used to replace a function in terms of an infinite summation. Taking only given number of low-order terms, and neglecting the rest of the higher-order terms, introduces a truncation error. Actually, this truncation determines the number of significant digits in a computation. For example, using the Taylor expansion of an exp(x) function as

$$e^x = 1 + \frac{x}{1!} + \frac{x^2}{2!} + \frac{x^3}{3!} + \frac{x^4}{4!} + \cdots + \frac{x^n}{n!} + \cdots \tag{2.1}$$

and keeping only the terms up to the third order yield an absolute truncation error of

$$\text{Absolute error} = \sum_{n=4}^{\infty} \frac{x^n}{n!}. \tag{2.2}$$

This absolute error is also an infinite series; therefore, its numerical evaluation is a challenge. In general, an upper bound for the error is given.

The number of significant digits helps to determine the uncertainty bounds (or the error) of a numerical value. For example, $x = 5000$, $y = 5000$, $z = 531$ as representations of x, y, and z show that these three quantities may take any value between $4500 < x < 5500$, $4999.5 < x < 5000.5$, and $530.5 < x < 531.5$. The reason for this is the number of significant digits in x, y, and z.

How many digits of a number obtained via a measurement or a calculation are meaningful? The answer resides in the degree of uncertainty, which means the amount of error. For example, the height of a building is measured to be $d = 5.280753$ m with a measurement uncertainty of 0.103 m. How will the result be conveniently presented? The first step is to specify the uncertainty as an absolute error. The error could be rounded off in two ways: $\Delta d = 0.10$ or $\Delta d = 0.1$. Then, the result for these two cases may be presented as $d \pm \Delta d = 5.28 \pm 0.10$ and $d \pm \Delta d = 5.3 \pm 0.1$, respectively. If the uncertainty has three significant digits, as given, then the result should be given as $d \pm \Delta d = 5.281 \pm 0.103$. An alternative way of presenting $d \pm \Delta d = 5.3 \pm 0.1$ is $d \pm \Delta d = 5.3\ (1.00 \pm 0.018)$. In summary, the number of significant digits and the error must be of the same order!

2.4 ERROR PROPAGATION

A model representing a physical (real-world) problem is built in terms of mathematical relations, which use measured and/or computed quantities. Error propagates because of model-based derived quantities. *Propagation error* is the error in the succeeding steps of a process due to an occurrence of an earlier error. A method or a result may become *unstable* due to the propagated error if the errors are magnified continuously during multiple derivations or iterative processes.

Consider a result of a measurement or a computation given as $A \pm \Delta a$. Here, A is the value, $\Delta a > 0$ is the absolute error, and $\Delta a/A$ is the relative error. If a quantity is expressed in terms of the addition or subtraction of two other quantities, either measured or computed, then the total (propagated) error is equal to the addition of absolute errors:

$$C = (A \pm B) \pm \Delta c \Rightarrow \Delta c = \Delta a + \Delta b. \tag{2.3}$$

If the quantity is given as the multiplication/division of two measured/computed quantities, then the total (propagated) relative error is equal to the addition of relative errors:

$$C = AB \quad \text{or} \quad C = A/B \Rightarrow \frac{\Delta c}{C} = \frac{\Delta a}{A} + \frac{\Delta b}{B}. \tag{2.4}$$

For example, what is the total distance of a trip if the journey lasts 3±0.1 h and the average speed is 60 ± 0.5 km/h? Answer: Since distance = velocity × time, the minimum distance is 2.9 × 59.5 = 172.55 km (rounded off to 172.6), and the maximum distance is 3.1 × 60.5 = 187.55 km (rounded off to 187.6). Similarly, the distance can be given as 180 ± 7.6 km if calculated from (2.4).

In general, the total (propagated) error is obtained from these two properties as

$$C = A^m B^n \Rightarrow \frac{\Delta c}{C} = |m|\frac{\Delta a}{A} + |n|\frac{\Delta b}{B}. \tag{2.5}$$

Finally, if $y = f(x_1, x_2, x_3, \ldots)$ is a multivariable function, then the total error is calculated from

$$\Delta y = \left(\frac{\partial f}{\partial x_1}\right)\Delta x_1 + \left(\frac{\partial f}{\partial x_2}\right)\Delta x_2 + \left(\frac{\partial f}{\partial x_3}\right)\Delta x_3 + \cdots. \tag{2.6}$$

2.5 ERROR AND CONFIDENCE LEVEL

Suppose the voltage across the terminals of a resistor is going to be measured using a high-precision voltmeter. Each time you repeat the measurement, you come out with a different result, basically because of randomly changing environmental factors. Now, suppose the problem at hand is to predict the average height of the students in a school, the average income of the workers in a factory, and so on using *sampling*. The characteristic difference between natures of these two cases is the size of the value space, called the *population*. The population in the voltage measurement is infinite, but the number of students in the school, or the number of workers at the factory, is finite. No matter how you increase the number of voltage measurements, the results have a distribution that can be represented via the *mean* and *variance* σ^2 (or its positive square root, the *standard deviation* σ).

The mean of a multiple measurement, repeated N times, is calculated as

$$x_{\text{avg}} = \frac{1}{N}\sum_{k=1}^{N} x_k = \frac{x_1 + x_2 + x_3 + x_4 + \cdots + x_N}{N}. \tag{2.7}$$

The standard deviation σ (or the error Δx of a measurement) is given as

$$\sigma = \Delta x = \sqrt{\frac{\sum_{k=1}^{N}\left(x_k - x_{\text{avg}}\right)^2}{N-1}} \tag{2.8}$$

(N is also used instead of $N - 1$ in the denominator, but the difference is negligible as N gets larger). The mean is a measure of the central tendency, while the standard deviation is a measure of dispersion.

Contrary to the resistor measurement, if all the students in a school or the workers in a factory are used in the measurements (i.e., if the sample space is equal to the population), the average height or income is determined without a distribution (of course, within the measurement error limits). In practice, the size of the population is often too large to consider completely, and therefore sampling is applied: the tests or measurements are conducted with a small, randomly selected part of the population. For example, suppose there are 2450 students in the school, and 100 of them are selected randomly for the measurements. The average height of these 100 students and the standard deviation are found to be 1.73 m and 25 cm, respectively. What can you say about the average height of the students in the school? Can you predict this average value? How accurate will that prediction be? How does your prediction change if the size of the sample space is increased to 500 or 1000?

It can easily be shown that in most physical problems where measurements or tests are repeated N times, the measured values are accumulated around an average value with a certain dispersion called a Gaussian distribution. The Gaussian distribution or the Gauss function is determined with an average value x_0 and a standard deviation σ,

$$g(x) = \frac{1}{\sigma\sqrt{2\pi}} e^{-(x-x_0)^2/2\sigma^2}. \tag{2.9}$$

A special case of the Gaussian distribution, where the average is zero ($x_0 = 0$) and the standard deviation is 1 ($\sigma = 1$) is called the normal distribution,

$$f(x) = \frac{1}{\sqrt{2\pi}} e^{-x^2/2}. \tag{2.10}$$

Figure 2.1 shows the normal distribution, which plays an important role in probabilistic calculations. The function $f(x)$ in (2.10) is also called the *probability density function* (PDF) for random variables with normal distributions, since $f(x) \geq 0$ for all x, and the area under the curve in the figure is equal to 1. It is customary to use uppercase letters for random variables and lowercase for their values. The horizontal axis shows the random variable x, and the area under this curve between $\pm x$

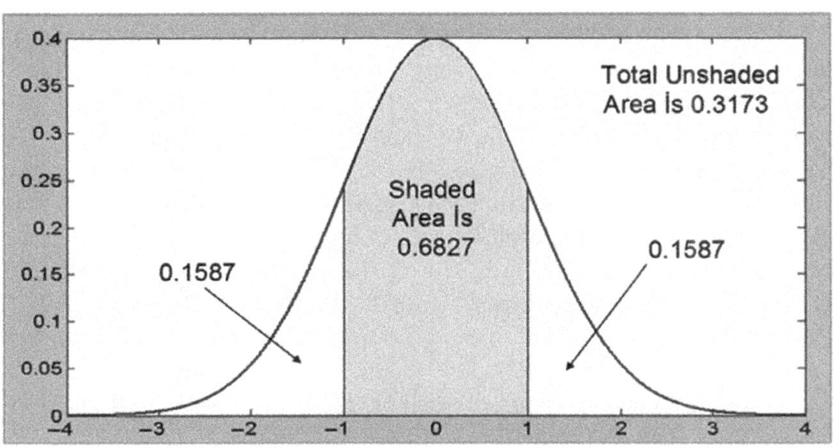

Figure 2.1. The normal distribution function.

values is the probability of occurrence. The total area under the curve is equal to 1, meaning that the probability of measuring any value is 100%.

The probability of measuring X between $-1 \leq x \leq +1$ (i.e., $P(-1 \leq x \leq +1)$) can be calculated via the integration of $f(x)$ between $-1 \leq x \leq +1$ (the shaded area in Fig. 2.1):

$$P(-1 \leq x \leq +1) = \frac{1}{\sqrt{2\pi}} \int_{-1}^{+1} e^{x^2/2} dx = 0.6827.$$

This is phrased either as "the probability of getting the result between $x_{avg} - \Delta x$ and $x_{avg} + \Delta x$ is 68.27%" or as "we are 68.27% sure that the result is between $(x_{avg} - \Delta x)$ and $(x_{avg} + \Delta x)$." The probability of having X outside this range is $1 - 0.6827 = 0.3173$. The probability of having X greater than $x_{avg} + \Delta x$ ($P(x \geq +1)$) is $0.3173/2 = 0.1587$.

It is quite easy to compute this probability integral numerically. For example, the trapz(x,y) command of *MATLAB* can be used via the three-line script to be shown later, and the integral within the given limits can be computed with a given accuracy (here 1000 samples are used, but the accuracy increases as the number of samples increases).

```
> x = -1.0:0.001:1.0;
> y = 1/sqrt(2.*pi)*exp(-x.^2/2);
> trapz(x,y)
> 0.6827
```

Here, x is an array containing 1000 elements between ± 1, and y is the 1000-element array of $f(x)$ values corresponding to those x values. The trapz(x,y) command computes the numerical integral using the well-known *trapezoidal rule*. As shown earlier, the probability of having X between $-1 \leq x \leq +1$ is 0.6827. Running the script again and again with different intervals yields the probabilities of other values. For example, the probabilities of having X between the intervals of $-1.96 \leq x \leq +1.96$, $-2 \leq x \leq +2$, $-2.58 \leq x \leq +2.58$, and $-3 \leq x \leq +3$ are, respectively, 0.9500, 0.9545, 0.9901, and 0.9973. Here, $\sigma = 1.96$, $\sigma = 2.58$, or $\sigma = 3.0$ are called the *critical values* of the desired confidence levels.

If X is a Gaussian random variable with average x_0 and standard deviation σ, then the transformation

$$Y = \frac{X - x_0}{\sigma} \tag{2.11}$$

yields a random parameter Y with a normal distribution. The horizontal axis also specifies *dispersion* around the mean, that is, the error after N measurements. If, for example, an N-element sample space has the average measured distance d_{avg} with a standard deviation (absolute error) Δd, then it may be speculated that the probability of having the true value in the interval $d_{avg} \pm \Delta d$ is 68.27% ($d_{avg} \pm \Delta d$ in (2.9) corresponds to ± 1 in (2.10) if (2.11) is used). This probability value is called the *confidence level.* If a 99% confidence level is desired, then the interval of the integral of (2.10) that yields 0.99 should be computed numerically. As done earlier, a 99% confidence level requires an interval of $\sigma = \pm 2.58$ (or $\pm 2.58\Delta d$). In this case, it may be speculated that with the 99% confidence level, the true value falls in the region of $d_{avg} \pm 2.58\Delta d$.

2.5.1 Predicting the Population's Confidence Interval

The confidence interval (CI) for the population mean may be calculated in two ways. If the size of the population is infinite and the sample size is N, then

$$\Delta d_N = \alpha \frac{\Delta d_\infty}{\sqrt{N}} \, (N \geq 30) \tag{2.12}$$

is used to derive the population mean. Otherwise,

$$\Delta d_N = \alpha \frac{\Delta d_M}{\sqrt{N}} \frac{\sqrt{M-N}}{\sqrt{M-1}} \, (N \geq 30) \tag{2.13}$$

should be used for the population mean (here, M is the size of the population). For example, assume there are $M = 4186$ students in a school and only the heights of $N = 200$ students are measured. Suppose the average value and the standard deviation are found to be $h_{avg} = 1.72$ m and $\Delta h = 0.23$ m, respectively. What can be said about the mean height of all the students in the school (population) with a 95% confidence level? As found after the measurements with the sample space, the mean height with a 68.27% confidence level is 1.72 ± 0.23 m. It can be said that with a 95% confidence level, the mean height of all students at the school will be

$$h_{avg} \pm \Delta h = h_{avg} \pm 1.96 \times \frac{0.23}{\sqrt{200}} = 1.72 \pm 0.03 \text{ m.}$$

Table 2.1.
Some Confidence Levels and Corresponding Critical Values

Confidence level	70%	80%	90%	95%	99%
Normal distribution (α)	1.040	1.280	1.645	1.960	2.580
Student's distribution (for $v = N = 15, \beta$)	0.536	0.866	1.340	1.750	2.600

Finally, if the number of measured samples is less than $N < 30$, then the distribution is more likely a *t-distribution* (Student's distribution) [2]. In this case, the critical values α in (2.12) and (2.13) are replaced with the critical values β of the t-distribution. The difference between these two distributions is almost negligible if the sample space is greater than 30. Table 2.1 lists some widely used values of confidence levels and the corresponding critical values of both distributions. Note that the β-values of the t-distribution are for $N = 15$ measurements.

EXAMPLE

A group of students is required to predict the received power of a transmitter at a given frequency and at a distance d. The transmitted power P_t, the transmitting antenna gain G_t, the signal wavelength λ, and the distance d are to be measured, and the received power is to be calculated from

$$P_r = \frac{P_t G_t \lambda^2}{(4\pi d)^2} \text{ [W]}. \tag{2.14}$$

After a series of measurements are conducted, the transmitted power P_t is found to be 120 kW with 1% accuracy. The wavelength is found to be $\lambda = 1.5 \pm 0.02$ cm. The transmitting antenna gain is given as 13 dB, with a relative measurement error of -20 dB. Finally, after 15 trials, the measurement results for the distance in kilometers are {7.54, 7.12, 7.09, 7.37, 7.86, 7.43, 7.03, 6.97, 8.04, 7.96, 7.61, 7.52, 7.80, 7.16, 7.77}.

1. What is the average distance d_{avg} and its standard deviation Δd (i.e., what are the error bounds)?
2. What will the distance d be if a 95% confidence level is required?
3. What value does the range error Δd approach with the same 95% confidence level if the number of range measurements tends to infinity?
4. What is the received power, $P_r \pm \Delta P_r$?
5. What is the lowest level of the received power under the 99% confidence level?

ANSWERS

1. The average transmitted power is the mean of 15 measurements:

$$d_{\text{avg}} = \frac{1}{N}\sum_{k=1}^{N} d_k = \frac{d_1 + d_2 + d_3 + d_4 + \cdots + d_N}{N}.$$

Adding all 15 measured values and dividing the total by 15 yields, $d_{\text{avg}} = 7.4847$ km. It should be noted that the number of significant digits here is two, so the result is rounded to $d_{\text{avg}} = 7.48$ km.

The distance error Δd is the standard deviation σ, which can be calculated via

$$\Delta d = \sqrt{\frac{\sum_{k=1}^{N}\left(d_k - d_{avg}\right)^2}{N-1}}.$$

In this case, it is found to be $\Delta d = 0.3544$ km. Rounding off to two digits yields the distance $d_m = d \pm \Delta d = (7.48 \pm 0.35) \times 10^3$ m.

2. The distance error, $\Delta d = 0.3544$ (which reduces to $\Delta d = 0.35$ km after rounding off), corresponds to the standard deviation of $\sigma = \pm 1$, and shows a 68.27% confidence level. In order to reach a confidence level of 95% ($\sigma = \pm 1.96$), Δd must be multiplied by 1.96. Therefore, the distance with 95% confidence level is $d_m = d \pm 1.96\Delta d = (7.48 \pm 0.67) \times 10^3$ m.

This would have been true if the number of measurements was greater than 30. According to sampling theory, for an experiment with a normal distribution, the mean of the population can be predicted via (2.12) as $d_m = d \pm 1.96\Delta d/\sqrt{N}$. If, for example, $N = 200$ with the same standard deviation, then we are 95% confident that the distance is in the range of $d_m = (7.48 \pm 0.09) \times 10^3$ m. Since $N = 15$ in this example, the critical value of the Student's distribution (t-distribution) should be used. It is seen from Table 2.1 that for $v = 15$ (i.e., for 15 degrees of freedom [df]), using a 95% CI, one should use $d_m = d \pm 1.75\Delta d/\sqrt{15}$ with $\Delta d = 0.35$, and obtain $d_m = (7.48 \pm 0.16) \times 10^3$ m (the range error is $\Delta d = 0.16$ km).

3. According to (2.12), if N goes to infinity, the range error tends to zero.

4. The decibel gain of the transmitting antenna is given by $G_{\text{dB}} = 10 \log_{10}(G)$. Therefore, $G_{\text{dB}} = 13$ corresponds to $G = 10^{G_{\text{dB}}/10} = 20$. The relative gain in decibels is calculated from $\Delta G_{\text{dB}} = 10 \log_{10}(\Delta G/G)$. A relative gain error of -20 dB corresponds to $\Delta G = 0.20$. As a result, the gain of the transmitting antenna that will be used in (2.14) is $G_t = 20.00 \pm 0.20$.

The mean received power is calculated using

$$P_r = \frac{P_t G_t \lambda^2}{(4\pi d)^2} = \frac{120 \times 10^3 \times 19.95 \times 0.015}{\left(4 \times \pi \times 7.48 \times 10^3\right)^2} = 4.06\ \mu\text{W}.$$

The error in the received power can be calculated via

$$\frac{\Delta P_r}{P_r} = \frac{\Delta P_t}{P_t} + \frac{\Delta G_t}{G_t} + 2\frac{\Delta \lambda}{\lambda} + 2\frac{\Delta d}{d}.$$

In our case, $\Delta P/P_r = 0.01 + 0.02 + 0.026 + 0.04 = 0.10$, and the absolute error in the received power will be 0.61 μW. Therefore, the received power is $P_r = 4.06 \pm 0.04$ μW.

5. As mentioned earlier, a 99% confidence level corresponds to $\pm 3\sigma$ (equivalently, three times the absolute error). Therefore, the calculations should be repeated with the Student's distribution for $v = N = 15$ (i.e., for 15 dfs), using a 99% CI (from Table 2.1, this value corresponds to 2.60).

Range error: $(2.6 \times 0.35/\sqrt{15})\ \Delta d = 0.23$ km.

Received power error: $(\Delta P_r/P_r = 0.01 + 0.02 + 0.026 + 0.06 = 0.12)$ $\Delta P_r = 0.12$ μW.

Received power: $P_r = 4.06 \pm 0.49$ μW.

The lowest level of received power will then be $P_r \geq 3.57$ μW.

2.6 HYPOTHESIS TESTING

Hypothesis testing is nothing but giving a *statistical decision* [4] whether or not a claim is true. *Null hypothesis* is the claim; any hypothesis that is different is called *alternative hypothesis*; it consists of the procedure to decide whether to accept or reject this null hypothesis. Sometimes, the null hypothesis is rejected when it is true (type-I error, *false rejection*), sometimes, it is accepted when it is false (type-II error, *false acceptance*).

If X is a Gaussian distributed stochastic parameter with mean μ and standard deviation σ, then $Z = (X - \mu)/\sigma$ is normally distributed (with zero mean and unit standard deviation). If n-samples are tested from an infinite population with mean μ_s and standard deviation σ_s, then $Z = \sqrt{n}(X - \mu_s)/\sigma_s$ is also normally distributed. The probability of $z_1 \leq Z \leq z_2$ (in the limit as $z_1 \to -\infty$, $z_2 \to +\infty$,) in these cases can be calculated from

$$P(z_1 \leq Z \leq z_2) = \frac{1}{\sqrt{2\pi}} \int_{z_1}^{z_2} \exp\left(-\frac{v^2}{2}\right) dv. \quad (2.15)$$

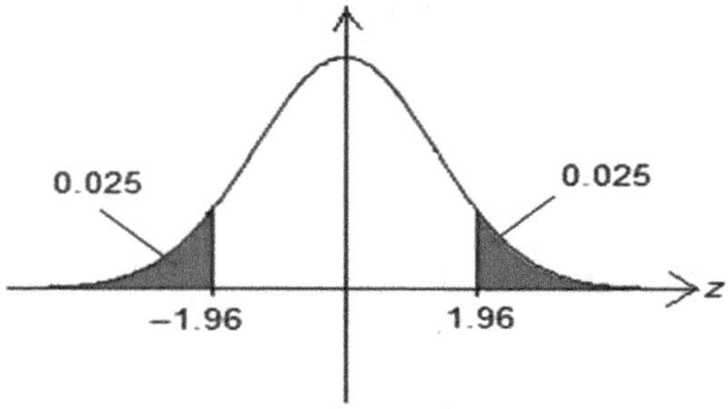

Figure 2.2. Critical value of a two-sided Z-test (95% CI, $\alpha = 0.5$).

This probability is equal to the area of unshaded region in Fig. 2.2 for $z_{1,2} = \pm 1.96$; therefore, the calculation of the probability is nothing but a numerical integration of the specified area under the PDF curve bounded by given two values z_1 and z_2.

This probability for $z_1 = -1.96$ and $z_2 = 1.96$ is 0.95. The verbal statement is either of the following:

- The probability of getting a z between $z_{1,2} = \pm 1.96$ is 0.95.
- We are 95% confident (sure) that z would be in the range of $-1.96 \leq z \leq 1.96$.

Designing a hypothesis testing starts with the choice of the *level of significance* (α), since it directly affects whether to accept or reject a null hypothesis. In practice, a significance level of $\alpha = 0.05$ or $\alpha = 0.01$ is customary. A significance level of $\alpha = 0.05$ means 95% CI (one is 95% sure that the result is true) about the outcome; .that is, 95 out of 100 tests would lead to the correct decision. In other words, one could be wrong with probability of 0.05. The p-value measures how likely it was true that one would have gotten his sample results if the null hypothesis were true. If for example, "there is no difference between any two processes" is the null hypothesis, then it will be accepted (rejected) if $p \geq \alpha$ ($p < \alpha$) (see Ref. 4 for the definitions of α, CI, and p-value).

Note that, *the higher the CI, the wider the range of uncertainty.* One may be 100% confident only when z is in the range of $-\infty \leq z \leq \infty$. The hypothesis testing is then reduced to finding critical values z_1 and/

Table 2.2.
Critical *Z* Values for the Hypothesis Testing

Level of significance (α)	0.10	0.05	0.01	0.005
Left-tailed test	−1.280	−1.645	−2.330	−2.580
Right-tailed test	+1.280	+1.645	+2.330	+2.580
Two-tailed test	±1.645	±1.96	±2.580	±2.810

or z_2 for a given CI or level of significance $(1 - P(z_1 \leq z \leq z_2))$. Figure 2.2 shows the critical values for 95% CI (or $\alpha = 0.5$).

Three types of hypothesis testing are possible: two tailed, left tailed, and right tailed. If, for example, the hypothesis test is about whether or not one process is better or worse than the other, such a test is called one-tailed test. If, on the other hand, the test is about whether or not two processes are the same, this is a two-tailed test. Table 2.2 lists some common level of significances and the corresponding critical values.

Categorically, hypothesis testing problems can be grouped into five as will be given later [4].

2.6.1 Testing Population Mean

Claim: Doğuş University students study at least half an hour in average every night.
Test:

Step 1. Get a random sample with n students and ask them how long they study every night in the average.

Step 2. Calculate the mean (μ_s) and standard deviation (σ_s) of the sample space.

Step 3. Calculate standard error ($\mathrm{SE} = \sigma_s / \sqrt{n}$).

Step 4. Calculate *Z*-test score ($z_t = (x - \mu_s)/\mathrm{SE}$).

Step 5. Compare test score z_t with the critical value z_c determined from the given CI.

Null hypothesis H_0: $\mu = \mu_s = 30$ min $\Rightarrow (z_t \geq z_c)$.
Alternative hypothesis H_1: $\mu = \mu_s \Rightarrow (z_t < z_c)$.

2.6.2 Testing Population Proportion

Claim: At least 30% of Doğuş University students carry a cell phone but not a calculator. (Is this claim clear for you?)

Test:

Step 1. Get a random sample with n students having cell phones and ask them if they have a calculator with them; say k of the n students have a calculator.

Step 2. Calculate sample proportion ($p_s = k/n$).

Step 3. Calculate standard error in terms of stated proportion p_0 ($\text{SE} = \sqrt{p_0(1 - p_0)/n}$).

Step 4. Calculate Z-test score ($z_t = (P_s - P_0)/\text{SE}$).

Step 5. Compare test score z_t with the critical value z_c determined from the given CI.

Null hypothesis H_0: $p_s \geq p_O \Rightarrow (z_t \geq z_c)$.
Alternative hypothesis H_1: $p_s < p_O \Rightarrow (z_t \geq z_c)$.

2.6.3 Testing Two Population Averages

Claim: Doğuş University female students spend more money on meals than male students.

Test:

Step 1. Get randomly chosen two sets of samples; n_1 female and n_2 male students; ask them how much they spend on food.

Step 2. Calculate sample means (μ_x, μ_y) and standard deviations (σ_x, σ_y); x for female and y for male.

Step 3. Calculate standard error ($\text{SE} = \sqrt{\sigma_x^2/n_1 + \sigma_y^2/n_2}$).

Step 4. Calculate Z-test score ($z_t = (\mu_x - \mu_y)/\text{SE}$).

Step 5. Compare test score z_t with the critical value z_c determined from the given CI.

Null hypothesis H_0: $\mu_x = \mu_y \Rightarrow (z_t \geq |z_c|)$.
Alternative hypothesis H_1: $\mu_x \neq \mu_y \Rightarrow (z_t < |z_c|)$.

2.6.4 Testing Two Population Proportions

Claim: Average numbers of smokers among male and female student at Doğuş University are the same.

Test:

Step 1. Get randomly chosen two samples; n_1 female and n_2 male students; ask them if they smoke; say number of female and male students are k_1 and k_2.

Step 2. Calculate sample proportions ($p_1 = k_1/n_1$ and $p_2 = k_2/n_2$).

Step 3. Calculate overall proportion ($p = (k_1 + k_2)/(n_1 + n_2)$).

Step 4. Calculate standard error ($\mathrm{SE} = \sqrt{p(1-p)(1/n_1 + 1/n_2)}$).

Step 5. Calculate Z-test score ($z_t = (p_1 - p_2)/\mathrm{SE}$).

Step 6. Compare test score z_t with the critical value z_c determined from the given CI.

Null hypothesis H_0: $p_1 = p_2 \Rightarrow (z_t \geq |z_c|)$.
Alternative hypothesis H_1: $p_1 \neq p_2 \Rightarrow (z_t < |z_c|)$.

2.6.5 Testing Paired Data

Claim: Doğuş University students are more successful in a classical exam than in a multichoice test.

Test:

Step 1. Get randomly chosen n students and give them both tests and list their scores.

Step 2. Calculate the differences (d_i, $i = 1, \ldots, n$) in the individual scores, and the mean (μ_d) and the standard deviation (σ_d) of the sample group.

Step 3. Calculate standard error ($\mathrm{SE} = \sigma_d/\sqrt{n}$).

Step 5. Calculate Z-test score:

$$z_t = \mu_d/\mathrm{SE} = \frac{\sum d}{\dfrac{\sqrt{n\sum d^2 - \left(\sum d\right)^2}}{\sqrt{n-1}}}.$$

Step 6. Compare test score z_t with the critical value z_c determined from the given CI.

Null hypothesis H_0: $(z_t \leq |z_c|)$.
Alternative hypothesis H_1: $(z_t > |z_c|)$.

Note that MATLAB has its own Statistics Toolbox and the *ztest* module can be used for the hypothesis testing:

$$[\text{h}, pvalue, \text{ci}] = ztest(x, mu, sigma, \text{alpha} = 0.05, \text{'}both\text{'}).$$

For n-element x-sample set, this command tests whether or not the sample set and the population (or the other set) with mean μ and standard deviation σ are the same [1].

2.7 HYPOTHETICAL TESTS ON CELL PHONES

Test 1:

A single blind test was performed among Doğuş University students: 26 out of 374 students who stayed all day in a room with a base station antenna inside (who were exposed to EM fields at 900 MHz) reported to experience dizziness. On the other hand, 8 students out of 210 who think they were exposed (placebo trial) also reported dizziness.

QUESTION

Do you think (with 95% CI or $\alpha = 0.5$) EM signals cause dizziness?

Categorically, this is testing two populations' proportions. Following the steps of test 4, the proportions of the two populations are found as $p_1 = 26/374 = 0.07$and $p_2 = 8/210 = 0.04$. The overall proportion is $p = (26 + 8)/(374 + 210) = 0.058$. Then, the standard error will be SE $= 0.02$. Finally, the Z-test score is $z_t = (0.07 - 0.04)/0.02 = 1.5$.

This is a single-side (right-side) test (why?), so put the null and alternative hypotheses as follows.

Null hypothesis: H_0: $p_1 = p_2$.

Alternative hypothesis: H_1: $p_1 > p_2$.

This means accept H_0 if $z_t < z_c$, otherwise reject the claim. The test Z-score and the critical value are $z_t = 1.5$ and $z_c = 1.645$, respectively; therefore, the null hypothesis should be rejected. *In other words, there is no enough evidence that RF exposure (i.e., 900 MHz GSM signals) causes dizziness.*

What if the results of this hypothesis testing are requested with 90% CI (or $\alpha = 0.1$)? In this case, the critical value (see Table 2.1) would be $z_c = 1.28$; therefore, the null hypothesis should be accepted. *In other words, there is enough evidence that RF exposure (i.e., 900 MHz GSM signals) causes dizziness.*

Is not that interesting? The result totally changes when the CI changes. Do you believe in this test? Is it that simple to do a test like this one?

Test 2:

Recently, Doğuş University students conducted a case-control study on EM signals and headache. A total of 975 male and female students sampled randomly from all the departments were considered. The case group was established with 200 students who have migraine, out of 975. The control group consisted of the rest (775 students) who did not have migraine.

QUESTION

Does a 30 min (or more) continuous use of the cell phone increase a headache of people with migraine?

The case-control experiment is designed as follows.

- The case group used their cell phones intentionally at least 30 min every morning for the whole month. The rest of the day/month, they used their phones regularly.
- The control group used their cell phones regularly every day for whole month.
- A total of 205 students, 96 from the case group and 109 from the controls, reported a headache after a month.
- The other 770 students, 104 from the case group and 666 from the controls, reported no headache after a month.
- The 2 × 2 table of this case-control study is given in Table 2.3.

Table 2.3 displays exposed cases (a_1), the exposed controls (b_1), unexposed cases (a_2), and the unexposed controls (b_2). The statistics generated to measure association is the *odds ratio* (OR), which is the ratio of the odds of exposure in the cases (a_1/a_2) to the odds of exposure

Table 2.3.
The 2 × 2 Table of the Case-Control Study

	Case	Control	Total
Exposed	96 (a_1)	109 (b_1)	205 (n_1)
Unexposed	104 (a_2)	666 (b_2)	770 (n_1)
Total	200 (m_1)	775 (m_1)	975 (N)

in the controls (b_1/b_2). This is equal to (a_1b_2/a_2b_1). The OR much greater than 1 means those with the migraine are much more vulnerable to the use of cell phones when headache is of interest. The OR close to 1 means cell phone use and headache are not likely associated. The OR far less than 1 suggests that the migraine is a protective factor in the causation of the headache when using a cell phone.

The OR from Table 2.3 is OR = (96 × 666)/(109 × 104) = 5.64. This OR suggests that headache is 6.64 times as frequent in the students with a migraine in Doğuş University. Since the student population of Doğuş University (nearly 5000) is much more than 975, there should be an interval in which this OR might change. This interval is stated in terms of the requested CI. The $100(1-\alpha)\%$ CI of this test interval (uncertainty) is calculated from

$$\Psi = e^{(\ln(\mathrm{OR}) \pm z_\alpha \mathrm{SE_{OR}})}, \tag{2.16}$$

where z_α is the critical value for the Z-test corresponding to the given CI (e.g., $z_\alpha = 1.65$ for 90% CI, $z_\alpha = 2.58$ for 99% CI) and $\mathrm{SE_{OR}}$ is the standard error calculated from

$$\mathrm{SE_{OR}} = \sqrt{\frac{1}{a_1} + \frac{1}{a_2} + \frac{1}{b_1} + \frac{1}{b_2}}. \tag{2.17}$$

The $\mathrm{SE_{OR}}$ from the values listed in Table 2.3 is calculated to be 0.1752; therefore, 95% CI interval of this case-control study is [4.00–7.95].

The OR = 5.64 (95%, [4.00–7.95]) therefore means that it does not matter how many times you specify the case and control groups and do the hypothesis test, the resulting OR would be around [4.00–7.95] for a given 95% CI.

Test 3:

The correlation between a cell phone usage and reaction times is going to be investigated (with 95% CI or $\alpha = 0.5$)?

Claim:

Using a cell phone for a long time (let us say 20 min and more) affects the reaction time of a person.

This is the fifth (double-side) test listed earlier (testing paired data). Ten students are randomly selected. Their reaction times are recorded before and after using a cell phone 20 min or more. Table 2.4 shows the test data.

Note that since there are only a few students (less than 30), the Student (t)-test would be used instead of Gaussian Z-test [10]. The df of this test is $n - 1 = 9$. The critical value for 95% CI or $\alpha = 0.5$ with df $= 9$ is read (from t-test tables) [10] to be $t_c = 2.262$. The test Z-score (or t-score in this case) is

$$t_t = \mu_d/\mathrm{SE} = \frac{\sum d}{\frac{\sqrt{n\sum d^2 - \left(\sum d\right)^2}}{\sqrt{n-1}}} = \frac{-26}{\frac{\sqrt{10\times 164 - (-26)^2}}{\sqrt{9}}} = -2.512$$

The null and alternative hypotheses are $t_t \leq |t_c|$ and $t > |t_c|$, respectively. Since t_t $(= -2.512)$ is less than t_c $(= -2.262)$, the null hypothesis should be rejected. In other words, *the use of cell phone for 20 min*

Table 2.4.
Data Table for the Reaction Time Test

Subject	Reaction Time Before, TB (ms)	Reaction Time After, TA (ms)	d (ms)	d^2
1	32	27	5	25
2	21	24	−3	9
3	23	29	−6	36
4	25	29	−4	16
5	18	22	−4	16
6	30	29	1	1
7	31	33	−2	4
8	30	35	−5	25
9	27	31	−4	16
10	16	20	−4	16
	$\mathrm{TB_{ave}} = 25.3$	$\mathrm{TA_{ave}} = 25.3$	$\sum d = -26$	$\sum d^2 = 164$

or more reduces the reaction time of a person. But, if we want to express the result of this test with a 99% CI, then t_t (= −2.512) would be within the interval of $-3.25 \leq t_c \leq 3.25$, which concludes that the null hypothesis should be accepted. In other words, we are 99% sure that *the use of cell phone for 20 min or more does not affect the reaction time of a person* (remember, this is a hypothetical test and the numbers are dummy!).

2.8 CONCLUSIONS

Everybody uses numbers: therefore, public understanding of innumeracy is essential [5]. Society must be prepared accordingly, especially in terms of the meanings of the numbers used. To some extent, everybody should be equipped to deal with *data acquisition*, *correlation*, *using a model*, *statistics*, *randomness in nature*, *risk management*, *information-based decision making*, *uncertainty*, and the *bounds of science*.

REFERENCES

1. L. Sevgi, "Speaking with Numbers: Public Understanding of Science," *ELEKTRIK, Turkish Journal of Electrical Engineering and Computer Sciences*, **14** (1), 2006, pp. 33–40.
2. M. Spiegel, J. Schiller, A. Srinivasan, *Schaum's Outline of Probability and Statistics—Crash Course*, McGraw-Hill, New York, 2000.
3. P. F. Dunn, *Measurement and Data Analysis for Engineering and Science*, McGraw-Hill, New York, 2005.
4. D. Rumsey, *Statistics for Dummies*, Wiley Publishing, Hoboken, NJ, 2003.
5. J. A. Paulos, *Innumeracy*, Hill and Wang, New York, 2001.

CHAPTER 3

Numerical Analysis in Electromagnetics

EM modeling and simulation (EM-MODSIM) necessitates a good understanding and use of numerical analysis [1,2]. Built-in MATLAB functions such as ***roots***(), ***diff***(), ***gradient***(), ***trapz***(), and ***inv***() can be used for most of numerical analysis jobs. Zero crossing of a function may be directly obtained via ***roots***(), numerical derivative of a function can be extracted by ***diff***() or ***gradient***(), the integral of a function between specified limits can be found by ***trapz***(), and an inverse of a square $N \times N$ matrix can be obtained via ***inv***(). Although MATLAB makes life easy for engineers, EM-MODSIM people should always depend on their own codes; at least aware of and well understand what is going on behind these built-in commands. This chapter reviews briefly some numerical analysis fundamentals which are essential for EM-MODSIM studies.

3.1 TAYLOR'S EXPANSION AND NUMERICAL DIFFERENTIATION

Any continuous function $f(x)$ having all order derivatives can be represented exactly and completely if its value and all order derivatives are known at a point x_0 as

Electromagnetic Modeling and Simulation, First Edition. Levent Sevgi.

$$f(x) = f(x_0) + \frac{(x-x_0)}{1!}\frac{df(x_0)}{dx} + \frac{(x-x_0)^2}{2!}\frac{d^2 f(x_0)}{dx^2} + \cdots + \frac{(x-x_0)^n}{n!}\frac{d^n f(x_0)}{dx^n} + \cdots. \tag{3.1}$$

This sum is called Taylor's expansion and is used to replace mathematical derivative with a finite-difference approximation. The sum is called MacLaurin series if $x_0 = 0$. Taylor's expansion is used to approximate a function for a given accuracy. If x is close to x_0, then a few terms would be adequate within a specified accuracy (e.g., less than a relative error of 1–2%). More and more terms would be required for the same specified accuracy as x goes further and further away from x_0. This is illustrated in the example given in Fig. 3.1.

Suppose y = cos(x) is taken into account. Its Taylor expansion at $x_0 = 0$ (i.e., MacLaurin series) will be

$$f(x) = f(0) + x\frac{df(0)}{dx} + \frac{x^2}{2}\frac{d^2 f(0)}{dx^2} + \frac{x^3}{6}\frac{d^3 f(0)}{dx^3} + \frac{x^4}{24}\frac{d^4 f(0)}{dx^4} + \cdots,$$

which yields

$$f(x) = 1 - \frac{x^2}{2!} + \frac{x^4}{4!} - \frac{x^6}{6!} + \frac{x^8}{8!} \pm \cdots.$$

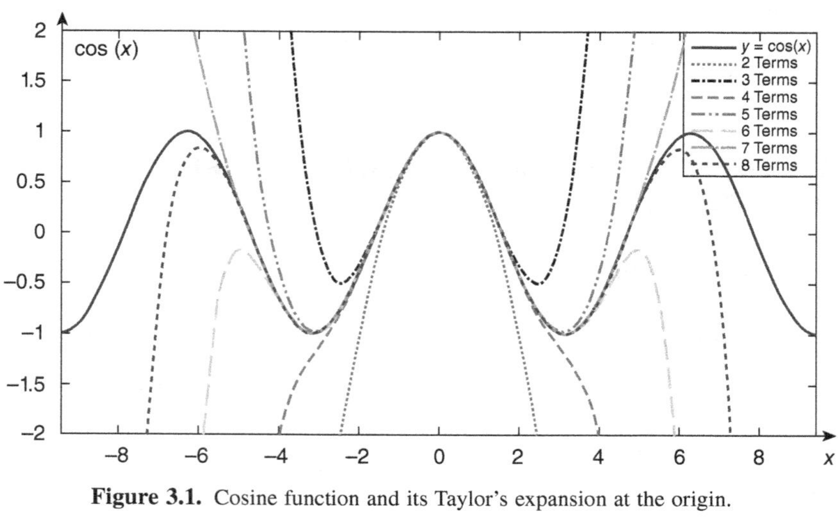

Figure 3.1. Cosine function and its Taylor's expansion at the origin.

Table 3.1.
Code for the Taylor's Expansion of Cosine Function

```
%=============================================
% Program: Taylor's Expansion
%=============================================
clear all; clc; eps=1e-10;
x=-2*pi:0.01:2*pi; fn=cos(x); fn1=1;
error=eps+1; n=1; m=1;
while n>=1
   plot(x,fn,'k',x,fn1, 'r-','LineWidth',2);
   grid; xlabel('x'); ylabel('Function');
   xlim([-2*pi 2*pi]);  %ylim([-1.2 1.2])
   legend('Function','Taylor Expansion');
   title(['Number of Terms:    ', num2str(n),]);
   pause;
   m=2*n; fn1=fn1+(-1)^n*x.^m/factorial(m);
   n=n+1;
end
%------------------------ END ------------------------
```

The MATLAB code which plots Fig. 3.1 is given in Table 3.1. Mathematical definition of a derivative is given as

$$f'(x_0)=\frac{df}{dx}=\mathrm{Lim}_{\Delta x\to 0}\frac{f(x_0+\Delta x)-f(x_0)}{\Delta x}. \tag{3.2}$$

Numerically, (3.2) can be replaced with

$$f'(x_0)\cong\frac{f(x_0+\Delta x)-f(x_0)}{\Delta x}+O(\Delta x), \tag{3.3}$$

which may easily be obtained from (3.1) if $\Delta x = x - x_0$. Here, $O(\Delta x)$ refers to the order of error (the most significant term of the residue). This is called the *forward difference* scheme, since only the two points x_0 and $x_0 + x_0$ are used. Similarly,

$$f'(x_0)\cong\frac{f(x_0)-f(x_0-\Delta x)}{\Delta x}+O(\Delta x) \tag{3.4}$$

is the *backward difference* scheme, while

$$f'(x_0) = \frac{f(x_0+\Delta x) - f(x_0-\Delta x)}{2\Delta x} + O(\Delta x^2) \tag{3.5}$$

is the *central difference* scheme. It should be noted that central difference scheme uses three points for the numerical derivative but the error is in the order of O(Δx^2), which is better than the order of the error of O(Δx) in forward and backward difference schemes. The order of error may be decreased if the number of points used in numerical derivation increases.

Similarly, finite-difference representations of second derivatives are

$$f''(x_0) = \frac{f(x_0+2\Delta x) - 2f(x_0-\Delta x) + f(x_0)}{\Delta x^2}, \tag{3.6}$$

$$f''(x_0) = \frac{f(x_0+\Delta x) - 2f(x_0) + f(x_0-\Delta x)}{\Delta x^2} \tag{3.7}$$

for forward and central difference approximations, yielding the error in the orders of O(Δx) and O(Δx^2), respectively.

3.1.1 Taylor's Expansion and Ordinary Differential Equations

Taylor's expansion is also one of a few important methods for the numerical solution of ordinary differential equations (ODEs). In order to explain the Taylor's expansion method in the ODE solution, an example problem is taken into account.

For example, let us solve

$$\frac{dy}{dx} = -2x - y \tag{3.8a}$$

for the initial condition of

$$y(0) = -1, \tag{3.8b}$$

whose analytical exact solution is given as

$$y(x) = -3e^{-x} - 2x + 2. \tag{3.9}$$

It is known that the Taylor's expansion for the first three terms is equal to

$$f(x) = f(x_0) + \frac{\Delta x}{1!}\frac{df(x_0)}{dx} + \frac{\Delta x^2}{2!}\frac{d^2 f(x_0)}{dx^2} + \frac{\Delta x^3}{3!}\frac{d^3 f(x_0)}{dx^3} + \text{error}. \quad (3.10)$$

In order to determine $y(x)$, one should start from (3.8a) with the initial condition given (3.8b). Using $x = 0$ and $y(0) = -1$ in (3.8a), one derives the second equation:

$$\frac{dy}{dx}_{|x=0} = y'(0) = -2(0) - (-1) \Rightarrow y'(0) = 1. \quad (3.11)$$

The second order derivative $y''(x)$ can be obtained from (3.8a) directly as

$$\frac{d^2y}{dx^2} = -2 - \frac{dy}{dx}, \quad (3.12)$$

and for $x = 0$, this is equal

$$\frac{d^2y}{dx^2}_{|x=0} = y''(0) = -2 - y'(0) = -2 - (-1) \Rightarrow y''(0) = -3. \quad (3.13)$$

Similarly, the values for the third and fourth derivatives can be obtained. First, get the derivate (3.12) once and twice again and get

$$\frac{d^3y}{dx^3} = -\frac{d^2y}{dx^2}, \frac{d^4y}{dx^4} = -\frac{d^3y}{dx^3}. \quad (3.14)$$

Then, find their values for $x = 0$:

$$\frac{d^3y}{dx^3}_{|x=0} = y'''(0) = -y''(0) = -(-3) \Rightarrow y'''(0) = 3, \quad (3.15)$$

$$\frac{d^4y}{dx^4}_{|x=0} = -y''''(0) = -y'''(0) = -(3) \Rightarrow y''''(0) = -3. \quad (3.16)$$

Finally, the solution is written as

$$f(x) = -1 + \Delta x - \frac{3}{2}\Delta x^2 + \frac{1}{2}\Delta x^3 - \frac{1}{8}\Delta x^4 + \text{error}. \quad (3.17)$$

Table 3.2.
Numerical Solutions Using (3.9) and (3.17)

X	y_{Taylor}	$y_{Analytic}$
0.0	−1.00000	−1.00000
0.1	−0.91451	−0.91451
0.2	−0.85620	−0.85619
0.5	−0.82031	−0.81059

Table 3.2 lists the numerical solutions of (3.8) calculated with the Taylor's expansion (3.17) and analytical exact solution (3.9) for different x-values.

3.1.2 Poisson and Laplace Equations

Elliptic-type Poisson equation is a partial differential equation widely used to represent physical problems in electrical and mechanical engineering, theoretical physics, and others. It represents various physical problems: temperature distribution in thermodynamics, gravitational fields and a vibrating string in mechanics, potential distribution in electrostatics, and so on. It is given as

$$\nabla^2 U(x, y, z) = f(x, y, z), \tag{3.18}$$

where, in three-dimensional (3D) Cartesian coordinates, the Laplace operator ∇^2 (or, as widely used Δ) is given as

$$\nabla^2 = \frac{\partial^2}{\partial x^2} + \frac{\partial^2}{\partial y^2} + \frac{\partial^2}{\partial z^2}. \tag{3.19}$$

If the right-hand side of (3.18) is the Dirac delta function $\delta(x - x')(y - y')(z - z')$, then the Green's function $g(x,y,z;x',y',z')$ replaces the function U. The Poisson equation is not a basic equation, but follows directly from the Maxwell equations if all time derivatives are zero, that is, for electrostatic conditions. In electrostatics, Poisson equation expresses Gauss's law; U is the scalar potential function and is obtained from the solution of (3.18) under a given charge density distribution function f (i.e., $f = -\rho/\varepsilon$, where ρ and ε are the volume charge density and permittivity of the region, respectively). In elasticity, on the other hand, U in one-dimensional (1D) denotes the transverse displacement along a direction of a stretched string under a nonzero tension T, with the right-hand side of (3.18) as $f = w/T$, w being force density. Equation (3.18) becomes Laplace equation for vanishing of f.

The wave equation, on the other hand,

$$\left(\nabla^2 - \frac{1}{c^2}\frac{\partial^2}{\partial t^2}\right)U(x, y, z; t) = 0, \tag{3.20}$$

represents waves (spherical in 3D, cylindrical in two dimensions [2D]) in the time domain in EMs and $c = 1/\sqrt{\varepsilon_0 \mu_0}$ is the velocity of light in free space. Note that for a specified geometry, (3.18) represents a boundary value problem (BVP), but (3.20) is an initial-BVP (IBVP). Wave equation will be treated numerically in details in future chapters.

3.1.3 An Iterative (Finite-Difference) Solution

Both Poisson and Laplace equations can be solved numerically. One way is to discretize and put them into iterative forms. Suppose the Poisson equation is taken into account in 1D along z-direction in a finite region $0 \leq z \leq Z_m$. The differential equation reduces to

$$\frac{d^2}{dz^2}U(z) = f(z) \tag{3.21}$$

and can uniquely be solved if the boundary conditions $U(0)$ and $U(Z_m)$ are supplied. The geometry of the problem is given in Fig. 3.2. The finite-difference solution of (3.21) is

$$U(k) = \frac{1}{2}[U(k+1) + U(k-1) - f(k)] \tag{3.22}$$

and is applied to all inner nodes (i.e., k = 2,3,4, . . ., KE-1, KE being the number of nodes). It states that the potential value at a node is

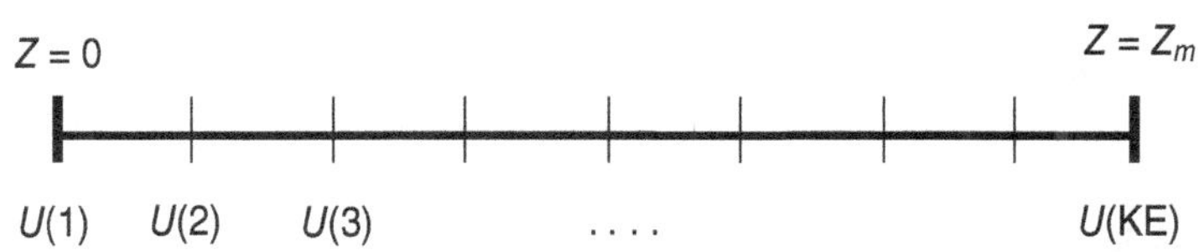

Figure 3.2. One-dimensional finite z-space.

determined by the values of two neighboring nodes and a given charge distribution. The terminating values $U(1)$ and $U(\mathrm{KE})$ should be supplied as boundary conditions. In 2D, (3.21) and (3.22) become

$$\left(\frac{\partial^2}{\partial x^2}+\frac{\partial^2}{\partial y^2}\right)U(x,y)=f(x,y), \tag{3.23}$$

$$U_{i,j}=\frac{1}{4}\left[U_{i+1,j}+U_{i-1,j}+U_{i,j+1}+U_{i,j-1}-f_{i,j}\right]. \tag{3.24}$$

It states that the potential at a node can be calculated from the potentials of four neighboring nodes plus a given charge distribution. Figure 3.3 shows the results of a Laplace problem with $Z_m = 1$, $U(0) = -2$, and $U(1) = 1$. It means the potential at left and right are -2 and 1 V, respectively. The problem is to find out $U(x)$ function which satisfies these boundary conditions and (3.21). Initially, all the inner nodes are assumed zero and (3.22) is applied over and over again in an infinite loop. The iterations cause the boundary potentials to propagate node by node from both ends. As the number of iterations increases, the potential values at the inner nodes change because of the

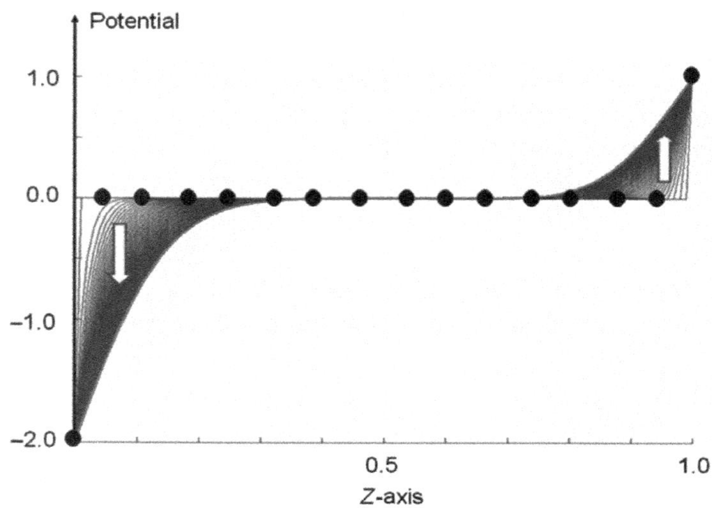

Figure 3.3. Numerical solution of Laplace equation for $Z_m = 1$, KE $= 100$, $U(0) = -2$, $U(1) = 1$, $\gamma = 62$, $\gamma \leq$ le -3.

contributions from left and right and then converge. The difference between the iteration and the previous one becomes less and less, and finally settles to a fix value. A kind of convergence criteria such as "the iterations may be terminated if the sum of the differences between two consecutive iterations of all nodes is less than a given value" may be set. In 1D, this is equivalent to setting

$$\sum_{k=1}^{\mathrm{KE}} \left|U_k^{n+1} - U_k^n\right| \leq \gamma, \tag{3.25}$$

where γ is the total error limit. The arrows in Fig. 3.3 show the propagation of the potential values at nodes as the number of iteration increases. The thick line is the final potential distribution which satisfy a given error criteria.

Figure 3.4 shows the progress of iterations for Poisson equation for the same space and boundary conditions given in Fig. 3.3 but with a Gaussian-type charge distribution. The dots show the given charge distribution. The left and right vertical axes correspond to potential and normalized charge densities (i.e., ρ_1/ε), respectively. Figure 3.5 shows another simulation result of the Poisson equation for the same parameters except the charge distribution. Again, the dots show normalized charge distribution and left and right vertical axes correspond to the potential and charge densities, respectively. Finally, two examples for

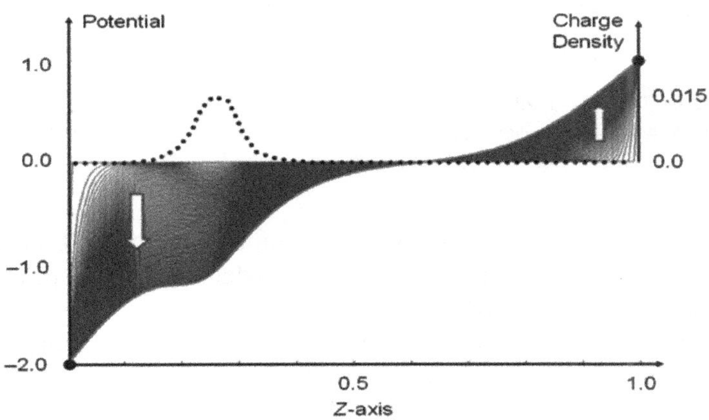

Figure 3.4. Numerical solution of Poisson equation for $Z_m = 1$, KE $= 100$, $U\ (0) = -2$, $U\ (1) = 1$, $\gamma = 146$, $\gamma \leq le\ -3$. The dots show the charge distribution.

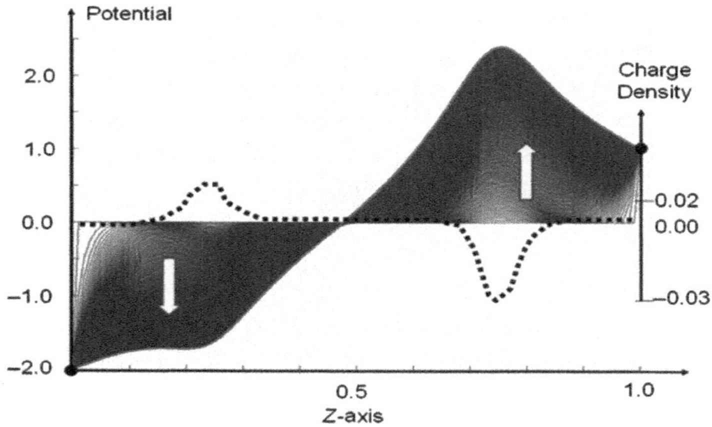

Figure 3.5. Convergence of the Poisson equation for for $Z_m = 1$, KE $= 100$, $U(0) = -2$, $U(1) = 1$, $\gamma = 256$, $\gamma \leq$ le -3. The dots show the charge distribution.

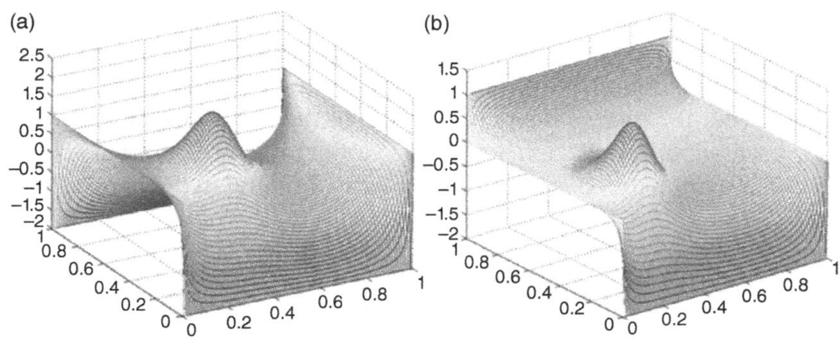

Figure 3.6. The solution of 2D Poisson equation for 1 m^2 plate (100 × 100 space) for the boundary values of (left, right, top, bottom) (a) (1,1,−2,−2), $\gamma = 942$, (b) (0,0,1,−2), $\gamma = 865$, with similar Gaussian charge distributions inside ($\gamma \leq$ le -3).

the solution of Poisson equation in 2D computed from (3.24) are illustrated in Fig. 3.6. A short MATLAB code which is used in producing the figures earlier is listed in Table 3.3.

Note that wave propagation and numerical propagation may sometimes be used improperly. People who deal with simulations in electrostatics and electrodynamics might refer to different concepts with the same term *numerical propagation*. This is because the word *iteration* has different meanings in the numerical simulation of Poisson and wave

Table 3.3.
A Short MATLAB Script for 1D Poisson Equation

```
%% ------------------------------------------------------
%   Program: LS_Poisson1D.m
%% ------------------------------------------------------
clc; clear all; close all;
LZ = 1;          % input('Length of 1D Medium [m] : ?');
NZ = 100;        % input('Number of nodes along Z : ?');
NI = 1000;       % input('Number of iterations (NI) : ');
u(1) = input('Potential at left boundary (z=0): ');
u(NZ) = input('Potential at right boundary (z=LZ): ');
% Parameters
Eps=1.e-3; dz=LZ/(NZ-1); i0=NZ/4; a0=1/sqrt(2.*pi*i0^2);
for i=1:NZ      % set initial values for z() and u()
     z(i)=(i-1)*dz;   u(i)=0; Ui(i)=a0*exp(-(i-i0)^2/
(NZ/4)));
end
figure (1); hold on; plot(z,u); xlim([0 LZ]);
xlabel('Z-axis'); ylabel('Potetial [V]');
% Main iteration loop
k=0; ueps=10;
while ueps>Eps
    k=k+1;   ueps=0;
    for i=2:NZ-1
         old=u(i); u(i)=(u(i+1)+u(i-1)-Ui(i))/2;
new=u(i);
         ueps=ueps+(abs(new-old))^2;
    end
    if k>NI
      display('Maximum iteration number is exceeded !..');
break
    end
    plot(z,u);   pause(0.002);
end    % End of iteration loop
plot(z,u,'r', 'Linewidth', 2);
fprintf('No of Iterations = %3d ', k); figure(2);
subplot(2,1,1); plot(z,Ui,'k-','Linewidth',3);
ylabel('Charge distribution')
subplot(2,1,2);plot(z,u,'r','Linewidth',2);
xlabel(ZX-axis');ylabel('Potetial [V]');
% ------------------------ END ------------------------
```

equations. It refers to the *convergence of the data* for Poisson equation, but yields the *time progress* in the wave equation as will be illustrated in later chapters.

3.2 NUMERICAL INTEGRATION

In EM-MODSIM, we usually meet with finite and/or infinite integrals; therefore, it is appropriate to include a simple numerical integration code. Basically, numerical integration is nothing but calculating the area of a function under its curve surrounded by the horizontal axis and both integration limits. If the integrand decays to zero as the integration limits tend to infinity, then the integration is carried out as in the integration with finite limits. In this case, the user only needs to observe the finite limits at both ends where the contributions become less than a chosen threshold.

3.2.1 Rectangular Method

Numerical integration is nothing but dividing the area under the curve between limits into small rectangular areas as depicted in Fig. 3.7. The

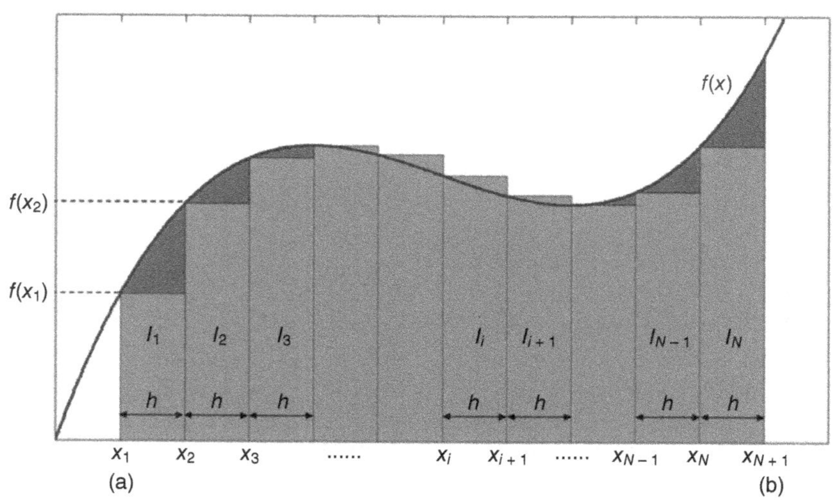

Figure 3.7. The curve of a function and rectangular representation of the area.

height of each rectangle is the value of the function at some point of the rectangle base. The basic question is how one chooses this point. If the left-bottom corner of the rectangle is chosen, then the area of that rectangle will be less (higher) than the real value for monotonically increasing (decreasing) function in that region. There are many methods to reduce the error introduced in this way, but the most important one is to reduce the base length of the rectangles (i.e., increase number of integration steps).

Table 3.4 lists a simple MATLAB code for the numerical integration. The code is based on replacing the mathematical integration with the addition of the areas of neighboring rectangular areas:

Table 3.4.
A Simple MATLAB Code for Numerical Integration

```
%% ---------------------------------------------
%   Program : LS_NumInt.m
%% ---------------------------------------------
clear all; clc; a=2; b=10; h=0.2; I1=0; I2=0; I3=0;
% MATLAB int command
x=a:(b-a)/200:b; fx=x.^3-18*x.^2+96*x; IA=trapz(x,fx);
for x=a:h:b-h  % 1. Rectangular Rule
     y=x^3-18*x^2+96*x;  I1=I1+y*h;
end
for x=a+0.5*h:h:b-0.5*h % Midpoint Rule
     y=x^3-18*x^2+96*x;  I2=I2+y*h;
end
for x=a:h:b-h % Trapezoidal Rule
     y1=x^3-18*x^2+96*x; d=x+h; y2=d^3-18*d^2+96*d;
I3=I3+h*(y1+y2)/2;
end
e1=abs(IA-I1); e2=abs(IA-I2); e3=abs(IA-I3);
fprintf('Rectangle     MidPoint      Trapezoidal \n');
fprintf('-------------------------------------- \n ');
fprintf('   \n');
fprintf(' %10.4f  %10.4f   %10.4f \n', I1, I2, I3);
fprintf('   \n');
fprintf(' Errors    \n');
fprintf('   \n');
fprintf(' %8.6f   %8.6f   %8.6f \n', e1, e2, e3);
% --------------------- END ---------------------
```

$$\int_a^b f(x)dx = f(a)\Delta x + f(a+\Delta x)\Delta x + f(a+2\Delta x)\Delta x + \cdots + f(b)\Delta x, \tag{3.26}$$

where $\Delta x = (b - a) \mid /N$ is the base length (width) of each rectangle. Two other integration models are also included in the code.

A polynomial $x^3 - 18x^2 + 96x$ is chosen as an example. The integral of this polynomial between $x = 1$ and $x = 11$ is 1440 and is calculated analytically ($\int_1^{11} x^3 - 18x^2 + 96x dx = 1440$). Table 3.5 shows absolute and relative errors with respect to the number of rectangles used in numerical integration. As shown there, as many as 20,000 or more rectangles are necessary to approach the exact integration value. This is the simplest numerical integration approach. An improved one is to use trapezoidals instead of rectangles. Built-in MATLAB command ***trapz***() uses this approach. As little as 10 trapezoidals are enough to approach the exact numerical integration value (check with this three line command: *x = 1.0 : 1.0 : 11.0; fx = x.^3-18.*x^2 + 96.*x;* ***trapz(x,fx)***).

Direct numerical integration is based on the calculation of the area under the function between specified boundaries by using tiny

Table 3.5.
Numerical Integration Error Versus Number of Rectangles

Rectangular Integration Method (Exact Value = 1440)			
N	Num Int.	Abs. Error	Rel. Error %
2	1115.0	325	22.569
5	1277.5	162.5	11.285
10	1375.0	65	4.5139
20	1407.5	32.5	2.2569
100	1433.5	6.5	0.45139
1000	1439.3	0.65	0.045139
10,000	1439.9	0.065	0.0045139
20,000	1440.0	0.0325	0.0022569
100,000	1440.0	0.0065	0.00045139

rectangles. The same approach may also be used to find out one of the boundaries if the value of the integration is given. This is important in statistical computations and discussed in Chapter 2. As described there, the probability is the area under the probability density function for the specified critical values. This is the direct numerical integration procedure. The inverse procedure is to find out the critical value once the probability is given. This is known as confidence level calculations in statistics. Table 3.6 lists a sample code for the calculation of critical values (i.e., integration boundaries) for a given confidence level.

Table 3.6.
A Simple MATLAB Code for Numerical Integration of Gauss Function

```
%% -----------------------------------------------
%     Program     : Confidence.m (Normal Distribution)
%% -----------------------------------------------
clc; clear all;
fprintf(1,'----------------------------------\n');
fprintf(1,'Which would you like to calculate? \n');
fprintf(1,' [1] CL of a value interval \n');
fprintf(1,' [2] Critical Values for a given CL \n');
fprintf(1,'---------------------------------- \n');
qq=input(' Your Choice? ');
if qq==1
  sigma=input('Give standard deviation (0-3): ');
  x=0.0:0.0001:sigma;
  normx=1/sqrt(2.*pi)*exp(-x.^2/2);
  result=trapz(x,normx)*200;
  fprintf(1,'The CL is %% %6.2f \n',result);
else
  conf=input('CL as a percentage (0-100)? ');
  conf=conf/200; area=0.0; del=0.0001; y=-del;
  while area&#x003C;=conf
    y=y+del;
    area=area+1/sqrt(2.*pi)*exp(-y^2/2)*del;
  end
  fprintf(1,'The standard deviation is %5.2f \n',y);
end
% ---------------------- End -----------------------
```

3.3 NONLINEAR EQUATIONS AND ROOT SEARCH

Another important topic in numerical analysis is the root finding of a given nonlinear function. Unfortunately, there is no generally applicable root finding method that can be used for all kinds of functions. A method may be best, for example, for a polynomial, maybe worse for a trigonometric function. Therefore, two of the methods are included here.

The simplest root search model is the fixed-point iteration method. Here, the function $y = f(x)$ is written in the form of $x_{i+1} = g(x_i)$. The intersection points of the graphs of $y = x$ and $y = g(x)$ gives the solution (i.e., the root). Table 3.7 lists a short MATLAB code for the root

Table 3.7.
Fixed-Point Iteration MATLAB Code

```
%% ------------------------------------------------
%    Program: LS_FixedPI.m
%    Purpose: Root search with fixed point iteration
%% ------------------------------------------------
function LS_FixedPI
clc; clear all;
ep=1e-5; %input('Error ?');
x0=-1.0; %input('First guess ?');
error=1;
n=0;
fprintf('    Root              Fuction           Error\n');
fprintf('===================================\n');
fprintf('         \n');
while abs(error) > ep,
   n=n+1;
   fx=ff(x0);
   x1=fx;
   fun=x1^2-4*x1-6;
   error=x1-x0;
   x0=x1;
   fprintf ('%12.8f %12.8f %12.8f\n', x1, fun, error);
   pause
end
function f=ff(x)
    f=sqrt(4*x+6);
%------------------------- END ---------------------
```

search using fixed-point iteration method. Here, roots of the function $y = x^2 - 4x - 6$ are searched. The roots of this function are $x_1 = 5.1623$ and $x_2 = -1.1623$. Note that built-in MATLAB command ***roots***() can be used as ***roots***([1 −4 −6]) and these roots can be obtained.

The function is put into $x_{i+1} = \sqrt{4x_i + 6}$ form. The initial guess is $x_0 = -1.0$ and one of the roots is searched up to four digits (i.e., error is set to 1×10^{-5}). The results are shown in Table 3.8.

The Newton–Raphson method uses an iterative process to approach one root of a function. The specific root that the process locates depends on the initial, arbitrarily chosen x-value:

$$x_{n+1} = x_n - \frac{f(x_n)}{f'(x_n)}. \tag{3.27}$$

Table 3.9 lists a simple MATLAB code for this process.

Table 3.8.
Fixed-Point Root Search of $y = x^2 - 4x - 6$ ($x_0 = -1.0$)

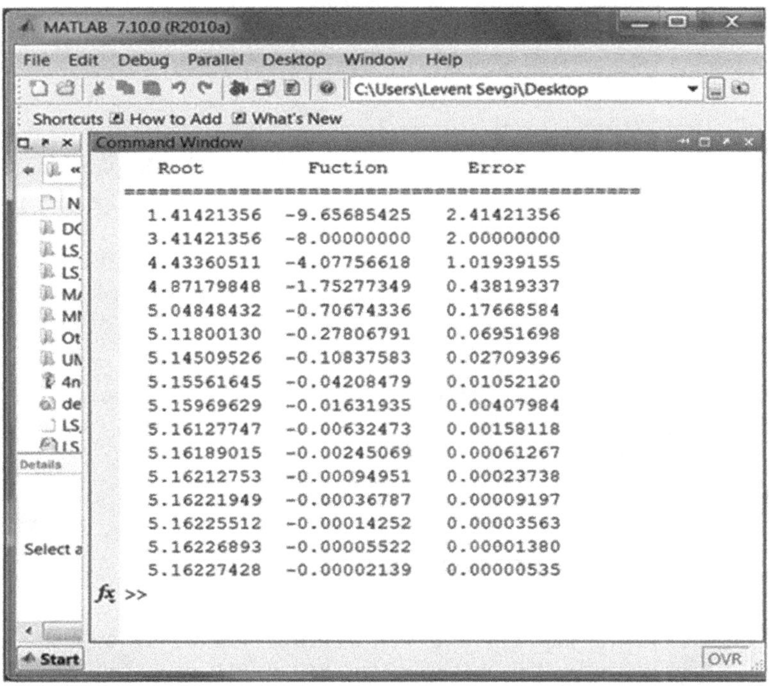

Root	Fuction	Error
1.41421356	-9.65685425	2.41421356
3.41421356	-8.00000000	2.00000000
4.43360511	-4.07756618	1.01939155
4.87179848	-1.75277349	0.43819337
5.04848432	-0.70674336	0.17668584
5.11800130	-0.27806791	0.06951698
5.14509526	-0.10837583	0.02709396
5.15561645	-0.04208479	0.01052120
5.15969629	-0.01631935	0.00407984
5.16127747	-0.00632473	0.00158118
5.16189015	-0.00245069	0.00061267
5.16212753	-0.00094951	0.00023738
5.16221949	-0.00036787	0.00009197
5.16225512	-0.00014252	0.00003563
5.16226893	-0.00005522	0.00001380
5.16227428	-0.00002139	0.00000535

Table 3.9.
Newton–Raphson Root Search MATLAB Code

```
% -----------------------------------------------
%    Program : LS_NewtonRaphson.m
% -----------------------------------------------
function LS_Newton
clc;  clear all; fprintf('\n');
fprintf('It # – Root   -- Fuction -- Error -\n');
fprintf('=============================\n');
x0=5; ep=1e-7; newton('fonk','dfonk',x0,ep);
function newton(f,df,x0,ep)
   format short e;  err=1; k=1;
   while abs(err) > ep,
      fx=feval(f,x0);   dfx=feval(df,x0);
      if dfx==0;    'df(x)=0.  Stop' ;   return; end
      x1=x0-fx/dfx;   err=x1-x0; x0=x1;
      fprintf ('%3d %12.8f %12.8f %12.8f\n',k,x0,fx,err);
      k=k+1; pause
end
function f=fonk(x); f=x^2-4*x-6;
function df=dfonk(x); df=2*x-4;
% -------------------- END -------------------
```

Take $f(x) = \exp(-x) - \sin(x)$ as an example. The derivative of this function is going to be $f'(x) = \exp(-x) - \cos(x)$; therefore, the iterative equation will be

$$x_{n+1} = x_n - \frac{e^{-x_n} - \sin(x_n)}{-e^{-x_n} - \cos(x_n)},$$

and the first three steps will be $x_0 = 0.7$, $x_1 = 0.582964$, $x_2 = 0.588520$, and $x_3 = 0.588520$, which shows a three digit convergence.

Now, take the function $y = x^2 - 4x - 6$ used with the fixed-point iteration. The Newton–Raphson iterations and the result are shown in Table 3.10. As observed, convergence is much faster.

3.4 LINEAR SYSTEMS OF EQUATIONS

As presented in Chapter 1 and will be implemented in later chapters (e.g., Chapter 13), method of moments (MoM) [3] solves linear systems

Table 3.10.
Newton–Raphson Root Search of $y = x^2 - 4x - 6$ ($x_0 = 5.0$)

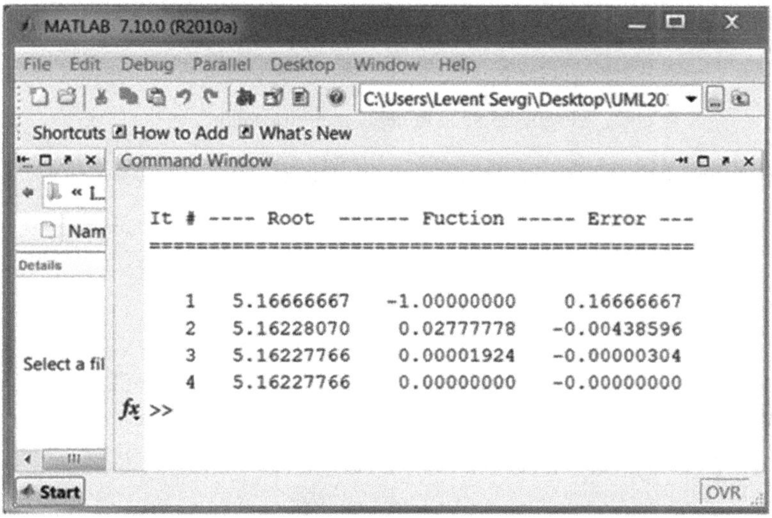

It #	Root	Fuction	Error
1	5.16666667	-1.00000000	0.16666667
2	5.16228070	0.02777778	-0.00438596
3	5.16227766	0.00001924	-0.00000304
4	5.16227766	0.00000000	-0.00000000

of equations. It starts with either *electric field integral equation* (EFIE) or *magnetic field integral equation* (MFIE), and the unknown function (usually, it is EM-field-induced currents on and/or in interacting objects) is approximated by a finite series of known expansion functions with unknown expansion coefficients. This results into a number of simultaneous algebraic equations for the unknown coefficients. These systems of equations are then solved using matrix calculus.

Multiloop circuits are also solved using linear systems of equations. Suppose the loop currents in the circuit in Fig. 3.8 are of interest. Kirchoff's loop current (KLC) method yields a 3 × 3 linear system in the form $[Z][I] = [V]$, where $[Z]$ is the resistor matrix, $[V]$ is the source vector, and $[I]$ is the unknown loop currents vector. A 3 × 3 linear system of equations is in the form of

$$\begin{aligned} Z_{11}I_1 + Z_{12}I_2 + Z_{13}I_3 &= V_1 \\ Z_{21}I_1 + Z_{22}I_2 + Z_{23}I_3 &= V_2 \\ Z_{31}I_1 + Z_{32}I_2 + Z_{33}I_3 &= V_3, \end{aligned} \tag{3.28}$$

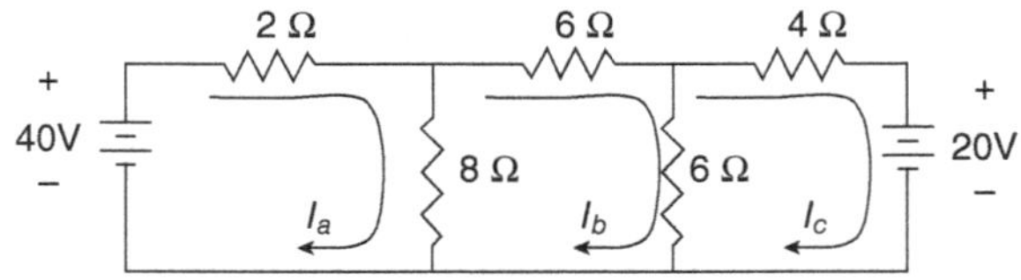

Figure 3.8. A three-loop electric circuit.

which can be written in the matrix format as

$$\begin{bmatrix} Z_{11} & Z_{12} & Z_{13} \\ Z_{21} & Z_{22} & Z_{23} \\ Z_{31} & Z_{32} & Z_{33} \end{bmatrix} \begin{bmatrix} I_1 \\ I_2 \\ I_3 \end{bmatrix} = \begin{bmatrix} V_1 \\ V_2 \\ V_3 \end{bmatrix}. \tag{3.29}$$

Here, Z_{ij} is the coefficient matrix, I_i is the unknown vector, and V_i is the known (usually, excitation) vector ($i = 1,2,3$, $j = 1,2,3$). If the equations are linearly independent (i.e., if the determinant of the coefficient matrix is not zero), the solution of this system will be

$$[I] = [Z]^{-1}[V]. \tag{3.30}$$

Matrix inversion is the critical issue in numerical analysis. There are many methods; only Gaussian elimination is included here. Gaussian elimination is a process which puts the coefficient matrix in a form where elements on the diagonal (Z_{ii}) are "1" and below diagonal elements are all "0":

$$\begin{bmatrix} 1 & a_{12} & a_{13} \\ 0 & 1 & a_{23} \\ 0 & 0 & 1 \end{bmatrix} \begin{bmatrix} I_1 \\ I_2 \\ I_3 \end{bmatrix} = \begin{bmatrix} b_1 \\ b_2 \\ b_3 \end{bmatrix}. \tag{3.31}$$

This can be achieved by applying the following procedure on the extended matrix:

$$\begin{bmatrix} Z_{11} & Z_{12} & Z_{13} & V_1 \\ Z_{21} & Z_{22} & Z_{23} & V_2 \\ Z_{31} & Z_{32} & Z_{33} & V_3 \end{bmatrix}. \tag{3.32}$$

First, divide both sides of the first equation in (3.28) by

$$Z_{11} \rightarrow \left(1 \quad \frac{Z_{12}}{Z_{11}} \quad \frac{Z_{13}}{Z_{11}} \quad \frac{V_1}{Z_{11}} \right).$$

This will make the first element on the diagonal "1." Then, multiply first equation by Z_{21}, and subtracting this equation from the second equation, this will make the first element on the second row,

$$"0" \to \left(0 \quad \frac{Z_{12}Z_{21}}{Z_{11}} - Z_{22} \quad \frac{Z_{13}Z_{21}}{Z_{11}} - Z_{23} \quad \frac{V_1 Z_{21}}{Z_{11}} - V_2 \right).$$

Then, repeat the last step for the third row (i.e., multiply first equation by Z_{31} and subtract this equation from the third equation. This will make the first element on the third row,

$$"0" \to \left(0 \quad \frac{Z_{12}Z_{31}}{Z_{11}} - Z_{32} \quad \frac{Z_{13}Z_{31}}{Z_{11}} - Z_{33} \quad \frac{V_1 Z_{31}}{Z_{11}} - V_3 \right).$$

Repeat the whole procedure starting from the second element on the diagonal Z_{22}. This will make the second element on the diagonal "1" and all underneath "0." Do this until the last element on the diagonal will be "1."

This elimination procedure assumes that all diagonal elements are nonzero. If some of them are zero, then the equations may be reordered (e.g., replace mth row with nth row if Z_{mm} or Z_{nn} is zero). This is called *pivoting*.

This upper triangular matrix is directly solved by back-substitution method. Note that the value of an equation does not change if both sides are multiplied/divided with the same constant. Gauss elimination applies this property only. Table 3.11 lists a simple MATLAB Gauss elimination code. Here, a 3 × 3 linear system of equations is specified by $A = [4\ 2\ 1;\ -1\ 1\ 4;\ 1\ 4\ 3]$. The output of the code is pictured in Table 3.12.

Table 3.11.
A Simple MATLAB Code for Gaussian Elimination Method

```
%% -----------------------------------------------------
% Program:  LS_GaussElim12
% Purpose:  Solve a system of equations by using Gauss
%             elimination method (including pivoting)
% [A] [x] = [r]    => [x] = [A](-1) [r]
%% -----------------------------------------------------
clc; clear all;
% n=input('the number of unknowns=  ');
% A=random('unif',0,30,n,n)-15; b=random('unif',0,2
5,n,1)-15;
A=[4 2 1; -1 1 4; 1 4 3]; b=[9; 5; 12];
row,col]=size(A); n=row;
disp(' Input Matrix A ='); disp(A);
```

(*Continued*)

Table 3.11.
(Continued)

```
s=inv(A)*b; disp(s);   % Solution with MATLAB's INV()
command
%-------------------------------------------------------------
t1=clock;
for i=1:n-1
   [maxi,row]=max(abs(A(:,i))); Dum=A(i,:);
A(i,:)=A(row,:);
   A(row,:)=Dum; dd=b(i); b(i)=b(row); b(row)=dd;
   ap=A(i,i); A(i,:)=A(i,:)/ap;  b(i)=b(i)/ap;
   disp('Triangularizatiopn steps ='); disp(A); pause;
   for k=i+1:n
       dd=A(k,i); A(k,:)=A(k,:)-dd*A(i,:);
       b(k)=b(k)-dd*b(i);
       disp('A='); disp(A); pause;
   end
end
b(n)=b(n)/A(n,n);
A(n,n)=A(n,n)/A(n,n);
disp('The Upper Triangular Matrix ='); disp('A=');
disp(A);
%-------------------------------------------------------------
x(n)=b(n);
for i=n-1:-1:1
   sum=0;
   for j=i+1:n
       sum=sum+A(i,j)*x(j);
   end
   x(i)=b(i)-sum;
end
t2=clock; time=etime(t2,t1)
disp(' The unknowns are :'); disp(x');
% ------------------- END ---------------------
```

Table 3.12.
The Output of the Code Listed in Table 3.10

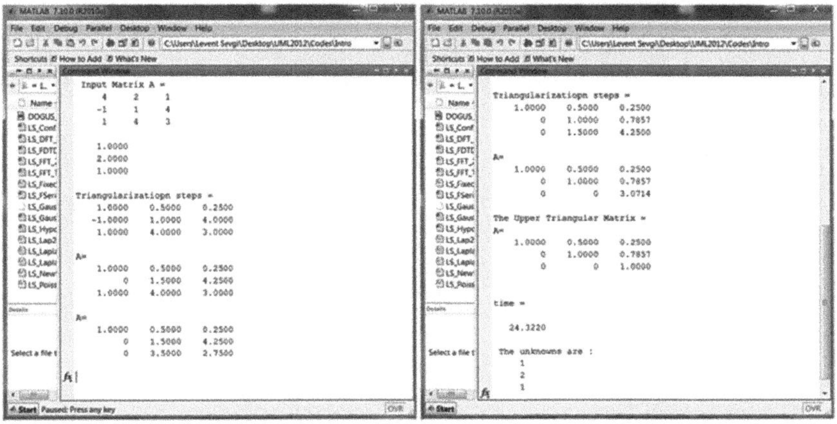

REFERENCES

1. C. F. Gerald, P. O. Wheatley, *Applied Numerical Analysis*, 5th ed., Addison-Wesley, Reading, MA, 1994.
2. L. V. Fausett, *Applied Numerical Analysis Using MATLAB*, Prentice Hall, Upper Saddle River, NJ, 1999.
3. R. F. Harrington, *Field Computation by Moment Method*, 1st ed., 1968, IEEE Press, New York, 1993.

CHAPTER 4

Fourier Transform and Fourier Series

4.1 INTRODUCTION

Analysis of a real world signal is a fundamental problem for engineers and scientists, especially for electrical engineers since almost every real world signal is transformed into an electrical signal by means of transducers, for example, accelerometers in mechanical engineering, electroencephalography (EEG) electrodes, and blood pressure probes in biomedical engineering, seismic transducers in Earth Sciences, antennas in electromagnetic (EM), microphones in communication engineering, and others.

Traditional way of observing and analyzing signals is to view them in time domain. Baron Jean Baptiste Fourier [1], more than a century ago, showed that any waveform that exists in the real world can be represented (i.e., generated) by adding up sinusoidal waves. Since then, we have been able to build (break down) our real world time signal in terms of (into) these sinusoidal waves. It is shown that the combination of sinusoidal waves is unique; any real world signal can be represented by only one sinusoidal combination [2].

The Fourier transform (FT) has been widely used in circuit analysis and synthesis, from filter design to signal processing, image reconstruction, and so on. The reader should keep in mind that the time domain–frequency domain relations in EM are very similar to the relations

Electromagnetic Modeling and Simulation, First Edition. Levent Sevgi.

between spatial and wave number domains. A simplest propagating (e.g., along z) plane wave is in the form of $\Phi(r, t) \propto e^{j(\omega t - kz)}$ (where k and z are the wave number and position, respectively) and, whatever characteristics $\exp(j\omega t)$ have are also applicable to $\exp(-jkz)$. Some characteristics are outlined as follows.

- A rectangular time (frequency) window corresponds to a beam-type (Sinc(.) function) variation in frequency (time) domain.
- Similarly, a rectangular aperture (array) in spatial domain corresponds to a beam-type (Sinc(.) function) variation in wave number domain.
- The wider the antenna aperture, the narrower the antenna beam; or the narrower the pulse in time domain, the wider the frequency band.

Therefore, FT has also been used in EM from antenna analysis to imaging and nondestructive measurements, even in propagation problems. For example, the split-step parabolic equation method (which is nothing but the beam propagation method in optics) has been in use more than decades and is based on sequential FT operations between the spatial and wave number domains. Two and three dimensional propagation problems with nonflat realistic terrain profiles and inhomogeneous atmospheric variations earlier have been solved with this method successfully [3–5]. A quick Internet search in the AP publications with the IEEE Explore will yield a number of different applications.

4.2 FOURIER TRANSFORM

The principle of a transform in engineering is to find a different representation of a signal under investigation. The FT is the most important transform widely used in electrical engineering.

4.2.1 Fourier Transform (FT)

The transformations between the time and the frequency domains are based on the FT and its inverse (IFT). They are defined via

$$S(\omega) = \int_{-\infty}^{\infty} s(t) e^{-j2\pi ft} dt, \; s(t) = \int_{-\infty}^{\infty} S(f) e^{j2\pi ft} df. \tag{4.1}$$

Here, $s(t)$, $s(\omega)$, and f are the time signal, the frequency signal, and the frequency, respectively, and $j=\sqrt{-1}$. We, the physicists and engineers, sometimes prefer to write the transform in terms of angular frequency $\omega = 2\pi f$ as

$$S(\omega)=\int_{-\infty}^{\infty} s(t)e^{-j\omega t}dt,\ s(t)=\frac{1}{2\pi}\int_{-\infty}^{\infty} S(\omega)e^{j\omega t}d\omega, \tag{4.2}$$

which, however, destroys the symmetry. To restore the symmetry of the transforms, the convention is to divide $1\ /\ 2\pi$ term into two and use $1\,/\sqrt{2\pi}$ during both FT and IFT. The FT is valid for real or complex signals, and, in general, is a complex function of ω (or f).

The FT is valid for both periodic and nonperiodic time signals that satisfy certain conditions. Almost all real world signals easily satisfy these requirements (it should be noted that the Fourier series is a special case of the FT). Mathematically, FT is defined for continuous time signals, and in order to go to frequency domain, the time signal must be observed from an infinite-extend time window. Under these conditions, the FT defined earlier yields frequency behavior of a time signal at every frequency, with zero frequency resolution. Some functions and their FT are listed in Table 4.1.

Table 4.1.
Some Functions and Their FTs

Time Domain	Fourier Domain
Rectangular window	Sinc function
Sinc function	Rectangular window
Constant function	Dirac delta function
Dirac delta function	Constant function
Dirac comb (Dirac train)	Dirac comb (Dirac train)
Cosine function	Two real-even delta function
Sine function	Two imaginary-odd delta functions
Exp function $\{\exp\{j\omega t\}\}$	One positive-real delta functions
Gauss function	Gauss function

4.2.2 Discrete Fourier Transform (DFT)

To compute the FT numerically on a computer, discretization plus numerical integration are required. This is an approximation of the true (i.e., mathematical), analytically defined FT in a synthetic (digital) environment, and is called the DFT. There are three difficulties with the numerical computation of the FT:

- *Discretization* (introduces periodicity in both time and frequency domains)
- *Numerical integration* (introduces approximation and numerical round-off and truncation errors)
- *Finite time duration* (introduces maximum frequency and resolution limitations).

The DFT of a continuous time signal sampled over a record period of T, with a sampling rate of Δt can be given as

$$S(m\Delta f) = \frac{T}{N}\sum_{n=0}^{N-1} s(n\Delta t)e^{-j2\pi m\Delta f n\Delta t}, \tag{4.3}$$

where $\Delta f = 1/T$ and is valid at frequencies up to $f_{max} = 1/2\Delta t$. Table 4.2 lists a simple MATLAB *m-file* that computes (4.3) for a time record $s(t)$ of two sinusoids whose frequencies are user specified. The record length and sampling time interval are also supplied by the user, and DFT of this record is calculated inside a simple integration loop.

Table 4.2.
A MATLAB Module for DFT Calculations

```
%% -------------------------------------------
% Program: LS_DFT.m
% Purpose: To get Discrete Fourier transform of a
function
%% -------------------------------------------
clear all; close all; clc;
% Get the input parameters
fr1=input('Frequency of the first sinusoid [Hz] = ? ');
fr2=input('Frequency of the second sinusoid [Hz] = ? ');
```

Table 4.2.
(Continued)

```
a1=input('Amplitude of the first sinuzoid = ? ');
a2=input('Amplitude of the second sinuzoid = ? ');
T=input ('Time record length [s] = ? ');
dt=input ('sampling time interval [s] = ? ');
fmax=input('maximum frequency for DFT [Hz] = ? ');
df=input('frequency sampling interval for DFT [Hz] = ?
');
% Calculate paprameters
a1=1; a2=0.5; N=T/dt; M=fmax/df;
w1=2*pi*fr1; w2=2*pi*fr2;
for k=1:N  % build the input time series
    t(k)=k*dt;
    st(k)=a1*sin(w1*dt*k)+a2*sin(w2*dt*k);
end
for k=1:M  % Apply the DFT with M points
    Sf(k)=complex(0,0);
    for n=1:N
        Sf(k)=Sf(k)+st(n)*exp(-i*2*pi*n*dt*k*df);
    end
    Sf(k)=Sf(k)*dt;
end
for k=1:M % Prepare the frequency samples
    f(k)=k*df;
end
% plot the output in the time domain
figure(1);
plot(t,st));
title('The Sum of two Sinusoids')
xlabel('Time [s]'); ylabel('Amplitude')% plot the output
in the frequency domain
figure(2);
plot(f,abs(Sf));
title('The DFT of the Sum of two Sinusoids')
xlabel('Frequency [Hz]'); ylabel('Amplitude')
% -------------------- END ----------------------
```

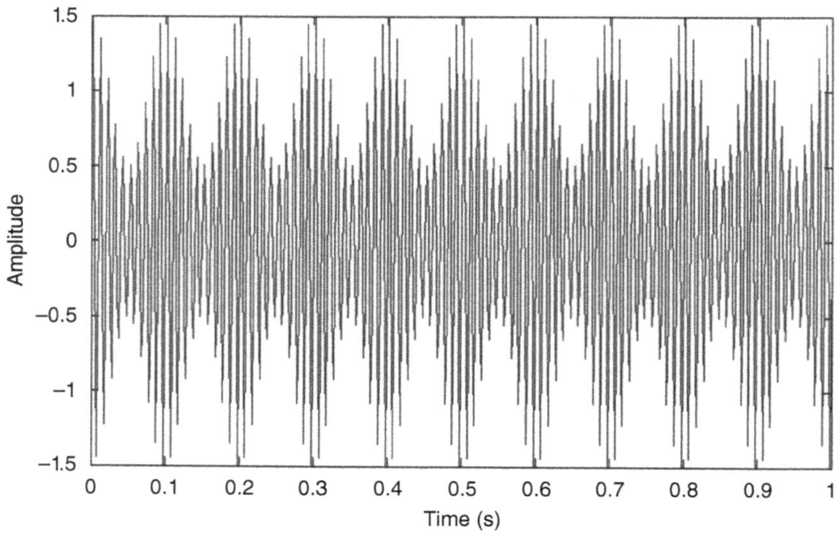

Figure 4.1. Time variation of two sinusoids.

Let us plot two sinusoids with 100 Hz/1 V and 110 Hz/0.5 V both in time and frequency domain. Choose $f_{max} = 150$ Hz, $\Delta f = 2$ Hz, $T = 1$ s, and $\Delta t = 2.5$ ms. Note that these parameters may be chosen arbitrary in DFT but frequency resolution and maximum frequency will be $\Delta f = 1 / T$ and $f_{max} = 1 / 2\Delta t$, respectively. Results are shown in Fig. 4.1 and Fig. 4.2.

4.2.3 Fast Fourier Transform (FFT)

The DFT requires excessive amount of computation time, particularly when the number of samples N is high. The FFT is an algorithm to speed up DFT computations. The FFT forces one further assumption that N is an integer multiple of 2. This allows certain symmetries to occur, reducing the number of calculations.

To write an FFT routine is not as simple as DFT routine, but there are many internet addresses where one can supply FFT subroutines (including source codes) in different programming languages, from FORTRAN to C++. Therefore, the reader does not need to go into details, rather to include them in their codes by simply using "include" statements or "call" commands. In MATLAB, the calling command is

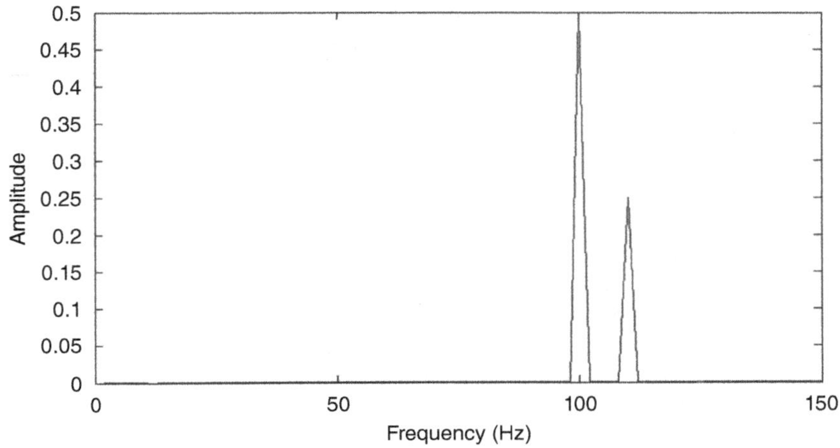

Figure 4.2. Frequency variation of two sinusoids obtained with DFT.

fft(s, N) for the FFT and *ifft*(S, N) for the inverse FFT, where s and S are the recorded N-element time array and its FT, respectively. The DFT code listed in Table 4.2 can be used for FFT. In order to do that, one needs to replace the loop for DFT in the table with "Sf=***fftshift***(***fft***(st)*dt)." This m-file is included in Table 4.3. Note that one needs to scale the results in the frequency domain (i.e., multiply the result by Δt) since MATLAB *fft*(s, N) command assumes $\Delta t = 1$; also, swap the first N / 2 samples with the second half using the *fftshift*(s) command. Figure 4.3 shows the FFT of the same signal.

4.2.4 Aliasing, Spectral Leakage, and Scalloping Loss

As stated earlier, performing FT in a discrete environment introduces artificial effects. These are called aliasing effects, spectral leakage, and scalloping loss [2]. The following should be kept in mind when dealing with discrete FT.

- Multiplication in the time domain corresponds to a convolution in the frequency domain.
- The FT of an impulse train in the time domain is also an impulse train in the frequency domain with the frequency samples separated by $T_0 = 1 / f_0$.

Table 4.3.
A MATLAB Module for FFT Calculations

```
%% --------------------------------------------------------
%   Program: LS_FFT.m
%   Purpose: To get Fast Fourier transform of a function
%% --------------------------------------------------------
clear all; close all; clc;
% Get the input parameters
f1=input('Frequency of the first sinusoid [Hz] = ? ');
f2=input('Frequency of the second sinusoid [Hz] = ? ');
a1=input('Amplitude of the first sinuzoid = ? ');
a2=input('Amplitude of the second sinuzoid = ? ');
dt=input ('Time sampling interval (Delta_t) [s] : ');
T=input ('Time duration (T) [s] : ');
N=floor(T/dt); df=1/T; fmax=1/(2*dt); a1=1.0; a2=0.5;
for n=1:N
    t(n)=(n-1)*dt;
    f(n)=-fmax+(n-1)*df;
    st(n)=a1*sin(2*pi*f1*t(n))+a2*sin(2*pi*f2*t(n));
end
Sf=fftshift(fft(st)*dt); % Apply FFT Swap the left and
right halves
% plot the output in the time domain
plot(t,st); title('The Sum of two Sinusoids')
xlabel('Time [s]'); ylabel('Amplitude')
% plot the output in the frequency domain
figure(2);
plot(f,abs(Sf)); title('The DFT of the Sum of two
Sinusoids')
xlabel('Frequency [Hz]'); ylabel('Amplitude')
% ---------------------- END --------------------
```

- The narrower the distance between impulses (T_0) in the time domain, the wider the distance between impulses (f_0) in the frequency domain (and vice versa).
- The sampling rate must be greater than twice the highest frequency of the time record, that is, $\Delta t \geq 1 / (2f_{max})$ (Nyquist sampling criterion).

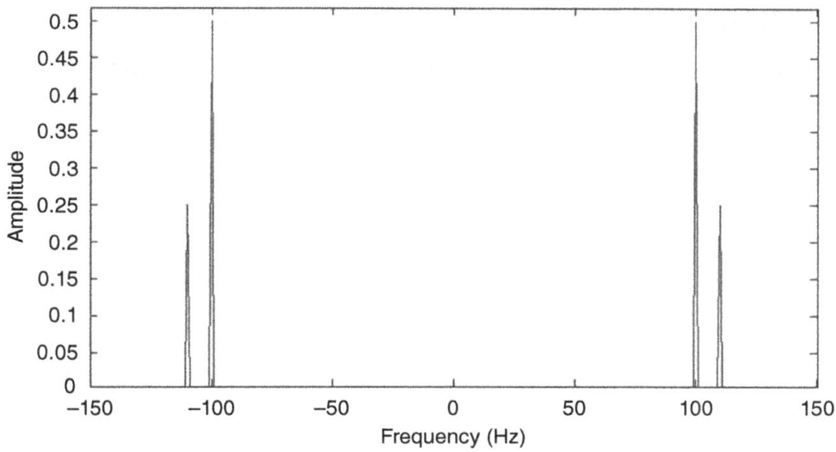

Figure 4.3. Frequency variation of two sinusoids obtained with FFT.

- Since *time–bandwidth* product is constant, narrow transients in the time domain possess wide bandwidths in the frequency domain.
- In the limit, the frequency spectrum of an impulse is constant and covers the whole frequency domain (that is why an impulse response of a system is enough to find out the response of any arbitrary input).

If the sampling rate in the time domain is lower than the Nyquist rate, *aliasing* occurs [2]. Two signals are said to alias if the difference of their frequencies falls in the frequency range of interest, which is always generated in the process of sampling (aliasing is not always bad; it is called mixing or heterodyning in analog electronics, and is commonly used in tuning radios and TV channels). It should be noted that although obeying Nyquist sampling criterion is sufficient to avoid aliasing, it does not give high quality display in time domain record. If a sinusoid existing in the time signal is not bin centered (i.e., if its frequency is not equal to any of the frequency samples) in the frequency domain, *spectral leakage* occurs. In addition, there is a reduction in coherent gain if the frequency of the sinusoid differs in value from the frequency samples, which is termed *scalloping loss*.

4.2.5 Windowing and Window Functions

Using a finite-length discrete signal in the time domain in FT calculations means to apply a rectangular window to the infinite-length signal [6]. This does not cause a problem with the transient signals which are time bounded inside this window. But, what happens if a continuous time signal (e.g., a sinusoidal wave) is of interest? If the length of the window (i.e., the time record of the signal) contains an integral number of cycles of the time signal, then periodicity introduced by discretization makes the windowed signal exactly same as the original. In this case, the time signal is said to be periodic in the time record. On the other hand, there is a difficulty if the time signal is not periodic in the time record, especially at the edges of the record (i.e., window). If the DFT or FFT could be made to ignore the ends and concentrate on the middle of the time record, it is expected to get much closer to the correct signal spectrum in the frequency domain. This may be achieved by a multiplication with a function that is zero at the ends of the time record and large in the middle. This is known as *windowing*.

The time record is tempered and perfect results should not be expected if windowing is applied. For example, windowing reduces spectral leakage, but does not totally eliminate it. Note that windowing is introduced to force the time record to be zero at the ends; therefore, transient signals which occur (starts and ends) inside this window do not require a window. They are called *self-windowed* signals, and examples are impulses, shock responses, noise bursts, sinusoidal bursts, and so on.

The FT of a rectangular pulse is the *Sinc*() function. It has the narrowest mainlobe, but the highest sidelobe level as opposed to other window functions. Therefore, there is a compromise between a narrow main lobe (for high resolution) and low sidelobes (for low spectral leakages). High frequency resolution provides accurate estimation of the frequency of an existing sinusoid and results in the separation of two sinusoids that are closely spaced in the frequency domain. Low spectral leakage improves the detectability of a weak sinusoid in the presence of a strong one that is not bin centered. A few examples of common window functions are ($n = 0, 1, 2, .., N - 1$) the following.

Rectangular:

$$W(n) = 1. \tag{4.4a}$$

Hanning:

$$W(n) = \frac{1}{2} - \frac{1}{2}\cos\left(\frac{2\pi n}{N}\right). \tag{4.4b}$$

Hamming:

$$W(n) = 0.54 - 0.46\cos\left(\frac{2\pi n}{N}\right). \tag{4.4c}$$

All of these window functions act as a filter with a very rounded top. If a sinusoid in the time record is centered in the filter, then it will be displayed accurately. Otherwise, the filter shape (i.e., the window) will attenuate the amplitude of the sinusoid (scalloping loss) by up to a few decibels (15–20%) when it falls midway between two consecutive discrete frequency samples. The solution of this problem is to choose a flattop window function:

Flattop:

$$W(n) = 0.281064 - 0.520897\cos\left(\frac{2\pi n}{N}\right) + 0.19804\cos\left(\frac{4\pi n}{N}\right), \tag{4.5}$$

which reduces the amplitude loss to a value less than 0.1 dB (1%). However, this accuracy improvement does not come without its price; it widens the mainlobe in the frequency domain response (i.e., a small degradation in frequency resolution). It should be remembered that there is always a pay off between accuracy and resolution when applying a window function.

4.3 BASIC DISCRETIZATION REQUIREMENTS

The mathematical FT is used to calculate the frequency spectrum of a time function without maximum frequency and resolution limitations. These limitations belong to DFT and FFT. The maximum frequency in the DFT or FFT depends on the sampling interval, and the frequency resolution is determined by the record length of the signal. That is, N samples of a time signal recorded during a finite duration of T with a sampling period of Δt ($N = T / \Delta t$) can be transformed into N samples in the frequency domain between $-f_{\max}$ and $+f_{\max}$ according to

$$f_{\max} = \frac{1}{2\Delta t}, \Delta f = \frac{1}{T}. \tag{4.6}$$

Since the sampling interval and the record length must be finite in computer calculations, maximum frequency and the resolution must also be finite. This means the following.

- Any frequency component f_c beyond $+f_{max}$ cannot be observed in its actual frequency; instead, it enters from left because of rotational symmetry and periodicity and appears at $-f_{max} + f_D$, where $f_D = f_c - f_{max}$.
- Similarly, any frequency component $-f_c$ beyond $-f_{max}$ cannot be observed in its actual frequency; instead, it enters from right because of rotational symmetry and periodicity and appears at $f_{max} - f_D$, where $f_D = |f_c - f_{max}|$.

This is pictured in Fig. 4.4. There are four sinusoids with different amplitudes. The first sinusoid f_1 is within $\pm f_{max}$; therefore, it appears at their correct locations. The other two (f_2 and f_3) are beyond $+f_{max}$. The dashed arrows show their ghost replicas rotationally shifted and entered from the left in the frequency domain. Finally, f_4 which is beyond $-f_{max}$ enters the spectrum from the right.

It should be noted that the MATLAB code in Table 4.2 directly integrates (4.3) numerically, so any number of frequency samples ($n \times \Delta f$) can be used to plot a frequency spectrum. Unfortunately, (4.6) still holds for the DFT. In other words, for example, if one wants to

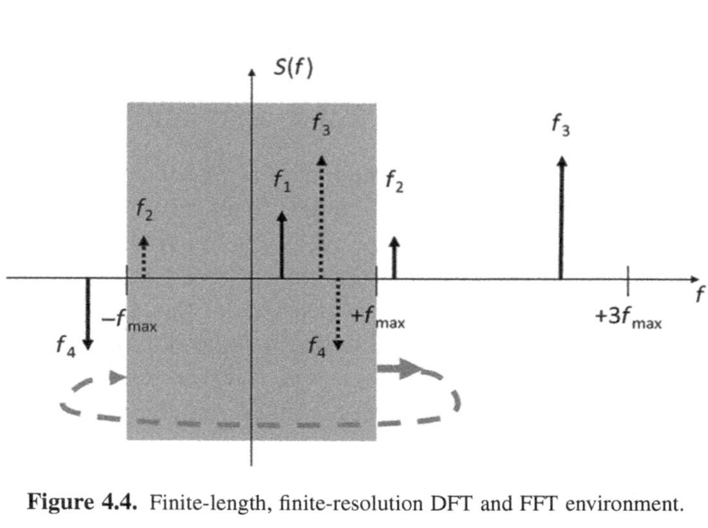

Figure 4.4. Finite-length, finite-resolution DFT and FFT environment.

discriminate two sinusoids with 50 and 52 Hz in the frequency domain, the frequency resolution Δf must be much less than their difference (2 Hz). This necessitates T to be minimum 0.5 s if $\Delta f = 2$ Hz is chosen. On the other hand, the DFT of a signal can be taken using any Δf, but the capability of discriminating two nearby sinusoids in the frequency domain is limited by the length of the record (T).

QUESTION

How do you choose T and Δt if you want to observe the spectrum of

(a) $s(t) = \sin(100\pi t) + \sin(70\pi t)$?

(b) $s(t) = \cos(98\pi t) + \exp(-206\pi t)$?

ANSWER

(a) The two frequencies of these two sinusoids are $f_1 = 50$ Hz and $f_1 = 35$ Hz. In the frequency domain, they appear at ± 50 and ± 35 Hz, respectively. Maximum frequency f_{max} in the frequency domain should be greater than 50 Hz, and frequency resolution be less than 15 Hz. It is appropriate, for example, to choose $f_{max} = 100$ Hz and $\Delta f = 1$ Hz; therefore, discrete time domain values will be $T = 1$ s and $\Delta t = 0.005$ s.

(b) The sinusoids in the frequency domain appear at ± 48 Hz and ± 103 Hz, respectively. It is therefore convenient to choose $f_{max} = 125$ Hz and $\Delta f = 1$ Hz; therefore, discrete time domain values will be $T = 1$ s and $\Delta t = 0.004$ s.

QUESTION

At which frequencies and why do you observe the signal if $\Delta t = 10$ ms and $T = 1$ s for

(a) $s(t) = \sin(100\pi t) + \sin(70\pi t)$?

(b) $s(t) = \cos(86\pi t) + \exp(-206\pi t)$?

(c) $s(t) = \exp(j104\pi t) + \sin(104\pi t)$?

ANSWER

Figure 4.5, Figure 4.6, and Figure 4.7 show the frequency spectra one should observe with the given parameters.

At which frequencies do you observe the signal (and why) if $\Delta t = 40$ ms and $T = 0.21$ s for

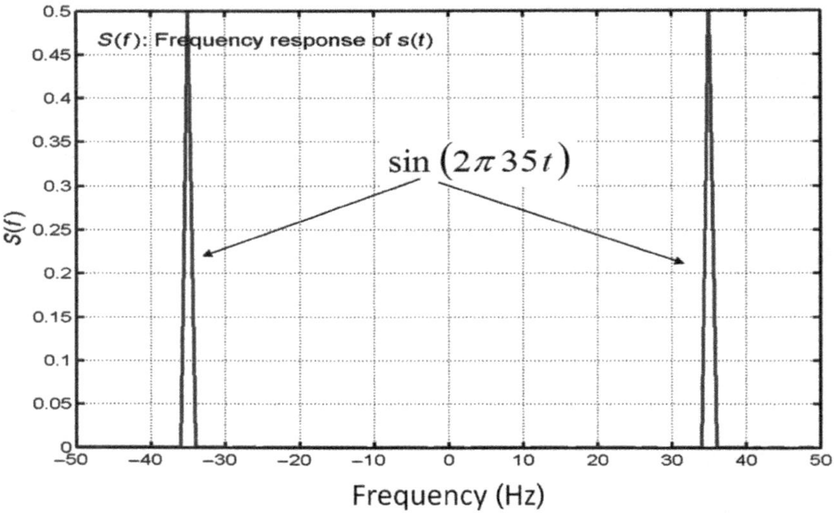

Figure 4.5. The frequency spectrum for question 2a (±50 Hz cannot be observed since they exactly appear at $\pm f_{\max}$).

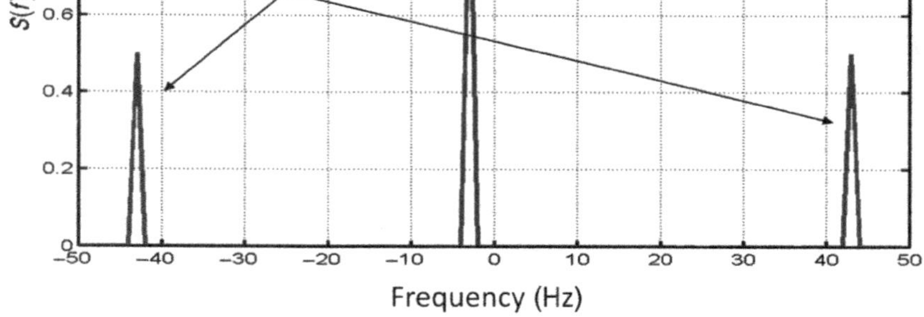

Figure 4.6. The frequency spectrum for question 2b ($\exp(-206\pi t)$ enters the spectrum from the right and is 53 Hz away from the right boundary; note the amplitude differences).

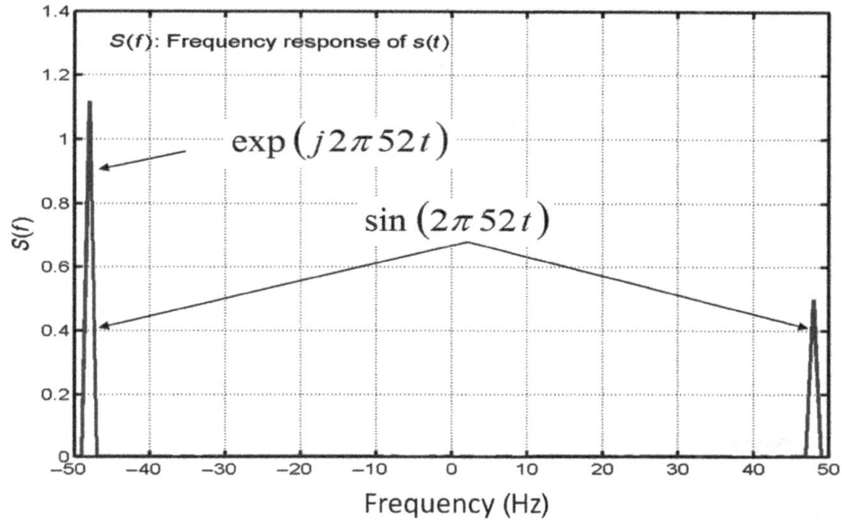

Figure 4.7. The frequency spectrum for question 2c ($\exp(j104\pi t)$ enters the spectrum from the left and is 2 Hz away from the left boundary).

(a) $s(t) = \sin(100\pi t) + \exp(j70\pi t)$?

(b) $s(t) = \exp(j32\pi t) + \sin(86\pi t)$?

(c) $s(t) = \exp(j120\pi t) + \exp(j206\pi t)$?

The answers of these three are left to the reader. Use any of the two MATLAB scripts listed in Table 4.2 and Table 4.3, and observe frequency spectra by yourselves.

4.4 FOURIER SERIES REPRESENTATION

Fourier transform is used for *energy signal* which contain finite energy. This means $\int_{-\infty}^{\infty} |s(t)|^2\, dt$ is finite. Periodic functions do not satisfy this property and are called *power signals*. Power signals have finite power in one period (i.e., $(1/P)\int_{-P/2}^{P/2} |s(t)|^2\, dt$ is finite, where P is the period of the signal). Power signals are represented in terms of Fourier series.

A function $f(x)$ is periodic, with period P, if

$$f(x) = f(x+P) = f(x+2P) + f(x+3P) + \cdots. \tag{4.7}$$

A periodic function $f(x)$ can be approximated by using Fourier series expansion as

$$f(x) \approx \frac{A_0}{2} + \sum_{n=1}^{\infty} A_n \cos\left(\frac{2\pi nx}{P}\right) + B_n \sin\left(\frac{2\pi nx}{P}\right), \tag{4.8a}$$

where

$$A_0 = \frac{2}{P}\int_{-P/2}^{P/2} f(x)dx,\; A_n = \frac{2}{P}\int_{-P/2}^{P/2} f(x)\cos\left(\frac{2n\pi x}{P}\right)dx, \tag{4.8b}$$

$$B_n = \frac{2}{P}\int_{-P/2}^{P/2} f(x)\sin\left(\frac{2n\pi x}{P}\right)dx. \tag{4.8c}$$

Nonperiodic functions may also be approximated by Fourier series inside a finite region by assuming the finite region as the period of that function. In this case, it should be remembered that the Fourier series representation no longer represents the function outside the region. Equation (4.8) shows that one needs to multiply the function with sine and cosine functions along the whole period and then integrate in order to find out Fourier coefficients A_n and B_n. Here, completeness and orthogonality

$$\frac{2}{P}\int_{-P/2}^{P/2} \sin\left(\frac{2n\pi x}{P}\right)\sin\left(\frac{2m\pi x}{P}\right)dx = \begin{cases} 0 & m = n \\ 1 & m \neq n \end{cases}, \tag{4.9a}$$

$$\frac{2}{P}\int_{-P/2}^{P/2} \cos\left(\frac{2n\pi x}{P}\right)\cos\left(\frac{2m\pi x}{P}\right)dx = \begin{cases} 0 & m = n \\ 1 & m \neq n \end{cases}, \tag{4.9b}$$

$$\frac{2}{P}\int_{-P/2}^{P/2} \sin\left(\frac{2n\pi x}{P}\right)\cos\left(\frac{2m\pi x}{P}\right)dx \equiv 0, \tag{4.9c}$$

properties of the trigonometric functions are used to extract Fourier coefficients.

Table 4.4 lists a simple MATLAB code for the Fourier series representation of a given function. A few examples plotted with this MATLAB code are given in Fig. 4.8, Fig. 4.9, and Fig. 4.10.

The function used in this example is $f(x) = (x^3 - 4x^2)\sin(5x^2 - 2x) - 3$ and the interval (i.e., the period) is $0 \leq x \leq 4$. First, only the first five terms in (4.8a) are used and the result is plotted in Fig. 4.8. The solid

Table 4.4.
A MATLAB Module for Fourier Series Representations

```
%% ---------------------------------------------------------
% Program: LS_FSeries.m
% Purpose: To represent a function with Fourier Series
%% ---------------------------------------------------------
clear all; clc; format long;
N=400; xf=0; xl=10; P=abs(xl-xf); delx=P/N; x0=2.5;
xT=.4;
tr=0.1;    %input ('Rise/Fall time [s] : ');
tau=2;     %input ('Pulse duration [s] : ');
fprintf('This program calculates FSeries of the
following: \n');
fprintf(' \n');
fprintf('=========================================
======\n');
fprintf(' \n');
fprintf('[1] Gauss Function \n');
fprintf(' [2] Modulated Gauss Function \n');
fprintf(' [3] 2 Gauss Functions (+) \n');
fprintf(' [4] 2 Gauss Functions (-) \n');
fprintf(' [5] Sine Function \n');
fprintf(' [6] Rectangular Pulse \n');
fprintf(' [7] Trapezoidal Pulse \n');
fprintf(' [8] Triangular Pulse \n');
fprintf(' \n');
fprintf('=========================================
======\n');
n=input('Your Choice [1-7] = ?'); fprintf(1, ' \n');
while n&lt;1 | n>7
   fprintf('\n Wrong!\n'); n=input('Your Choice [1-7] =
?');
end
for j=1:N     % Generate the function at N points
  xx(j)=xf+(j-1)*delx;
  if n == 1
    fnr(j)=1/sqrt(2*pi)*exp(-(xx(j)-x0).^2/(2*xT^2));  %
Gauss Function
  elseif n==2
    fnr(j)=sin(8*pi*xx(j))/sqrt(2*pi)*exp(-
(xx(j)-x0).^2/(2*xT^2));
  elseif n==3
```

(Continued)

Table 4.4.
(Continued)

```
    fnr(j)=1/sqrt(2*pi)*(exp(-(xx(j)-x0).^2/
(2*xT^2)). . .
             +exp(-(xx(j)-x0-P/2).^2/(2*xT^2)));
  elseif n==4
    fnr(j)=1/sqrt(2*pi)*(exp(-(xx(j)-x0).^2/
(2*xT^2)). . .
             -exp(-(xx(j)-x0-P/2).^2/(2*xT^2)));
  elseif n==5
    fnr(j)=(xx(j).^2)*sin(3*pi*xx(j));
  elseif n==6
    if xx(j)&lt;=2*tau    % Rectangular Pulse
       fnr(j)=0;
    elseif xx(j)&lt;=3*tau
       fnr(j)=1;
   else
       fnr(j)=0;
    end
  elseif n==7
    if xx(j)&lt;=2*tau    % Trapezoidal Pulse
       fnr(j)=0;
    elseif xx(j)&lt;=2*tau+tr
       fnr(j)=(xx(j)-2*tau)/tr;
    elseif xx(j)&lt;=3*tau+tr
       fnr(j)=1;
    elseif xx(j)&lt;=3*tau+2*tr
       fnr(j)=3-(xx(j)-3*tau+tr)/tr;
    else
       fnr(j)=0;
    end
  elseif n==8
    if xx(j)&lt;=2*tau    % Triangular Pulse
      fnr(j)=0;
    elseif xx(j)&lt;=2.5*tau
      fnr(j)=(xx(j)-2*tau)/(0.5*tau);
    elseif xx(j)&lt;=3*tau
      fnr(j)=-(xx(j)-3*tau)/(0.5*tau);
    else
      fnr(j)=0;
    end
```

Table 4.4.
(Continued)

```
   end
end
A0=trapz(xx,fnr)*2./P; % Calculate A0
for j=1:N; fn(j)=A0/2; end;
OK=1; k=0; Nt=0;
fprintf(1, 'Coefficient No --- An --- Bn --- \n');
fprintf(1, '-------------------------------- \n ');
while OK>0
   Nt=Nt+1;
   for k=1:N % Calculate A(k) and B(k)
    g1(k)=fnr(k)*cos(2*pi*Nt*xx(k)/P); g2(k)=fnr(k)*sin(2
*pi*Nt*xx(k)/P);
   end
   An=trapz(xx,g1)*2./P;
   Bn=trapz(xx,g2)*2./P;
   fprintf(1, '  %3d        %12.8f     %12.8f    \n', Nt,
An, Bn);
   for j=1:N     % Calculate F Series of the function at
M points
      fn(j)=fn(j)+An*cos(2*Nt*pi*xx(j)/P)+Bn*sin(2*Nt*pi
*xx(j)/P);
   end
% Plot both the function and its FS representation
   plot(xx,fnr,'k',xx,fn,'r-','LineWidth',2); grid;
   xlabel('Time [s])'); ylabel('f(t)');
   legend('Function','FSeries');
   title(=['Term: ', num2str(Nt), '  An:  ', num2str(An),
        '  Bn:  ', num2str(Bn),];);
   pause;
end
% -------------------- END --------------------
```

and dashed lines in the figures correspond to the function and its Fourier series approximation, respectively. Figure 4.9 corresponds to the same scenario but with the first 10 terms in the series summation in (4.8a). Finally, Fig. 4.10 belongs to the same comparisons with the first 25 terms. The agreement in curves in Fig. 4.10 shows that 25 terms are adequate for this function in this interval (period).

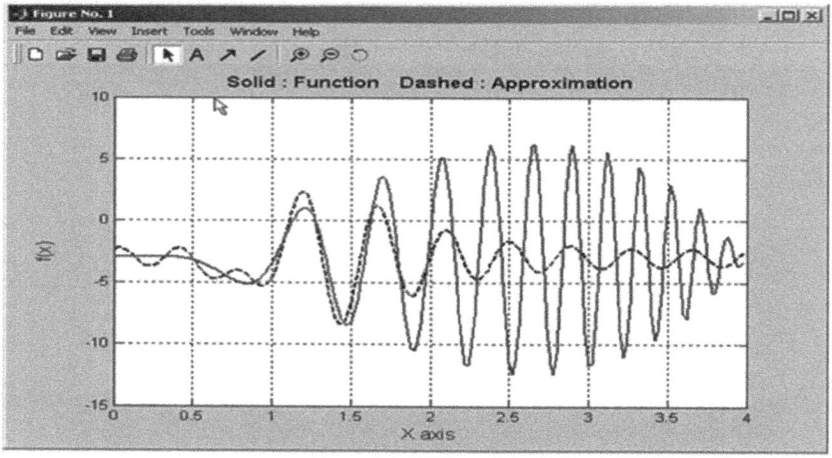

Figure 4.8. The function and its Fourier series approximation with five terms.

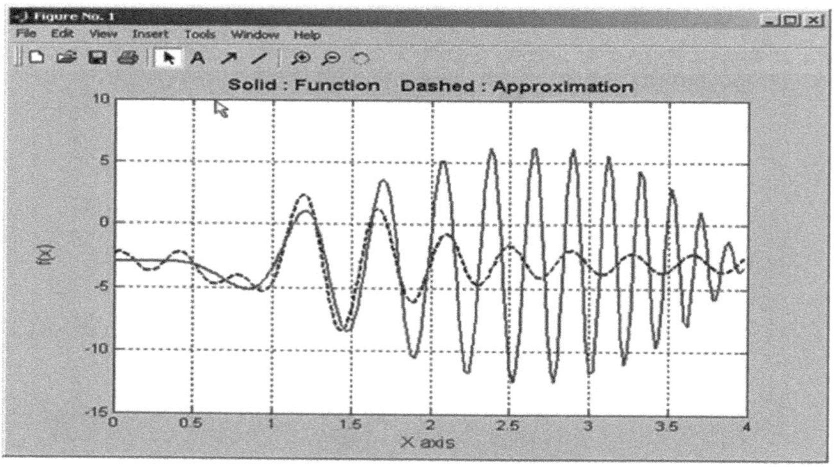

Figure 4.9. The function and its Fourier series approximation with 10 terms.

It should be noted that the function $f(x) = (x^3 - 4x^2)\sin(5x^2 - 2x) - 3$ is not periodic, but its Fourier series representation is periodic with a period of 4. This is well understood if the function and its Fourier representation are plotted for $4 \leq x \leq 8$. Figure 4.11 belongs to these plots.

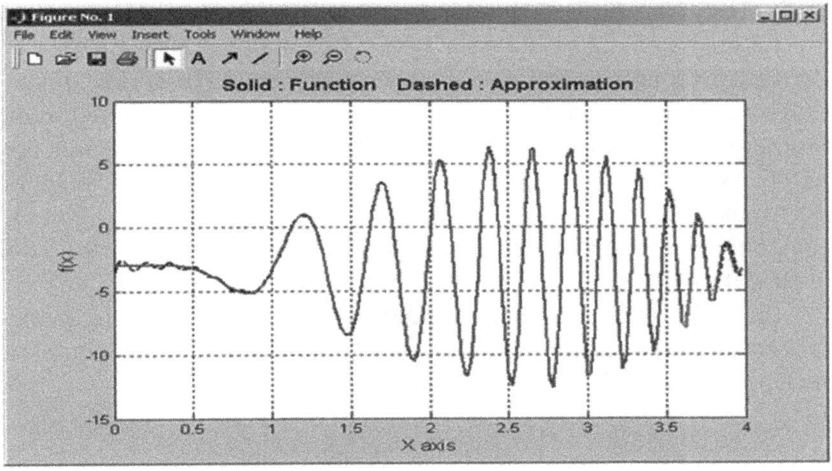

Figure 4.10. The function and its Fourier series approximation with 25 terms.

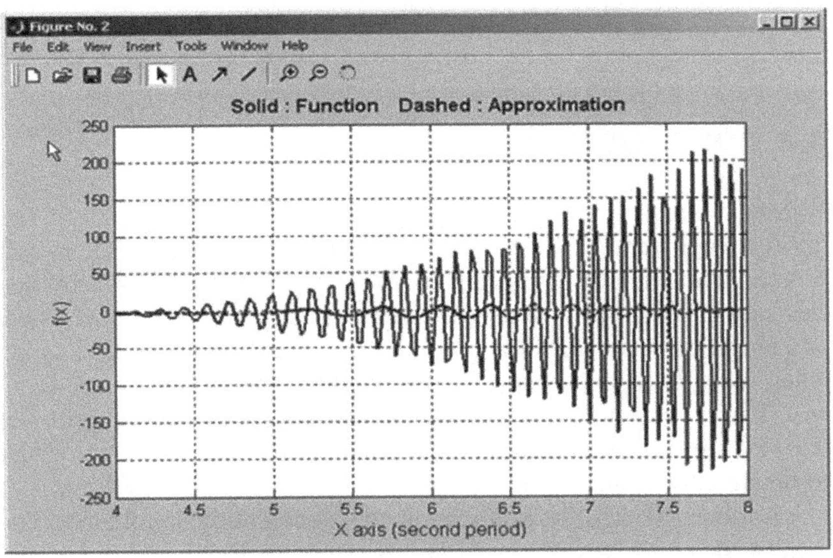

Figure 4.11. The function and its 25-term Fourier series approximation for $4 \leq x \leq 8$.

4.5 RECTANGULAR PULSE AND ITS HARMONICS

As shown earlier, any function may be approximated by series summation of sine and cosine functions. The number of terms required in the Fourier series representation depends on the smoothness of the function and the specified accuracy. The degree of smoothness of the function determines the number of terms in its Fourier representation. In addition, only sine or cosine terms contribute to the function if it is odd or even symmetric.

In understanding digital communication concepts, the frequency content of the rectangular pulse should be well analyzed. A symmetric rectangular pulse is defined as

$$\mathrm{Rect}\left(\frac{t}{T}\right)=\begin{cases}1 & \text{if} \quad -\dfrac{T}{2}\leq t\leq +\dfrac{T}{2} \\ 0 & \text{if} \quad t>\left|\dfrac{T}{2}\right|\end{cases}, \tag{4.10}$$

and an infinite number of terms is required to fully represent this function with the Fourier series summation. The terms are called harmonics. It is interesting to visualize term by term contributions in the Fourier series representation. This is illustrated in Fig. 4.12 for a 2 s rectangular pulse between $4\leq t\leq 6$ s.

4.6 CONCLUSIONS

Fourier transform and Fourier series expansion are important procedures in engineering [7]. The DFT and FFT are discrete tools to analyze time domain signals. One needs to know the problems caused by discretization and specify the parameters accordingly to avoid nonphysical and nonmathematical results. Moreover, extra attention should be paid when using built-in commands in different computer languages (e.g., MATLAB). As pointed out earlier, one needs to multiply the results of the FFT taken using MATLAB's *fft(s, N)* command with Δt in order to obtain correct amplitude values.

Another MATLAB example is the convolution command. The *conv(A, B)* convolves vectors *A* and *B*. The result is a vector with the size of *length(A)+length(B)–1*. The reader naturally assumes that this

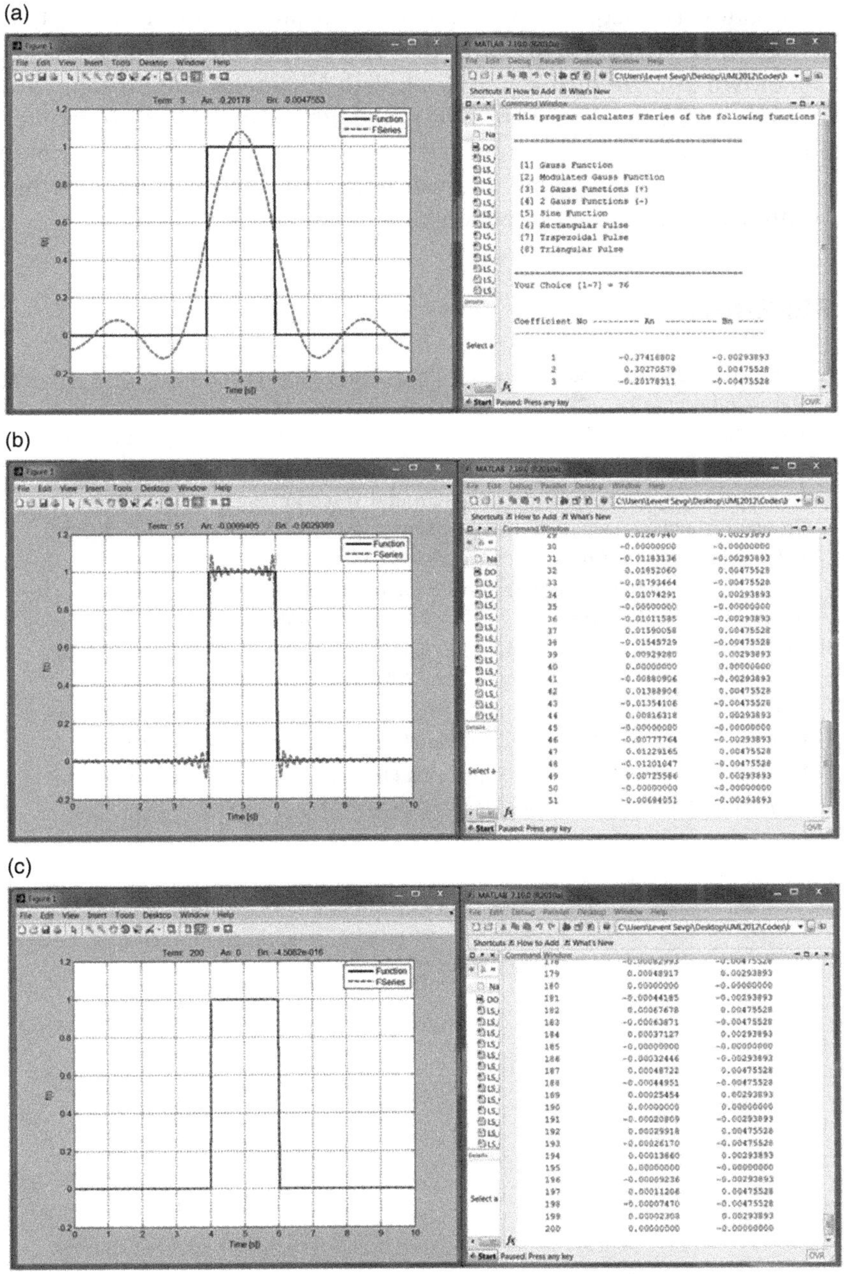

Figure 4.12. A rectangular pulse and its FS representations.

information is adequate. Again, when you attempt to convolve two discrete time signals, you realize that you are in trouble. The convolution of two discrete time signals each with N samples and Δt sampling intervals, one extending from t_1 to t_2, the other from t_3 to t_4, is another time signal which extends from $t_1 + t_3$ to $t_2 + t_4$ having $2N - 1$ samples with the same Δt sampling interval. Again, one needs to multiply the results with Δt in order to obtain correct amplitude values.

REFERENCES

1. Baron Jean Baptiste Fourier (see, e.g., http://en.wikipedia.org/wiki/Baron_Fourier).
2. HP, "The Fundamentals of Signal Analysis," Hewlett Packard Application note 243, 1994.
3. L. Sevgi, F. Akleman, L. B. Felsen, "Ground Wave Propagation Modeling: Problem-Matched Analytical Formulations and Direct Numerical Techniques," *IEEE Antennas and Propagation Magazine*, **44** (1), 2002, pp. 55–75.
4. L. Sevgi, *Complex Electromagnetic Problems and Numerical Simulation Approaches*, IEEE Press – John Wiley and Sons, 2003.
5. M. Levy, *Parabolic Equation Methods for Electromagnetic Wave Propagation*, IEE, Institution of Electrical Engineers, 2000.
6. F. J. Harris, "On the Use of Windows for Harmonic Analysis with the Discrete Fourier Transform," *Proceedings of the IEEE*, **66** (1), 1978, pp. 51–83.
7. L. Sevgi, "Numerical Fourier Transforms DFT and FFT," *IEEE Antennas and Propagation Magazine*, **49** (3), 2007, pp. 238–243.

CHAPTER 5

Stochastic Modeling in Electromagnetics

5.1 INTRODUCTION

Performance tests of multisensor systems with tens even hundreds of targets can only be done in a synthetic environment so computer simulations are almost mandatory [1–4]. A typical multisensor surveillance system is pictured in Fig. 5.1. The country may be Italy or any other country such as Turkey, Greece, Italy, Egypt, England, Germany, Spain, France in the same (Mediterranean) region, or the United States, Canada, China, Japan, Sri Lanka, India, and so on, somewhere else. The mission may be national or international, commercial or military. The aim is to monitor surface and air activity continuously under all weather conditions.

Suppose you want to do real tests for the performance of such a system:

- How do you design characteristic scenarios?
- How many air and surface test targets should you use?
- How do you specify their routes and why?
- Under what environmental conditions do you repeat tests and why?

Electromagnetic Modeling and Simulation, First Edition. Levent Sevgi.

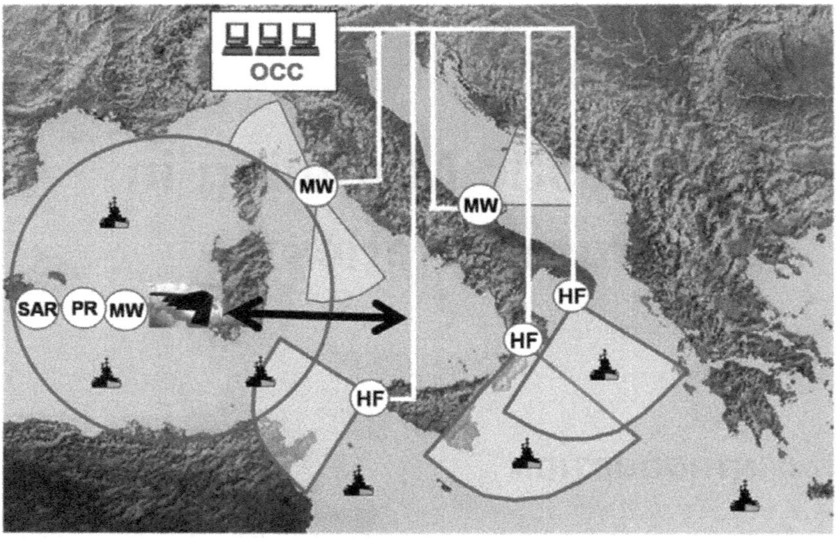

Figure 5.1. A typical multisensor system. HF, high-frequency; MW, microwave; PR, profiling; SAR, synthetic aperture radar; OCC, operation control center.

- How do you synchronize targets and personnel?
- How do you design and switch plan B when something goes wrong?
- Finally, how do you do all these with a limited budget and time?

Obviously, a real-world test of this problem is almost an impossible task; therefore, numerical simulations are of great importance.

Once everything is moved on to a computer, problems related to studying in a nonphysical, numerical environment become important. Table 5.1 and Table 5.2 list fundamental system-level simulation requirements and challenges.

Even if these requirements are satisfied and challenges are taken care of, there are still important questions left for clarification [5]:

- How do you design characteristic/typical realistic test scenarios?

Table 5.1.
System-Level Simulation Requirements

- Understanding and modeling (deterministic or stochastic, mathematical or numerical) of different types of sensors with different sets of parameters (such as power, frequency, coverage, capacity, accuracy, etc.)
- Repetitious runs in order to gain confidence and/or establish the sensitivity of the simulator output data on variations in each of the assumed parameters or conditions
- A trade-off between simulation run time and model accuracy in an intelligent way to balance model fidelity and intended applications
- A design decision on a choice between online or offline simulations partially as standalone sensors or subsystems.

Table 5.2.
System-Level Simulation Challenges

Design	Available sensors, optimum sensors, optimum locations, etc.
Software	Detection and tracking algorithms, level of sensor or data fusion, etc.
Communications	Types of communication, data flow over communication lines, etc.
Tests	Calibration, verification, and validation
Performance evaluation	Performance, success/failure criteria.
System integration	Communication between subsystems, input/output relations, etc.

- How many times should you repeat these tests?
- How do you define evaluation/performance criteria?
- How and when do you say the system works successfully?
- How do you judge on the performance difference between 80% success and 85% success? What does this mean?

Keeping in mind all these issues and questions, this short tutorial aims to present stochastic modeling (random generation) of the signal environment.

5.2 RADAR SIGNAL ENVIRONMENT

A radar signal usually contains echoes from objects under investigation (targets), unwanted objects' echoes (clutter), noise, and other interfering signals [6,7]. All fluctuate randomly with time so the signal environment is *stochastic*. Usually, the target signal is embedded within a background (noise + clutter) signal and/or obscured by interference; that is, its power level is much less than the others. The process of extracting target information from the total echo is called (stochastic) signal processing and performed via powerful, intelligent algorithms. The power of these algorithms comes from (1) the physical and statistical understanding of the target, noise, clutter, and interfering signals and (2) computer generation of the synthetic signal environment. The first step is to generate random numbers with a probability density function (pdf) that represents statistical characteristics of the physical event.

5.2.1 Random Number Generation

Random processes with different statistical characteristics (such as PDF, mean, variance, etc.) are numerically represented by using random number generation [8]. The four well-known probability distributions in radar and communication simulations are uniform, Gaussian, Rayleigh, and exponent distributions. Table 5.3 lists short scripts which create an array of random numbers that obey these four distributions. Note that an array of random numbers with Gaussian, Rayleigh, and exponential distributions can be generated using uniform distribution. This is illustrated in Table 5.3.

In other words, all other distributions can be realized using ***rand*** command. The ***rand*** command generates random numbers uniformly distributed between 0 and 1. A transformation of $a + (b - a)$ rand is required if random numbers uniformly distributed between a and b are of interest. Similarly, Rayleigh distributed random numbers with standard deviation σ can be generated using ***rand*** command via $\sqrt{-2\sigma \ln(\text{rand})}$ transformation [8].

Similar to ***rand***, a random array with a normal distribution can be generated using ***randn*** command. Normal distribution is Gaussian distribution with *zero mean* and *unit standard deviation*. A normally distributed random array and its PDF are given in Fig. 5.2 (on the right and left, respectively).

Table 5.3.
Scripts for Random Number Generations

function out = randUNIFORM(a,b,n) **for** k=1:n out(k) = a + (b − a) * ***rand***; **End**	Crates of *n*-element uniformly distributed random variable between *a* and *b*.
function out = randGAUSS(mean,var,n) **for** k=1:n out(k) = sqrt(var)****randn*** + mean; **End**	Creates *n*-element Gaussian random variable with mean *mean* and variance *var*.
function out = randRALEIGH(var,n) **for** k=1:n out(k) = sqrt(var)*sqrt(−2*log(***rand***)); **End**	Creates *n*-element Raleigh random variable with variance *var*.
function out = randEXPO(a,n) **for** k=1:n out(k) = −log(***rand***)/a; **End**	Creates *n*-element exponentially distributed random variable with parameter *a*.

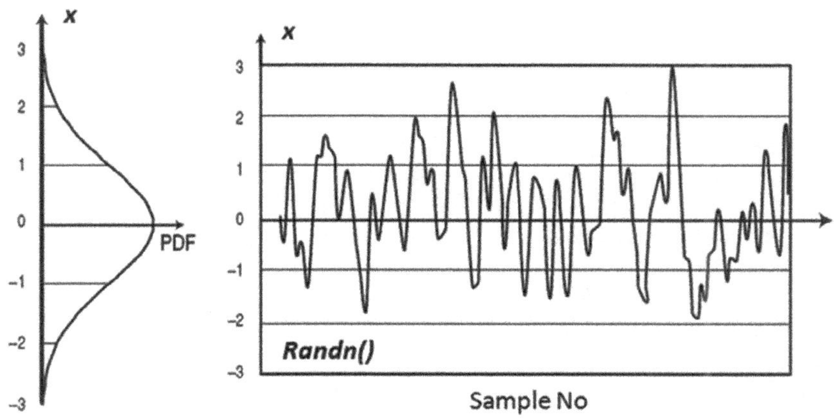

Figure 5.2. Normally distributed random array and its histogram.

The PDF of a random array can be plotted directly by using MATLAB ***hist()*** command. The Gaussian PDF is maximum at zero value and falls off away from zero, as shown (rotated 90°) on the left. Any Gaussian random array with a given *mean* and *standard deviation* may then be produced by shifting and rescaling as given in "rand-GAUSS" function in Table 5.3.

The reader may produce similar figures using this three line MATLAB script: x = ***randn***(1,5000); ***plot***(x); ***hist***(x). Here, the size of the array with normal distribution is 5000. Table 5.4 lists a short MATLAB code for the generation of an array of random numbers with three different probability distribution functions (PDFs). Arrays generated with this code and their histograms are plotted in Fig. 5.3, Fig 5.4, and Fig 5.5.

Table 5.4.
A MATLAB Code for Random Number Generation

```
%% ------------------------------------------------------
%   Program Name      : LS_Random.m
%% ------------------------------------------------------
function LS_Random
clear all; clc;
fprintf('===================================== \n');
fprintf(' \n');
fprintf('(1) Uniform distributions \n');
fprintf('(2) Gaussian distribution \n');
fprintf('(3) Rayleigh distribution \n');
fprintf('===================================== \n');
fprintf(' \n');
x = input ('Your Choice ? : '); fprintf(' \n');
ns = input ('Number of samples? : ');
if x == 1              % Uniform
   lower = input ('Give the lower limit : ');
   upper = input ('Give the upper limit : ');
   out = randunifc(lower,upper,ns);
elseif x == 2              % Gaussian
   ave = input ('Give the average : ');
   var = input ('Give the variance : ');
   out = randgauss(ave,var,ns);
elseif x == 3              % Rayleigh
   var = input ('Give the variance : '); out =
randray(var,ns);
end
figure(1); plot(out); xlim([0,ns]); xlabel('Sample No');
ylabel('Signal'); title('Time history of the random
array')
figure(2); hist(out,50);
%----------- END of MAIN -----------
```

Table 5.4.
(Continued)

```
function out = randunifc(a,b,n)
  for k=1:n
    out(k) = a + (b-a) * rand;
  end
%----------------------------------------
function out = randgauss(a,d,n)
  f = sqrt(d);
  for k=1:n
    out(k) = f * randn + a;
  end
%-----------------------------------------
function out = randray(a,n)
  f = sqrt(a);
  for k=1:n
    u=rand; out(k) = f * sqrt(-2*log(u));
end
%---------------------- END ------------------------
```

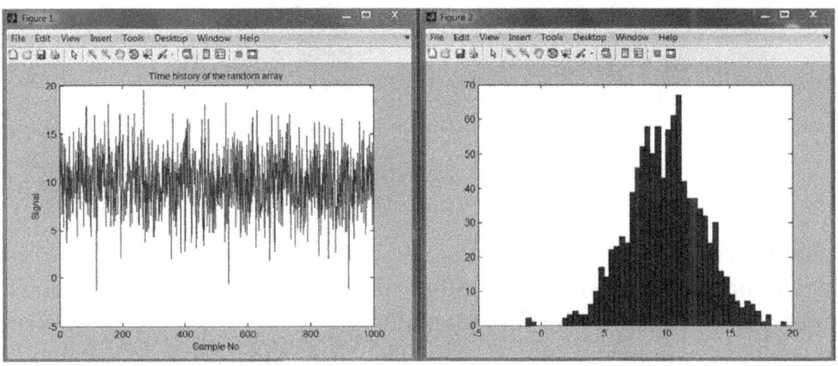

Figure 5.3. An array of Gaussian random numbers and its histogram ($n = 1000$).

5.2.2 Noise Generation

Noise is a random signal and handled in a stochastic manner [9]. Its amplitude distribution characteristics, average value, deviation, frequency characteristics (power spectrum), and so on must be well understood when dealing with communication/radar theory. The distinguishing

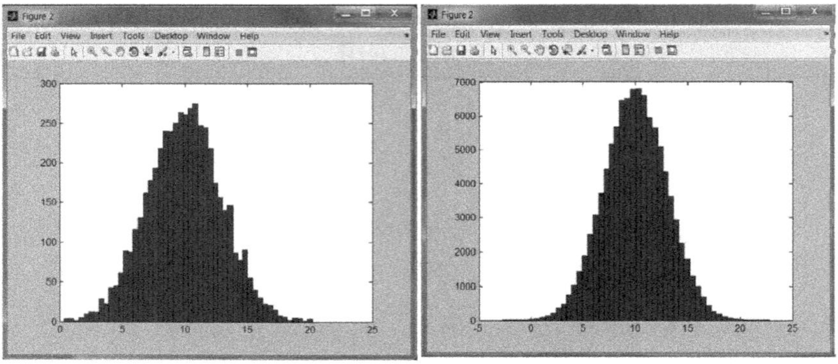

Figure 5.4. Histograms of Gaussian random numbers (L) $n = 5000$, (R) $n = 100{,}000$.

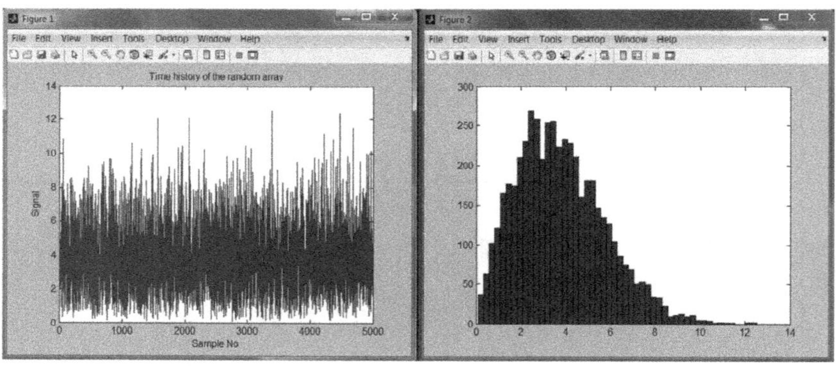

Figure 5.5. An array of Rayleigh random numbers and its histogram ($n = 5000$).

characteristic of the noise is that it is a sample-to-sample (pulse-to-pulse) uncorrelated signal (usually called as *White* or *Gaussian* noise). Signal correlation in the time domain and power spectrum in the frequency domain form a Fourier pair. This means that the autocorrelation function of a random array, for example, generated by the ***rand*** command will be a delta function. A delta function in the time domain corresponds to a constant value in the frequency domain (noise is present in all frequencies). Therefore, the wider the (receiver) bandwidth, the higher the noise level in radar receivers.

Statistical description of the noise (i.e., a one-dimensional random variable) can be specified and numerically generated by using its PDF and/or pdf. Some well-known random variables are uniform, Gaussian, and Rayleigh distributed, and they are listed as follows.

Uniform

(mean: $m = (a + b)/2$; variance: $\sigma^2 = (b - a)^2/12$); variable range:

$$a < n < b, \tag{5.1a}$$

$$\text{PDF} \quad F(n) = \begin{cases} 0 & n < a \\ \dfrac{n-a}{n-b} & a \le n \le b, \\ 1 & n > b \end{cases} \tag{5.1b}$$

$$\text{pdf} \quad f(n) = \frac{1}{b-a}. \tag{5.1c}$$

Gaussian (normal)

m: mean (m^2: DC power); σ: variance (σ^2: AC power), variable range:

$$-\infty < n < +\infty, \tag{5.2a}$$

$$\text{PDF} \quad F(n) = \frac{1}{\sigma\sqrt{2\pi}} \int_{-\infty}^{n} \exp\left\{-\frac{(t-m)^2}{2\sigma^2}\right\} dt, \tag{5.2b}$$

$$\text{pdf} \quad f(n) = \frac{1}{\sigma\sqrt{2\pi}} \exp\left\{-\frac{(n-m)^2}{2\sigma^2}\right\} dt. \tag{5.2c}$$

Rayleigh

(σ: variance) Variable range:

$$0 < n < +\infty, \tag{5.3a}$$

$$\text{PDF} \quad F(n) = \begin{cases} 0, & n \le 0 \\ 1 - \exp\left\{\dfrac{-n^2}{2\sigma^2}\right\}, & n > 0, \end{cases} \tag{5.3b}$$

$$\text{pdf} \quad f(n) = \frac{n}{\sigma^2} \exp\left\{-\frac{n^2}{2\sigma^2}\right\}. \tag{5.3c}$$

The commercial software packages and compilers (e.g., MATLAB, Fortan, C++, etc.) have standard built-in functions that generate a random variable with uniform distribution. If U is a random variable with uniform distribution (between 0 and 1), then random variables with Gaussian and Rayleigh distributions may be generated as follows:

Uniform (with the lower and upper limits of a and b, respectively)

Step 1. Give the lower and upper boundaries (a, b).

Step 2. Generate a random number with uniform distribution (u).

Step 3. Transform to the new uniform number via $n = (b - a)u + a$.

Gaussian

Step 1. Give the mean and variance (m, σ).

Step 2. Generate two random numbers with uniform distribution (u_1 and u_2).

Step 3. Transform to standard Gaussian random number n_0 via $n_0 = \sqrt{-2\ln(u_1)}\cos(2\pi u_2)$ (zero mean, unit variance).

Step 4. Transform to Gaussian random number n_g with mean m and variance σ $n_g = \mathrm{m} + \sigma\, n_0$.

Rayleigh

Step 1. Give the variance (σ).

Step 2. Generate a random number with uniform distribution (u).

Step 3. Transform to standard Rayleigh random number n_0 via $n_0 = \sqrt{-2\ln(u)}$.

Step 4. Transform to Rayleigh random number n_r with variance σ $n_r = \sigma n_0$.

It is often required to generate a sequence that contains both the signal and the noise. The amount of noise added to the signal can be specified with a signal to noise ratio (SNR). The SNR is defined in terms of power of a signal. The mean power in a (of a sequence of) signal s can be calculated from

$$\text{power} \approx \frac{1}{N}\sum_{k=1}^{N} S_k^2. \tag{5.4}$$

If a signal is a sinusoid, then the signal power is equal to one-half the square amplitude of the sinusoid (e.g., the power of $10 \times \sin(\omega t)$ is 50). SNR is the ratio of the power in a signal to the power in the noise. For example SNR = 1 specifies that the noise and the signal have equal powers.

An array of random numbers with a sinusoid, band limited white noise, and a given SNR (dB) may be generated as follows.

Step 1. Give SNR (dB).

Step 2. Calculate SNR from SNR = 10 ^ (SNR_{dB}/10).

Step 3. Calculate signal power from $s(t) = V_m \sin(\omega_o t)$, where $P_s = (V_m)^2/2$.

Step 4. Calculate noise power P_n from SNR $= P_s/P_n$, where $P_n = (V_m)^2/(2\text{SNR})$.

Step 5. Calculate the fluctuation limits of the uniformly distributed noise from$P_n = (2N_m)^2/12$, where

$$N_m = \sqrt{\frac{3V_m^2}{2\text{SNR}}}.$$

Step 6. Generate an N-element array of uniformly distributed random numbers u_k ($k = 1,2, \ldots, N$).

Step 7. Change the variance of the noise: $n_k = 2N_m u_k - N_m$.

Step 8. Generate the signal as $s_k = V_m \sin(\omega_o k \Delta t)$.

Step 9. Add noise to the signal $v_k = s_k + n_k$.

Table 5.5 lists a short MATLAB script which generates a 192-element array containing a sinusoidal signal between $0 - 6\pi$ and Gaussian distributed noise with an SNR of 10 dB. The signal level is set to 0 dB (1 W).

Figure 5.6 shows time variation of signal plus noise for SNR = 10 dB. Solid and dashed curves correspond to uniformly and normally distributed noises, respectively. Note that signal plus noise for a given SNR can be performed with a single MATLAB command AWGN. Table 5.6 explains the use of AWGN command.

Noise is uncorrelated; meaning that it can be reduced or filtered out by simply *oversampling* and *averaging*. This is illustrated in Fig. 5.7. Roughly speaking, the SNR increases by a factor of $\sqrt{m}$, where m is the number of samples averaged (in this example (SNR is improved 10 dB).

It should be noted that SNR yields relative power difference between the signal and the noise levels; absolute noise level (the

Table 5.5.
A MATLAB Script for Signal Plus Noise Generation

```
% -----------------------------------------------------
% Program: LS_SNR.m
% Purpose: Generate signal and noise for a given SNR
% -----------------------------------------------------
clear all; clc;
SNRdB=input('SNR: Signal to noise ratio [dB] = ');
SNR=10^(SNRdB/10); Vs=sqrt(2);  Psignal=0.5*Vs^2;
X = Vs*sin(0:pi/32:6*pi);     % Signal array
Pnoise=Psignal/SNR;           % Noise Power
Vn=sqrt(Pnoise);              % Noise Voltage
[nn nSample]=size(X);
Nk=randn(1,nSample);          % Random numbers
Noise=Vn*Nk;                  % Noise array
Sigtot=X+Noise;               % Total signal
plot(Y,'r-') ;  hold on;
plot(Sigtot); xlabel('Sample No');
ylabel('Signal+Noise Voltage [V]')
% --------------------- END --------------------
```

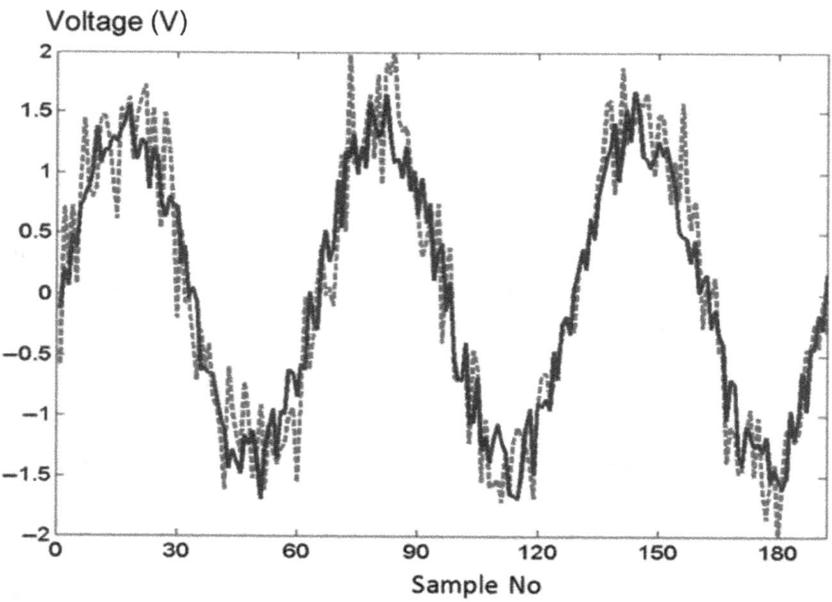

Figure 5.6. Synthetically generated sinusoidal signal + noise (SNR = 10 dB); solid: uniform noise; dashed: Gaussian noise.

Table 5.6.
MATLAB's AWGN Command

AWGN(X,SNR)	Adds white Gaussian noise to *X*. The SNR is in decibels. The power of *X* is assumed to be 0 dBW. If *X* is complex, then AWGN adds complex noise.
AWGN(X,SNR,XPOW)	When XPOW is numeric, it represents the signal power in dBW. When SIGPOWER is "measured," AWGN measures the signal power before adding noise.
Example: X = sqrt(2)*sin(0:pi/8:6*pi); Y = AWGN(X,13,0);	Specifies the power of *X* to be 0 dBW and add noise to produce SNR= 13 dB

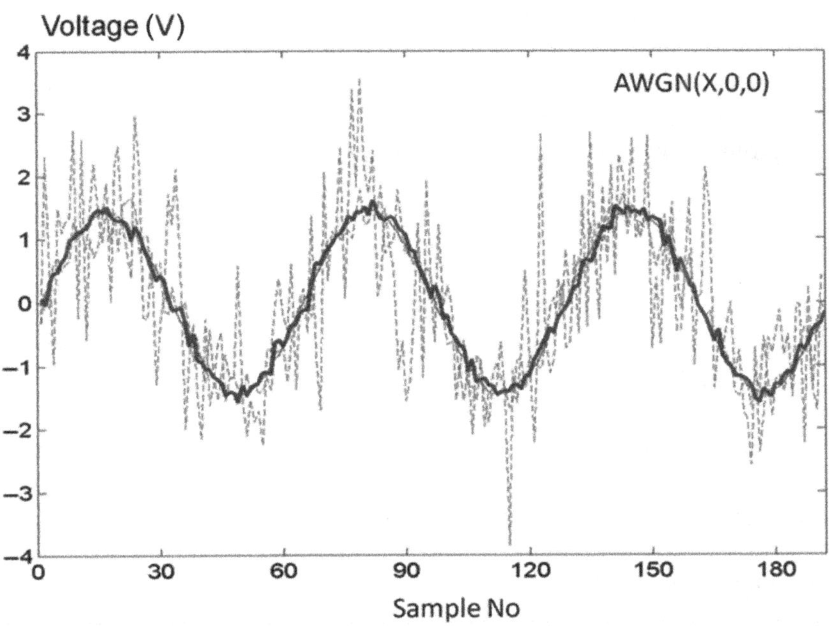

Figure 5.7. Signal + noise (SNR = 0 dB). Dashed: single generation; solid: oversampled and averaged.

detection threshold) is determined from the physical characteristics of the noise. For example, if the thermal noise is dominant, then average noise floor can be calculated from $N_0 = kTB$, where $k = 1.38 \times 10^{-23}$ Joules/K is the Boltzman's constant, T is the temperature in degrees Kelvin (K), and B (Hz) is the signal bandwidth. The effect of added noise is also shown in Fig. 5.7 for a single rectangular pulse signal. Here, the amplitude and the duration of the pulse are 1 V and 0.25 s, respectively. The noise amplitude (i.e., variance) is chosen 0.5 V.

5.2.3 Signal Generation

Useful signal can be generated as explained earlier if the carrier signal of a communication/radar system is of interest. When the radar cross section (RCS) of a radar target is of interest, there may be spatial or temporal fluctuations (or both) which differs from target to target [6]. The fluctuations may be slow (correlated within seconds, even minutes, e.g., a tanker ship navigating on a calm ocean) or fast (correlated only in milliseconds, e.g., a maneuvering high speed fighter aircraft), the fluctuating RCS values may be distributed or may accumulate around a few dominant scatterers. These are categorized in terms of PDFs and are grouped into three as [5–7] follows.

- Steady targets such as mountains, buildings, and so on (TYPE 0).
- Group of small scatterers (without dominant contribution) (slow: TYPE 1 and fast: TYPE 2).
- Group of scatterers with a few dominant ones (slow: TYPE 3 and fast: TYPE 4).

These types are also called as Swerling types (SW) and determine radar waveform as well as echo integration process in radar receivers. For example, since SW 1 and SW 3 targets (which have a decorrelation time of seconds) are sample-to-sample stationary, pulse-to-pulse coherent integration can be applied.

5.2.4 Clutter Generation

Clutter is a word used to describe all unwanted echoes in a radar receiver [6,10]. Unlike noise, which inherently is present at the receiver, clutter is a target like echo signal that come from many small scatterers, such as rain droplets, birds, ocean waves, terrain irregularities, vegetation, aurora, meteors, and others, therefore it is parameter and range dependent. When clutter power dominates over noise, the radar is said to operate in a *clutter-limited* condition.

Unwanted echoes usually occur as distributed clutter, as surface clutter (such as land and sea echoes) or as volume clutter (such as rain, chaff, etc.). They depend on many factors, such as type of the terrain observed, direction of illumination and observation, radar wavelength, and polarization. For example, ocean clutter (usually described as RCS density) depends on ocean wave height, wind characteristics (direction and speed), and wave direction [6]. Land clutter depends on soil type, surface roughness, foliage cover, and so on. Both are sample-to-sample coherent signals, so elimination by just averaging (postdetection integration as done for noise) is not possible. Usually, clutter elimination is performed in the frequency domain. Land clutter occupies a very low frequency range around zero Doppler frequency. On the other hand, ocean clutter has different Doppler characteristics at different radar frequencies.

Ocean clutter is the result of the interaction of the radiated electromagnetic wave with ocean waves [5–7] and has a quite interesting Doppler spectrum. The dominant contribution is produced by scatter from ocean waves having a wavelength half that of the radar wavelength and moving radially to and away from the radar site. This first order resonant scatter results in two dominant peaks called Bragg lines [10–12] at two distinct Doppler frequencies $f_b = \pm\sqrt{g/\pi\lambda} \approx 0.102\sqrt{f_{\text{MHz}}}$ in hertz corresponding to the velocities of the propagation of these ocean waves [10]. Here, λ is the radar wavelength, f is the radar frequency in megahertz, and g is the acceleration due to gravity. A typical ocean spectrum is pictured in Fig. 5.8.

Ocean parameters and their effects on the spectra are (A) *surface wind direction*, (B) *radial current*, (C,D) *wind speed*, and (E,F) *scalar wave spectrum.* Ocean spectra are also used in inverse problems, and surface current and wind maps are prepared especially near economically important harbors and narrow straits (see, for example, Ref. 11).

Ocean clutter is modeled as an ensemble of N complex elements that have a specified amplitude distribution and are pulse-to-pulse correlated. This correlation is established through imposition of a specified shape of the power spectrum. A few available empirical ocean wave

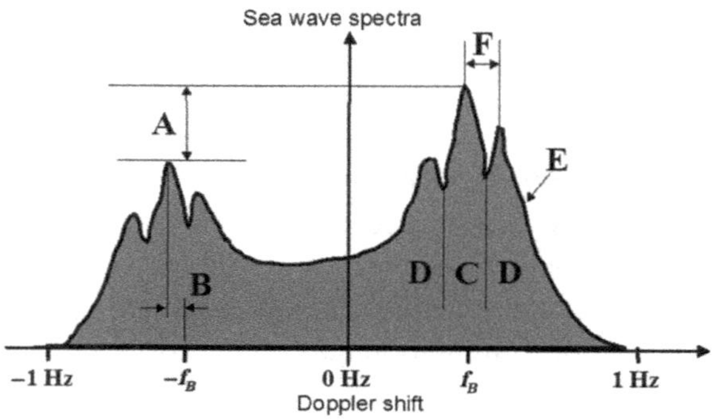

Figure 5.8. A typical HF radar ocean wave spectra [5].

models are Neumann–Pierson spectrum, Phillips spectrum, Pierson–Moskowitz spectrum, and JONSWAP [10–13].

The spectral density (power spectrum) $S_\mu(\omega)$ of a stationary random process $\mu(t)$ and its correlation function $K_\mu(\tau)$ are Fourier pairs [9],

$$S_\mu(\omega) = \int_{-\infty}^{\infty} K_\mu(\tau) e^{-j\omega t} dt \leftrightarrow K_\mu(\tau) = \int_{-\infty}^{\infty} S_\mu(\omega) e^{j\omega t} d\omega, \tag{5.5}$$

for example,

$$S_\mu(\omega) = \frac{\sigma^2}{\pi} \frac{\alpha}{\omega^2 + \alpha^2} \leftrightarrow K_\mu(\tau) = \sigma^2 e^{-\alpha|\tau|}. \tag{5.6}$$

Once the input parameters for this example (σ, α, and Δt) are supplied, signal coherency can be imposed by using either of the expressions in (5.6).

Generation of a Gaussian stationary process with a correlation function given in (5.2) starts with the generation of the first normally distributed random number z_0. This could be done in MATLAB by using either ***randn*** or ***rand*** commands:

$$z_0 = \sqrt{-2\ln(\text{rand})}\cos(2\pi\ \text{rand}) \quad \text{or} \quad z_0 = \text{randn}. \tag{5.7}$$

The first clutter sample is then produced by imposing the correlation via

$$\mu_0 = \sigma\sqrt{1 - e^{-2\alpha\Delta t}}\, z_0. \tag{5.8}$$

The other samples z_k, μ_k, $k = 1, 2, \ldots, N$ are obtained iteratively via

$$z_k = \sqrt{-2\ln(\text{rand})}\cos(2\pi\ \text{rand}), \tag{5.9}$$

$$\mu_k = e^{-\alpha\Delta t}\mu_{k-1} + \sigma\sqrt{1-e^{-2\alpha\Delta t}}\,z_k. \tag{5.10}$$

Alternative procedure for the clutter generation is as follows [7]:

- Two Gaussian distributed unit amplitude phasor arrays g_{1k}, g_{2k} are generated as $\xi_k = g_{1k} + ig_{2k}$, $|\xi| = 1$.
- The Gaussian-shaped power spectrum $S_\mu(f_k)$ of the ocean clutter around dominant Bragg frequencies are generated at K_{FFT} points (at $f_1, f_2, \ldots, f_{\max}$, where $f_{\max}$ is the maximum frequency in the Doppler spectrum).
- The width of the Gaussian-shaped power spectrum is related to the ocean wave height. The ensemble of K_{FFT} elements X_k are generated via $X_k = \xi_k\sqrt{S_\mu(f_k)}$ $(k = 1,2, \ldots, K_{\text{FFT}})$.
- Inverse Fourier transform is utilized via fast Fourier transform (FFT) to obtain the complex elements $\mu(t_k)$.

5.3 TOTAL RADAR SIGNAL

Total radar signal is the addition of signal, noise, and clutter. This is best illustrated with the following example [14–16].

QUESTION

Specify a threshold level for a constant false alarm rate (CFAR) detector if the total signal recorded contains a normally distributed, uncorrelated environmental noise with 13 dB SNR, a coherent ocean clutter having 30 dB clutter to noise ratio (CNR) with a Gaussian-type amplitude distribution and a Gaussian-shaped frequency spectrum with 0.05 Hz bandwidths around 0.25 Hz Bragg peaks, and a Swerling-II-type fluctuating target echo with a 5 dB fluctuation margin and a Doppler shift of 0.75 Hz.

The Bragg peaks ($f_b \approx \pm 0.102\sqrt{f_{\text{MHz}}}$) around ± 0.25 Hz correspond to the radar carrier frequency of 6 MHz. Assume this carrier frequency is removed at the front end of the receiver after a demodulation. The frequency spectrum should cover ±0.25 Hzof Bragg peaks and 0.75 Hz of Doppler shift caused by the radial component of the moving target. Therefore, maximum frequency (f_{max}) may be set to, for example, ±1.0 or ±2.0 Hz. These two yields, for $N = 512$, the frequency resolutions of $\Delta f \approx 0.004$ Hz and $\Delta f \approx 0.008$ Hz, respectively. The signal integration time (i.e., recording period) is inversely proportional with the frequency resolution and is equal to ($T = 1 / \Delta f$) 250 s for $\Delta f \approx 0.004$ Hz (see Ref. 14 for a DFT and FFT tutorial).

The target is Swerling-II type with a fluctuation margin of 5 dB. A 5 dB power fluctuation corresponds to approximately 2.5 V fluctuations in amplitude (i.e., voltage). Swerling-II-type target means a target with small scatterers, without a dominating scatterer; therefore, either of *uniform* or *Gaussian* distributions may be used (actually, this example belongs to a high-frequency radar, where radar signal, i.e., electromagnetic waves interact with a target as a whole). It is the resonance RCS regime; therefore, target amplitude fluctuations may be neglected. But, for the microwave radars, that is, for the optical RCS regime, target fluctuations must be taken into account.

The total received echo signal is synthetically generated as follows.

- Produce the environmental noise first. Produce an array of N-element (e.g., $N = 512$) normally distributed noise samples. Set the noise power (i.e., variance of the array) to a specified value (e.g., −30 dB).
- Add N-discrete target samples using a sinusoidal signal V_m $\sin(2\pi f_{\text{Doppler}} n\Delta t)$ for $n = 1, 2, 3, \ldots, 512$ and $\Delta t = T / 512$.
- Normalize the array as explained in Table 5.3 in order to obtain an SNR of 13 dB.
- Generate N-element clutter array using either of the methods explained earlier. Do it either directly in the time domain using the autocorrelation function or in the frequency domain using a specified power spectrum. Form two Gaussian-shaped power spectra around the two Bragg peaks.
- Normalize the array as explained in Table 5.4 and set the CNR to 30 dB.

Figure 5.9 and Figure 5.10 present the total signal environment in the time and frequency domains, respectively. Since SNR = 13 dB and

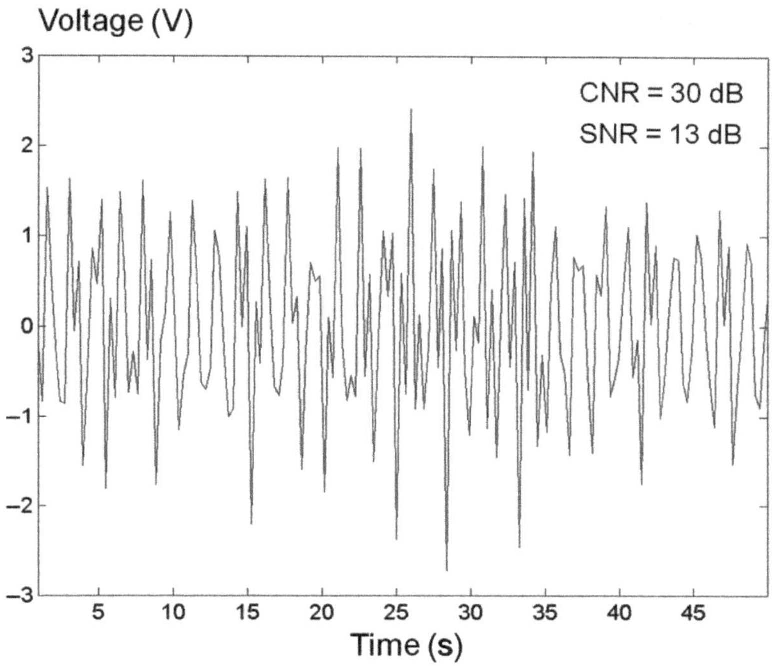

Figure 5.9. Signal + noise + clutter vs. time (I channel).

CNR = 30 dB, the target is embedded in the clutter signal. As observed in Fig. 5.9, it is impossible to detect target in the time domain in this case. Therefore, the detection procedure is carried out in the frequency domain as observed in Fig. 5.10 (note that horizontal scale is normalized to the Brag Frequency).

In Fig. 5.10, the two Bragg peaks around ± 1 and the target around ± 3 normalized frequencies, respectively, are clearly identified. If the CFAR threshold in the detection algorithm is set, let us say, around -15 dB, the target will be detected. Note that there may be a few ghost detections also because of the Gaussian noise (e.g., the noise sample denoted with a question mark will falsify the detection algorithm).

Fortunately, noise is uncorrelated from sample to sample as well as from spectrum to spectrum. Therefore, ghost targets may easily be eliminated if the detection decision is given after averaging a few

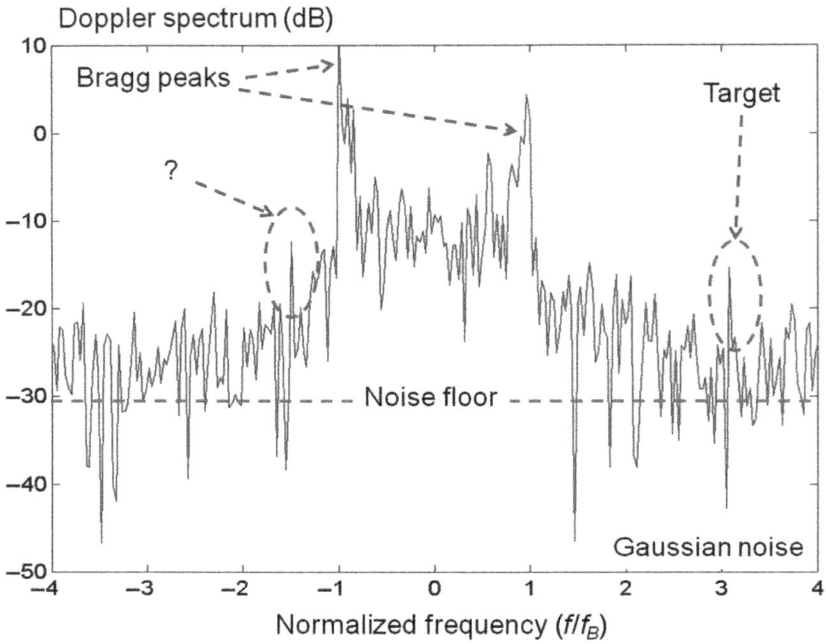

Figure 5.10. A typical HF radar spectrum (SNR = 13 dB, CNR = 30 dB, noise = −30 dB).

Doppler spectra. This is illustrated in Fig. 5.11. Although Fig. 5.10 and Fig 5.11 belong to same parameter sets, ghost target in Fig. 5.10 does not appear in Fig. 5.11.

On the other hand, Bragg peaks and ocean clutter are present at all Doppler spectra so spectra averaging do not work in the elimination of the clutter. Targets with slow radial velocities (i.e., low Doppler shifts) fall in this high clutter region which make them invisible [3,5].

5.4 DECISION MAKING AND DETECTION

Detection is the first task in establishing a communication link or tracking a target with a radar system. This is a decision making problem and is performed through a hypothesis testing procedure [17]. The hypothesis testing which is reviewed in Chapter 2 (Section 2.6) is nothing but CFAR detection.

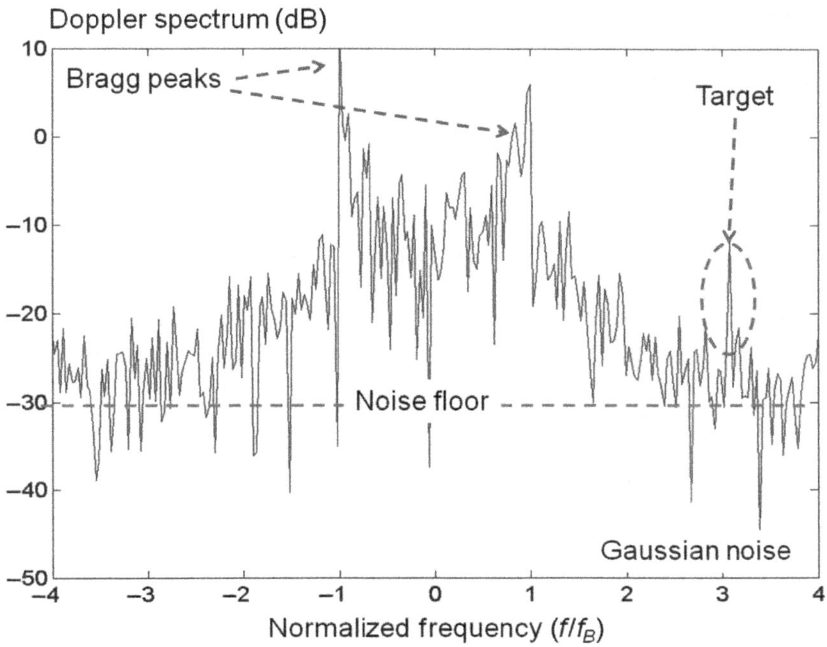

Figure 5.11. Another spectrum with the same parameters given in Fig. 5.10.

5.4.1 Hypothesis Operating Characteristics (HOCs)

Reliable research usually includes data acquisition and its statistical evaluation. Decision making results from *statistical evaluations* based on *hypothesis testing*. This will take us to study the derivation of *receiver operating characteristic* (ROC) and *hypothesis operating characteristic* (HOC) which are actually different phrasing of the same question [18–20].

Null hypothesis is the basic assumption and hypothesis testing is the procedure to decide whether to accept or reject this null hypothesis. Any hypothesis that differs from a given null hypothesis is named as the *alternative hypothesis*. The error is called type I (*false rejection*) when the null hypothesis is rejected when it happens to be true. If the null hypothesis is accepted when it should be rejected, type-II error (*false acceptance*) is made (see Fig. 5.14). Designing a hypothesis testing starts with the choice of the *level of significance* (α), since it directly affects whether to accept or reject a null hypothesis. In practice, a significance level of $\alpha = 0.05$ or $\alpha = 0.01$ is customary. A significance level of $\alpha = 0.05$ means 95% confidence interval (CI) (one is 95% sure that the result is true) about the outcome; that is, 95 out of 100 tests would lead to the correct decision. In other words, one could be wrong with probability of 0.05. The p-value measures how likely it was true that one would have gotten his sample results if the null hypothesis were true. If, for example, "there is no difference between any two processes" is the null hypothesis, then it will be accepted (rejected) if $p \geq \alpha$ ($p < \alpha$).

If X is a Gaussian distributed stochastic parameter with mean μ and standard deviation σ, then $Z = (X - \mu)/\sigma$ is normally distributed (with zero mean and unit standard deviation). If n-samples are tested from an infinite population with mean μ_s and standard deviation σ_s, then $Z = \sqrt{n}(X - \mu_s)/\sigma_s$ is also normally distributed. The probability of $z_1 \leq Z \leq z_2$ (in the limit as $z_1 \to -\infty$, $z_2 \to +\infty$) in these cases can be calculated from

$$P(z_1 \leq Z \leq z_2) = \frac{1}{\sqrt{2\pi}} \int_{z_1}^{z_2} \exp\left(-\frac{v^2}{2}\right) dv. \tag{5.11}$$

The hypothesis testing is then reduced to finding critical values z_1 and/or z_2 for a given CI or level of significance α. In terms of numerical analysis, it is a numerical integration problem (see Fig. 5.14). Three types of hypothesis testing are possible: two tailed, left tailed, and right tailed. If, for example, the hypothesis test is about whether or not one process is better or worse than the other, such a test is called one-tailed test. If, on the other hand, the test is about whether or not two processes are the same, this is a two-tailed test. Table 5.7 lists some common level of significances and the corresponding critical values.

A simple MATLAB code which calculates a critical value for a given CI (or vice versa) is given in Table 5.8. Finding the probability for a given critical values is simply a numerical integration procedure. The integration intervals for left-, right-, and two-sided hypothesis testing are $(z_c \leq z \leq +\infty)$, $(-\infty \leq z \leq -z_c)$, and $(z_c \leq z \leq z_c)$, respectively. On the other hand, finding the critical value for a given probability is a numerical integral search procedure.

Quiz 1:

A diet pill is tested on 64 people in Istanbul. The mean weight of people in Istanbul is known to be 68 kg with a standard deviation of

Table 5.7.
Critical *z* Values for the Hypothesis Testing

Level of significance α	0.10	0.05	0.01	0.005
Left-tailed test	-1.28	-1.645	-2.33	-2.58
Right-tailed test	$+1.28$	$+1.645$	$+2.33$	$+2.58$
Two-tailed test	±1.645	±1.96	±2.58	±2.81

Table 5.8.
A MATLAB Code for the Calculation of CI or α

```
%% --------------------------------------------------------
%  Program: LS_Confidence.m
%  Purpose: Calculate the confidence level of a data
%               being inside a given interval
%% --------------------------------------------------------
clc; clear all;
% This script is used for Normal distribution
fprintf(1,'-----------------------------------\n');
fprintf(1,' Which would you like to calculate? \n');
fprintf(1,' [1] CI of a value interval \n');
fprintf(1,' [2] The value interval for a given CI \n');
fprintf(1,'---------------------------------- \n');
qq=input(' Your Choice? ');
%
if qq==1
   sigma=input('Give the standard deviation value (0-3):
');
   x=0.0:0.0001:sigma; normx=1/sqrt(2.*pi)*exp(-x.^2/2);
   result=trapz(x,normx)*200;
   fprintf(1,'The CI is %% %6.2f \n',result);
else
   conf=input('Give the CI as a percentage (0-100): ');
   conf=conf/200;
   area=0.0; del=0.0001; y=-del;
   while area&lt;=conf
      y=y+del; area=area+1/sqrt(2.*pi)*exp(-y^2/2)*del;
   end
   fprintf(1,'The standard deviation of your CI is
%5.2f \n',y);
end
% -------------------- End ----------------------
```

12 kg. Design a hypothesis test and find out whether the diet pill is effective with a given significance level.

Solution:

This is a left-tailed hypothesis testing because one needs to show whether or not, after a certain period, those who take diet pills loose significant weight. Suppose the level of significance is $\alpha = 0.05$ (i.e., 95% CI). From Table 5.6, the critical value is found to be $z_c = -1.645$ (this could also be calculated by using the MATLAB code listed in Table 5.7). If the mean weight of these 64 people who take the diet pill after a given period is μ_s, then one needs to compare $z = \sqrt{n}\,(\mu_s - \mu)/\sigma$ with the critical value $z \leq z_c$. If $z_c = -1.645$, then the diet pill is accepted to be effective; otherwise it is not.

The hypothesis is set as follows:

H_0 (null hypothesis): The diet pill works.
H_1 (alternative hypothesis): The diet pill does not work.

Accept H_0 if $\mu_s \leq 65.35$ kg with the level of significance $\alpha = 0.05$; otherwise reject H_0.

Quiz 2:

A coin is tossed 64 times. For a level of significance $\alpha = 0.05$, how many heads (or tails) one should expect to conclude that "the coin is fair"?

Solution:

Tossing a coin is independent and called Bernoulli trials [1]. Its distribution is binomial. The probability that an event (head or tail) will happen x times out of n trials (x: success, $n - x$: failure) is

$$P(X = x) = \frac{n!}{x!(n-x)!} p^x q^{n-x}, \tag{5.12}$$

where p and $q = 1 - p$ are the probabilities of success and failure, respectively. The mean and standard deviation of binomial distribution are $\mu = np$ and $\sigma = \sqrt{npq}$. According to central limit theorem [5], if n is large, binomial distributed random event turns into a normally distributed event via $Z = (X - \mu)/\sigma$.

This is a double-side test with the critical value $z_c = 1.96$. The number of heads (or tails) should be between $\mu \pm 1.96\sigma$. The probability of getting head (or tail) with a fair coin is $p = 0.5$. Here, $\mu = np = 32$ and $\sigma = \sqrt{npq} = 4$. Therefore, the number of heads (or tails) should be between $24.16 \leq X \leq 39.84$.

The hypothesis is set as follows:

H_0 (null hypothesis): $p = 0.5$ (the coin is fair).
H_1 (alternative hypothesis): $p \neq 0.5$ (the coin is not fair).

Accept H_0 if the number of heads (or tails) is between 24 and 39; otherwise reject H_0.

Hypothesis operation curves (HOCs) are the curves of type-II error versus distance ($d = (\mu - \mu_0)/\sigma\sqrt{n}$) for a given number of samples. A short MATLAB code is listed in Table 5.9 which prepares HOC. Running this code yields the HOC shown in Fig. 5.12.

As explained earlier, type-II error occurs if the null hypothesis H_0 ($\mu = \mu_0$) is accepted when the alternative hypothesis H_1 ($\mu \neq \mu_0$) is true. Suppose $\mu = \mu_0 + \delta$ ($\delta > 0$); then the test statistic and distance are $z_t = (x - \mu_0 - \delta)/(\sigma / \sqrt{n}) + \delta\sqrt{n}/\sigma$ and $d = \delta\sqrt{n}/\sigma$, respectively. The calculation of the number of samples required for a hypothesis testing can be done with the MATLAB code given in Table 5.10.

5.4.2 A Communication/Radar Receiver

The first task of a receiver is to decide whether or not a recorded sample is a useful signal/target or noise for a given detection probability P_D and

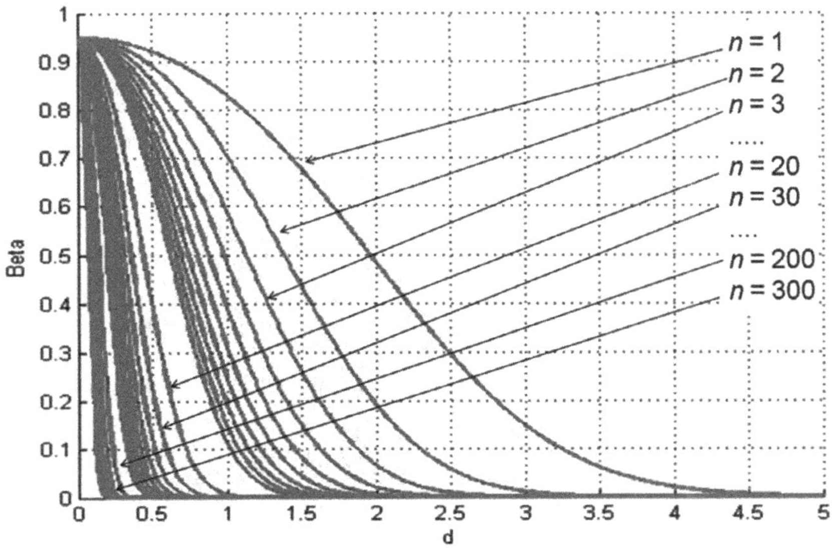

Figure 5.12. HOC prepared via MATLAB code listed in Table 5.3.

Table 5.9.
A Short MATLAB Code for Plotting HOC

```
%% ---------------------------------------------------------
%  Program: LS_HOC.m
%  Purpose: To plot Hypothesis Operating Characteristics
Curves
%% ---------------------------------------------------------
clc; clear all; figure(1); hold on; grid on;
CI=95;    %input('Required Confidence Level [%]= ?');
kk=1;   %input('Two-sided test [1], Single-sided test [2]
= ?' );
if kk==1; alpha=(1-CI/100)/2;   else   alpha=1-CI/100;
end
z=-10; del=10/10000; area=0;
while area&lt;= alpha   % Calculate z_critical
     z=z+del; area=area+1/sqrt(2*pi)*exp(-z^2/2)*del;
end
zc=-z;  n=1;
while n>0      % Number of samples loop
     k=1;
     for d=0.01:.01:5
        dd(k)=d; z1=zc-d*sqrt(n); zz=z1-10:del:z1;
        fz=1/sqrt(2*pi)*exp(-zz.^2/2);
beta1=trapz(zz,fz);
        z2=-zc-d*sqrt(n); zz=z2-10:del:z2;
        fz=1/sqrt(2*pi)*exp(-zz.^2/2);
beta2=trapz(zz,fz);
        beta(k)=beta1-beta2; k=k+1;
     end
     plot(dd,beta,'r','LineWidth',2); xlabel('d');
ylabel('Beta');
     pause;
     if n&lt;10; n=n+1; elseif n&lt;100; n=n+10; else
n=n+100;  end
end
% ------------------ END --------------------
```

Table 5.10.
A MATLAB Code Calculating the Required Number of Samples

```
%% -----------------------------------------------------
%   Program: LS_HYPO.m
%   Purpose: To find out the number of samples required
for
%            a hypothesis testing for a given CI
%% -----------------------------------------------------
clc; clear all;
mu=input(' Mean value of the statistical test = ?');
sigma=input('Standard deviation of the statistical test
= ?');
CI=input('Required Confidence Level [%]= ?');
Error=input('Error = ?');
kk=input('Two-sided test [1], Single-sided test [2] = ?'
);
if kk==1;
  alpha=(1-CI/100)/2;
else
  alpha=1-CI/100;
end
z=-10; del=10/10000; area=0;
while area&lt;=alpha
    z=z+del;
    area=area+1/sqrt(2*pi)*exp(-z^2/2)*del;
end
alphac=-z;  n=(alphac*sigma/Error)^2;
fprintf('The number of samples required = %4i \n',
ceil(n));
fprintf('Critial Value = %8.5f  \n', alphac);
% ------------------- END ---------------------
```

false alarm rate P_{FA}. The generation of a received signal with and without random noise was discussed in Section 5.2 with sample MATLAB codes. Time sequences of signal, noise, and signal plus noise generated with the MATLAB code given there are given in Fig. 5.13. Here a 3 V sinusoidal signal and a white noise with 3 dB SNR are shown.

The problem, at any instant, is then to decide whether the next sample is noise only or noise plus signal at the radar receiver. A classical

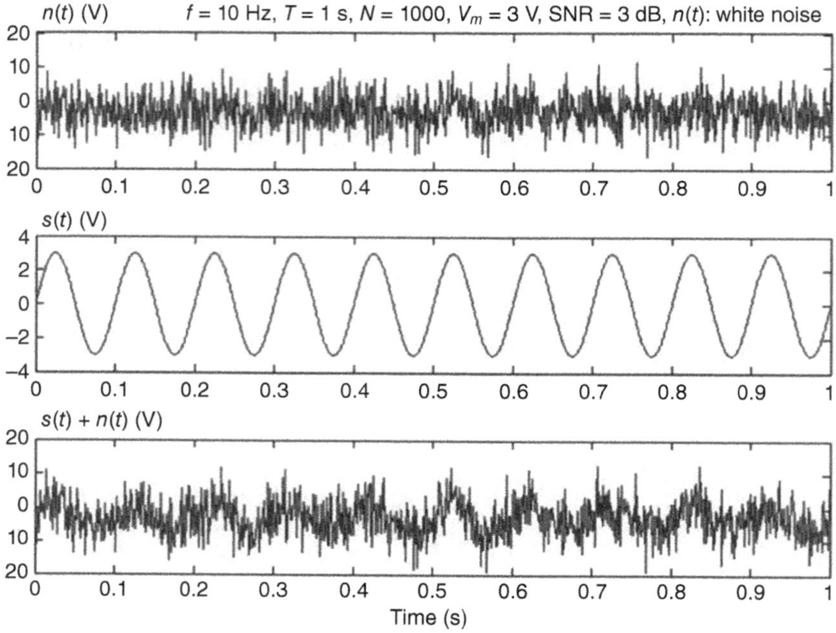

Figure 5.13. A computer generated white noise, sinusoidal signal, and both.

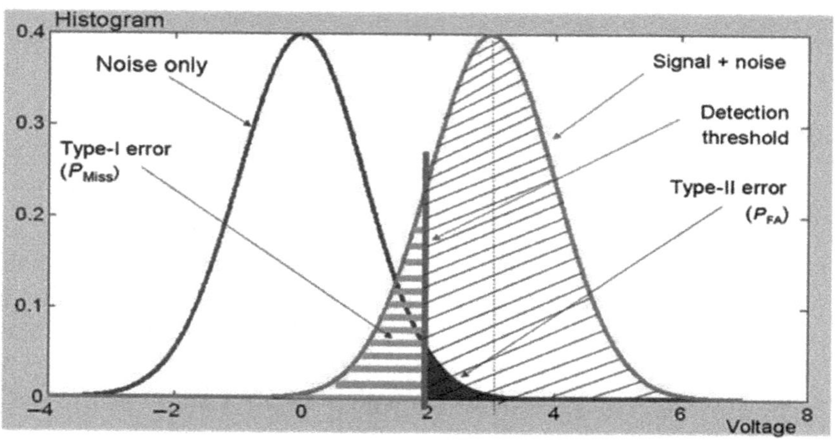

Figure 5.14. Detection probability and false alarm rates for white noise receivers.

approach is to use CFAR procedure and find out the threshold voltage above (under) which detection (no detection) decision is given. This is illustrated with Fig. 5.14. The Gauss functions on the left and right show the pdfs for the "noise only" and "signal plus noise" cases, respectively.

For a single channel representation, once the threshold voltage is specified, the false alarm rate and the detection probabilities can be calculated as follows (*rms*–power of the noise is σ_n^2):

$$P_{\mathrm{FA}} = \int_{V_{\mathrm{TH}}}^{\infty} \frac{1}{\sqrt{2\pi\sigma_n^2}} \exp\left(-\frac{v^2}{2\sigma_n^2}\right) dv, \tag{5.13}$$

$$P_{\mathrm{D}} = \int_{V_{\mathrm{TH}}}^{\infty} \frac{1}{\sqrt{2\pi\sigma_n^2}} \exp\left(-\frac{v^2+V_0^2}{2\sigma_n^2}\right) dv. \tag{5.14}$$

These two equations can be used to prepare ROCs. The signal and noise powers are given as $P_{\mathrm{noise}} = \sigma_n^2$, $P_{\mathrm{signal}} = V_0^2/2$. ROCs represent detection probabilities versus SNR for a given false alarm rate. These curves are prepared as follows.

- First, specify the false alarm rate P_{FA}.
- Use (5.13) in an iterative form and calculate the threshold voltage V_{TH} for this P_{FA}.
- Fix noise power $P_{\mathrm{noise}} = \sigma_n^2$; use $\mathrm{SNR} = V_0^2/2\sigma_n^2$. For a specified SNR, calculate V_0. Use the threshold voltage in (5.14) and calculate detection probability versus SNR.

A short MATLAB code listed in Table 5.11 is prepared for this purpose. The ROC prepared using this code is given in Fig. 5.15.

Most of the radar and/or communication receivers do the detection in the frequency domain; this necessitates tracing the phase information also. Therefore, detection is based on I + Q (in phase and quadrature) channel representations.

The PDF for signal plus noise in this case is given with

$$f(v) = \frac{v}{\sigma_n^2} I_0\left(\frac{vV_0}{\sigma_n^2}\right) \exp\left(\frac{-\left(v^2+V_0^2\right)}{2\sigma_n^2}\right). \tag{5.15}$$

For noise-only ($V_0 = 0$) detection, this reduces to

$$f(v) = \frac{v}{\sigma_n^2} \exp\left(-\frac{v^2}{2\sigma_n^2}\right). \tag{5.16}$$

Table 5.11.
A Short MATLAB Code for the Generation of ROC

```
%% ---------------------------------------------------
%    Program: LS_ROC.m
%    Purpose: To plot PD vs. SNR for fixed PFA
%% ---------------------------------------------------
clc; clear all; figure(1)
xlabel('SNR [dB]'); ylabel('Probability of Detection
(PD)');
grid; hold on
for k=2:2:10
     PFA=10^(-k); area=0.0; del=0.00001; y=-10;
     while area&lt;=PFA
       y=y+del;
       area=area+1/sqrt(2.*pi)*exp(-y^2/2)*del;
     end
     Vth=-y; j=0;
     for snr=1:0.5:100
       j=j+1; snrdB(j)=10*log10(snr); Pnoise=1; Vn=1;
       Psignal=snr*Pnoise; Vs=sqrt(Psignal);
       x=Vth:0.0005:Vs+10;
       fx=1/sqrt(2.*pi)*exp(-(x-Vs).^2/2);
       PD(j)=trapz(x,fx);
   end
%  plot(snrdB,PD,'r','LineWidth',2)
set(gca,'ytick',[.1 .2 .3 .4 .5 .6 .7 .75 .8 .85 .9 .95
.9999])
set(gca,'xtick',[1 2 3 4 5 6 7 8 9 10 11 12 13 14 15 16
17 18 19 20])
semilogy(snrdB, PD,'r-','LineWidth',2);
title(['Receiver Operation Characteristic (ROC)']); pause
end
% --------------------- END ----------------------
```

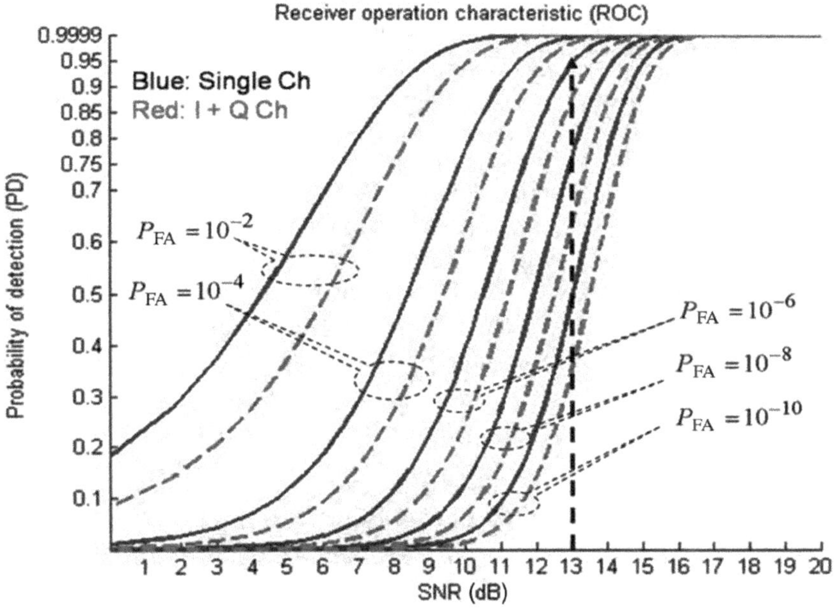

Figure 5.15. ROC prepared using MATLAB code in Table 5.11 (solid: single channel; dashed: I + Q channel).

The false alarm rate P_{FA} is then calculated from

$$P_{FA} = \int_{V_{TH}}^{\infty} \frac{v}{\sigma_n^2} \exp\left(-\frac{v^2}{2\sigma_n^2}\right) dv = \exp\left(-\frac{v^2}{2\sigma_n^2}\right). \tag{5.17}$$

On the contrary, the threshold voltage can be calculated from

$$V_{TH} = \sqrt{2\sigma_n^2 \ln\left(\frac{1}{P_{FA}}\right)} \tag{5.18}$$

if false alarm rate is given. The detection probability, using this threshold voltage, can be found from

$$P_D = \int_{V_{TH}}^{\infty} \frac{v}{\sigma_n^2} I_o\left(\frac{vV_0}{\sigma_n^2}\right) \exp\left(\frac{-(v^2 + V_0^2)}{2\sigma_n^2}\right) dv. \tag{5.19}$$

Since $\mathrm{SNR} = V_0^2/2\sigma_n^2$, (5.19) reduces to

$$P_\mathrm{D} = \int_{V_\mathrm{TH}}^{\infty} \frac{v}{\sigma_n^2} I_o(2v\mathrm{SNR})\exp\left(-\frac{v^2}{2\sigma_n^2}\right)\exp(-\mathrm{SNR})\,dv. \qquad (5.20)$$

A short MATLAB code listed in Table 5.12 belongs to the preparation of ROC for I + Q channel detection procedure. Curves for this case are also plotted in Fig. 5.15.

Both ROC and HOC can be validated numerically; this can be achieved by using the MATLAB code presented in Table 5.13. This code produces a number of random transmitted bits for a given false

Table 5.12.
A MATLAB Code Which Plots ROC for I + Q CHANNEL

```
%-------------------------------------------------------
%   Program   : IQ_ROC.m
%   Purpose   : To plot PD vs. SNR for fixed PFA
%-------------------------------------------------------
clc; clear all; figure(1); grid; hold on
xlabel('SNR [dB]'); ylabel('Probability of Detection
(PD)');
for k=2:2:10
   PFA=10^(-k); Vth=sqrt(2*log(1/PFA));    j=0;
   for snr=1:0.5:100              % Calculate detection
probability
      j=j+1;   snrdB(j)=10*log10(snr);   Pnoise=1; Vn=1;
      Psignal=snr*Pnoise; Vs=sqrt(2*Psignal);
      x=Vth:0.002:Vth+15;
      fx=x.*BESSELI(0,x.*Vs).*exp(-(x.^2+Vs^2)/2);
      PD(j)=trapz(x,fx);
   end
set(gca,'ytick',[.1 .2 .3 .4 .5 .6 .7 .75 .8 .85 .9 .95
.9999])
set(gca,'xtick',[1 2 3 4 5 6 7 8 9 10 11 12 13 14 15 16
17 18 19 20])
semilogy(snrdB, PD,'r','LineWidth',2);
title(['Receiver Operation Characteristic (ROC)']);  pause
end
% -------------------- END ---------------------
```

Table 5.13.
A Short MATLAB Code for the ROC Test

```
%% -------------------------------------------------------
%   Program: LS_ROC_Test.m
%   Purpose: To test ROC via N transmitted bits
%% -------------------------------------------------------
clc; clear all;
snrdB = input ('SNR[dB]?: '); snr=10^(snrdB/10);
NS=20000;
nsb=0; nnb=0;          % number of signal and noise bits
for k=1:NS
  bits(k)=rand(1);
  if (bits(k)>=0.5)
     bits(k)=1; nsb = nsb +1;
  else
     bits(k)=0; nnb =nnb +1;
  end
end
Psignal=1; Vs=sqrt(Psignal); Pnoise=Psignal/snr;
Vnoise=sqrt(Pnoise); noise= Vnoise*randn(1,NS); %Noise
samples
PFA=10^(-2); area=0.0; del=0.00001; y=-10;
while area&lt;=PFA
y=y+del;
area=area+1/(sqrt(2.*pi)*Vnoise)*exp(-y^2/
(2*Vnoise^2))*del;
end
vth=-y;
% Calculate PD
x=vth:0.00001:Vs+6;
fx =1/(sqrt(2.*pi)*Vnoise)*exp(-(x-Vs).^2/(2*Vnoise^2));
PDC =trapz(x,fx); Vk = bits+noise;
for k=1:NS    % DO the Detection
  if (Vk(k)>= vth); dbits(k)=1; else dbits(k)=0; end
end
ndet=0; nghost=0;
for k=1:NS     % Calculate PFA and PD
    if bits(k)==1
       if dbits(k)==1; ndet=ndet+1; end
    else
       if dbits(k)==1; nghost=nghost+1; end
```

(*Continued*)

Table 5.13.
(Continued)

```
    end
end
PFA=nghost/nnb;  PD=ndet/nsb;
fprintf(1,'CFAR Detection Simulation Results \n');
fprintf(1,'---------------------------------- \n');
fprintf(1,'PD (Calculated)  = %6.3f \n',PDC);
fprintf(1,'Number of Noise bits     = %6d \n',nnb);
fprintf(1,'Number of Signal bits    = %6d \n',nsb);
fprintf(1,'Number of Ghost bits     = %6d \n',nghost);
fprintf(1,'Number of detected bits = %6d \n',ndet);
fprintf(1,'PFA = %6.3f \n',PFA);
fprintf(1,'PD  = %6.3f \n',PD);
% -------------------- END -----------------------
```

alarm rate and SNR, and adds random noise accordingly. It then calculates the threshold voltage and detection probability mathematically using (5.13) and (5.14).

Note that $\mathrm{SNR} = V_0^2/\sigma_n^2$ is used for binary CFAR detection; instead of $\mathrm{SNR} = V_0^2/2\sigma_n^2$, which is used for sinusoidal signal. Finally, number of detected and ghost bits are counted numerically. The results are compared with the ones obtained mathematically.

Two test-run results for SNR = 0 dB and SNR = 10 dB with $P_{\mathrm{FA}} = 0.01$ are given in Fig. 5.16; a total of 20,000 bits are produced. As shown, mathematical results agree very well with the numerical ones. Note that the detection probability for SNR = 10 dB is 0.798 in binary CFAR detection. Check from Fig. 5.16 that the CFAR detection probability for the sinusoidal signal is around 0.98.

Note that the MATLAB code listed in Table 5.13 may also be used in the preparation of ROC/HOC under different noises and/or clutter distributions. If, for example, noise and/or clutter has Rayleigh, Rician, Weibul, or some other distribution, then these curves can only be prepared numerically. Note also that MATLAB has its own Statistics Toolbox and the *ztest* module can be used for the hypothesis testing:

$$[\mathrm{h}, \mathit{pvalue}, \mathrm{ci}] = \mathit{ztest}(x, \mathit{mu}, \mathit{sigma}, \mathrm{alpha} = 0.05, \text{'}\mathit{both}\text{'}).$$

For n-element x-sample set, this command tests whether or not the sample set and the population (or the other set) with mean μ and stan-

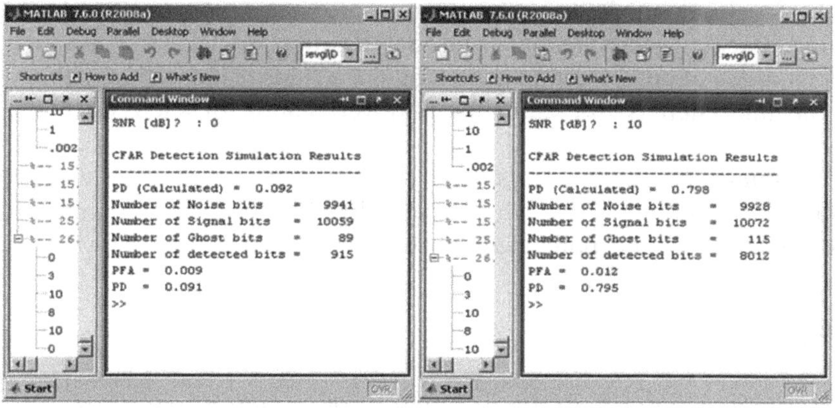

Figure 5.16. CFAR detection tests performed using the code in Table 5.13.

dard deviation σ are the same. For example, if the mean weight of the samples given in Quiz 1 reduces to 65 kg after taking the pill for a suggested period, the hypothesis test can be done via

```
> pmu = 68;
> n = 64;
> psigma = 12;
> smu = 65
> pmu = ssigma = psigma/sqrt(n);
> wkg = ssigma*rand(1,n) + smu;
> hist(wkg);
> [h,pvalue,ci] = ztest(wkg,pmu,psigma,0.05,'l
eft')
```

5.5 CONCLUSIONS

Realistic communication, radar, multisensor systems necessitates the inclusion of randomness of signals. This is achieved via stochastic modeling and simulation [7]. Stochastic modeling and simulation, in turn, necessitate synthetic generation of radar signal environment as realistic as possible. This could only be achieved by good understanding of physical as well as statistical behaviors of radar target, noise, clutter, and interference.

Stochastic modeling and simulation also necessitate reliable detection, tracking, classification, and identification models. But, the first step is detection. Detection is nothing but a statistical decision making. The final step is the validation and verification of such stochastic systems. This is achieved via performance evaluation processes. Unfortunately, there is no generally accepted performance evaluation criteria for the tests of such systems; therefore, not only a simulator design but also its application are real challenges.

REFERENCES

1. N. Ince, E. Topuz, E. Panayirci, C. Isik, *Principles of Integrated Maritime Surveillance Systems*, Kluwer Academic, Boston, 2000.
2. L. Sevgi, A. M. Ponsford, H. C. Chan, "An Integrated Maritime Surveillance System Based on Surface Wave HF Radars, Part I—Theoretical Background and Numerical Simulations," *IEEE Antennas and Propagation Magazine*, **43** (4), 2001, pp. 28–43.
3. A. M. Ponsford, L. Sevgi, H. C. Chan, "An Integrated Maritime Surveillance System Based on Surface Wave HF Radars, Part II—Operational Status and System Performance," *IEEE Antennas and Propagation Magazine*, **43** (5), 2001, pp. 52–63.
4. L. Sevgi, "Target Reflectivity and RCS Interaction in Integrated Maritime Surveillance Systems Based on Surface Wave HF Radar Radars," *IEEE Antennas and Propagation Magazine*, **43** (1), 2001, pp. 36–51.
5. L. Sevgi, Chapter 1: Radars, in R. Bansal, ed., *Engineering Electromagnetics: Applications*, CRC Press–Taylor & Francis Group, New York, 2006.
6. M. I. Skolnik, *Introduction to Radar Systems*, McGraw Hill, New York, 1985.
7. L. Sevgi, *Complex Electromagnetic Problems & Numerical Simulation Approaches*, IEEE Press–John Wiley & Sons, Piscataway, NJ, 2003.
8. D. C. Montgomery, G. C. Runger, *Applied Statistics & Probability for Engineers*, John Wiley & Sons, Hoboken, NJ, 2003.
9. S. Haykin, M. Moher, *Introduction to Analog & Digital Communications*, 2nd ed., John Wiley & Sons, Inc., Prentice-Hall, NJ, 2007.
10. D. E. Barrick, "Theory of HF and VHF Propagation Across the Rough Sea, 1, The Effective Surface Impedance for a Slightly Rough Highly Conducting Medium at Grazing Incidence," *Radio Science*, **6** (5), 1971, pp. 517–526.
11. H. Gunther, et al., "The EuroROSE Project," Proceedings of the 16th International Conference of the American Meteorological Society on "Interactive Information and Processng Systems (IIPS) for Metereology, Oceanography and Hydrology," California, USA, January 2000, pp. 214–217.
12. B. Kinsman, *Wind Waves*, Prentice Hall, Englewood Cliffs, NJ, 1965.

13. O. M. Phillips, *Dynamics of the Upper Ocean*, Cambridge University Press, London, 1966.
14. L. Sevgi, "Numerical Fourier Transforms DFT and FFT," *IEEE Antennas and Propagation Magazine*, **49** (3), 2007, pp. 238–243.
15. L. Sevgi, "Synthetic Radar Signal Environment: Computer Generation of Signal, Noise, and Clutter," *IEEE Antennas and Propagation Magazine*, **49** (5), 2007, pp. 192–198.
16. L. Sevgi, "Hypothesis Testing and Decision Making: Constant-False-Alarm Rate," *IEEE Antennas and Propagation Magazine*, **51** (3), 2009, pp. 218–224.
17. L. Sevgi, "Biostatistics and Epidemiology: Hypothetical Tests on Cell Phone Users," *IEEE Antennas and Propagation Magazine*, **52** (1), 2010, pp. 267–273.
18. D. Rumsey, *Statistics for Dummies*, Wiley Publishing, Hoboken, NJ, 2003.
19. C. F. Gerald, P. O. Wheatley, *Applied Numerical Analysis*, 6th ed., Addison-Wesley, New York, 1999.
20. B. R. Mahafza, A. Z. Elsherbeni, *MATLAB Simulations for Radar Systems Design*, Chapman & Hall/CRC LLC, Boca Raton, FL, 2004.

CHAPTER 6

Electromagnetic Theory: Basic Review

Electromagnetics (EM) has increasingly gained attention for the last decade or so parallel to widespread usage of wireless systems; therefore (as proposed by my colleagues and dear friend Professor Anton Tijhuis from the Netherlands), we have to replace the motto *Engineering Electromagnetics* with *Electromagnetic Engineering*. This replacement also necessitates full review of electrical engineering education. EM engineers today should be well equipped with theory as well as practice including the classical EM field theory, wave scattering, antennas, circuit theories, transmission line theory, and microstrip circuits. Modern topics such as radar systems and wireless communication necessitate a good understanding of signal processing techniques together with deterministic and stochastic methods. This chapter roughly brings together fundamental EM concepts, mathematical relations, and physical meanings.

6.1 MAXWELL EQUATIONS AND REDUCTION

Any EM problem may be postulated in terms of Maxwell equations via

$$\nabla \times \vec{E} = -\frac{\partial}{\partial t}\vec{B}, \nabla \times \vec{H} = \vec{J} + \frac{\partial}{\partial t}\vec{D}, \tag{6.1a}$$

Electromagnetic Modeling and Simulation, First Edition. Levent Sevgi.

$$\nabla \cdot \vec{D} = \rho, \nabla \cdot \vec{B} = 0, \tag{6.1b}$$

where the vectors used are $\vec{E}$, electric field (V/m); $\vec{D}$, electric flux density (C/m^2); $\vec{H}$, magnetic field (A/m); and $\vec{B}$, magnetic flux density (Wb/m^2).

The theory behind EM fields is well established by Maxwell equations. The parameters σ, ε, and μ (non-negative real numbers) are used as constitutive parameters of the medium under interest, where σ is conductivity (S/m) (models losses dissipated as heat), ε is permittivity (dielectric constant) (F/m) (models E-field storage capability), and μ is permeability (magnetic constant) (H/m) (models H-field storage capability).

A medium is called to be "simple" when it is homogeneous (constitutive parameters do not vary with position), linear (constitutive parameters are not functions of applied/external EM fields), and isotropic (constitutive parameters are not functions of directions of applied/external fields). In a simple medium, EM vectors are related to each other and also to the excitation current density as

$$\vec{D} = \varepsilon \vec{E}, \vec{B} = \mu \vec{H}, \vec{J} = \sigma \vec{E} + \vec{J}_v. \tag{6.2}$$

The excitations of EM fields are expressed via electric current density $\vec{J}$ (A/m^2), and electric charge density ρ (C/m^3). Electric current density has conduction and convection components as $\vec{J} = \vec{J}_c + \vec{J}_v$ with the conduction and convection current densities expressed as $\vec{J}_c = \sigma \vec{E}$ and $\vec{J}_v = \rho \vec{v}$, where $\vec{v}$ is the velocity of the charge. Mathematical solution of Maxwell equations is problem dependent. Roughly speaking, one can either decompose equations into transverse and longitudinal components or use vector potentials.

6.2 WAVEGUIDING STRUCTURES

The strategy for solving EM guided waves is different from strategies of antenna/radiation and scattering problems [1,2]. In guided wave theory, the strategy is usually based on transverse and longitudinal decompositions. Under exp($j\omega t$) time dependence, one obtains

$$\vec{E}(\vec{r}, t) = \vec{E}(\vec{r}, \omega) e^{j\omega t}, \quad \vec{H}(\vec{r}, t) = \vec{H}(\vec{r}, \omega) e^{j\omega t}, \tag{6.3}$$

$$\nabla \times \vec{E}(\vec{r}) = -j\omega\mu \vec{H}(\vec{r}) - \vec{M}(\vec{r}), \tag{6.4}$$

$$\nabla \times \vec{H}(\vec{r}) = j\omega\varepsilon\vec{E}(\vec{r}) + \vec{J}(\vec{r}), \tag{6.5}$$

where $\vec{r} = (x, y, z)$, $\mu = \mu(z)$, and $\varepsilon = \varepsilon(z)$. The first step is to scalarize the equations given in vector form via the decomposition of the fields into *longitudinal* and *transverse* components. Equation (6.3) is first vector multiplied by $\vec{e}_z$, which is the unit vector with respect to z-axis to obtain

$$\vec{e}_z \times \left[\nabla \times \vec{E}\right] = -\left(-j\omega\mu\vec{H} - \vec{M}\right) \times \vec{e}_z, \tag{6.6a}$$

$$j\omega\mu\vec{H} \times \vec{e}_z + \vec{M} \times \vec{e}_z = \vec{e}_z \times \left[\nabla \times \vec{E}\right], \tag{6.6b}$$

$$j\omega\mu\vec{H}_t \times \vec{e}_z + \vec{M}_t \times \vec{e}_z = \nabla_t E_z - \frac{\partial}{\partial z}\vec{E}_t, \tag{6.7}$$

where $\nabla \equiv \nabla_t + \vec{e}_z\partial / \partial z$ and $\vec{E} \equiv \vec{E}_t + \vec{e}_z E_z$. Here, $\vec{E}_t$ and ∇_t denote the transverse components of electric field and transverse Nabla operator, respectively. Then, dot product by $\vec{e}_z$ is applied to (6.4) resulting in

$$\begin{aligned} -j\omega\mu H_z - M_z &= \vec{e}_z \cdot \nabla \times \vec{E} = -\nabla \cdot (\vec{e}_z \times \vec{E}) \\ &= -\nabla \cdot (\vec{e}_z \times \vec{E}_t) = -\nabla_t \cdot (\vec{e}_z \times \vec{E}_t). \end{aligned} \tag{6.8}$$

With the help of the duality property, the equations

$$j\omega\mu\vec{e}_z \times \vec{E}_t - \vec{J}_t \times \vec{e}_z = \nabla_t H_z - \frac{\partial}{\partial z}\vec{H}_t \tag{6.9}$$

and

$$j\omega\mu E_z + J_z = \nabla_t \cdot (\vec{H}_t \times \vec{e}_z) \tag{6.10}$$

are obtained from (6.4). Combining (6.7) and (6.10) yields

$$-\frac{\partial}{\partial z}\vec{E}_t = j\omega\mu\vec{H}_t \times \vec{e}_z - \nabla_t\left[\frac{1}{j\omega\varepsilon}\left[\nabla_t \cdot \vec{H}_t \times \vec{e}_z - J_z\right]\right] + M_t \times \vec{e}_z, \tag{6.11}$$

$$-\frac{\partial}{\partial z}\vec{E}_t = j\omega\mu\left[1 + \frac{\nabla_t\nabla_t}{k^2}\right] \cdot (\vec{H}_t \times \vec{e}_z) + \vec{M}_{\text{te}} \times \vec{e}_z, \tag{6.12}$$

$$\vec{M}_{\text{te}} = \vec{M}_t + \vec{e}_z \times \frac{\nabla_t J_z}{j\omega\varepsilon}. \tag{6.13}$$

Here, (6.12) denotes the transverse equation for electric field. $\vec{M}_{\text{te}}$ is the effective transverse magnetic (TM) current element. The magnetic field transverse equation can be reconstructed via the combination of (6.8) and (6.9),

$$-\frac{\partial}{\partial z}\vec{H}_t = j\omega\varepsilon\left[1+\frac{\nabla_t\nabla_t}{k^2}\right]\cdot(\vec{e}_z\times\vec{E}_t)+\vec{e}_z\times\vec{J}_{\text{te}}, \tag{6.14}$$

where

$$\vec{J}_{\text{te}} = \vec{J}_t - \vec{e}_z\times\frac{\nabla_t M_z}{j\omega\mu} \tag{6.15}$$

denotes the effective transverse electric (TE) current. Equations (6.12) and (6.14) form the transverse field equations together. The longitudinal field components can be obtained by using the transverse field equations with Maxwell equations as

$$E_z = \frac{1}{j\omega\varepsilon}\left\{\nabla_t\cdot\left(\vec{H}_t\times\vec{e}_z\right)-J_z\right\}, \tag{6.16}$$

$$H_z = \frac{1}{j\omega\mu}\left\{\nabla_t\cdot\left(\vec{e}_z\times\vec{E}_t\right)-M_z\right\}. \tag{6.17}$$

6.3 RADIATION PROBLEMS AND VECTOR POTENTIALS

Electromagnetic problems such as antenna radiation or radar cross-section prediction are better modeled via vector (and scalar) potentials. Under the same $\exp(j\omega t)$ time dependence, the complex permittivity is defined as $\varepsilon' = \varepsilon - j\sigma / \omega$. The vector potential is introduced by using

$$\nabla\cdot\vec{H} = 0, \tag{6.18}$$

which illustrates that the magnetic field has only circulation (that is why it is sometimes called a solenoidal field). Because it possesses only a circulation, it can be represented by the curl of some other vector function as

$$\vec{H} = \frac{1}{\mu}\nabla\times\vec{A}. \tag{6.19}$$

Here, $\vec{A}$ is the magnetic vector potential. Equation (6.19) is the result of the vector property *the divergence of the curl of any vector is equal to zero*:

$$\nabla \cdot \left(\nabla \times \vec{A}\right) = 0. \tag{6.20}$$

Using (6.19) in the curl of electric field equation gives

$$\nabla \times \left(\vec{E} + j\omega\vec{A}\right) = 0. \tag{6.21}$$

The term in the parenthesis in (6.21) is an electric field and represents the static form (since its curl is equal to zero). The electric scalar potential Φ is defined from

$$\vec{E} + j\omega\vec{A} = -\nabla\Phi \tag{6.22}$$

since it satisfies (6.21) identically, that is, for any Φ,

$$\nabla \times (\nabla\Phi) = 0. \tag{6.23}$$

Solving the electric field from (6.22), one obtains

$$\vec{E} = -j\omega\vec{A} - \nabla\Phi. \tag{6.24}$$

The fields $\vec{E}$ and $\vec{H}$ are now represented in terms of Φ and $\vec{A}$. Substituting (6.18) into Maxwell magnetic curl equation results in

$$\nabla \times \vec{H} = \frac{1}{\mu}\nabla \times \nabla \times \vec{A} = j\omega\varepsilon\vec{E} + \vec{J}. \tag{6.25}$$

Using the vector identity

$$\nabla \times \nabla \times \vec{A} = \nabla\left(\nabla \cdot \vec{A}\right) - \nabla^2\vec{A} \tag{6.26}$$

and (6.24) in (6.25) yields

$$\nabla\left(\nabla \cdot \vec{A}\right) - \nabla^2\vec{A} = j\omega\mu\varepsilon\left(-j\omega\vec{A} - \nabla\Phi\right) + \mu\vec{J} \tag{6.27}$$

or

$$\left\{\nabla^2 + \omega^2\varepsilon\mu\right\}\vec{A} - \nabla\left(j\omega\mu\varepsilon\Phi + \nabla \cdot \vec{A}\right) = -\mu\vec{J}, \tag{6.28}$$

where the divergence of $\vec{A}$ is yet to be specified. A convenient choice would be Lorenz condition,

$$\nabla \cdot \vec{A} = -j\omega\mu\varepsilon\Phi, \tag{6.29}$$

that eliminates the third term in (6.28) and results in

$$\left\{\nabla^2 + \omega^2\varepsilon\mu\right\}\vec{A} = -\mu\vec{J}, \tag{6.30}$$

the *vector wave equation*. In summary, for a given source distribution $\vec{J}$, the vector field $\vec{A}$ can be calculated from (6.30) and then the field $\vec{H}$ is derived from (6.19), the field $\vec{E}$ is derived from

$$\vec{E} = \frac{1}{j\omega\varepsilon}\nabla\times\vec{H}. \tag{6.31}$$

Although only $\vec{A}$ is required to specify the fields $\vec{E}$ and $\vec{H}$, the scalar potential function can also be obtained by solving the scalar wave equation,

$$\{\nabla^2 + \omega^2\varepsilon\mu\}\Phi = -\frac{\rho}{\varepsilon}. \tag{6.32}$$

Note that $\vec{J}$ and ρ are related via (6.29) or via the continuity relation,

$$\nabla\cdot\vec{J} = -j\omega\rho. \tag{6.33}$$

In rectangular coordinates, the vector wave equation (6.30) can be reduced to the components

$$\{\nabla^2 + k^2\}A_x = -\mu J_x, \tag{6.34a}$$

$$\{\nabla^2 + k^2\}A_y = -\mu J_y, \tag{6.34b}$$

$$\{\nabla^2 + k^2\}A_z = -\mu J_z, \tag{6.34c}$$

where $k^2 = \omega^2\varepsilon\mu$ is the amplitude of the wave vector. These three equations are identical in form; therefore, solving one of them automatically will yield the solutions of the other two.

6.4 THE DELTA DIRAC FUNCTION

The concept of delta function is used in EM source representations. Mathematically, it is an operator which transforms a continuous function $f(x)$ into $f(x_0)$ at the point x_0 via

$$\int_{-\infty}^{+\infty} f(x)\delta(x - x_0)dx = f(x_0), \tag{6.35a}$$

therefore, is represented as

$$\delta(x - x_0) = \begin{cases} 0 & x \neq x_0 \\ \int_{-\infty}^{+\infty} \delta(x - x_0)dx = 1 & x = x_0 \end{cases}. \tag{6.35b}$$

Note that the second equation in (6.35a) shows that the delta Dirac is actually not a function, it is a functional. Mathematically, it represents sampling when multiplied by any function.

6.5 COORDINATE SYSTEMS AND BASIC OPERATORS

The well-known three coordinate systems are depicted in Fig. 6.1. Here r is the distance measured from the origin, ρ is its projection on xy-plane, ϕ is the angle from x-axis on xy-plane, and θ is the angle measured from z-axis. Usually, ϕ and θ are called horizontal and vertical angles, respectively. The relations among the three coordinate systems are the following.

Cylindrical

$$x = \rho\cos(\phi),\ y = \rho\sin(\phi),\ z = z. \tag{6.36a}$$

Spherical

$$x = r\sin(\theta)\cos(\phi),\ y = r\sin(\theta)\sin(\phi),\ z = r\cos(\theta). \tag{6.36b}$$

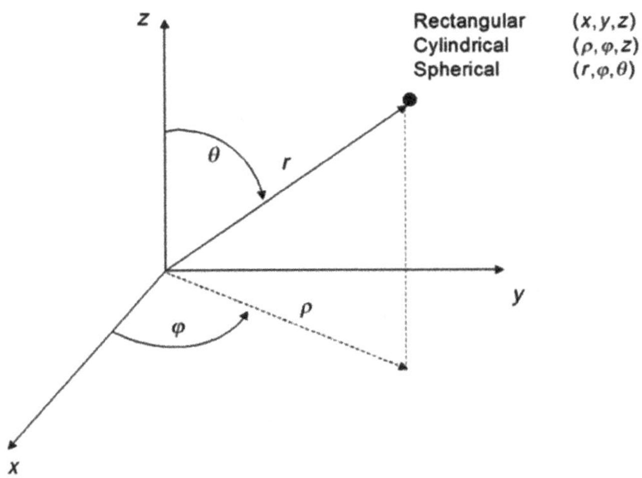

Figure 6.1. Rectangular, cylindrical, and spherical coordinates.

The Gradient

The *gradient* of a scalar quantity is the space rate of change of that quantity:

$$\nabla V = \text{grad} V = \frac{\partial V}{\partial x}\vec{e}_x + \frac{\partial V}{\partial y}\vec{e}_y + \frac{\partial V}{\partial z}\vec{e}_z.$$

In cylindrical coordinates,

$$\nabla V = \text{grad} V = \frac{\partial V}{\partial \rho}\vec{e}_\rho + \frac{1}{\rho}\frac{\partial V}{\partial \phi}\vec{e}_\varphi + \frac{\partial V}{\partial z}\vec{e}_z. \tag{6.37}$$

In spherical coordinates,

$$\nabla V = \text{grad} V = \frac{\partial V}{\partial r}\vec{e}_r + \frac{1}{r}\frac{\partial V}{\partial \theta}\vec{e}_\theta + \frac{1}{r\sin(\theta)}\frac{\partial V}{\partial \phi}\vec{e}_\phi \tag{6.38}$$

The Divergence

The divergence of any vector is the net flow out per unit volume:

$$\nabla \cdot \vec{A} = \text{div}\vec{A} = \frac{\partial(A_x)}{\partial x} + \frac{\partial A_y}{\partial y} + \frac{\partial A_z}{\partial z}.$$

In cylindrical coordinates,

$$\nabla \cdot \vec{A} = \text{div}\vec{A} = \frac{1}{\rho}\frac{\partial(\rho A_\rho)}{\partial \rho} + \frac{1}{\rho}\frac{\partial A_\phi}{\partial \phi} + \frac{\partial A_z}{\partial z}. \tag{6.39}$$

In spherical coordinates,

$$\nabla \cdot \vec{A} = \text{div}\vec{A} = \frac{1}{r^2}\frac{\partial(r^2 A_r)}{\partial r} + \frac{1}{r\sin(\theta)}\frac{\partial(\sin(\theta)A_\theta)}{\partial \theta} + \frac{1}{r\sin(\theta)}\frac{\partial A_\phi}{\partial \phi}. \tag{6.40}$$

The Curl

The curl of a vector is the degree of its rotation:

$$\nabla \times \vec{A} = \text{curl}\vec{A} = \vec{e}_x\left(\frac{\partial A_z}{\partial y} - \frac{\partial A_y}{\partial z}\right) + \vec{e}_y\left(\frac{\partial A_x}{\partial z} - \frac{\partial A_z}{\partial x}\right) + \vec{e}_z\left(\frac{\partial A_y}{\partial x} - \frac{\partial A_x}{\partial y}\right).$$

In cylindrical coordinates,

$$\nabla \times \vec{A} = \vec{e}_\rho\left(\frac{1}{\rho}\frac{\partial A_z}{\partial \phi} - \frac{\partial A_\phi}{\partial z}\right) + \vec{e}_\varphi\left(\frac{\partial A_\rho}{\partial z} - \frac{\partial A_z}{\partial \rho}\right) + \vec{e}_z\left(\frac{1}{\rho}\left[\frac{\partial(\rho A_\phi)}{\partial \rho} - \frac{\partial A_\rho}{\partial \phi}\right]\right). \tag{6.41}$$

In spherical coordinates,

$$\nabla \times \vec{A} = \vec{e}_r \left(\frac{1}{r\sin(\theta)} \left[\frac{\partial(\sin(\theta)A_\theta)}{\partial\theta} - \frac{\partial A_\theta}{\partial\phi} \right] \right) + \vec{e}_\varphi \left(\frac{1}{r} \left[\frac{\partial(rA_\theta)}{\partial r} - \frac{\partial A_r}{\partial\theta} \right] \right)$$
$$+ \vec{e}_\theta \left(\frac{1}{r\sin(\theta)} \frac{\partial A_r}{\partial\phi} - \frac{1}{r} \frac{\partial(rA_\phi)}{\partial r} \right). \tag{6.42}$$

The Laplacian Operator

The operator $\nabla^2 = \nabla \cdot \nabla$ is the divergence of the gradient of a scalar quantity:

$$\nabla^2 V = \left\{ \frac{\partial^2}{\partial x^2} + \frac{\partial^2}{\partial y^2} + \frac{\partial^2}{\partial z^2} \right\} V.$$

In cylindrical coordinates,

$$\nabla^2 V = \left\{ \frac{1}{\rho} \frac{\partial}{\partial\rho} \left(\rho \frac{\partial}{\partial\rho} \right) + \frac{1}{\rho^2} \frac{\partial^2}{\partial\phi^2} + \frac{\partial^2}{\partial z^2} \right\} V. \tag{6.43}$$

In cylindrical coordinates,

$$\nabla^2 V = \left\{ \frac{1}{r^2} \frac{\partial}{\partial r} \left(r^2 \frac{\partial}{\partial r} \right) + \frac{1}{r^2 \sin(\theta)} \frac{\partial}{\partial\theta} \left(\sin(\theta) \frac{\partial}{\partial\theta} \right) + \frac{1}{r^2 \sin^2(\theta)} \frac{\partial^2}{\partial\phi^2} \right\} V. \tag{6.44}$$

6.6 THE POINT SOURCE REPRESENTATION

In EM sources such as an antenna in free space, an impulse in time domain may be represented in terms of delta Dirac function as shown in Fig. 6.2.

In rectangular coordinates ($dV = dxdydz$),

$$\delta(\vec{r} - \vec{r}') = \delta(x - x')\delta(y - y')\delta(z - z'). \tag{6.45}$$

In cylindrical coordinates ($dV = \rho d\rho d\phi dz$),

$$\delta(\vec{r} - \vec{r}') = \frac{1}{\rho}\delta(\rho - \rho')\delta(\phi - \phi')\delta(z - z'). \tag{6.46}$$

In spherical coordinates ($dV = r^2\sin(\theta)drd\theta d\phi$),

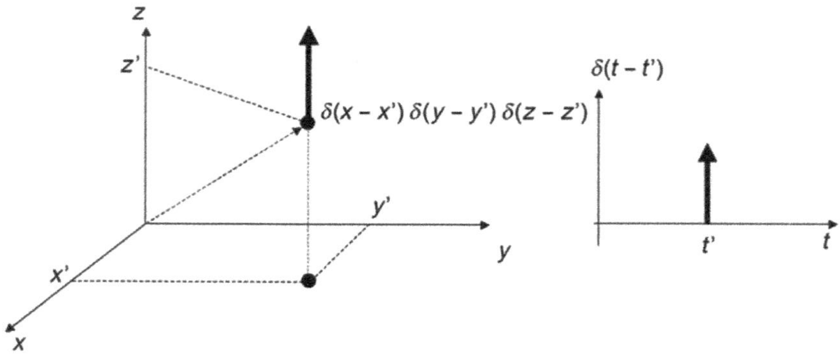

Figure 6.2. Delta Dirac function (left) of a point source and (right) an impulse in time.

$$\delta(\vec{r}-\vec{r}')=\frac{1}{r^2\sin(\theta)}\delta(r-r')\delta(\phi-\phi')\delta(\theta-\theta'). \tag{6.47}$$

In some cases, the problem at hand may have symmetry with respect to some axes. For example, a cylindrical waveguide located along z-axis, an antenna above smooth earth, are these types of problems. Moreover, sometimes it is difficult to work out a problem in a three-dimensional (3D) space directly; instead, we reduce it into two dimensional (2D) and deal with a simplified version. In order to achieve all these, a *line source* representation is used. A line source is the combination of infinite number of point sources aligned with respect to an axis. For example, a line source along z-axis is the following.

In rectangular coordinates,

$$\delta(\vec{r}-\vec{r}')=\delta(x-x')\delta(y-y'). \tag{6.48}$$

In cylindrical coordinates,

$$\delta(\vec{r}-\vec{r}')=\frac{1}{\rho}\delta(\rho-\rho')\delta(\phi-\phi'). \tag{6.49}$$

6.7 FIELD REPRESENTATION OF A POINT/LINE SOURCE

The Green's function (GF) problem associated with the vector potential problem for a point source at the origin is ($A_z=\psi$)

$$\left\{\nabla^2+k^2\right\}\psi=-\delta(x)\delta(y)\delta(z). \tag{6.50}$$

Since the source is zero everywhere except at the origin, (6.50) becomes

$$\{\nabla^2 + k^2\}\psi = 0 \tag{6.51}$$

away from the origin. Equation (6.51) is known as *Helmholtz equation*. Because of spherical symmetry, the Laplacian may better be written in spherical coordinates so that ψ has only radial dependence. The two solutions of (6.51) are

$$\psi(r) = \frac{e^{-jkr}}{4\pi r} \quad \text{and} \quad \psi(r) = \frac{e^{jkr}}{4\pi r}, \tag{6.52}$$

which correspond to waves propagating radially outward and inward, respectively. If the source were located at an arbitrary position $\vec{r}'$, (6.52) becomes

$$\psi(r) = \frac{e^{-jkR}}{4\pi R} \quad \text{and} \quad \psi(r) = \frac{e^{jkR}}{4\pi R}, \tag{6.53}$$

where $R = |\vec{r} - \vec{r}'|$ is the distance between the source and observer. The GF of (6.50) may then be built in terms of (6.52).

For an arbitrary z-directed current density, the vector potential is also z-directed. If the source is assumed to be a collection of point sources with z-directed current densities J_z, then the z-directed vector potential will be the sum of the responses of these point sources and can be expressed by an integral over the source volume as

$$A_z = \iiint_{v'} \mu J_z \frac{e^{jkR}}{4\pi R} dv'. \tag{6.54}$$

Similar equations hold for the x- and y-components. Finally, the GF for a z-directed line source can be given as

$$\psi(\rho) = \frac{e^{-jk\rho}}{2\pi\sqrt{\rho}} \quad \text{and} \quad \psi(\rho) = \frac{e^{jk\rho}}{2\pi\sqrt{\rho}}. \tag{6.55}$$

The $1/r$ and $1/\sqrt{\rho}$ in (6.51) and (6.53), respectively, represent the spherical and cylindrical wave spreads.

6.8 ALTERNATIVE FIELD REPRESENTATIONS

The analysis and synthesis of source-driven EM propagation problems is effectively done in the complex configuration (space–time)–spectrum (wave number–frequency) phase space and it decomposes a space–time dependent wave functions into their spectral constituents. Having $\vec{r} = (x, y, z)$ and $\vec{k} = (k_x, k_y, k_z)$ as the *spatial* and *wave number* vectors, respectively, the configuration space function $G(\vec{r}, t)$ (i.e., transient point source response which is called Green's function) can be represented via phase space components $\overline{G}(\vec{r}, t, \vec{k}, \omega)$ as

$$G(\vec{r}, t) = \int \iiint \overline{G}\left(\vec{r}, t, \vec{k}, \omega\right) d\vec{k} d\omega. \tag{6.56}$$

The configuration (space–time)–spectrum (wavenumber–frequency) phase space of a problem can be accessed via continuous and discrete local and global transforms that are derived from (6.54). Coordinate separable canonical geometries provide great flexibility and opportunity for visual illustrations to the analytical modeler.

A generic problem for a source-driven structure is the GF problem (see Chapter 7). The GF is the response of a linear system to a point source of unit strength. A source of unit strength at a position z' along, for example, z-coordinate can be represented by the Dirac delta function $\delta(z - z')$. The general form of a GF problem may be postulated via a second-order nonhomogeneous differential equation with a delta function on the right-hand side, together with boundary conditions (BCs) that represent the geometry of the problem. Omitting the delta function on the right-hand side (i.e., using homogeneous second-order differential equation and BC) yields the source-free (i.e., an eigenvalue) problem. An eigenvalue problem is to find out a set of complete functions that characterize the geometry of the problem (and yields eigenmodes and eigen spectral wave numbers that describe source-free wave fields which individually satisfy BC in wave-confining environments). Many of EM problems, either source-free or source driven, can be represented by a unique second-order differential equation: the Sturm–Liouville (SL) equation. Solving SL equation under certain conditions yields nothing but the GF. GF solution may also be built from eigenfunction solutions. Alternatively, eigenfunctions may also be derived from the GF solution. It illustrates that a guided wave problem may be handled as either an eigenvalue problem or a GF problem, since the former also yields the latter and vice versa. These are discussed in the next chapter.

6.9 TRANSVERSE ELECTRIC/MAGNETIC FIELDS

Guided wave propagation problems in telecommunications media, such as transmission lines, waveguides with rectangular and/or circular cross sections, infinite-extent parallel plate waveguides, dielectric slab, and so on, are best treated using reduced equations obtained from the longitudinal and transverse decomposition of Maxwell equations. This yields the well-known transverse electric (TE), transverse magnetic (TM), and transverse EM (TEM) field representations under different polarizations. These are classical topics taught in almost every introductory as well as advanced level EM wave courses/books/lecture notes [1–4]. But, it might be confusing issue if different books, lecture notes, or papers are used for the following reasons.

- Some use perpendicular and parallel polarizations, while others prefer horizontal or vertical polarizations. How do we define polarization and which corresponds to which in terminology?
- Some use, for example, H_x, E_y, and H_z components for TE-type representation, but some others use H_x, H_y, and E_z for the same TE representation. Which set is correct? How should TE/TM representations be classified?
- Suppose a TE-type finite-difference time-domain (FDTD) simulator is based on H_x, E_y, and H_z components. Can you use this simulator with the same three field components and simulate a TM-type problem? Can the *duality principle* be used for this purpose?

6.9.1 The 3D TE/TM Waves

A rectangular cross-section metallic waveguide is a classical 3D structure in treating/teaching TE and TM waves in *Guided Wave Theory*. The procedure is as follows.

- Start with the two Maxwell equations in 3D:

$$\nabla \times \vec{E} = -\mu \frac{\partial \vec{H}}{\partial t} - \vec{M}, \tag{6.57a}$$

$$\nabla \times \vec{H} = \varepsilon \frac{\partial \vec{E}}{\partial t} + \vec{J}. \tag{6.57b}$$

- Write down open form of these vector equations:

$$\vec{e}_x\left(\frac{\partial E_z}{\partial y}-\frac{\partial E_y}{\partial z}\right)+\vec{e}_y\left(\frac{\partial E_x}{\partial z}-\frac{\partial E_z}{\partial x}\right)+\vec{e}_z\left(\frac{\partial E_y}{\partial x}-\frac{\partial E_x}{\partial y}\right)$$
$$=-\vec{e}_x\left(\mu\frac{\partial H_x}{\partial t}+M_x\right)-\vec{e}_y\left(\mu\frac{\partial H_y}{\partial t}+M_y\right)-\vec{e}_z\left(\mu\frac{\partial H_z}{\partial t}+M_z\right), \tag{6.58a}$$

$$\vec{e}_x\left(\frac{\partial H_z}{\partial y}-\frac{\partial H_y}{\partial z}\right)+\vec{e}_y\left(\frac{\partial H_x}{\partial z}-\frac{\partial H_z}{\partial x}\right)+\vec{e}_z\left(\frac{\partial H_y}{\partial x}-\frac{\partial H_x}{\partial y}\right)$$
$$=\vec{e}_x\left(\varepsilon\frac{\partial E_x}{\partial t}+J_x\right)+\vec{e}_y\left(\varepsilon\frac{\partial E_y}{\partial t}+J_y\right)+\vec{e}_z\left(\varepsilon\frac{\partial E_z}{\partial t}+J_z\right), \tag{6.58b}$$

- Locate the waveguide longitudinally along a chosen direction (z-axis is chosen as a usual notation),
- Define transverse TE_z and TM_z cases by assuming $E_z = 0$ and $H_z = 0$, respectively, and write down the sets of equations for each case.

This is straightforward. It is based on the fact that the wave is trapped transversely and propagates longitudinally. The derivation of all field components for TE_z and TM_z cases then reduces to looking separately for the solutions for only H_z (when $E_z = 0$) and E_z (when $H_z = 0$) components, respectively. The other field components can be derived in terms of these two. Of course, alternative representations can be used for partially filled inhomogeneous waveguides (i.e., mathematically transverse propagation and longitudinal trapping may also be assumed) [2].

6.9.2 The 2D TE/TM Waves

Although a little bit tricky and confusing, the transverse field representation is best understood in 2D [5,6,9]. A plane wave with a chosen polarization hitting an infinite homogeneous dielectric medium is a classical/canonical problem, but it has to be treated with care because the 2D environment may be either of the yz-, xy-, or xz-plane. Also, one is faced with the question of "transverse to what?" The only way to overcome these difficulties is to remember that the electric field vector, the magnetic field vector, and the Poynting vector form a *right-hand system* like the Cartesian coordinate system itself.

6.9.2.1 TE/TM Solutions on the *yz*-Plane

Assume *yz*-plane is the 2D environment as shown in Fig. 6.3. As mentioned earlier, the propagation direction is determined by the Poynting vector ($\vec{E} \times \vec{H} \Rightarrow \vec{k}$). Therefore, to create a 2D scenario, first locate the medium, then specify the Poynting vector and either electric or magnetic field vector. The other vector is automatically determined from the right-hand system assumption.

The infinite extent of the dielectric medium along *x*-direction implies $\partial / \partial x \equiv 0$, and using this in (6.58) results in

$$-\mu \frac{\partial H_x}{\partial t} - M_x = \frac{\partial E_z}{\partial y} - \frac{\partial E_y}{\partial z}, \tag{6.59a}$$

$$-\mu \frac{\partial H_y}{\partial t} - M_y = \frac{\partial E_x}{\partial z}, \tag{6.59b}$$

$$-\mu \frac{\partial H_z}{\partial t} - M_z = -\frac{\partial E_x}{\partial y}, \tag{6.59c}$$

$$\varepsilon \frac{\partial E_x}{\partial t} + J_x = \frac{\partial H_z}{\partial y} - \frac{\partial H_y}{\partial z}, \tag{6.60a}$$

$$\varepsilon \frac{\partial E_y}{\partial t} + J_y = \frac{\partial H_x}{\partial z}, \tag{6.60b}$$

$$\varepsilon \frac{\partial E_z}{\partial t} + J_z = -\frac{\partial H_x}{\partial y}. \tag{6.60c}$$

Note that the dielectric medium may be located either vertically (as shown in Fig. 6.3), which yields the air-dielectric boundary on the

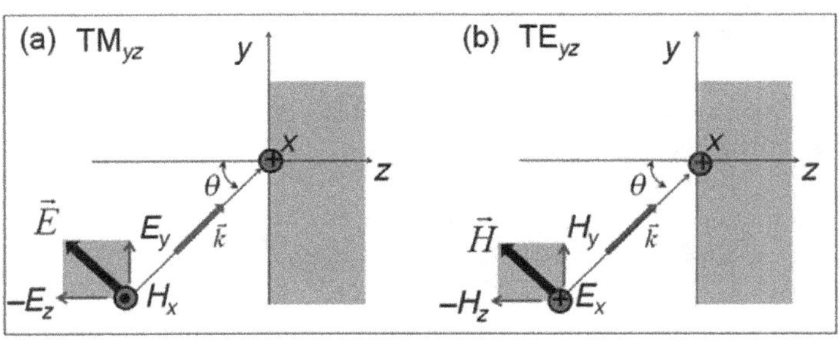

Figure 6.3. Transverse field components in 2D on the *yz*-plane, (a) TM$_{yz}$ case, and (b) TE$_{yz}$ case.

xy-plane or horizontally having the boundary on the xz-plane. The TE and TM field components and the equations remain the same. The choice of the "transverse" depends on the choice of the "longitudinal." In 2D (as will be shown in Fig. 6.4 and Fig. 6.5), the longitudinal (propagation) direction may be either of the two axes; that is why two letters are used to name TE and TM cases (note that the problem reduces to 1D plane wave propagation through multilayered media if $\theta = 0°$: normal incidence). Note also that (6.60) may also be obtained from (6.59) using the duality principle summarized in Table 6.1.

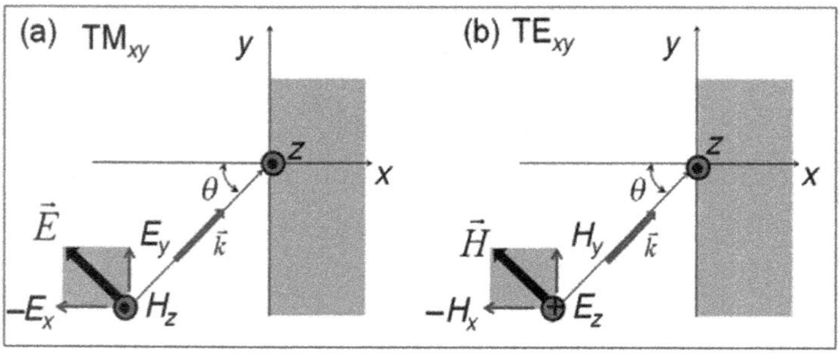

Figure 6.4. Transverse field components on the xy-plane: (a) TM_{xy} (= TE_z) case and (b) TE_{xy} (= TM_z) case.

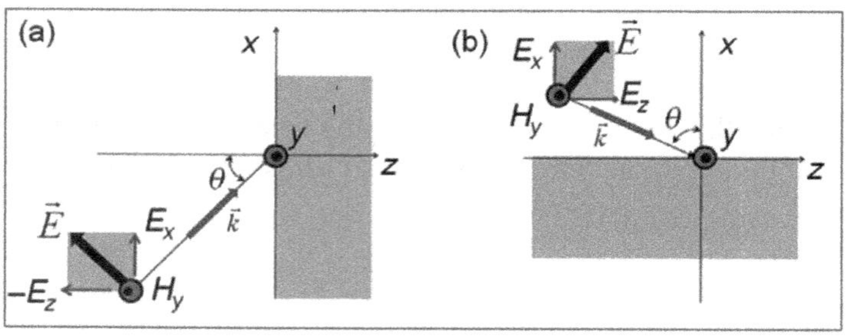

Figure 6.5. Transverse field components in 2D on the xz-plane, TM_{xz} case, (a) on the xy interface and (b) on the yz-interface.

Table 6.1.
Duality Relations in Maxwell Equations

Fields	E → H	H → −E
Sources	J → M	M → −J
Media	$\varepsilon \to \mu$	$\mu \to \varepsilon$

Table 6.2.
The *TE* and *TM* Sets of Equations for Figure 6.3.

TM_{yz} case (H_x, E_y, E_z)	TE_{yz} case (E_x, H_y, H_z)
$\frac{\partial H_x}{\partial z}=\varepsilon\frac{\partial E_y}{\partial t}+J_y$	$\frac{\partial E_x}{\partial z}=-\mu\frac{\partial H_y}{\partial t}-M_y$
$\frac{\partial H_x}{\partial y}=-\varepsilon\frac{\partial E_z}{\partial t}-J_z$	$\frac{\partial E_x}{\partial y}=\mu\frac{\partial H_z}{\partial t}+M_z$
$\frac{\partial E_z}{\partial y}-\frac{\partial E_y}{\partial z}=-\mu\frac{\partial H_x}{\partial t}-M_x$	$\frac{\partial H_z}{\partial y}-\frac{\partial H_y}{\partial z}=\varepsilon\frac{\partial E_x}{\partial t}+J_x$

Table 6.2 lists both TE_{yz} and TM_{yz} sets of equations on the yz-plane. The easiest way to check whether or not everything (i.e., the field components, field vectors, and the wave vector $\vec{k}$) is consistent is to calculate the Poynting vector and to compare it with the wave vector. They must be the same except a multiplicative factor $\vec{E}\times\vec{H}=\left(\vec{H}\cdot\vec{H}/(\omega\mu)\right)\vec{k}$ $(=\left(\vec{E}\cdot\vec{E}/(\omega\varepsilon)\right)\vec{k})$! For the scenario shown in Fig. 6.3, the Poynting vector is obtained from $\vec{E}\times\vec{H}$, which is equal to $(E_y\vec{e}_y-E_z\vec{e}_z)\times(H_x\vec{e}_x)$ or $E_zH_x\vec{e}_y+E_yH_x\vec{e}_z$ for TM_{yz} case. The wave vector in Fig. 6.3 is equal to $\vec{k}=k_y\vec{e}_y+k_z\vec{e}_z$, therefore, is consistent with the field components and the specified directions.

Some prefer to name TM_{yz} or TE_{yz} in Fig. 6.3 as TE_x or TM_x, respectively. As shown in, for example, (6.59a), TM_{yz} automatically yields TE_x, but this would be appropriate notation as long as x-axis is the longitudinal propagation direction. In 2D, this is certainly not the case.

6.9.2.2 TE/TM Solutions on the *xy*-Plane

Now, assuming the xy-plane is the 2D environment as shown in Fig. 6.4, six relations grouped as TE/TM equations, derived from Maxwell equations under $\partial / \partial z\equiv 0$, are listed in Table 6.3 (check whether or not the choice for vectors as in Fig. 6.2 is consistent).

6.9.2.3 TE/TM Solutions on the *xz*-Plane

Finally, assume xz-plane as the 2D environment (the figure for this case is left to the reader). The TM/TE equations under $\partial / \partial y\equiv 0$ (on the xy-plane) are listed in Table 6.4. It should be noted that the scenario for

Table 6.3.
The Two Sets of Equations for Fig. 6.4.

TM_{xy} case (E_x, E_y, H_z)	TE_{xy} case (H_x, H_y, E_z)
$\frac{\partial H_z}{\partial y} = \varepsilon \frac{\partial E_x}{\partial t} + J_x$	$\frac{\partial E_z}{\partial y} = -\mu \frac{\partial H_x}{\partial t} - M_x$
$\frac{\partial H_z}{\partial x} = -\varepsilon \frac{\partial E_y}{\partial t} - J_y$	$\frac{\partial E_z}{\partial x} = \mu \frac{\partial H_y}{\partial t} + M_y$
$\frac{\partial E_x}{\partial y} - \frac{\partial E_y}{\partial x} = \mu \frac{\partial H_z}{\partial t} + M_z$	$\frac{\partial H_y}{\partial x} - \frac{\partial H_x}{\partial y} = \varepsilon \frac{\partial E_z}{\partial t} + J_z$

Table 6.4.
Sets of Equations for TE/TM Cases on the *xz*-Plane

TM_{xz} case (E_x, H_y, E_z)	TE_{xz} case (H_x, E_y, H_z)
$\frac{\partial H_y}{\partial x} = \varepsilon \frac{\partial E_z}{\partial t} + J_z$	$\frac{\partial E_y}{\partial x} = -\mu \frac{\partial H_z}{\partial t} - M_z$
$\frac{\partial H_y}{\partial z} = -\varepsilon \frac{\partial E_x}{\partial t} - J_x$	$\frac{\partial E_y}{\partial z} = \mu \frac{\partial H_x}{\partial t} + M_x$
$\frac{\partial E_x}{\partial z} - \frac{\partial E_z}{\partial x} = -\mu \frac{\partial H_y}{\partial t} - M_y$	$\frac{\partial H_x}{\partial z} - \frac{\partial H_z}{\partial x} = \varepsilon \frac{\partial E_y}{\partial t} + J_y$

a plane wave with a specified polarization (e.g., TM_{xz}) hitting an infinite dielectric medium may be either of the two cases shown in Fig. 6.5.

Different notations and terms are used for TE and TM representations. Traditionally, propagation engineers use the terms horizontal and vertical polarizations. This may be understood easily because they are accustomed to dealing with EM wave propagation over earth's surface in 2D as field versus height/range variations. Long- and medium-wave communications (signals in the range of kilohertz to megahertz) use vertically polarized waves excited by vertically erected wire antenna systems. First TV broadcasts were horizontally polarized and receive antennas were roof-mounted *Yagi–Uda*-type systems with conducting wires placed parallel to the ground.

Early analytical solutions belong to plane waves hitting 2D (stratified) plane earth as shown in Fig. 6.3 and Fig. 6.4. For example, in Fig.

6.3b, the 2D plane earth extends infinitely with earth–atmosphere boundary on the xy-plane (the boundary normal vector is the negative z-axis). The plane of incidence in this case is the plane containing the incident wave (wave vector $\vec{k}$) and the surface normal, $\vec{n} = -\vec{e}_z$. Horizontal polarization is the case where electric field vector is parallel to the boundary surface (i.e., the xy-plane). In other words, electric field vector is perpendicular to the plane of incidence (i.e., the yz-plane). This is why, in physics, perpendicular polarization is also used for horizontal polarization. In microwave engineering, horizontal or perpendicular polarization is nothing but the TE representations.

Similarly, vertical polarization is used for the case where magnetic field vector is parallel to the boundary surface (i.e., the xy-plane). In other words, magnetic field vector is perpendicular to the plane of incidence. In physics, parallel polarization is also used for vertical polarization. In microwave engineering, vertical or parallel polarization is nothing but the TM representations.

6.10 THE TE/TM SOURCE INJECTION

In 2D, TE/TM field simulations wave injection is important. Waves in these problems may be injected using either of three field components. Figure 6.6 illustrates source excitation through the three different field components (either Ref. 7 or 8 can be used to reproduce these figures). The three field components for the TM_{yz} case on the yz-plane (as in Fig. 6.1a) are H_x, E_y, and E_z. Injection of the waves via only H_x is equivalent to using a magnetic line source, and this produces cylindrical wave spread on the yz-plane as shown in Fig. 6.6a and b. On the other hand, E_y and E_z injection correspond to vertical and horizontal short electric dipole excitations, respectively (as shown in Fig. 6.6c and d). Note that all four plots in Fig. 6.4 are produced from the line source (H_x) excitation. This is consistent with (6.59a); the time variation of magnetic field component changes with the spatial variations of both of the electric field (E_y and E_z) components.

The situation is not the same for the same problem if the fields are excited via electric field components. As can be seen from (6.60b) or (6.60c), time rate of change of vertical component of the electric field (i.e., E_y) determines the horizontal variation of the magnetic field (i.e.,H_x), or time injection of horizontal component of the electric

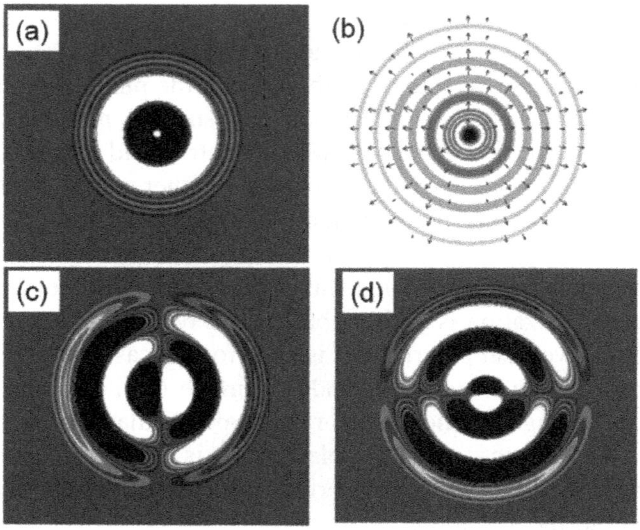

Figure 6.6. Field excitation on *yz*-plane for the TM_{yz} case: (a) H_x is excited, H_x is observed, (b) H_x is excited, Poynting vector is observed, (c) H_x is excited, E_y is observed, (d) H_x is excited, E_z is observed (produced by packages presented in Refs. 7 and 8).

field (i.e., E_z) determines the vertical variation of the magnetic field (i.e., H_x). This is illustrated in Fig. 6.7.

Note that the three field components for the TE_{yz} case on the *yz*-plane are E_x, H_y, and H_z. Injection of the waves via only E_x is equivalent to using an electric line source, and (as H_x does in the TM_{yz} case) this also produces a cylindrical wave on the *yz*-plane as shown in Fig. 6.6a and b. On the other hand, H_y and H_z injection correspond to vertical and horizontal short magnetic loop excitations, respectively.

To excite a plane TM_{yz} wave incident upon an infinite-extent planar dielectric layer with an angle θ with respect to the plane normal $\vec{n} = -\vec{e}_z$ (see Fig. 6.3), one needs to inject

$$H_x = H_0 \sin(\omega t), \tag{6.61}$$

$$E_y = \eta_0 H_0 \sin(\omega t) \sin\left(\theta + \frac{\pi}{2}\right), \tag{6.62}$$

$$E_z = \eta_0 H_0 \sin(\omega t) \cos\left(\theta + \frac{\pi}{2}\right). \tag{6.63}$$

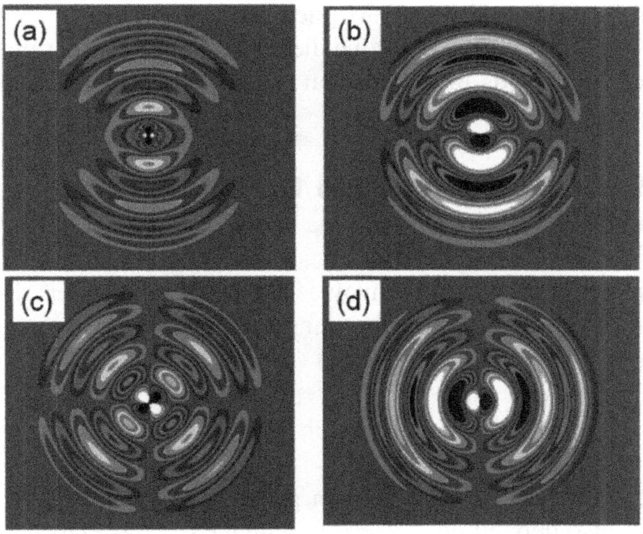

Figure 6.7. Field excitation on xy-plane for the TM_{yz} case: (a) E_z is excited, E_z is observed; (b) E_z is excited, H_x is observed, (c) E_y is excited, E_z is observed, (d) and E_y is excited, H_x is observed (produced by the virtual tools presented in [Refs. 7 and 8).

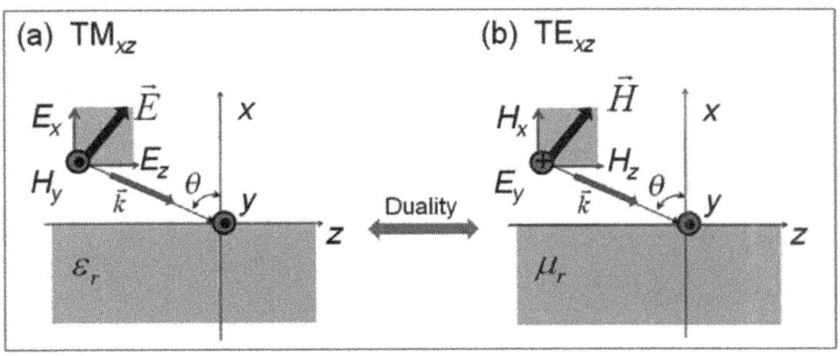

Figure 6.8. Field components on the xz-plane and the duality between TE and TM cases.

Injecting the source via only some of the three components excites plane waves propagating along both $\vec{k}$ and $-\vec{k}$ directions.

Note that duality does not hold when the problem is not anymore in free space and contain boundaries. For example, a TE wave hitting an infinite-extent homogeneous dielectric plane with an oblique incidence is equivalent to the TM wave hitting an infinite-extent homogeneous medium with a constant permeability (see Fig. 6.8). If the problem is about observing EM phenomena for TE and TM types of plane wave incidence on the same dielectric slab with the same oblique incidence, one needs to write down the governing equations for each case separately.

6.11 SECOND-ORDER EM DIFFERENTIAL EQUATIONS

Electromagnetic problems involve linear, second-order differential equations. In general, a second-order partial differential equation (PDE) is given by

$$\left(a\frac{\partial^2}{\partial x^2}+b\frac{\partial^2}{\partial x\partial y}+c\frac{\partial^2}{\partial y^2}+d\frac{\partial}{\partial x}+e\frac{\partial}{\partial y}+f\right)\psi=g. \tag{6.64}$$

The coefficients a, b, and c, in general, are functions of x and y; they may also depend on φ itself, in which case, the PDE is said to be *nonlinear*. A PDE in which $g\ (x, y)$ *equals* zero is termed homogeneous; it is inhomogeneous if $g\ (x, y)$ *is not equal to* 0.

A PDE, in general, can have both *boundary* values and *initial* values. PDEs whose BCs are specified are called *steady-state equations*. If only initial values are specified, they are called *transient equations*. Any linear, second-order PDE can be classified as *elliptic*, *hyperbolic*, or *parabolic* depending on the coefficients a, b, and c.

Elliptic: $(b^2 - 4ac) < 0$.

Hyperbolic: $(b^2 - 4ac) > 0$.

Parabolic: $(b^2 - 4ac) = 0$.

The terms hyperbolic, parabolic, and elliptic are derived from the fact that the quadratic equation

$$\left(ax^2+bxy+cy^2+dx+ey+f\right)=0. \tag{6.65}$$

represents a hyperbola, parabola, or ellipse if $b^2 - 4ac$ is positive, zero, or negative, respectively. For example, Laplace equation is elliptic type, wave equation is hyperbolic type, and diffusion equation is parabolic type. In EM, elliptic-type equations models boundary-value problems; in terms of numerical modeling and simulation, this means the solution is in closed form. One needs to solve a linear system of equations with (up to) billions of equations. On the other hand, hyperbolic- and parabolic-type equations model initial-boundary value problems and their solutions are in closed forms. Iterative models with quite a few equations are possible. Remember, closed-form solutions are always stable, but open-form solutions are conditionally stable; one needs to pay attention to stability conditions.

6.12 EM WAVE–TRANSMISSION LINE ANALOGY

There is a close analogy between the propagation of plane waves in a homogeneous medium and the propagation of voltage and current waves along a uniform transmission line. This analogy can be used not only as an aid to understand these concepts but also to obtain the solutions to realistic problems. A background in EM wave theory, for example, certainly helps to write down the solutions to EM wave problems or vice versa.

For a uniform transmission line having primary parameters R, L, C, and G per unit length, the coupled transmission line equations of the voltage and current in differential forms may be written as

$$\frac{\partial v(x,t)}{\partial x} + L\frac{\partial i(x,t)}{\partial t} + Ri(x,t) = 0, \tag{6.66a}$$

$$\frac{\partial i(x,t)}{\partial x} + C\frac{\partial v(x,t)}{\partial t} + Gv(x,t) = 0, \tag{6.66b}$$

while Maxwell equations for a uniform plane wave propagating along the $+x$ direction inside a homogeneous medium, having only E_y and H_z components, can be written as

$$\frac{\partial E_y(x,t)}{\partial x} + \mu\frac{\partial H_z(x,t)}{\partial t} = 0, \tag{6.67a}$$

$$\frac{\partial H_z(x,t)}{\partial x} + \varepsilon\frac{\partial E_y(x,t)}{\partial t} + \sigma E_y(x,t) = 0. \tag{6.67b}$$

Inspecting Equations (6.66) and (6.67) shows that the following quantities are analogous:

V (V)	E (V/m)
I (A)	H (A/m)
C (farad/m)	ε (farad)
L (H/m)	μ (H)
G (S/m)	σ (S/m)
R (Ω/m)	

If the voltages and currents, and electric and magnetic fields vary sinusoidally with time (i.e., under phasor notation), (6.66) and (6.67) become

$$\frac{dV(x)}{dx}+(R+j\omega L)I(x)=0, \quad \frac{dE_y(x)}{dx}+(0+j\omega\mu)H_z(x)=0, \tag{6.68a}$$

$$\frac{dI(x)}{dx}+(G+j\omega C)V(x)=0, \quad \frac{dH_z(x)}{dx}+(\sigma+j\omega\varepsilon)E_y(x)=0, \tag{6.68b}$$

and the second-order, uncoupled equations are

$$\frac{d^2V(x)}{dx^2}-\gamma^2V(x)=0, \quad \frac{d^2E_y(x)}{dx^2}-\gamma^2E_y(x)=0, \tag{6.69a}$$

$$\frac{d^2I(x)}{dx^2}-\gamma^2I(x)=0, \quad \frac{d^2H_z(x)}{dx^2}-\gamma^2H_z(x)=0, \tag{6.69b}$$

where

$$\gamma^2=(R+j\omega L)(G+j\omega C). \tag{6.70a}$$

in transmission line representation, and

$$\gamma^2=(j\omega\mu)(\sigma+j\omega\varepsilon). \tag{6.70b}$$

in EM plane wave representation. The solution of (6.69) can be given as

$$\begin{aligned} V(x)&=A_1\cosh(\gamma x)+B_1\sinh(\gamma x),\\ E_y(x)&=A_1\cosh(\gamma x)+B_1\sinh(\gamma x), \end{aligned} \tag{6.71a}$$

$$\begin{aligned} I(x)&=A_2\cosh(\gamma x)+B_2\sinh(\gamma x),\\ H_z(x)&=A_2\cosh(\gamma x)+B_2\sinh(\gamma x). \end{aligned} \tag{6.71b}$$

The characteristic impedance of the transmission line (Z_0) and characteristic or intrinsic impedance of the propagation medium (η_0) are

$$Z_0=\sqrt{\frac{R+j\omega L}{G+j\omega C}} \quad \text{and} \quad \eta_0=\sqrt{\frac{j\omega\mu}{\sigma+j\omega\varepsilon}}. \tag{6.72}$$

For the special case of a *lossless* line ($R = G = 0$) or lossless (*nonconducting*) ($\sigma = 0$) medium,

$$Z_0 = \sqrt{\frac{L}{C}}, \quad \eta_0 = \sqrt{\frac{\mu}{\varepsilon}}, \tag{6.73a}$$

$$\gamma = \alpha + j\beta \Rightarrow (\alpha = 0) \Rightarrow \beta = \omega\sqrt{LC} \quad \text{or} \quad \beta = \omega\sqrt{\varepsilon\mu}. \tag{6.73b}$$

Finally, wave propagation along a lossless and in a lossless medium will be

$$v = \frac{1}{\sqrt{LC}} = 310^8 \text{ m/s} \quad \text{or} \quad v = \frac{1}{\sqrt{\varepsilon_0\mu_0}} = 310^8 \text{ m/s}. \tag{6.74}$$

6.13 TIME DEPENDENCE IN MAXWELL EQUATIONS

We have two alternatives but the right-hand system is chosen in EM. We also have two alternatives for the choice of time dependence. Traditionally, exp($j\omega t$) is used in engineering society, but exp($-i\omega t$) is used in physics society. This results in two sets of kernels for waves propagating along $\vec{r}$: $\exp\left(j\omega t - \vec{k}\cdot\vec{r}\right)$ and $\exp\left(-i\omega t + \vec{k}\cdot\vec{r}\right)$. Complex relative permittivities in these cases will be $\overline{\varepsilon}_r = \varepsilon_r - j\sigma/\omega\varepsilon_0$ and $\overline{\varepsilon}_r = \varepsilon_r + i\sigma/\omega\varepsilon_0$, respectively. Finally, complex wave vectors will be $k = k_r - jk_i$ and $k = k_r + ik_i$ with positive k_r and k_i, respectively. Usually, solutions are expressed in terms of radiation and/or scattering integrals with infinite integration contours (i.e., all possible wave number values), therefore typical integration contours will be as in Fig. 6.9.

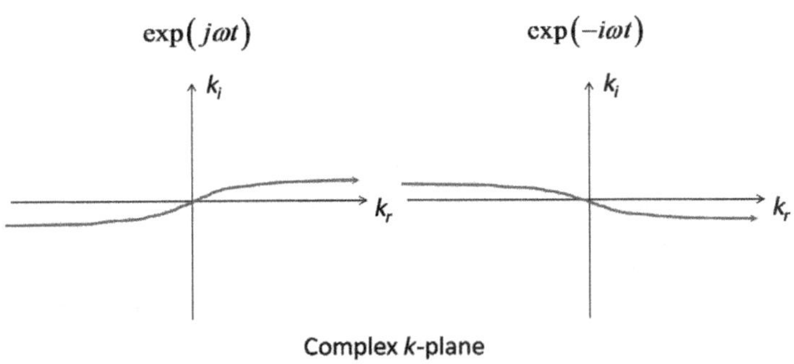

Figure 6.9. Typical scattering integral contours.

6.14 PHYSICAL FUNDAMENTALS

Always remember physics behind mathematics (i.e., equations).

- EM field is the space (volume) where we "feel" EM phenomena.
- Flux is the rate of transfer of fluid, particles, or energy across a given surface.
- A (closed) magnetic field intensity is produced by a conductive or displacement current and vice versa.
- The magnetic flux is continuous (the net flux out of a closed surface is zero).
- The electric flux exiting from a closed surface is equal to the (inside) charges.
- The magnetic flux is always closed. It has no beginning and no end.
- Electric flux is produced by a charge density. It starts at a positive charge and ends at a negative one.
- A rotating magnetic field strength is produced around a current (and vice versa)
- A rotating electric field strength is produced by a time changing magnetic flux density.

REFERENCES

1. D. G. Dudley, *Mathematical Foundations for Electromagnetic Theory*, IEEE Press (Series on Electromagnetic Waves), New York, 1994.
2. L. B. Felsen, N. Marcuvitz, *Radiation and Scattering of Waves*, Prentice-Hall, Englewood Cliffs, NJ, 1973, classic reissue IEEE Press, 1994.
3. R. E. Collin, *Field Theory of Guided Waves*, McGraw-Hill, Inc., New York, 1960.
4. E. C. Jordan, K. G. Balmain, *Electromagnetic Waves and Radiating Systems*, Prentice-Hall, Englewood Cliffs, NJ, 1968.
5. L. B. Felsen, F. Akleman, L. Sevgi, "Wave Propagation inside a Two-dimensional Perfectly Conducting Parallel Plate Waveguide: Hybrid Ray-Mode Techniques and Their Visualizations," *IEEE Antennas and Propagation Magazine*, **46** (6), 2004, pp. 69–89.

6. L. Sevgi, "Guided Waves and Eigenvalue Extraction from Propagation Characteristics," *IEEE Antennas and Propagation Magazine*, **50** (2), 2008, pp. 222–234.
7. G. Çakır, M. Çakır, L. Sevgi, "A Multipurpose FDTD-Based Two Dimensional Electromagnetic Virtual Tool," *IEEE Antennas and Propagation Magazine*, **48** (4), 2006, pp. 142–151.
8. M. Çakır, G. Çakır, L. Sevgi, "A Two-dimensional FDTD-Based Virtual Metamaterial—Wave Interaction Visualization Tool," *IEEE Antennas and Propagation Magazine*, **50** (3), 2008, pp. 166–175.
9. L. Sevgi, "Guided Waves and Transverse Fields: Transverse to What?," *IEEE Antennas and Propagation Magazine*, **50** (6), 2008, pp. 221–225.

CHAPTER 7

Sturm–Liouville Equation: The Bridge between Eigenvalue and Green's Function Problems

Guided wave theory (GWT) is well established with some canonical examples and classical lecture notes. The beauty and strength come from mathematical derivations. Guided wave problems can be grouped into two: eigenvalue (EV) problems (i.e., the effects of the environment on to the propagating waves) and the Green's function (GF) problem (the effects of the excitation in the environment). This chapter reviews GWT in one dimension (1D) which is represented by the Sturm–Liouville (SL) equation.

7.1 INTRODUCTION

Suppose you are staring at the water in front of a river. Can you predict where water comes from without knowing the route it follows by just analyzing its biological/chemical content? Alternatively, can you predict biological/chemical content of the water if you know the river bed (route)? The answers of both questions are *yes* if there are characteristic

Electromagnetic Modeling and Simulation, First Edition. Levent Sevgi.

features for the river bed (waveguide) and the water (source). If, for example, different parts of the river bed have different chemical/biological contents, then analyzing the content gives you a possibility to predict where it comes from.

In GWT, SL equation establishes a bridge between source-free (homogeneous) and source-driven (nonhomogeneous) wave equations. This means a waveguiding structure can be characterized by its distinctive features (i.e., EVs and eigenfunctions). These features contain signatures of wave shaping characteristics. That is, if you know the EVs and eigenfunctions of a waveguide, then you can predict the source and how electromagnetic (EM) energy will propagate through. Alternatively, if you observe EM waves somewhere inside, emanating from a kind of a known source, then you can extract EVs and eigenfunctions of the waveguide. The aim of this chapter is to review these connections via the SL equation. There are excellent books on GWT which become classical (see, for example, Refs. 1–3), and the reader is strongly referred to them for the subjects which are only summarized here.

7.2 GUIDED WAVE SCENARIOS

The strategy of solving Maxwell equations under given boundary/initial conditions and sources is problem dependent. In most EM problems, one can either reduce and solve the problem in terms of *transverse* and *longitudinal* components, or simplify full wave equations at the beginning by using *vector* and *scalar potentials*. In the former, the problem not only is reduced and scalarized but also solved; in the latter, the scalarization of these auxiliary functions is the second step. In this case, one needs to use equations that relate vector/scalar potentials to the electric and magnetic fields. GWT is based on the former approach [1,2]. Chapter 8 is reserved for this problem.

A few two dimensional (2D) guided wave scenarios are pictured in Fig. 7.1. These are all canonical problems thought in undergraduate as well as graduate level EM courses. Propagation inside a parallel plate waveguide with nonpenetrable boundaries (see Fig. 7.1a), for example, is given to teach ray/mode solutions and their alternative hybrid combinations. The modes of the parallel plate waveguide are called normal modes (NMs). The paper by Kamel and Felsen [4] is an excellent source to understand mathematics behind the physics of this problem.

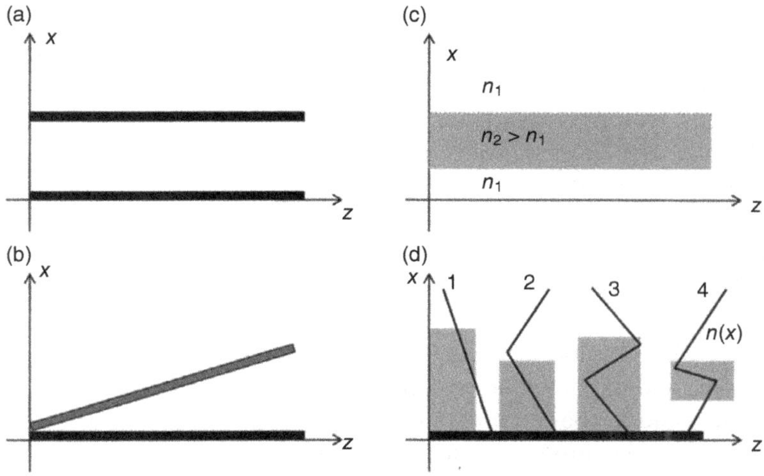

Figure 7.1. Canonical waveguiding problems (a) 2D parallel plate waveguide with nonpenetrable boundaries, (b) wedge waveguide with penetrable and/or nonpenetrable boundaries, (c) 2D dielectric (optical) film, and (d) surface/elevated ducting and antiducting on flat earth.

Also, the follow-up paper by Felsen et al. [5] introduced MATLAB-based virtual tools, which can be used to observe individual as well as collective contributions of ray and mode pieces and their various hybrid combinations on to the propagating fields.

The scenario in Fig. 7.1b belongs to the wedge-waveguide problem. The propagation inside a wedge waveguide with nonpenetrable boundaries can be modeled via NMs in the cylindrical coordinate system because the wave equation is separable there. Alternatively, NM-like wave objects can be introduced in the Cartesian coordinates if the wedge angle is small (i.e., a gradual change in the width of the wedge). These are called adiabatic modes (AMs) [6]. Finally, extension of AM to some critical regions where AM fails brought intrinsic mode (IM) representations, all of which are well reviewed in a recent paper by Sevgi et al. [7]. A MATLAB-based virtual tool was also introduced there to visualize NM, AM, and IM contributions (Chapter 9 is reserved for propagation modeling inside a wedge waveguide in terms of modal summation).

The transverse confinement of EM energy, sometimes, occurs due to the environmental variations of EM parameters such as refractive index (permittivity). Dielectric (optical) thin film shown in Fig. 7.1c is an example. Although the transverse domain extends infinitely at both ends, (mostly) energy is trapped inside the film because of its refractive index (this problem is revisited in Section 15.11).

Finally, Fig. 7.1d shows propagation above penetrable or nonpenetrable layer with different transverse refractivity profiles. These represent EM propagation above plane earth under various surface and/or elevated ducting refractivity profiles [8] (see Sections 15.6 and 15.7).

The strategy of guided wave propagation modeling (i.e., transverse and longitudinal decomposition of Maxwell equations) is pictured in Fig. 7.2. The solution can be obtained from transverse, source excited equations, and the boundary condition (BC); the longitudinal fields are dependent and can be extracted from the transverse solutions. Transverse solutions can be grouped into three depending on three—Dirichlet, Neumann, and Cauchy—type BCs. Scalar wave equations for these cases are obtained from source-free representations of the transverse vector Maxwell equations [1,2,8]. Three equations then yield modal (eigen) solutions and source information is extracted from the modal amplitudes which are derived from the solution of transmission line (TL) equations.

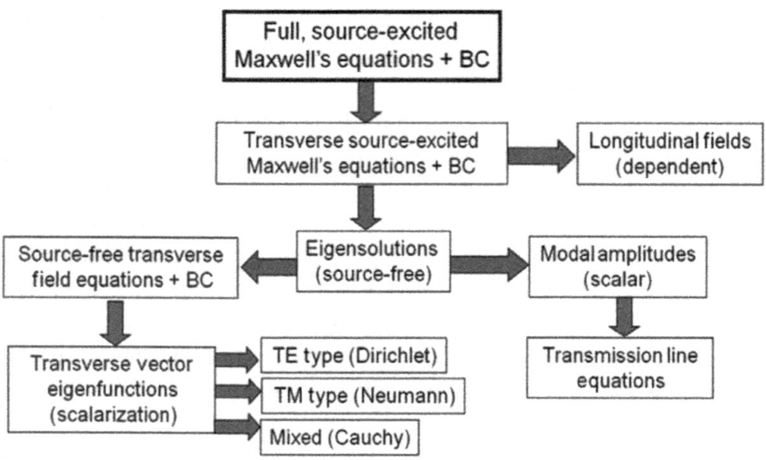

Figure 7.2. The transverse and longitudinal decomposition strategy for GWT.

7.3 THE STURM–LIOUVILLE EQUATION

Many of the boundary value problems in EM lead to the 1D SL equation (see Table 7.1 for different forms of the SL equation). The SL equation is of the form of

$$\left[\frac{d}{dz}p(z)\frac{d}{dz}-q(z)+\lambda w(z)\right]g(z,z';\lambda)=-\delta(z-z'), \tag{7.1}$$

and the BC at z_1 and z_2 ($z_1 \leq (z, z') \leq z_2$). Here, p, q, and w are real (in general continuous functions of z) that correspond to geometrical quantities of the problem, z', z are the locations of source and observer, and λ is a spectral parameter (not the wavelength). The g (z, z', λ) is called the Green's function (GF). The three common BCs are

$$g=0 \quad \text{or} \quad \frac{dg}{dz}=0 \quad \text{or} \quad g+\alpha\frac{dg}{dz}=0 \quad \text{at} \quad z=z_{1,2}, \tag{7.2}$$

where α is a parameter. These BCs belong to Dirichlet-, Neumann-, and Cauchy-type problems, respectively. There are three SL problem types [2]. The SL1 corresponds to the finite interval and real λ (i.e., p, q, and w are real and continuous functions of z; z_1 and z_2 are finite, and λ is real and independent of z) case, while the SL2 belongs to the complex λ (i.e., p, q, and w are real and continuous functions of z, z_1 and z_2 are finite, and λ is real and independent of z). Finally, if any of these conditions is violated (e.g., p, q, or w is not a continuous function of z, either or both of z_1 and z_2 are infinite, etc.), it is a SL3 problem.

The analysis of the SL equation is pictured in Fig. 7.3. The SL models [3] (1) a TL system directly if λ is a given fixed parameter (GF problem), (2) an EV problem (when the equation is homogeneous, i.e., right-hand side is zero) where λ is a free parameter (**that cannot be an EV**). Various canonical problems that are represented by SL equation for different sets of parameter are listed in Table 7.1. 1D wave and Airy equations in Cartesian coordinates, Bessel equation in cylindrical, and Legendre polynomials in spherical coordinates are only a few examples. It is well known that Airy equation is met in many ground wave propagation problems above earth's surface with vertical linear refractivity profiles (e.g., surface and/or elevated ducs). Similarly, Bessel equation is the characteristic radial wave equation in cylindrical coordinates, for structures, for example, cylindrical geometries, wedge waveguides, optical fibers, radial and coaxial TLs, etc.

Table 7.1.
Reduced SL Equations Under Different Parameter Sets

SL Equation	Parameters	Coordinate System	Solution
$\left[\frac{d^2}{dz^2}+k_z^2\right]U_z(z)=f(z)$	$p=1, q=0$ $w=1, \lambda=k_z^2$	Cartesian	Oscillatory (sine or cosine function) for finite region, progressing (exponential functions) for infinite region
$\left[\frac{d}{dz}z\frac{d}{dz}-\nu^2+k_z^2z^2\right]U_z(z)=f(z)$	$p=z, q=\nu^2$ $w=z^2, \lambda=k_z^2$	Cylindrical	Bessel functions ($J_\nu(z)$, $N_\nu(z)$, $H_\nu^{1,2}(z)$)
$\left[\frac{d^2}{dz^2}-k_z^2z\right]U_z(z)=f(z)$	$p=1, q=0$ $w=z, \lambda=k_z^2$	Cartesian	Airy functions (Ai(z), Bi(z), $W_{1,2}(z)$)
$\left[-\frac{d}{dz}(1-z^2)\frac{d}{dz}-n(n-1)\right]U_z(z)=f(z)$	$p=1-z^2$, $q=0$ $w=1$, $\lambda=n(n-1)$	Spherical	Legendre polynomials (P_n^m)

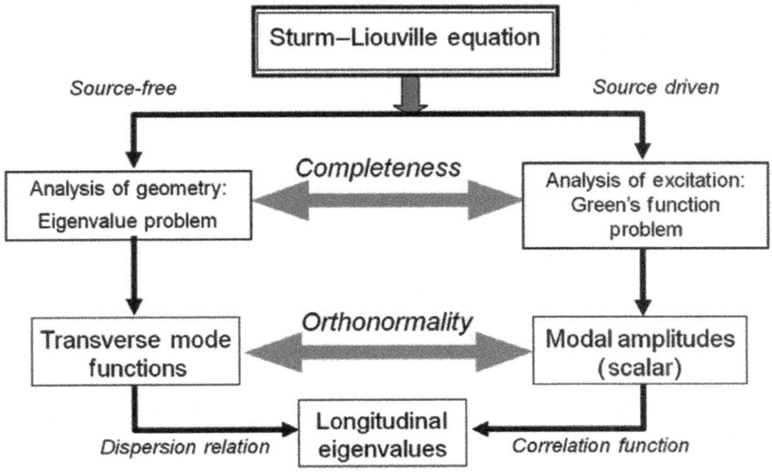

Figure 7.3. Interconnections for source-free and source-driven problems characterized by the SL equation.

7.3.1 The Eigenvalue Problem

As shown in Fig. 7.3, once BCs are specified for a given problem, the homogeneous form of the SL equation in (7.1) yields an infinite set of solutions $U(z) = U_n(z)$ that correspond to particular values of $\lambda = \lambda_n$ $(g\,(z, z'; \lambda) \rightarrow U(z))$. Here, $U_n(z)$ and λ_n are the *eigenfunctions* and *EVs*, respectively. The eigenfunctions form an orthogonal set with respect to the weighing function $w(z)$. Each of the functions in the orthogonal set (preferably) has unit energy, so they are also normalized. *Orthonormality* condition is then given as

$$\int_{z_1}^{z_2} w(z)U_n(z)U_m^*(z) = \delta_{mn}, \delta_{mn} = \begin{cases} 1 & n = m \\ 0 & n \neq m \end{cases}. \tag{7.3}$$

In EM, eigenfunctions are called normal modes (NM). NMs satisfy source-free wave equation in a coordinate system that yields separation of variables. They also satisfy transverse BCs on each transverse cross section. NM confinement may be due to fixed transverse boundaries and/or mode dependent virtual boundaries (refractive confinement). NMs propagate in longitudinal direction with distinct propagation constants. Each NM carries finite energy (can be normalized) and is independent of every other NM (orthogonality).

Any permissible transverse profile $f(z)$ can be replaced with an eigenfunction expansion,

$$f(z)=\sum_{n} K_n U_n(z), \tag{7.4}$$

and modal excitation coefficients K_n are derived from the orthonormality condition [3],

$$K_n = \int_{z=z_1}^{z=z_2} f(z)w(z)U_n^*(z)dz. \tag{7.5}$$

The limiting case is $f(z) = \delta(z - z')$ and yields the completeness relation

$$\frac{\delta(z-z')}{w(z')}=\sum_{n} U_n(z)U_n^*(z'). \tag{7.6}$$

It should be noted that eigenfunction expansion of a given source function for the SL1 and SL2 problems (i.e., finite spatial region with discrete EVs) is nothing but using a *Fourier series expansion* of an arbitrary function in a finite region.

7.3.2 The Green's Function (GF) Problem

The GF $g\ (z, z', \lambda)$ is the response of a linear system to a point source of unit strength [2] which can be represented by the Dirac delta function. The GF satisfies homogeneous SL equation everywhere except $z = z'$. At $z = z'$, GF must be continuous, but its first derivative is discontinuous by a unit amount, so that the second derivative yields a delta function singularity. The GF of a problem can be constructed via two different ways [1–3]: an eigenfunction expansion may be used,

$$g(z,z';\lambda)=\sum_{n=1}^{\infty}\frac{U_n(z)U_n^*(z')}{\lambda-\lambda_n}\ (\text{method I}), \tag{7.7}$$

or two solutions $f_1(z_<)$ and $f_2(z_>)$ may be matched at the source location,

$$g(z,z',\lambda)=-\frac{f_1(z_<)f_2(z_>)}{p(z')W(z')}\ (\text{method II}), \tag{7.8}$$

where $z_<$ is the lesser of z or z', and $z_>$ is the greater of z or z'. Here $f_1(z_<)$ and $f_2(z_>)$ are the solutions of homogeneous form of (7.1) and satisfy the BCs at the left and right ends, respectively; and $W(z')$ is called Wronskian, which is a global constant and can be calculated from

$$W = \left(f_1(z)\frac{df_2(z)}{dz} - f_2(z)\frac{df_1(z)}{dz} \right) \tag{7.9}$$

at any point between $z_1 - z_2$. The delta function is symmetric, that is, $\delta(z - z') = \delta(z' - z)$, whence $g(z, z', \lambda) = g(z', z, \lambda)$, thereby exhibiting reciprocity under the interchange of source and observation points.

7.3.3 Finite *z*-Domain Problem

Suppose the wave function $U(z)$ in a finite, bounded region $0 \leq z \leq a$ satisfies

$$\left(\frac{d^2}{dz^2} + k_z^2 \right) U(z) = 0 \tag{7.10}$$

and Dirichlet BC

$$U(0) = U(a) = 0. \tag{7.11}$$

The n-indexed eigensolutions (eigenfunctions) $U_n(z)$ corresponding to the EVs k_{zn} are given by

$$U_n(z) = A_n \sin(k_{zn} z), \quad k_{zn} = \frac{n\pi}{a}, \quad n = 1, 2, 3, \ldots, \tag{7.12}$$

where A_n is an arbitrary constant. It is readily verified that for any two eigenfunctions with indexes n_1 and n_2, respectively, the following relation—the *orthogonality relation*—applies:

$$A_{n1} A_{n2} \int_0^a U_{n1}(z) U_{n2}(z)\, dz = A_{n1} A_{n2} \delta_{n_1} \delta_{n2}, \tag{7.13}$$

where the Kronecker delta $\delta_{n_1, n_2} = 0, \quad n_1 \neq n_2$, while $\delta_{n_1, n_2} = 1, \quad n_1 = n_2$. It is convenient to normalize the eigenfunctions with respect to A_n so that

$$\int_0^a U_{n_1}(z) U_{n_2}(z)\, dz = \delta_{n_1, n_2}; \quad \text{i.e.,} \quad A_n = \sqrt{\frac{2}{a}}, \tag{7.14}$$

which renders the eigenset *orthonormal*. The implied *completeness* of the eigenset $\{U_n(z)\}$ admits the eigenexpansion of any "suitable" function $F_x(x)$ (for which the series converges),

$$F(z) = \sum_{n=0}^{\infty} a_n U_n(z), \quad 0 \le z \le a. \tag{7.15}$$

To determine the amplitudes a_n, multiply (7.15) by any other eigenfunction $U_m(z)$, perform the integration along z from 0 to a on both sides, interchange the order of integration and summation on the right-hand side, and invoke (7.14). This yields the result

$$\int_0^a F(z) U_m(z) dz = \sum_{n=0}^{\infty} a_n \int_0^a U_n(z) U_m(z) dz = \sum_{n=0}^{\infty} a_n \delta_{nm} = a_m. \tag{7.16}$$

This completes the expansion of $F(z)$ in terms of the $\{U_n(z)\}$ eigenbasis by setting $F(z) = \delta\,(z - z')$, which reduces the integral on the left-hand side of (7.16) to $U_m(z')$ via the properties of the delta function. Accordingly, $a_m = U_m(z')$, and from (7.15),

$$\delta(z - z') = \sum_{n=0}^{\infty} U_n(z) U_n(z'), \quad 0 \le (z, z') \le a. \tag{7.17}$$

The spectral representation of the delta function in (7.17) is referred to as *completeness relation* because any suitable function $F(z)$ can be expressed by performing the (identity) operation,

$$F(z) = \int_0^a F(z') \delta(z - z') dz', \tag{7.18}$$

on the summand of the right-hand side of (7.17), which recovers (7.15), with (7.16). Equations (7.6) and (7.14) form the basis for a z-domain Fourier series representation of the total field.

7.3.4 Infinite *z*-Domain Problem

Suppose the wave function $U(z)$ in an infinite, unbounded region $-\infty < z < +\infty$ satisfies the same wave (7.10) and the radiation condition for $z \to \pm\infty$. In this case, the EVs coalesce into a continuum

$k_{zn} \rightarrow k_z$, where k_z is a continuous variable $-\infty < k_z < +\infty$ (this can be verified by starting with a finite domain $z_1 \leq z \leq z_2$ and letting $z_{1,2} \rightarrow \pm\infty$). The corresponding trigonometric eigenfunctions $U(z, k_z)$ render the normalizing integral $\int_{-\infty}^{+\infty} U^2(z, k_z)dz$ divergent. To assign a proper "weight" to the divergence, one can systematically track the transition from the initial finite z-domain mentioned earlier to infinity to find for the *orthonormality condition*,

$$\int_{-\infty}^{+\infty} U(z, k_{z1})U^*(z, k_{z2})dz = \delta(k_{z1} - k_{z2}),$$
$$U(z, k_z) = \frac{1}{\sqrt{2\pi}} e^{-jk_z z}, \quad -\infty < (z, k_z) < +\infty, \tag{7.19}$$

where the complex conjugate has been introduced in view of the convenient exponential form for the eigenfunctions. Thus, the normalization expressed by δ_{n_1,n_2} for the discrete spectra in (7.14) is replaced for continuous spectra by the delta function $\delta(k_{z1} - k_{z2})$. By symmetry, interchanging $z \rightarrow k_z$, $k_z \rightarrow z$ leads to the *completeness relation*,

$$\frac{1}{2\pi} \int_{-\infty}^{+\infty} e^{-jk_z(z-z')} dk_z = \delta(z - z'), \quad -\infty < (z, z', k_z) < +\infty. \tag{7.20}$$

Any suitable function F(z) can thereby be represented via the operation

$$F(z) = \int_{-\infty}^{+\infty} F(z')\delta(z - z')dz'.$$

Equations (7.19) and (7.20) form the basis for z-domain Fourier integral formulations of the total waveguide field.

7.3.5 Relation between Eigenvalue and Green's Function Problems

The GF problem in the physical domain differs from the EV problem due to the presence of the delta function (i.e., source terms) on the right-hand side. A link between these two is established by integrating

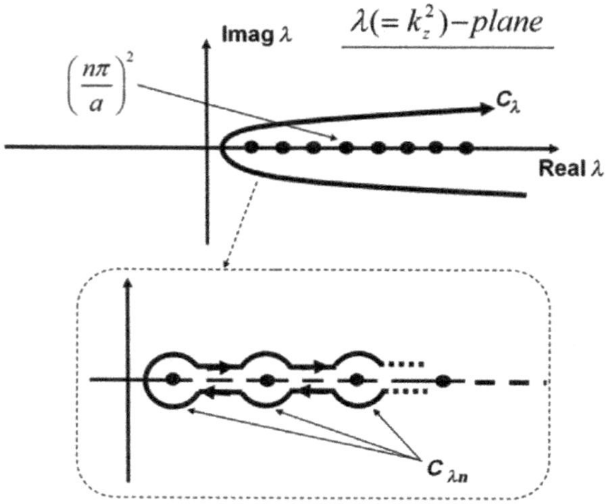

Figure 7.4. Singularities on complex λ-plane and the integration contour.

the eigenfunction expansion (7.11) along an infinite contour C_λ, enclosing all of the pole singularities of g, at $\lambda = \lambda_n$, on complex λ-plane (see, Fig. 7.4), and by invoking the residue theorem [3]:

$$\frac{-1}{2\pi j}\oint_{C_x} g_z(z,z';\lambda)d\lambda = \sum_{n=1}^{\infty} U_n(z)U_n(z') = \delta(z-z'). \tag{7.21}$$

This yields the desired *construction* of the completeness relation in terms of a contour integral over the closed form spectral GF, which plays a critical role in generating alternative representations for most of the field representations [5]. The link between GF and EV problem can also be established numerically [9,10]. Graduate level details of alternative GF representations may be found in References 4 and 5.

7.4 CONCLUSIONS

This brief chapter reviews the SL problem in 1D. Most of guided wave problems, either in 2D or in three dimensions (3D), start with transverse and longitudinal decomposition and then reduce related differential equations into 1D SL equations either in finite or infinite domains. Two canonical problems, propagation inside a parallel plate waveguide and inside wedge waveguide, are discussed in Chapter 8 and Chapter 9, respectively.

REFERENCES

1. L. B. Felsen, N. Marcuvitz, *Radiation and Scattering of Waves*, Prentice-Hall, Englewood Cliffs, NJ, 1973, classic reissue IEEE Press, 1994.
2. D. G. Dudley, *Mathematical Foundations for Electromagnetic Theory*, IEEE Press (Series on Electromagnetic Waves), New York, 1994.
3. R. E. Collin, *Field Theory of Guided Waves*, McGraw-Hill, Inc., New York, 1960.
4. L. B. Felsen, A. H. Kamel, "Hybrid Ray-Mode Formulation of Parallel Plate Waveguide Green's Functions," *IEEE Transactions on Antennas and Propagation*, **29** (4), 1981, pp. 637–649.
5. L. B. Felsen, F. Akleman, L. Sevgi, "Wave Propagation inside a Two-Dimensional Perfectly Conducting Parallel Plate Waveguide: Hybrid Ray-Mode Techniques and Their Visualizations," *IEEE Antennas and Propagation Magazine*, **46** (6), 2004, pp. 69–89.
6. A. D. Pierce, "Extension of the Method of Normal Modes to Sound Propagation in Almost Stratified Medium," *The Journal of the Acoustical Society of America*, **37**, 1965, pp. 19–27.
7. L. Sevgi, F. Akleman, L. B. Felsen, "Visualizations of Wave Dynamics in a Wedge-Waveguide with Non-Penetrable Boundaries: Normal, Adiabatic, and Intrinsic Mode Representations," *IEEE Antennas and Propagation Magazine*, **49** (3), 2007, pp. 76–94.
8. L. Sevgi, *Complex Electromagnetic Problems and Numerical Simulation Approaches*, IEEE & John Wiley Press, New York, 2003.
9. L. Sevgi, "Sturm-Liouville Equation: The Bridge Between Eigenvalue and Green's Function Problems," *ELEKTRIK, Turkish Journal of Electrical Engineering and Computer Sciences*, **14** (2), 2006, pp. 293–311.
10. L. Sevgi, "Guided Waves and Eigenvalue Extraction from Propagation Characteristics," *IEEE Antennas and Propagation Magazine*, **50** (2), 2008, pp. 222–234.

CHAPTER 8

The 2D Nonpenetrable Parallel Plate Waveguide

Guided wave propagation inside a two-dimensional (2D) parallel plate waveguide with nonpenetrable boundaries is a canonical problem [1–4]. It is also classical when ray, mode, and their hybrid forms are of interest in explaining waveguiding phenomena. This chapter deals with high frequency electromagnetic wave propagation in guiding environments. It is aimed at advanced senior and first-year graduate students, who are familiar with the usual engineering mathematics for wave equations, especially analytic functions, contour integrations in the complex plane, and so on, and also with rudimentary saddle point (HF) asymptotics. After an introductory overview of issues and physical interpretations pertaining to this broad subject area, detailed attention is given to the simplest canonical, thoroughly familiar, test environment: a (time-harmonic) line-source-excited 2D infinite waveguide with perfectly electrical conducting (PEC) plane-parallel boundaries. After formulating the Green's function (GF) problem within the framework of Maxwell equations, alternative field representations are presented and interpreted in physical terms, highlighting two complementary phenomenologies: progressing (ray-type) and oscillatory (mode-type), culminating in the self-consistent hybrid ray-mode scheme which is usually not included in conventional treatments at this level. This provides the analytical background for two educational MATLAB

Electromagnetic Modeling and Simulation, First Edition. Levent Sevgi.

packages which explore the dynamics of ray fields, mode fields, and the ray-mode interplay [5]. The first package RAYMODE serves as a tool to compute and display eigenray trajectories between specified source/observer locations and to analyze their contributions to wave fields individually. The second package HYBRID may be used to display range and/or height variations of the wave fields comparatively, calculated via ray summation, mode field summation, and hybrid ray-mode synthesis.

8.1 INTRODUCTION

When high frequency time-harmonic wave propagation is spatially (transversely) constrained by physical impenetrable boundaries or by "virtual" boundaries (ducts) established through wave-trapping (graded index) refraction in transversely inhomogeneous media, the resulting source-excited longitudinally guided (ducted) waves have traditionally been described in terms of two alternative phenomenologies: progressing and oscillatory. The progressing formulation views the wave fields as continuous spectra of waves which propagate (through the guiding environment) from the source to the receiver via multiple transverse reflections and (or) refractions that are localized around source–receiver-dependent "eigenray" trajectories. The oscillatory formulation views the wave fields as discrete (or discrete-continuous) superpositions of frequency-dependent (guided mode) "eigenspectra" which are matched to the entire transversely confining cross section, which are individually independent of source and receiver locations, but whose *amplitudes* of excitation and reception do depend on these locations.

In the "overmoded" high frequency regime, where the transverse dimensions span many wavelengths and many modes can propagate, the *oscillatory* formulation becomes unwieldy, and is not well matched to the wave physics, for *large* source–receiver separations since the forward-propagating lower mode eigenspectra (eigenangles) there are found to form closely spaced clusters. Here, the eigenray *progressing* wave spectra can be more efficiently adapted to the wave phenomenology. On the other hand, at *small* source–receiver separations, the progressing formulation becomes unwieldy because it now entails many closely spaced near-vertical multiple reflections, whereas the oscillatory lower eigenmode spectra are widely spaced and are therefore more efficient; while excited in their totality very near the source, these modes decay exponentially (are evanescent) longitudinally away from the source when the number of oscillations in their transverse profile exceeds a frequency-dependent cutoff threshold. These circumstances have motivated efforts to combine the two *complementary* spectral methodologies, neither of which is convenient for *all* source–receiver locations, in a manner that seeks to exploit the best features of each. The outcome has been a comprehensive rigorously based, self-consistent *hybrid ray-mode algorithm*, which has clarified the ray-mode interplay in the frequency and time domains through a series of spectral studies and wide-ranging applications to complex waveguiding environments.

The basic prototype problem by Felsen and Kamel [4] demonstrates alternative hybrid ray-mode partitionings of parallel plate waveguide GFs, with numerical examples that quantify all relevant dynamic constituents. Changes in the source–receiver locations give rise to intricate compensating wave mechanisms that account for smooth transitions when a mode field or a ray field enters or leaves a specified angular spectrum interval. These mechanisms can be systematically examined in this idealized environment, thus furnishing cogent physical insight that is essential for treating more complicated guiding configurations. This chapter revisits the problem tutorially so as to make it more easily accessible to the reader (especially advanced senior and first-year graduate students). For a more formal mathematical treatment of the parallel plate waveguide, GFs approach can be used especially for advanced graduate students, which does, however, not contain the hybrid ray-mode decomposition. Two MATLAB packages have been prepared for tutorial visualization of the hybrid ray-mode interplay.

8.2 PROPAGATION INSIDE A 2D-PEC PARALLEL PLATE WAVEGUIDE

The most general case deals with ray, mode, and hybrid field formulations for the 2D parallel plate waveguide with general *impedance* (Cauchy)-*type* boundary conditions. There, rays between the transmitter and observer encounter lateral shifts at each reflection when they interact with the boundaries. When PEC boundaries (i.e., Dirichlet-type for the TE_z problem and Neumann-type for the TM_z problem) are assumed as in our presentation here, the lateral shifts disappear and

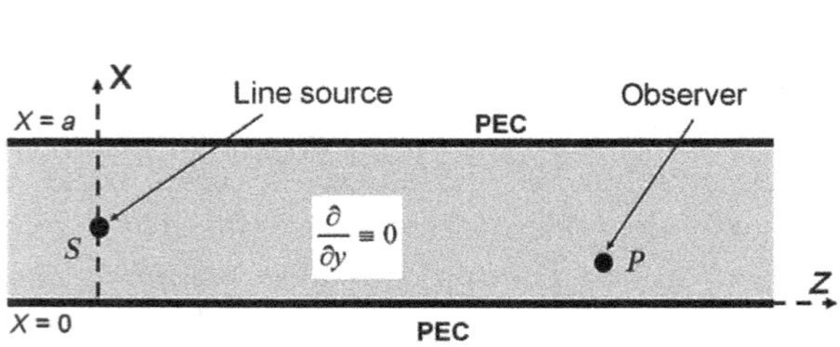

Figure 8.1. Physical configuration: 2D, homogeneously filled, parallel plate waveguide with PEC boundaries.

rays are reflected from the boundary at the angle of incidence (Snell's reflection law).

8.2.1 Formulation of the TE- and TM-Type Problems

The PEC parallel plate waveguide, with height a, has its propagation direction along the z-coordinate of a Cartesian system extending from $-\infty$ to $+\infty$ along the y- and z-directions (see Fig. 8.1). Assuming that the medium between the plates is free space, one can use Maxwell's curl equations in the form (a time dependence $\exp(j\omega t)$ is suppressed)

$$\nabla \times \vec{\mathrm{E}} = -j\omega\mu_0 \vec{H} - \vec{\mathrm{M}}, \tag{8.1a}$$

$$\nabla \times \vec{\mathrm{H}} = j\omega\varepsilon_0 \vec{\mathrm{E}} + \vec{\mathrm{J}}, \tag{8.1b}$$

where the arrows denote vector quantities. The nonphased source currents $\vec{J}$ and $\vec{M}$ extend along the y-direction, whence

$$\frac{\partial}{\partial y} \equiv 0 \tag{8.2}$$

throughout (i.e., the fields are functions of $[x,z]$).

Maxwell equations inside the parallel plate waveguide (in vacuum) are

$$\frac{\partial H_z}{\partial y} - \frac{\partial H_y}{\partial z} = j\omega\varepsilon_0 E_x, \quad \frac{\partial E_z}{\partial y} - \frac{\partial E_y}{\partial z} = -j\omega\mu_0 H_x, \tag{8.3a}$$

$$\frac{\partial H_x}{\partial z} + \frac{\partial H_z}{\partial x} = j\omega\varepsilon_0 E_y, \quad \frac{\partial E_x}{\partial z} - \frac{\partial E_z}{\partial x} = -j\omega\mu_0 H_y, \tag{8.3b}$$

$$\frac{\partial H_y}{\partial x} - \frac{\partial H_x}{\partial y} = j\omega\varepsilon_0 E_z, \quad \frac{\partial E_y}{\partial x} - \frac{\partial E_x}{\partial y} = -j\omega\mu_0 H_z, \tag{8.3c}$$

and the two wave equations will be

$$\left\{\frac{\partial^2}{\partial x^2}+\frac{\partial^2}{\partial y^2}+\frac{\partial^2}{\partial z^2}+k^2\right\}\vec{\mathrm{E}}=0, \quad \left\{\frac{\partial^2}{\partial x^2}+\frac{\partial^2}{\partial y^2}+\frac{\partial^2}{\partial z^2}+k^2\right\}\vec{\mathrm{H}}=0. \quad (8.4)$$

Assume $\exp(-j\beta z)$ dependence ($\beta = k_2$) along z (lossless waveguide walls and vacuum inside). In this case, a wave propagating in z-direction will be $\exp(j(\omega t - \beta z))$. The variation of each of the field components in the z-direction may be shown explicitly, for example, as $H_x = H_x^0 \exp(-j\beta z)$, therefore, $\partial H_x/\partial z = -j\beta H_x$. Making use of this and remembering $\partial/\partial y \equiv 0$, Maxwell equations become

$$\beta H_y = \omega\varepsilon_0 E_x, \quad j\beta E_y = -j\omega\mu_0 H_x, \quad (8.5a)$$

$$-j\beta H_x - \frac{\partial H_z}{\partial x} = j\omega\varepsilon_0 E_y, \quad -j\beta E_x - \frac{\partial E_z}{\partial x} = -j\omega\mu_0 H_y, \quad (8.5b)$$

$$\frac{\partial H_y}{\partial x} = j\omega\varepsilon_0 E_z, \quad \frac{\partial E_y}{\partial x} = -j\omega\mu_0 H_z. \quad (8.5c)$$

These equations may be solved to give

$$H_x = -\frac{j\beta}{k_x^2}\frac{\partial H_z}{\partial x}, \quad E_x = -\frac{j\beta}{k_x^2}\frac{\partial E_z}{\partial x}, \quad (8.6a)$$

$$H_y = -j\frac{\omega\varepsilon_0}{k_x^2}\frac{\partial E_z}{\partial x}, \quad E_y = j\frac{\omega\mu_0}{k_x^2}\frac{\partial H_z}{\partial x}, \quad (8.6b)$$

where $k_x^2 = k^2 - \beta^2$ is the transverse (x-domain) wave number. It is interesting to note that all four field components will be zero if the longitudinal fields are zero. Therefore, for the rectangular waveguide propagation, either E_z or H_z field components must exist. As in the parallel PEC plate waveguide in 2D, the solution can be grouped into two:

TE (transverse electric with respect to z-axis) type (i.e., $E_z = 0$*).*

TM (transverse magnetic with respect to z-axis) type (i.e., $H_z = 0$*).*

The two wave equations then reduce to

$$\left\{\frac{\partial^2}{\partial x^2}+k^2-\beta^2\right\}E_z=0, \quad \left\{\frac{\partial^2}{\partial x^2}+k^2-\beta^2\right\}H_z=0. \quad (8.7)$$

The reduced two sets of scalar equations (i.e., TM_z and TE_z) are given as follows:

SET 1: TM$_z$

$$\frac{\partial H_y}{\partial z} = -J_x - j\omega\varepsilon_0 E_x, \tag{8.8a}$$

$$\frac{\partial E_x}{\partial z} - \frac{\partial E_z}{\partial x} = -M_y - j\omega\mu_0 H_y, \tag{8.8b}$$

$$\frac{\partial H_y}{\partial x} = J_z + j\omega\varepsilon_0 E_z. \tag{8.8c}$$

SET 2: TE$_z$

$$\frac{\partial E_y}{\partial z} = M_x + j\omega\mu_0 H_x, \tag{8.9a}$$

$$\frac{\partial H_x}{\partial z} - \frac{\partial H_z}{\partial x} = J_y + j\omega\varepsilon_0 E_y, \tag{8.9b}$$

$$\frac{\partial E_y}{\partial x} = -M_z - j\omega\mu_0 H_z. \tag{8.9c}$$

These equations, in turn, can be combined into second-order partial differential equations, the wave equations, as follows:

TM$_z$ wave equation: The 2D TM$_z$ wave equation is

$$\left\{\frac{\partial^2}{\partial x^2} + \frac{\partial^2}{\partial z^2} + k^2\right\} H_y = j\omega\varepsilon_0 M_y + \frac{\partial J_z}{\partial x} - \frac{\partial J_x}{\partial z}, \tag{8.10}$$

with boundary conditions

$$\frac{\partial H_y}{\partial x} = 0 \quad \text{at } x = 0, a, \tag{8.11a}$$

$$\textit{outgoing wave (radiation) condition} \text{ at } z \to \pm\infty. \tag{8.11b}$$

Here, $k = 2\pi / \lambda = \omega\sqrt{\varepsilon_0\mu_0}$ is the free-space wave number and λ is the free-space wavelength. The remaining two TM$_z$ field components can be derived from

$$E_x = -\frac{1}{j\omega\varepsilon_0}\left\{\frac{\partial H_y}{\partial z} + J_x\right\}, \tag{8.12a}$$

$$E_z = \frac{1}{j\omega\varepsilon_0}\left\{\frac{\partial H_y}{\partial x} - J_z\right\}. \tag{8.12b}$$

TE_z wave equation: Similarly, the 2D TE_z wave equation is

$$\left\{\frac{\partial^2}{\partial x^2}+\frac{\partial^2}{\partial z^2}+k^2\right\}E_y = j\omega\mu_0 J_y - \frac{\partial M_z}{\partial x}+\frac{\partial M_x}{\partial z}, \tag{8.13}$$

with boundary conditions

$$E_y = 0 \quad \text{at } x = 0, a, \tag{8.14a}$$

$$\textit{outgoing wave (radiation) condition} \text{ at } z \to \pm\infty. \tag{8.14b}$$

The remaining two field components for this set are

$$H_x = \frac{1}{j\omega\mu_0}\left\{\frac{\partial E_y}{\partial z} - M_x\right\}, \tag{8.15a}$$

$$H_z = -\frac{1}{j\omega\mu_0}\left\{\frac{\partial E_y}{\partial x} + M_z\right\}. \tag{8.15b}$$

Since the TE_z and TM_z sets are decoupled, each can be excited independently of the other by appropriate selection of the sources, **J** and **M**. The line sources M_y, J_x, J_z excite the TM_z set, whereas the line sources M_x, M_z, J_y excite the TE_z set. Further simplification can be obtained by setting the source components $J_x = 0$, $J_z = 0$ for the TM_z set and $M_x = 0$, $M_z = 0$ for the TE_z set. These simplifications will be assumed in what follows.

8.2.2 The Green's Function Problem

The GF problem associated with both the TE_z set (when $M_x = M_z = 0$) and the TM_z set (when $J_x = J_z = 0$) is defined by the equation

$$\left\{\frac{\partial^2}{\partial x^2}+\frac{\partial^2}{\partial z^2}+k^2\right\}g(x,z;x',z') = -\delta(x-x')\delta(z-z'), \tag{8.16}$$

with boundary conditions

$$g(x,z;x',z') = 0 \quad \text{at } x = 0, a \quad (TE_z), \tag{8.17a}$$

$$\frac{\partial}{\partial x}g(x,z;x',z') = 0 \quad \text{at } x = 0, a \quad (TM_z), \tag{8.17b}$$

$$\textit{outgoing wave (radiation) condition} \text{ at } z \to \pm\infty. \tag{8.17c}$$

Here, (x',z') and (x,z) specify source and observation points, respectively, and $\delta(x)$, $\delta(z)$ are the Dirac delta functions, with

$$\delta(x)=0, \quad x\neq 0; \quad \int_a^b \delta(x)dx=1, \quad a<x<b. \tag{8.18}$$

The delta function is symmetric, i.e., $\delta(a-a')=\delta(a'-a)$, whence $g_x(x,z;x',z') = g_x(x',z';x,z)$, thereby exhibiting reciprocity under the interchange of source and observation points. The field $F(x,z)$ excited by a physically distributed source $S(x,z)$ can be synthesized in terms of the (symbolic) line source GF as follows:

$$F(x,z)=\int\!\!\int_{\substack{\text{Source}\\\text{region}}} dx'dz'S(x',z')g(x,z;x',z'). \tag{8.19}$$

8.2.3 Accessing the Spectral Domain: Separation of Variables

To solve the (source-free) wave equation

$$\left\{\frac{\partial^2}{\partial x^2}+\frac{\partial^2}{\partial z^2}+k^2\right\}U(x,z)=0 \tag{8.20}$$

subject to the boundary conditions

$$U(x,z)=0 \quad \text{at } x=0,a; \quad \text{radiation condition at } z\to\pm\infty, \tag{8.21}$$

assume that we can decompose

$$U(x,z)=U_x(x)U_z(z), \tag{8.22}$$

where U_x satisfies the first and U_z satisfies the second boundary condition in (8.17), respectively. Substituting (8.22) into (8.20), performing the partial derivatives, and rearranging the result, one finds

$$-\frac{U_x''}{U_x}=\frac{U_z''}{U_z}+k^2, \tag{8.23}$$

where a prime denotes the derivative with respect to the argument. Since the left-hand side in (8.23) is a function of x only, while the right-hand side is a function of z only, and the equation is valid for arbitrary (x,z) in the domains of (8.21), each side must be a constant which is denoted by k_x^2. Thus, (8.23) is reduced to two separate ordinary differential equations,

$$U_x'' + k_x^2 U_x = 0, \quad 0 \leq x \leq a, \quad U_x = 0 \text{ at } x = 0, a, \tag{8.24a}$$

$$\left\{\frac{d^2}{dz^2} + k_z^2\right\} U_z(z) = 0; \quad -\infty < z < +\infty, \text{ radiation condition at } z \to \pm\infty, \tag{8.24b}$$

$$k^2 = k_x^2 + k_z^2. \tag{8.24c}$$

The dispersion equation in (8.24c) relates x-domain and z-domain spectral wave numbers k_x and k_z, respectively, to the physical wave number k.

8.2.4 Spectral Representations: Eigenvalue Problems

Spectral representations of fields are structured around discrete (or continuous) sums (or integrals) extending over complete (preferably orthogonal) sets of eigenfunctions (basis) that are matched to the problem conditions. The 2D coordinate-separable problem in (8.20) and (8.21) admits separate one-dimensional (1D) eigenvalue problems based on (8.24a) and (8.24b), respectively.

EM waves inside the nonpenetrable parallel plate waveguide are trapped transversely and propagate longitudinally (see Fig. 8.1). Spectral representations are presented in Chapter 7; therefore, only results are given as follows.

***x*-Domain.** The finite-domain spectral representation (see Section 7.3.3 with (7.10)–(7.18)) is

$$U_{zn}(z) = \sqrt{\frac{2}{a}} \sin(k_{zn} z), \quad k_{zn} = \frac{n\pi}{a}, \quad n = 1, 2, 3, \ldots, \quad 0 \leq x \leq a, \tag{8.25}$$

where $U_{zn}(z)$ and k_{zn} correspond to the eigenfunctions and eigenvalues, respectively. These eigenfunctions are orthonormal and form a complete set. The implied *completeness* of the eigenset admits the eigen-expansion of any source function.

***z*-domain.** The infinite-domain spectral representation (see Section 7.3.4 with (7.19) and (7.20)) is

$$U_z(z, k_z) = \frac{1}{\sqrt{2\pi}} e^{-jk_z z}, \quad -\infty < (z, k_z) < +\infty. \tag{8.26}$$

8.2.5 Spectral Representations: 1D Characteristic Green's Functions

The 2D GF problem in the physical domain has been formulated in (8.16)–(8.18) and differs from the eigenvalue problem in (8.20) due to the presence of the (delta function) source terms on the right-hand side. The corresponding reduced 1D *spectral* problems in the x- and z-domains are modified accordingly.

***x*-Domain.** The spectral (also referred to as the "characteristic") GF g_x is defined as follows:

$$\left\{\frac{d^2}{dx^2}+k_x^2\right\}g_x(x,x';k_x)=-\delta(x-x'),\quad 0\leq(x,x')\leq a, \qquad (8.27)$$
$$g_x=0 \text{ at } x=0,a.$$

Now, k_x is a *specified* spectral wave number. For the unique specification of g_x in (8.27), it is necessary that $k_x \neq \pm k_{xn}$, for any m, because the eigenfunctions satisfy the source-free form of (8.27) and can therefore be added to the source-driven solution without violating (8.27). The solution for g_x can be constructed explicitly in the following form:

$$g_x(x,x';k_x)=\frac{\overleftarrow{g}_x(x_<,k_x)\overrightarrow{g}_x(x_>,k_x)}{-W_x(\overleftarrow{g}_x,\overrightarrow{g}_x)}, \qquad (8.28)$$

where $x_< = x$ for $x < x'$, but x' for $x > x'$; similarly, $x_> = x$ for $x > x'$, but x' for $x < x'$. Here, $\overleftarrow{g}_x$ and $\overrightarrow{g}_x$ are solutions of the homogeneous ($x \neq x'$) form of (8.27), which satisfy the boundary conditions at $x = 0$ and $x = a$, respectively. Thus, reciprocity is satisfied in (8.28). The Wronskian

$$W_x(\overleftarrow{g}_x,\overrightarrow{g}_x)\equiv\left[\overleftarrow{g}_x\frac{d\overrightarrow{g}_x}{dx}-\overrightarrow{g}_x\frac{d\overleftarrow{g}_x}{dx}\right]_{x=x'} \qquad (8.29)$$

is a function of k_x only, independent of x (or x'). The solution in (8.28) is continuous at $x = x'$, but has a discontinuous x-derivative there, which gives rise to the impulsive (delta function) second derivative that accounts for the source term on the right-hand side of (8.27). Specifically, the solution in (8.28) can be constructed via

$$\overleftarrow{g}_x(x,k_x)=\sin(k_x x), \qquad (8.30a)$$
$$\overrightarrow{g}_x(x,k_x)=\sin(k_x(a-x)), \qquad (8.30b)$$

whence $W_x = -k_x\{\sin(k_x x)\cos[k_x(a - x)] + \sin[k_x(a - x)]\cos(k_x x)\} = -k_x \sin(k_x a)$ from (8.35). Accordingly, characteristic GF will be

$$g_x(x, x'; k_x) = \frac{\sin(k_x x_<)\sin(k_x(a - x_>))}{k_x \sin(k_x a)}. \tag{8.31}$$

z-Domain. The characteristic GF g_z is defined as follows:

$$\left\{\frac{d^2}{dz^2} + k_z\right\} g_z(z, z'; k_z) = -\delta(z - z'), -\infty \leq (z, z', k_z) \leq +\infty, \quad \text{rad. cond. at } z \to \pm\infty, \tag{8.32}$$

with the solution

$$g_z(z, z'; k_z) = \frac{\overleftarrow{g}_z(z_<, k_z)\overrightarrow{g}_z(z_>, k_z)}{-W_z(\overleftarrow{g}_z, \overrightarrow{g}_z)}. \tag{8.33}$$

Here, $\overleftarrow{g}_z$ and $\overrightarrow{g}_z$ are outgoing wave solutions in $z < z'$ and $z > z'$, respectively, in accord with the radiation condition (Re $\{k_z\} > 0$, Im $\{k_z\} < 0$ for the assumed $\exp(j\omega t)$ time dependence)

$$\overleftarrow{g} = e^{+jk_z z}, \quad \overrightarrow{g} = e^{-jk_z z}, \quad W_z = -2jk_z. \tag{8.34}$$

Thus, since $(z_< - z_>) = -|z - z'|$, characteristic GF will be

$$g_z(z, z'; k_z) = \frac{e^{-jk_z|z-z'|}}{2jk_z}. \tag{8.35}$$

8.2.6 The 2D Green's Function Problem: Alternative Representations

The 2D GF problem in physical space is defined in (8.16) with (8.17). Alternative field representations for the 2D GF can be generated by various combinations of the 1D spectral domain eigenfunction expansions and characteristic GFs pertaining to the physical x- and z-domains.

Eigenfunction Expansion in x-domain, 1D GF in z-Domain. The GF can be written as

$$g(x, z; x', z') = \sum_{n=1}^{\infty} \overline{A}_n(z, z')U_{xn}(x)U_{xn}(x'), \tag{8.36}$$

where $U_{xn}(x)$ is given in (8.25), with $A_n = \sqrt{2/a}$. This form is suggestive in view of the delta function representation in (8.30), which is used to replace $\delta(x - x')$ on the right-hand side of (8.16). Substituting (8.36) into (8.16), and performing the partial differentiations on the summand in (8.36), one obtains, for the x-domain,

$$\frac{\partial^2}{\partial x^2} U_{xn} \rightarrow -k_{xn}^2 U_{xn}.$$

In view of (8.24a) and (8.24c). With this reduced x-domain formulation in mind, (8.16) becomes

$$\sum_{n=1}^{\infty}\left(\frac{d^2}{dz^2}+k_{zn}^2\right)\overline{A}_n(z,z')U_{xn}[x]U_{xn}(x') = -\sum_{n=1}^{\infty}U_{xn}(x)U_{xn}(x')\delta(z-z'), \tag{8.37}$$

where $k_{zn}^2 = k^2 - k_{xn}^2$. Equating the coefficients of the orthogonal eigenfunction expansions on both sides, one identifies $\overline{A}_n(z, z')$ as a spectral GF that satisfies

$$\left(\frac{d^2}{dz^2}+k_{zn}^2\right)g_{zn}(z,z') = -\delta(z-z'), \quad -\infty < (z,z') < +\infty \tag{8.38}$$

subject to the radiation condition at $|z| \rightarrow \infty$. The solution of (8.38) is given in (8.35), provided that $k_z \rightarrow k_{zn}$,

$$g_{zn}(z,z') = \frac{e^{-jk_{zn}|z-z'|}}{2jk_{zn}}, \quad k_{zn} = \sqrt{k^2 - \left(\frac{n\pi}{a}\right)^2}. \tag{8.39}$$

Thus, the desired representation for $g(x,z; x',z')$ is

$$g(x,z;x',z') = \sum_{n=1}^{\infty}\left\{\left(\frac{2}{a}\right)\sin\left(\frac{n\pi x}{a}\right)\sin\left(\frac{n\pi x'}{a}\right)\right\}\frac{e^{-jk_{zn}|z-z'|}}{2jk_{zn}}, \tag{8.40}$$

which is convenient for evaluating the waveguide field away from the source plane $z = z'$ when the number of *propagating* modes with real k_{zn} (i.e., $k > n\pi/a$) is not too large. Since $\mathrm{Im}\{k_{zn}\} \leq 0$, the higher-order nonpropagating modes with $k < n\pi/a$ decay along z.

Eigenfunction Expansion in z-Domain, 1D GF in x-Domain. Following the same procedure, one can write

$$\begin{aligned} g(x,z;x',z') &= \int_{-\infty}^{+\infty} dk_z \overline{A}(x,x';k_z)U(z,k_z)U^*(z',k_z), \\ U(z,k_z) &= \frac{1}{\sqrt{2\pi}}e^{-jk_z z}. \end{aligned} \tag{8.41}$$

Replacing $\delta(z - z')$ on the right-hand side of (8.16) by (8.32), substituting (8.41) into (8.16), and recalling that

$$\frac{\partial^2}{\partial z^2}U(z,k_z) \to -k_z^2 U(z,k_z),$$

one obtains

$$\begin{aligned}&\int_{-\infty}^{+\infty} dk_z U(z,k_z)U^*(z',k_z)\left(\frac{d^2}{dx^2}+k^2-k_z^2\right)\bar{A}(x,x';k_z)\\&=-\int_{-\infty}^{+\infty} dk_z U(z,k_z)U^*(z',k_z)\delta(x-x').\end{aligned} \tag{8.42}$$

Equating the (unique Fourier) integrands on both sides identifies $\bar{A}(x, x'; k_z)$ as the x-domain GF in (8.31), so that

$$\begin{aligned}&g(x,z;x',z') = \frac{1}{2\pi}\int_{-\infty}^{+\infty} \frac{\sin(k_x x_<)\sin(k_x(a-x_>))}{k_x \sin(k_x a)} e^{-jk_z(z-z')}dk_z,\\&k_x = \sqrt{k^2-k_z^2}.\end{aligned} \tag{8.43}$$

This representation is useful for field evaluations in the near zone of the source, $|z| \approx z'$, because the large number of closely spaced, weakly decaying *evanescent* modes, which cannot be neglected there, render the series in (8.40) poorly convergent. It is also useful at high frequencies when (8.40) encompasses many *propagating* modes. Note that the integrand in (8.43) contains pole singularities at $\sin(k_{xn}a) = 0$, $n = \pm1, \pm2, \ldots$, that is, $k_{xn} = n\pi/a$, which must be circumnavigated by analytic extension of the real-axis integrand into the complex k_z^2-plane (see Fig. 7.4 in Chapter 7). The residues at these poles generate precisely the x-domain eigenfunction expansion in (8.40).

8.3 ALTERNATIVE REPRESENTATION: EIGENRAY SOLUTION

The integral representation for eigenfunctions in the z-domain and propagation along x in (8.43) can be restructured in a manner that expresses the x-domain GF behavior in terms of multiply reflected traveling plane waves between the boundaries at $x = 0$, a, rather than the sinusoidal standing waves as shown. We expand the resonant denominator in the integrand into an infinite power series

$$(1-\Omega)^{-1} = \sum_{n=0}^{\infty} \Omega^n, \quad \Omega = \exp(-j2k_x a) \tag{8.44}$$

obtained by writing $\sin(k_x a) = [(1/2j)(1 - \Omega)\exp(jk_x a)]$. This series can be ordered into four (i)-indexed infinite periodic "image arrays" in the x-domain. Interchanging the orders of summation and integration, each of the integrals represents an exact 2D "image" line source GF $\hat{g}_n^{(i)}$ in *free space* (see (8.45b)), which can be identified asymptotically with a ray-optical field reaching the observer along a direct or multiply reflected ray path, and which is frequently referred to as an "eigenray" from a specified source point to a specified observation point. Following this procedure, one obtains the GF fields

$$g(x, z; x', z') = \sum_{i=1}^{4} \sum_{n=0}^{\infty} \hat{g}_n^{(i)}(x, z; x_n^{(i)}, z'), \tag{8.45a}$$

where ($k_x = \sqrt{k^2 - k_z^2}$)

$$\begin{aligned} \hat{g}_n^{(i)}\left(x, z; x_n^{(i)}, z'\right) &= -\frac{j}{4}\left[H_n^{(2)}\left(x, z; x_n^{(i)}, z'\right)\right]^{(i)} \\ &= \frac{1}{2\pi} \int_{-\infty}^{\infty} \frac{e^{-jk_x(x - x_n^{(i)})}}{2jk_x} e^{-jk_z(z-z')} dk_z \end{aligned} \tag{8.45b}$$

is the image line source GF in free space, whose closed-form and plane wave spectral representations are given by the first and second equalities in (8.45b), respectively. When the $\hat{g}_n^{(i)}$ integral is approximated asymptotically in the high frequency range, one obtains by the *stationary phase method*,

$$g(x, z; x', z') \approx \frac{j}{4\pi} \sum_{i=1}^{4} \sum_{n=0}^{\infty} \left[\frac{2j\pi}{k_0 q_n^{(i)''}(W_n^{(i)})}\right]^{1/2} \exp\{-jk_0 q_n^{(i)}(W_n^{(i)})\}, \tag{8.45c}$$

where a prime denotes the derivative with respect to the argument. In (8.45b), $H_n^{(2)}$ is the Hankel function of the second kind, with the argument

$$kr = k\left[\left(x - x_n^{(i)}\right)^2 + \left(z - z'\right)^2\right]^{1/2}.$$

The four different ray species are defined by the following phase accumulations:

$$q_n^{(1)} = (x_> - x_<)\cos W + z\sin W + 2an\cos W, \quad W = \sin^{-1}\left(\frac{k_x}{k}\right), \tag{8.46a}$$

$$q_n^{(2)} = (x_> + x_<)\cos W + z\sin W + 2an\cos W, \tag{8.46b}$$

$$q_n^{(3)} = -(x_> + x_<)\cos W + z\sin W + 2a(n+1)\cos W, \tag{8.46c}$$

$$q_n^{(4)} = -(x_> - x_<)\cos W + z\sin W + 2a(n+1)\cos W. \tag{8.46d}$$

Here, the *angular* spectrum variable W denotes the angle between the ray (i.e., the *local plane wave*) and the normal to the boundaries as shown in Fig. 8.2, whereas n is the number of reflections from both boundaries.

The eigenrays emanating from the source and reaching the observer are determined from the eigenray (stationary phase) condition

$$\frac{d}{dW} q_n^{(i)}(W) = 0 \Rightarrow W = W_n^{(i)}, \quad i = 1, 2, 3, 4 \tag{8.47}$$

The four eigenray species in (8.47) are pictured in Fig. 8.3, and their trajectories are organized according to their departure directions (up and down) at the source and arrival directions (up and down) at the observer (x_s = source, x_p = observer):

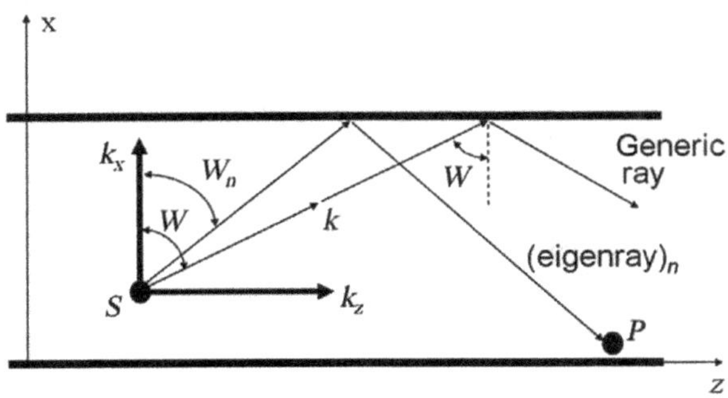

Figure 8.2. Spatial and spectral coordinates (transverse and longitudinal wave numbers); local plane wave (i.e., ray) trajectories: generic ray and n-indexed eigenray; spectral angles W and W_n (measured with respect to the x-direction).

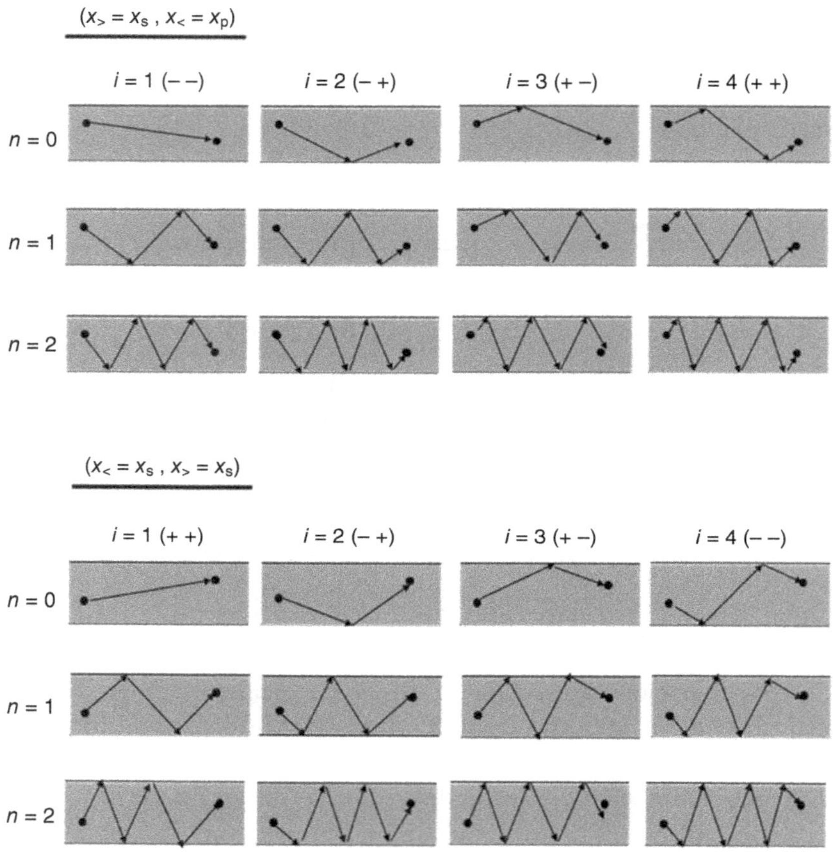

Figure 8.3. Trajectories for eigenrays, where n and i tag the number of reflections and the eigenray species, respectively. Relations between n and i are as follows: $i = 1$: n reflections at both boundaries ($- -$ for $x_> = x_s$, $x_< = x_p$ and $+ +$ for $x_> = x_p$, $x_< = x_s$); $i = 2$: n reflections at upper and $(n + 1)$ reflections at lower boundaries $(- +)$; $i = 3$: $(n + 1)$ reflections at upper and n reflections at lower boundaries $(+ -)$; $i = 4$: $(n + 1)$ reflections at both boundaries ($+ +$ for $x_> = x_s$, $x_< = x_p$ and $- -$ for $x_> = x_p$, $x_< = x_s$). This i-ordering corresponds to upward $(+)$ or downward $(-)$ departure directions at the source (first symbol inside the parenthesis), and the corresponding arrival directions at the observer (second symbol inside the parenthesis).

- $i = 1{:}n$ reflections at both boundaries (down/down for $x_> = x_s, x_< = x_p$; up/up for $x_> = x_p$, $x_< = x_s$),
- $i = 2{:}n$ reflections at upper and $(n + 1)$ reflections at lower boundary (down, up),
- $i = 3{:}(n + 1)$ reflections at upper and n reflections at lower boundary (up, down),
- $i = 4{:}(n + 1)$ reflections at both boundaries (up/up for $x_> = x_s$, $x_< = x_p$; down/down for $x_> = x_p$, $x_< = x_s$).

8.3.1 Relation between Eigenmode and Eigenray Representations

The generating function for either the eigenmode or the eigenray representation is the denominator shown in (8.44), pertaining to the spectral integral formulation in (8.43). The eigenmodes are parameterized by the mode eigenvalues W_m obtained from the resonance equation $\sin(k_{xm}a) = \sin(ka\sin W_m) = 0$, $m = \pm1,\pm2,\ldots$, which keeps the denominator intact. The eigenrays are parameterized via the series expansion in (8.45) by the eigenangles $W_n^{(i)}$, which satisfy the stationary phase condition $(d/dW)q_n^{(i)}(W, z, z') = 0$ at $W = W_n^{(i)}$ in (8.47). These alternative parameterizations are summarized in Fig. 8.4, which displays the eigenmode and eigenray "densities," respectively.

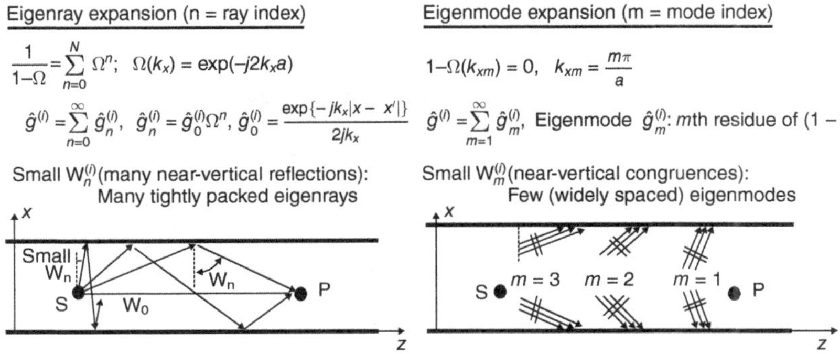

Figure 8.4. Schematizations of eigenrays and eigenmode ray congruences, with inclusion of the analytic generating function for eigenray and eigenmode expansions. (a) Eigenrays and (b) eigenmode ray congruences.

The complementary behavior of the mode and ray eigenangles plays an important role in the hybrid ray-mode representations.

8.3.2 2D GF and Hybrid Ray-Mode Decomposition

The computationally most useful and phenomenologically most intriguing representation of the 2D GF is the *hybrid ray-mode* decomposition which splits the propagating angular wave number spectral domain into sectors that can be "filled" sequentially either by ray fields or mode fields so as to take advantage of the "best" properties of each (see Fig. 8.5). This algorithm has been described conceptually earlier and will be discussed in detail later. By an entirely different approach, the hybrid ray-mode scheme can be derived directly through application of finite Poisson summation to convert a finite series of ray fields into a finite series of mode fields (plus remainders), and vice versa.

The hybrid ray-mode formulation can be justified by returning to the 2D GF form in (8.43) and expanding the resonant denominator $\sin(k_x a)$ in g_x not into an *infinite* power series as in (8.44) but into an exact decomposition with a *finite* power series plus a *resonant remainder*, according to

$$\frac{1}{1-\Omega} = \sum_{n=0}^{N} \Omega^n + \frac{\Omega^{N+1}}{1-\Omega} = \sum_{n=0}^{N} \Omega^n - \frac{\Omega^N}{2} + \left(\frac{1+\Omega}{1-\Omega}\right)\frac{\Omega^N}{2}, \qquad (8.48)$$
$$\Omega = \exp(-j2k_x a).$$

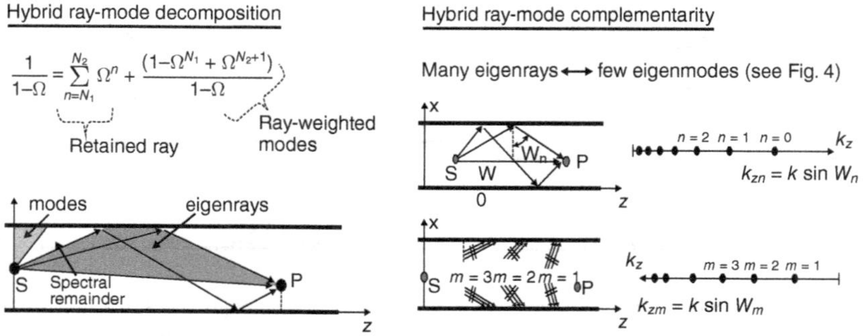

Figure 8.5. As in Fig. 8.4, but for hybrid ray-mode partitioning. (a) Partitioning in physical domain and (b) hybrid ray-mode schematization.

This can be shown (by algebraic manipulations in (8.43)) to lead instead of (8.45) to the following exact expansion:

$$g(x, z; x', z') = \sum_{i=1}^{4} \sum_{n=0}^{N} \Big[\hat{g}_n^{(i)}\left(x, z; x_n^{(i)}, z'\right) - \frac{1}{2} \hat{g}_N^{(i)}\left(x, z; x_N^{(i)}, z'\right) + I_N^{(i)}\left(x, z; x_n^{(i)}, z'\right) \Big], \tag{8.49}$$

where

$$I_N^{(i)}\left(x, z; x_N^{(i)}, z'\right) = \frac{1}{2\pi} \int_{-\infty}^{+\infty} \cot\left(k_x\, a\right) \frac{e^{-jk_x\left|x - x_N^{(i)}\right|}}{2jk_x} e^{-jk_z(z-z')} dk_z, \tag{8.50}$$

With $k_x = \sqrt{k^2 - k_z^2}$. The first two terms on the right-hand side of (8.49) represent traveling (i.e., ray-type) fields having undergone $n = 0$ to N reflections ($n = 0$, N being the limiting rays). Concerning (8.50), if in the integrand the term in brackets is ignored for the moment, the remaining integral has the form of $\hat{g}_N^{(i)}(x, z; x_N^{(i)}, z')$, that is, ray-type. Assuming that the bracketed term is slowly varying, then the entire integral behaves asymptotically (via the saddle point method) like a ray field with N reflections, with the addition of residue contributions due to the pole singularities at $k_{xm} = m\pi/a$, $m = 1,2,3,\ldots$ $\left(k_{zm} = \sqrt{k^2 - (m\pi / a)^2}\right)$ in the bracketed term that lie between $k_z = 0$ and $k_z = k_{zNs}^{(i)}$, where $k_{zNs}^{(i)}$ is the saddle point for the limiting ray N. This leads to the following asymptotic approximation:

$$g\left(x, z; x', z'\right) \approx \sum_{i=1}^{4} \sum_{n=1}^{N} \Bigg\{ \hat{g}_{ns}^{(i)}\left(x, z; x_n^{(i)}, z'\right) + \left[\frac{1}{2} + \cot\left(k_{xNs}^{(i)} a\right)\right] \hat{g}_{Ns}^{(i)}\left(x, z; x_N^{(i)}, z'\right) + \frac{1}{2} \sum_{n=1}^{M(N)} \left[\frac{\exp\left[-jq_N^{(i)}(x, z; x_n^{(i)}, z')\right]}{2jk_x}\right]_{k_{xn}} \mathrm{Res}\left\{\cot\left(k_{xm} a\right)\right\} \Bigg\}, \tag{8.51}$$

where $M(N)$ is the N-dependent last intercepted pole, $\hat{g}_{ns}^{(i)} \equiv \hat{g}_n^{(i)}\Big|_{k_z = k_{zns}^{(i)}}$, $k_{xNs}^{(i)} = k_x\left(k_{zNs}^{(i)}\right)$, and the saddle point $k_{zns}^{(i)}$ is defined by $(d / dk_z) q_n^{(i)}$ $(x, z; x_n^{(i)}, z) = 0$ at $k_z = k_{zns}^{(i)}$. (Note that n is the ray index and m is the mode index here). Moreover,

$$q_n^{(i)}\left(x, z; x_n^{(i)}, z\right) = k_x \left|x - x_n^{(i)}\right| + k_x\left(z - z'\right), \tag{8.52}$$

$$\hat{g}_{ns}^{(i)}\left(x, z; x_n^{(i)}, z'\right) \approx \frac{-j}{4}\sqrt{\frac{2\pi}{kr_n^{(i)}}}e^{-jkr_n^{(i)}+j\frac{\pi}{4}}, \tag{8.53}$$

with $r_n^{(i)} = \left[(x - x_n^{(i)})^2 + (z - z')^2\right]^{1/2}$ (cf. (8.46b)). Expression (8.51) is valid provided that $\cot(k_x a)$ is slowly varying, that is, k_{xm} does not lie near a saddle point $k_{zNs}^{(i)}$; when this is not the case, uniform asymptotics is required.

The result in (8.51) constitutes the desired hybrid ray-mode algorithm. The first term on the right-hand side represents the first N eigenrays (i.e., those with large k_z and small k_x) defined by $k_{zns}^{(i)}$, thereby implying that $k_{zNs}^{(i)} \leq k_{zns}^{(i)} < k$. The second term represents a weighting of the limiting eigenray $n = N$. The third term represents the mode contributions that compensate for the remaining eigenrays with $k_{zns}^{(i)} < k_{zNs}^{(i)}$ in the eigenray expansion. It should be noted that the decomposition of g_x in (8.43) into traveling wave constituents generates within each such constituent branch point singularities at $k_z = \pm k_x$ in the complex k_x-plane, which are not encountered when all of these constituents are recombined into complete mode fields [5]. In fact, g_x in (8.43) is an even function of k_x (replacing k_x by $-k_x$ does not change the result) so that its behavior for small k_x is proportional to $k_x^2 = k^2 - k_z^2$, which behaves regularly near $k_z = \pm k$.

Some fundamental and remarkably simple insights emerge from the mechanism that explains the proper mix of ray and modal fields: representing each mode in terms of two ray congruences that propagate transversely at the modal (characteristic) angles, the required modes have those values that fill the void left by a truncated angular spectrum of rays. Conversely, the required rays have orientation angles that fill the void left by an incomplete series of modes. The hybrid formulation is, in general, numerically more efficient than either the mode formulation alone or the ray formulation alone, when the number of contributing modes is very large. Selection of the best combination of rays and modes depends on the problem at hand.

8.4 A 2D-PEC PARALLEL PLATE WAVEGUIDE SIMULATOR

A MATLAB simulation package has been developed for the calculation of wave fields inside the PEC parallel plate waveguide subjected to time-harmonic line source excitation [5].

8.4.1 Representations Used for Mode, Ray, and Hybrid Solutions

The GF $g(x,z;\, x',z')$ of the TE$_z$ problem given in (8.40) with (8.39) is used for the mode solution. The line-source-excited electric field E_y is then given by $E_y = -j\omega\mu_0 g$ (see (8.13)) so that

$$E_y = -j\omega\mu_0 \sum_{m=1}^{\infty} \frac{2}{a} \frac{e^{-jk_{zm}|z-z'|}}{2jk_{zm}} \sin\left(\frac{m\pi x}{a}\right) \sin\left(\frac{m\pi x'}{a}\right),$$
$$k_{zm} = \sqrt{k^2 - \left(\frac{m\pi}{a}\right)^2}. \tag{8.54}$$

The eigenangle W_m of the mth mode is given by

$$W_{\mathrm{m}} = \tan^{-1}\left(\frac{k_{zm}}{k_{xm}}\right), \quad k_{\mathrm{xm}} = \frac{m\pi}{a}, \quad k_{zm} = \sqrt{k^2 - k_{xm}^2}. \tag{8.55}$$

For a distributed excitation, the E_y field in (8.13) that satisfies the boundary conditions in (8.14) may be obtained from the GF via (8.19). Three of the lower-order mode (LOM) profiles ($m = 1$, $m = 3$, and $m = 6$) are pictured in Fig. 8.6. Any transverse source profile can be synthesized with a complete set of these modes whose excitation amplitudes are determined from the orthonormality condition ((7.18) in

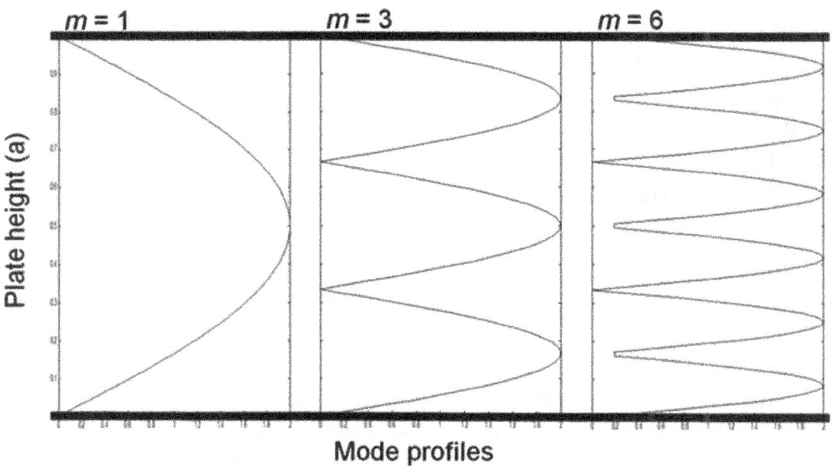

Figure 8.6. Vertical profiles of the first, third, and sixth modes.

Chapter 7). The propagating and nonpropagating (evanescent cutoff) modes have positive real and purely negative imaginary propagation constants k_{zm}, respectively. By equating k_{zm} to zero, the number of propagating modes can be determined once the frequency $\omega = k / \sqrt{\varepsilon_0 \mu_0}$ of the source is specified. Although an infinite number of modes (both propagating and nonpropagating) is required to reconstruct the line source field in the near zone of the source, higher-order nonpropagating modes are damped as the wave propagates longitudinally inside the waveguide, and their contribution becomes negligible after a few tens of wavelengths in range. Therefore, it is adequate to retain only the propagating modes in (8.54) at longer ranges. The modal sum in (8.54) tends to diverge for a parameter set (frequency–waveguide width) such that $k_{zm} \to 0$ (modal resonance condition). To obtain an acceptable result in this case, the mode at or near cutoff should be omitted.

A simple MATLAB program file *ModeSum.m* is given in Table 8.1, which computes mode fields versus height using (8.54) at a specified range when waveguide width and the frequency are supplied. Note that (8.54) belongs to a line source. Here, in the program, a smooth Gauss function $f(x,0) = \exp(-(x - hh)^2/2tt^2)$ is used as a vertical field distribution at the initial range (this is merely because line source necessitate infinite number of modes, but a finite number of modes is usually enough to reconstruct the source profile). Modal eigenvalues and longitudinal propagation constants are calculated in the first loop of the program. Then, orthonormality condition is used to extract modal excitation coefficients in the second loop and given initial Gauss profile is reproduced in terms of mode summation approach. Fields inside the waveguide at two different ranges are also calculated in terms of mode summation approach. Results are plotted in four different figures. From left to right, the first figure shows vertical mode shape and the second one plots both the initial Gauss profile and its mode summation equivalent. In this plot, one can observe how mode summation result approaches the Gauss profile as the number of modes included increases. The last two figures show vertical field distributions at the specified ranges. The pause command in the code makes it possible to observe mode contributions individually.

The first example produced with the code listed in Table 8.1 is given in Fig. 8.7. The width of the plate is 1 m. The number of propagating modes is 5 for $f = 800$ MHz. Note that propagating and nonpropagating (evanescent cutoff) modes have positive real and purely

Table 8.1.
A MATLAB Mode Summation Module

```
%% -------------------------------------------------
%    Program: ModeSum.m
%     d = Plate width; fr = frequency; nx = # of
height points
%     hh = source location; tt = source extend
%% -------------------------------------------------
clear all; clc;
c=3.E8; fr=18e8; k0=(2*pi*fr)/c; d=1; nx=500; hh=.3;
tt=.02;
dx=d/(nx-1); z1=2e3; z2=25e3;
for k=1:nx   % Transverse Gaussian Source profile
  xx(k)=(k-1)*dx; sgauss(k)=exp(-(xx(k)-hh)^2/
(2.*tt^2));
  nm(k)=0.0; nmfld(k)=0.0; nmfldz1(k)=0.0; nmfldz2(k)=0.0;
end
OK=1; n=0;
while OK>0   % Get propogation constants (beta)
  n=n+1;
  if k0>((n*pi)./d)
     beta(n)=sqrt(k0^2-(n*pi./d)^2);
  else
     beta(n)=i*sqrt((n*pi./d)^2-k0^2);
  end
  for k=1:nx % Use orthonormality & extract excitation
coeffs
    ff(k)=sgauss(k)*sqrt(2./d)*sin((n*pi*xx(k))/d);
  end
  kn(n)=trapz(ff,xx);
  for k=1:nx   % Calculate fields at three different
ranges
    nm(k)=kn(n)*sqrt(2./d)*sin((n*pi*xx(k))./d);
    nmfld(k)=nmfld(k)+nm(k);
    nmfldz1(k)=nmfldz1(k)+nm(k)*exp(i*beta(n)*z1);
    nmfldz2(k)=nmfldz2(k)+nm(k)*exp(i*beta(n)*z2);
  end
  %------------------ Plot output ------------------
  subplot(1,4,1); plot(abs(nm),xx,'LineWidth',2);
ylim([0,d]);
  xlabel('Mode Profile'); ylabel('Plate Width [m]');
```

(Continued)

Table 8.1.
(Continued)

```
   legend('Mode No:', num2str(n));
   subplot(1,4,2); xlim([0,1]);
   plot(sgauss,xx,abs(nmfld),xx,'r-','LineWidth',2,'MarkerS
ize',5)
   xlabel('Field at z=0'); legend('Gs Source','NM Sum');
   subplot(1,4,3); plot(abs(nmfldz1),xx,'k','LineWidth',2);
   xlim([0,1]); ylim([0,d]); xlabel('Field Profile');
   legend('z=z1');
   subplot(1,4,4); plot(abs(nmfldz2),xx,'k','LineWidth',2);
   xlim([0,1]); ylim([0,d]); xlabel('Field Profile');
   legend('z=z2'); pause
end
%--------------------- END ---------------------
```

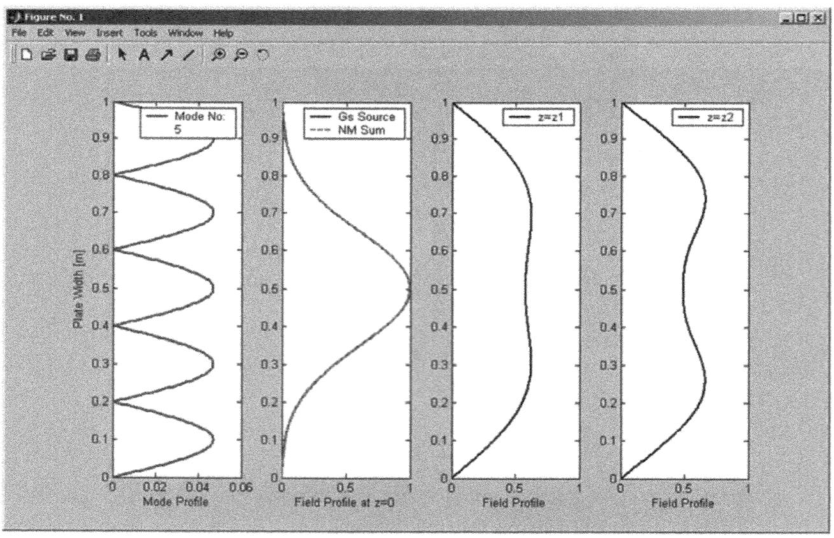

Figure 8.7. Fields inside the waveguide. Left to right: Vertical extend of the fifth mode, source profile and its mode sum equivalent, and mode fields at two ranges.

negative imaginary propagation constants, respectively. As observed in Fig. 8.7, the first five modes are enough to construct the given smooth Gaussian source profile. On the other hand, all of the first 41 modes are required for the source profile given in Fig. 8.8, but only the first five propagating modes contribute to the fields at ranges far from the

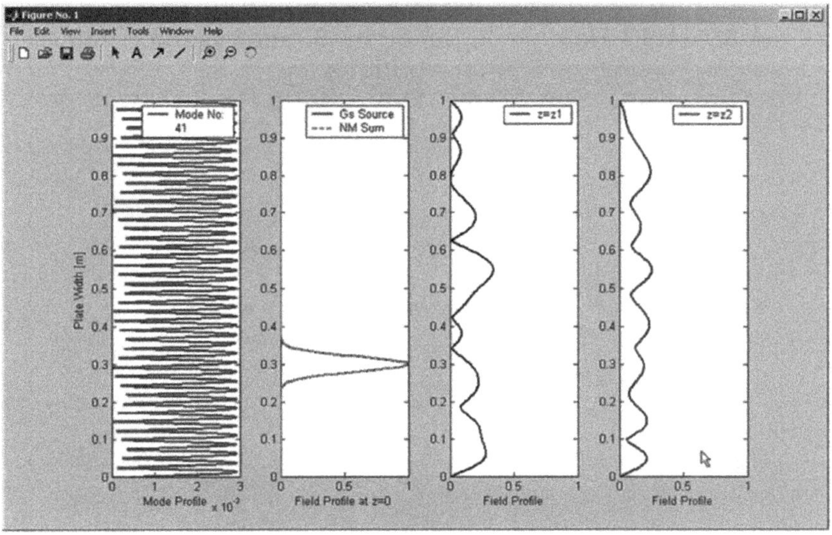

Figure 8.8. The same as in Fig. 8.7 but for another source profile.

source. It should be noted that an infinite number of modes (both propagating and nonpropagating) are required to reconstruct the line source field in the near zone of the source, higher-order nonpropagating modes are damped as the wave propagates longitudinally inside the waveguide, and their contribution becomes negligible after a few tens of wavelengths in range.

To calculate the eigenangles from the eigenray equation (8.47), with (8.46), for specified source/observer locations, some sort of root-search algorithm is required. Since the eigenrays are source/observer dependent, the root search must be repeated whenever one or both source/observer locations change. For example, for plots of wave field versus range at constant height, or wave field versus height at constant range, roots at every range or every height must be calculated and stored. More specifically, suppose that 100 height points are used to construct a wave field profile versus height at given range, with the number of ray reflections chosen as 10 (i.e., $n = 10$), then the number of eigenray angles that should be calculated by a root-search algorithm will be $100 \times 10 \times 4 = 4000$.

Equations (8.46c)–(8.47) are used for the ray field calculations and eigenray visualizations. A simple MATLAB program file *EigenRay.m* is given in Table 8.2, which computes eigenray paths and ray fields for

Table 8.2.
A MATLAB Field Code Based on Eigenrays and Their Contributions

```
function output=eigenraysummation
% ----------------------------------------------------
% Program: Eigenraysum.m
% EigenRay Summation inside a 2D-PEC parallel plate
waveguide
% for the calculation of
%    Field vs Range at a given height (xo)
%    Field vs Height at a given range (zo)
% E0: Field amplitude, nray: number of reflections
% fr: frequency [MHz], a: plate height [m], zmax: max
range[m]
% xs: source height [m], (zo,xo): observer range, height
 m]
% NX: number of height points, NZ: number of range
points
% nray: number of rays
% pol: polarization type, 0 for TEz case, 1 for TMz
case
%
% output.zz:     range points
% output.xx:     height points
% output.fx0:    vertical field at z=0
% output.fx:     vertical field at z=zo
% output.fz:     horizontal field at x=xo
%-----------------------------------------------------
E0=1; fr=2300; a=1; zmax=6; xs=.3; zo=5; xo=.7;
NX=100; NZ=100;
nray=4; pol=0; k0=2*pi*fr/300; xx=(0:a/(NX-1):a).';
zz=(0:zmax/(NZ-1):zmax).';
zz=0.5*(zz(1:end-1)+zz(2:end));
Fv0=zeros(NX,1); Fv1=zeros(NX,1);
Fh1=zeros(length(zz),1);
for k=1:NX  % Field vs Height at z=0
    g1=0; g2=0; g3=0; g4=0;
    for n=0:nray
      qq1= abs(xx(k)-xs)+2*a*n; qq2=
abs(xx(k)+xs)+2*a*n;
```

Table 8.2.
(Continued)

```
      qq3=-abs(xx(k)+xs)+2*a*(n+1);
qq4=-abs(xx(k)-xs)+2*a*(n+1);
      % Calculate four rays
      g1=g1+1i/(4*pi)*sqrt((2*pi*1i)/
(k0*-qq1))*exp(-1i*k0*qq1);
      g2=g2+1i/(4*pi)*sqrt((2*pi*1i)/
(k0*-qq2))*exp(-1i*k0*qq2);
      g3=g3+1i/(4*pi)*sqrt((2*pi*1i)/
(k0*-qq3))*exp(-1i*k0*qq3);
      g4=g4+1i/(4*pi)*sqrt((2*pi*1i)/
(k0*-qq4))*exp(-1i*k0*qq4);
    end
    if pol==0,
       Fv0(k)=Fv0(k)+g1-g2-g3+g4;
    elseif pol==1
       Fv0(k)=Fv0(k)+g1+g2+g3+g4;
    end
end
% Field vs Height z=zo
for k=1:NX
   g1=0; g2=0; g3=0; g4=0;
   for n=0:nray
     qq1=@(t)
abs(xx(k)-xs)*cos(t)+zo*sin(t)+2*a*n*cos(t);
     qq2=@(t)  abs(xx(k)+xs)*cos(t)+zo*sin(t)+2*a*n*co
s(t);
     qq3=@(t) -abs(xx(k)+xs)*cos(t)+zo*sin(t)+2*a*(n+1)*
cos(t);
     qq4=@(t) -abs(xx(k)-xs)*cos(t)+zo*sin(t)+2*a*(n+1)*
cos(t);
     % Calculate eigenray angles
     wn1=atan(zo/(abs(xx(k)-xs)+2*a*n));
     wn2=atan(zo/(abs(xx(k)+xs)+2*a*n));
     wn3=atan(zo/(-abs(xx(k)+xs)+2*a*(n+1)));
     wn4=atan(zo/(-abs(xx(k)-xs)+2*a*(n+1)));
     % Calculate four rays
     g1=g1+1i/(4*pi)*sqrt((2*pi*1i)/
(k0*(-qq1(wn1)))). . .
         *exp(-1i*k0*qq1(wn1));
```

(Continued)

Table 8.2.
(Continued)

```
     g2=g2+1i/(4*pi)*sqrt((2*pi*1i)/
(k0*(-qq2(wn2)))). . .
         *exp(-1i*k0*qq2(wn2));
     g3=g3+1i/(4*pi)*sqrt((2*pi*1i)/
(k0*(-qq3(wn3)))). . .
         *exp(-1i*k0*qq3(wn3));
     g4=g4+1i/(4*pi)*sqrt((2*pi*1i)/
(k0*(-qq4(wn4)))). . .
         *exp(-1i*k0*qq4(wn4));
  end
  if pol==0,
     Fv1(k)=Fv1(k)+g1-g2-g3+g4;
  elseif pol==1
     Fv1(k)=Fv1(k)+g1+g2+g3+g4;
  end
end
for k=1:length(zz)    % Field vs Range at x=xo
   g1=0; g2=0; g3=0; g4=0;
   for n=0:nray
      qq1=@(t)   abs(xo-xs)*cos(t)+zz(k)*sin(t)+2*a*n*co
s(t);
      qq2=@(t)   abs(xo+xs)*cos(t)+zz(k)*sin(t)+2*a*n*co
s(t);
      qq3=@(t)  -abs(xo+xs)*cos(t)+zz(k)*sin(t)+2*a*(n+1)
*cos(t);
      qq4=@(t)  -abs(xo-xs)*cos(t)+zz(k)*sin(t)+2*a*(n+1)
*cos(t);
      % Calculate eigenray angles
      wn1=atan(zz(k)/(abs(xo-xs)+2*a*n));
      wn2=atan(zz(k)/(abs(xo+xs)+2*a*n));
      wn3=atan(zz(k)/(-abs(xo+xs)+2*a*(n+1)));
      wn4=atan(zz(k)/(-abs(xo-xs)+2*a*(n+1)));
      % Calculate four rays
      g1=g1+1i/(4*pi)*sqrt((2*pi*1i)/
(k0*(-qq1(wn1)))). . .
         *exp(-1i*k0*qq1(wn1));
      g2=g2+1i/(4*pi)*sqrt((2*pi*1i)/
(k0*(-qq2(wn2)))). . .
         *exp(-1i*k0*qq2(wn2));
```

Table 8.2.
(Continued)

```
      g3=g3+1i/(4*pi)*sqrt((2*pi*1i)/
(k0*(-qq3(wn3)))). . .
          *exp(-1i*k0*qq3(wn3));
      g4=g4+1i/(4*pi)*sqrt((2*pi*1i)/
(k0*(-qq4(wn4)))). . .
          *exp(-1i*k0*qq4(wn4));
    end
    if pol==0,
        Fh1(k)=Fh1(k)+g1-g2-g3+g4;
    elseif pol==1
        Fh1(k)=Fh1(k)+g1+g2+g3+g4;
     end
end
output.xx=xx;     output.zz=zz; % Results
if pol==0,
   output.Fv0=E0*Fv0; output.Fv1=E0*Fv1; output.
Fh1=E0*Fh1;
elseif pol==1,
   output.Fv0=E0/(120*pi)*Fv0; output.Fv1=E0/
(120*pi)*Fv1;
   output.Fh1=E0/(120*pi)*Fh1;
end
plot(xx,abs(Fv1))
% ---------------------- END ----------------------
```

given source/receiver locations once the waveguide width and the frequency are specified [6].

Note that eigenray paths are derived from the solution of the eigenray equation (8.47). Alternatively, these rays and their paths can also be obtained by using image method. A simple MATLAB code which plots eigenrays using image method is listed in Table 8.3. Figure 8.9 shows a sample output of this code.

Wave field calculations in large waveguides have been performed by direct summation of either the modal or the ray field series. In both instances, when the number of contributing terms is exceedingly large, it is customary to truncate the series. Inefficient calculations involving a finite number of modes plus a remainder term, or a finite number of

Table 8.3.
A MATLAB Module for the Calculation of Eigenray Paths

```
function EigenRayPaths
% ------------------------------------------------------
% Program: EigenrayPaths.m
% Eigenray Paths inside a 2D Parallel Plate Waveguide
% a: plate height   [m], zmax: maximum range [m]
% xs: source height [m], xo: observer height [m]
% NZ: number of range points
% Case 1: nr reflections at both boundaries
% Case 2: nr reflections at Top and (nr+1) at BOTTOM
% Case 3: (nr+1) reflections TOP and nr at BOTTOM
% Case 4: (nr+1) reflections at BOTH boundaries
clc, close all, clear all; a=1; zmax=2; xs=0.3; xo=0.8;
NZ=1001; nr=25; zz=(0:zmax/(NZ-1):zmax).';
figure; axis([0 zmax 0 a]),hold on
for n=0:nr
    if xs>=xo      % Case 1
        wn1=((xo-xs)-2*a*n)/zmax;
    else
        wn1=((xo-xs)+2*a*n)/zmax;
    end
    findray(wn1,NZ,zz,xs,a);
    % Case 2
    wn2=-(abs(xo+xs)+2*a*n)/zmax;
findray(wn2,NZ,zz,xs,a);
    % Case 3
    wn3=(-abs(xo+xs)+2*a*(n+1))/zmax;
findray(wn3,NZ,zz,xs,a);
    % Case 4
    if xs>=xo
        wn4=((xo-xs)+2*a*(n+1))/zmax;
    else
        wn4=((xo-xs)-2*a*(n+1))/zmax;
    end
    findray(wn4,NZ,zz,xs,a);  pause
end
function findray(wn,NZ,zz,xs,a)
xx=zeros(1,NZ);
for kk=1:NZ,
    xx(kk)=wn*zz(kk)+xs;
```

Table 8.3.
(Continued)

```
    if xx(kk)>a
        nn=fix(xx(kk)/a);
        if mod(nn,2)==0,
            xx(kk)=-nn*a+wn*zz(kk)+xs;
        else
            xx(kk)=(nn+1)*a-wn*zz(kk)-xs;
        end
    elseif xx(kk)&#x003C;0
        nn=fix(abs(xx(kk)/a));
        if mod(nn,2)==0,
            xx(kk)=-nn*a-wn*zz(kk)-xs;
        else
            xx(kk)=(nn+1)*a+wn*zz(kk)+xs;
        end
    end
end
plot(zz,xx,'b-');
% ---------------------- END ---------------------
```

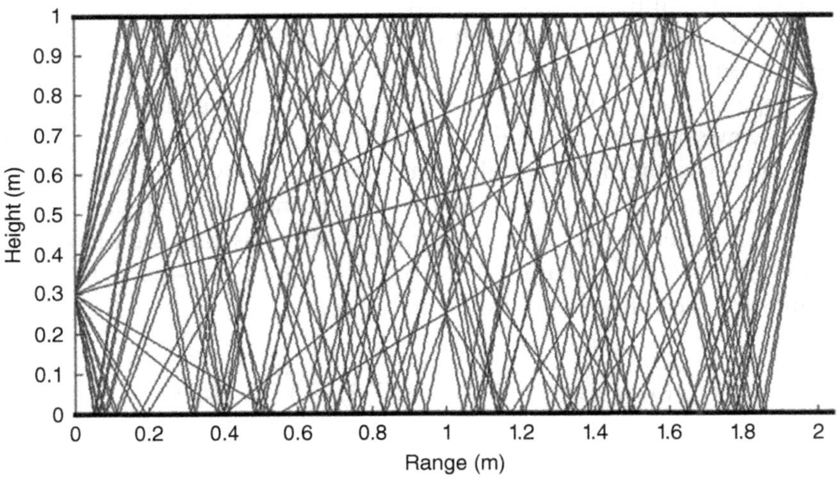

Figure 8.9. Eigenray paths obtained from image method.

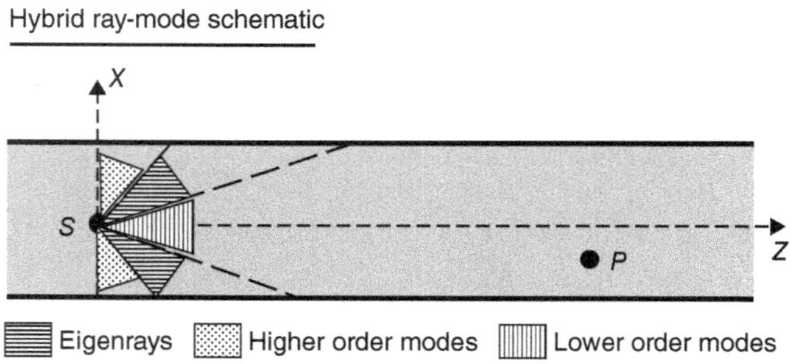

Figure 8.10. Multisector partitioning as in (8.51) showing groupings of eigenrays and modes. These intervals can be filled either with rays or modes depending on the problem conditions; for example, a few LOM and HOM plus a few eigenrays may be used in an efficient hybrid form to account for the field distribution.

eigenrays plus a remainder term, may be used for most of the source/observer locations, with truncations that render contributions from the remainder terms negligible.

The alternative self-consistent hybrid ray-mode approach (see (8.51)) utilizes a group of rays together with a group of lower and/or higher-order modes (HOMs) plus modified remainder terms, tuned so as to render the wave field computation more efficient. The physical interpretation of the process is pictured in Fig. 8.10. The propagating angular spectrum intervals which are inefficient for modes are filled by rays and vice versa. As the observer moves either longitudinally or transversely, the ray-mode combination must be readjusted whenever the departure angle of a limiting eigenray coincides with the constituent plane wave propagation angle of a limiting mode. The adjustment can be made by either retaining the same number of rays and altering the number of modes or by retaining the same number of modes and altering the number of rays. The former approach is used in the numerical packages that are explained later on. The hybrid ray-mode package comprises the following constituents:

$$g(x, z; x', z') = \sum_{m=0}^{M_1} G_m(\text{LOM}) + \sum_{n=N_1}^{N_2} G_n(\text{rays}) + \sum_{l=M_2}^{M_3} G_l(\text{HOM}) + \text{Remainder}, \tag{8.56}$$

where the first three terms on the right-hand side correspond to LOMs ($0 \leq m \leq M_1$), rays with finite number of reflections ($N_1 \leq n \leq N_2$), and HOMs ($M_2 \leq m \leq M_3$), respectively. Here, M_1 is the highest mode index of the LOMs, M_2 is the lowest mode index of the HOMs, and M_3 is the highest mode (truncation) index of the HOMs. Also, N_1 and N_2 are the included rays with minimum and maximum numbers of reflections, respectively. Setting N_1 equal to zero results in M_1 equal to zero, which means that the ray-mode mix involves a number of rays plus a few HOMs. If N_2 is large, then M_2 will be equal to M_3, which means that the ray-mode mix involves a few LOMs plus a number of rays.

The package may be used in one of two ways: (1) one can specify the number of eigenrays (i.e., N_1 and N_2) and use the algorithm to add LOM and HOM automatically for a given set of parameters (fix the number of rays and float the number of LOM and HOM); (2) one can specify a sector of angular spectrum (i.e., LOM and HOM) and use the algorithm to add eigenrays automatically (fix the number of LOM and HOM, float the number of eigenrays). Here, the latter option is used for the hybrid ray-mode calculations.

8.4.2 MATLAB Packages: RayMode and Hybrid

Two MATLAB packages have been prepared. The first is the RAYMOD package which is designed to investigate line-source-excited wave fields in terms of ray and mode solutions [5]. The run-time window is shown in Fig. 8.11, depicting the waveguide, eigenray plot, and wave field versus height plot. The geometrical parameters that should be supplied by the user are waveguide (plate) height (m), and distance from the source (m); the operational parameters are the frequency in megahertz and number of reflections (n) for the ray solution. The sliders at the left and right ends of the waveguide are used to specify the source and observation heights. There are three operational buttons. When the first button is clicked, the package calculates the number of propagating modes (i.e., the modes with positive real modal propagation constants) and displays it. When the second button, the "Calculate rays" button, is clicked, the package calculates eigenray angles by using a built-in root search algorithm "fsolve" of MATLAB. This may take time especially when the number of required rays is large. The last operational button, the "Add Ray" button, is reserved to plot eigenrays and wave

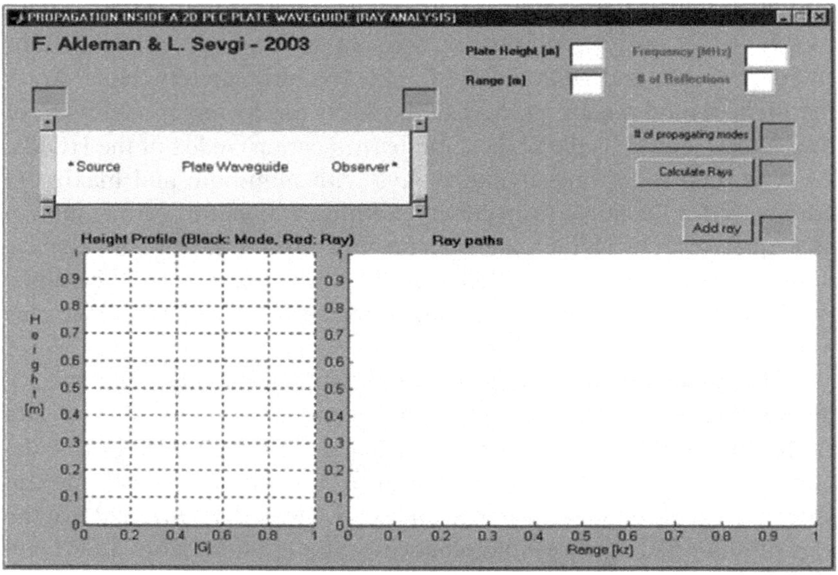

Figure 8.11. Front panel of the RAYMODE package. The user supplies plate height, source/observer heights, horizontal distance, frequency, and number of reflections (n) for the ray solution. The package calculates the modal (reference) solution and searches for the eigenrays for the specified geometrical parameters.

field variations. It should be noted that the "Add Ray" button must be used after the calculations.

Two graphics are reserved for the output. Eigenrays between source and observer can be plotted on the bottom right, whereas wave field versus height at a specified range can be plotted in terms of both mode solutions and ray solutions on the bottom left. The package is designed in such a way that eigenrays with different numbers of reflections, and their contributions, can be plotted separately so that the user may analyze the effects of the group of eigenray species. Each time the user changes any of the parameters, the calculations must be repeated by clicking the operation "Calculate Rays" button. This initiates both the mode and ray calculations.

The second package is the HYBRID package and is designed to calculate wave field versus range at a constant height and/or wave field versus height at a specified range via the mode representation, ray representation, and the hybrid form. The run-time window is shown in

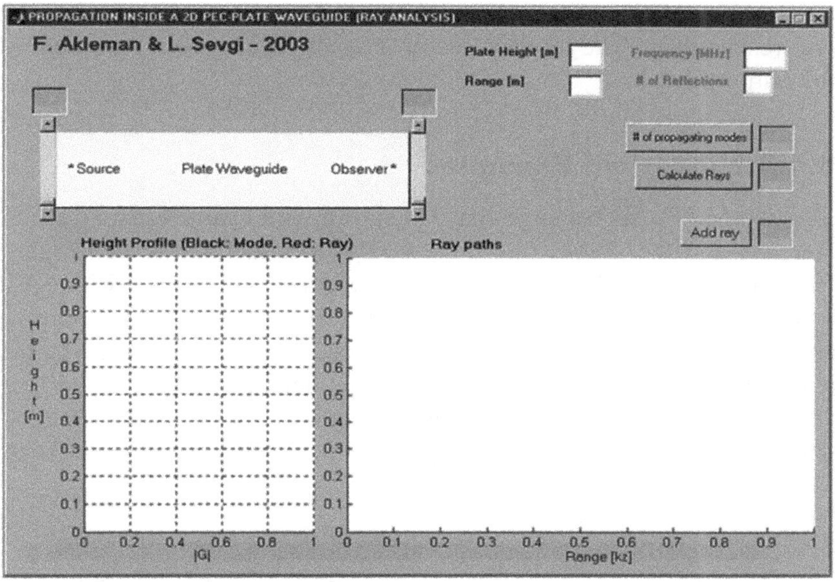

Figure 8.12. The front panel of the HYBRID package. Field versus height and/or range can be calculated via (i) mode summation, (ii) ray summation, and (iii) hybrid ray-mode formulation.

Fig. 8.12. It looks similar to the RAYMODE package. The waveguide schematic, input windows for waveguide height, frequency, and the sliders for the source/observer heights are exactly the same as in the RAYMODE package. The user must supply three parameters for the range calculations: initial range (m), final range (m), and the range increment (m). The last two parameters are required for the hybrid ray-mode analysis. The minimum and maximum numbers of reflected rays is stored to provide N_1 and N_2 in (8.51), respectively.

Two graphics are used in the HYBRID package, plotting either the range profile or the height profile of the line source wave field. The two "Calculate" buttons are used to perform the calculations first and then to plot the profiles. The range profile is plotted from the given initial range to the final range with the range increment supplied. The height profile is calculated for the last range only and is plotted on the right graphics. The number of LOM and HOM at each range for wave field versus range calculations, and at each height for wave field versus height calculations, is stored in the output files, *range.dat* and *height.dat*, respectively. The angular sector at each point is also recorded in the output files.

8.4.3 Numerical Examples

The two MATLAB packages have been run with characteristic parameter sets and the outputs are presented in this section. Tests are performed with the RAYMODE package; results are given in Fig. 8.13 and Fig. 8.14, for the same set of parameters. The waveguide width is chosen to be 1 m. The heights of source and observer are 0.3 and 0.7 m, respectively. Range is 5.6 m and the frequency is 2387 MHz (i.e., $ka = 50$). Rays with up to the tenth reflection are used in the ray representation. The number of propagating modes for these parameters is

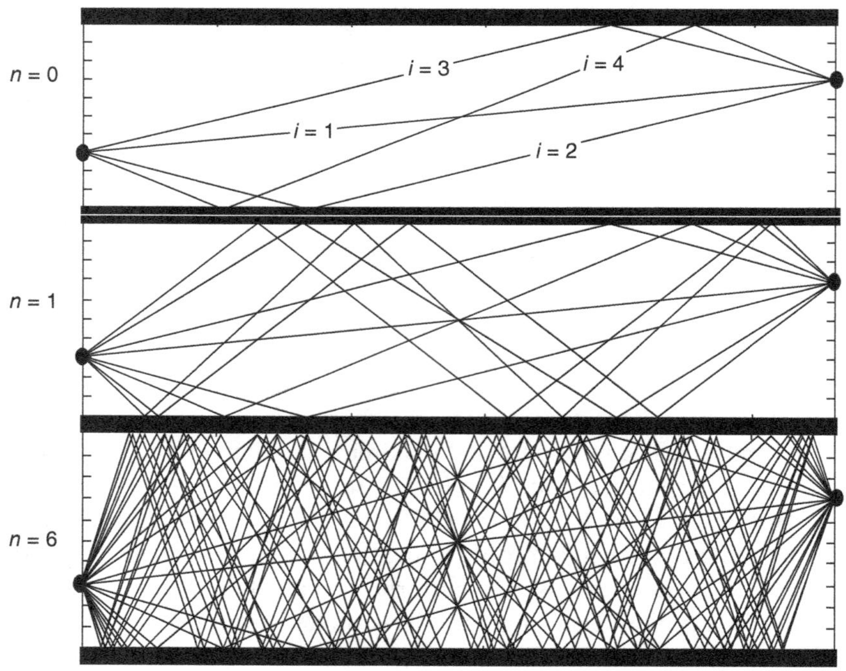

Figure 8.13. Eigenray paths calculated via RAYMODE for the following parameters: $a = 1$ m, distance = 5.6 m, source height = 0.3 m, observer height = 0.7 m, $f = 2387$ MHz ($ka = 50$), $n = 6$. Top: $n = 0$, middle: $n = 1$, bottom: $n = 6$.

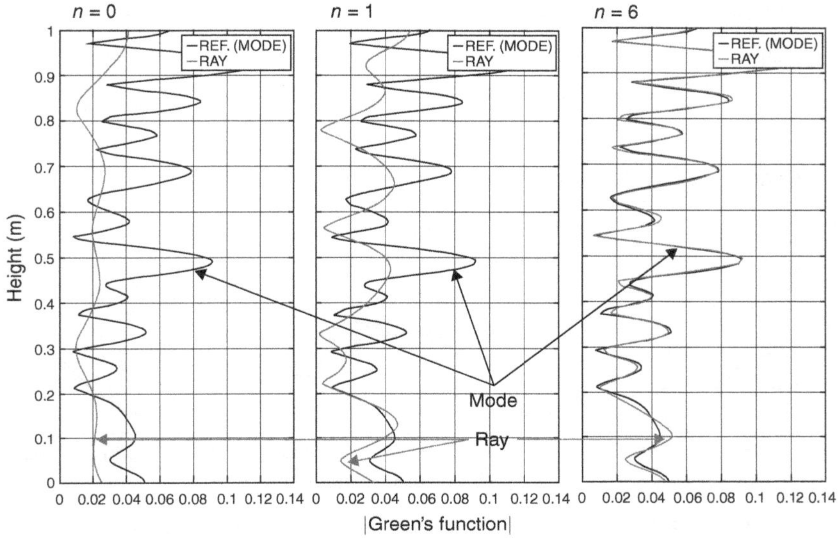

Figure 8.14. Field versus. height calculated via RAYMODE. The parameters are $a = 1$ m, distance = 5.6 m, source height = 0.3 m, $f = 2387$ MHz ($ka = 50$). Left: $n = 0$, middle: $n = 1$, right: $n = 6$.

15. Figure 8.13 shows eigenray photos with different numbers of reflections ($n = 0$, $n = 1$, $n = 6$).

As observed from the figure, eigenrays with $n = 0$ and $n = 1$ are far from filling the entire angular spectrum. On the other hand, eigenrays with up to six reflections (i.e., $n = 6$) cover the angular spectrum almost adequately. Eigenrays with $n = 0$ include the direct ray, the ray with a single reflection from the bottom boundary, the ray with a single reflection from the top boundary, and the ray with a single reflection from both boundaries.

Height profiles of the wave field at 5.6 m range with different numbers of eigenrays are plotted in Fig. 8.14. Summation of 15 propagating modes may be considered as the reference solution since all higher-order nonpropagating modes contribute negligibly at this range (which corresponds to a distance of a few hundred wavelengths). Whenever the "Add ray" button is clicked, a group of eigenrays with one higher reflection number is added. The figure again displays results for three different n values ($n = 0$, $n = 1$, $n = 6$). When the number

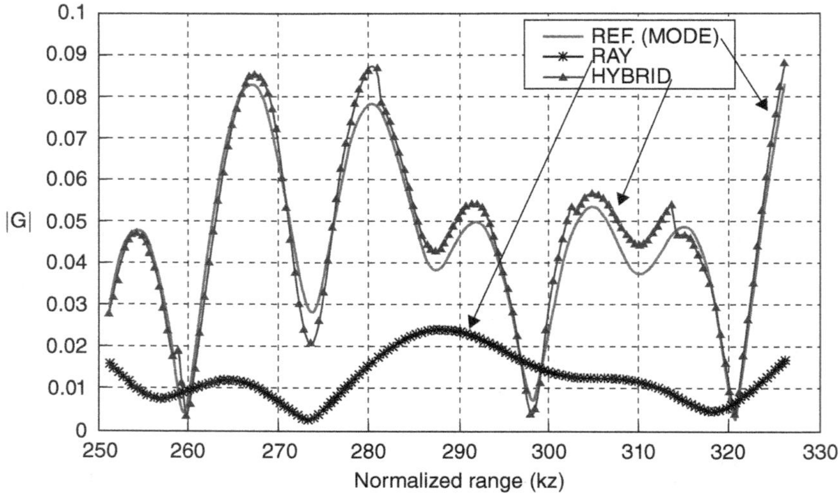

Figure 8.15. Field versus range calculated via the HYBRID package.

of eigenrays increases to 24 or higher, the ray summation results approach the mode solution. The difference between field plots for $n = 6$ and $n > 6$ is almost indistinguishable for the set of chosen parameters.

The package HYBRID is designed to test the sensitivity of various hybrid ray-mode combinations to frequency and source/observer locations. Figure 8.15 and Figure 8.16 display two typical results for the following parameters: waveguide width = 1 m, source height = 0.2 m, observer height = 0.5 m, frequency = 2400 MHz, first range = 5 m, last range = 6.5 m, and range increment = 0.01 m. Eigenrays with $n = 2$, 3, and 4 ($N_1 = 2$, $N_2 = 4$) have been chosen for inclusion in the hybrid ray-mode formulation. In this case, the number of propagating modes is 15 and all higher-order nonpropagating mode contributions are negligible in the range of interest. The range profile is calculated first and is plotted in Fig. 8.15, the height profile at the final range is calculated next, and both profiles are plotted in Fig. 8.16. Reasonably good agreement is obtained between the mode sum (the reference solution) and the hybrid solution. Ranges and heights where discrepancies occur require the addition of the remainder term.

The hybrid ray-mode results presented here are based on filling the angular spectrum partially with modes and rays in a few different ways.

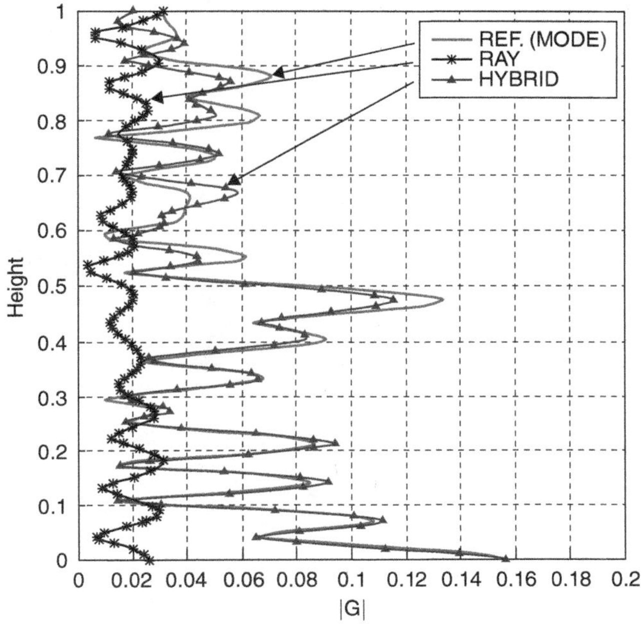

Figure 8.16. Field versus height calculated via the HYBRID package.

The modal eigenangles are determined from the waveguide width (a) and the frequency (f), and are constant once they are specified. On the other hand, eigenray spectral angles are source/observer dependent. The longer the range, the larger is the ray eigenangle for a fixed height and frequency (recall that the eigenray angle is measured from the vertical axis, not the horizontal axis). Therefore, different numbers of LOM and HOM may be required as the range changes in order to fill the angular spectrum left empty by the eigenrays. To illustrate this, the range in the first example (range $=$ 5.6 m) is more than doubled (range $=$ 12 m) and eigenray contributions with 10 reflections and 19 reflections are plotted in Fig. 8.17 and Fig. 8.18. Although eigenrays with up to 10 reflections give quite good results at the range of 6 m for the given parameters, they are incapable of reproducing the reference solution calculated from the first 15 propagating modes. Even increasing the number of reflections of the eigenrays to 19 (i.e., total of 72 eigenrays) does not meet the requirement. It is obvious that more than 20 eigenrays are required to construct acceptable results.

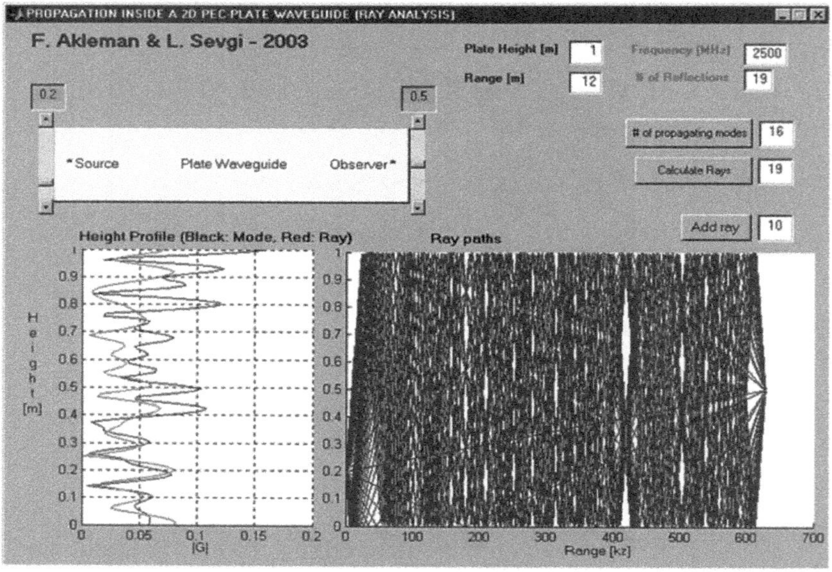

Figure 8.17. Eigenrays and wave field versus height obtained via RAYMODE package. All parameters are the same as in Fig. 8.15, except the range is more than doubled. The number of reflections $n = 10$.

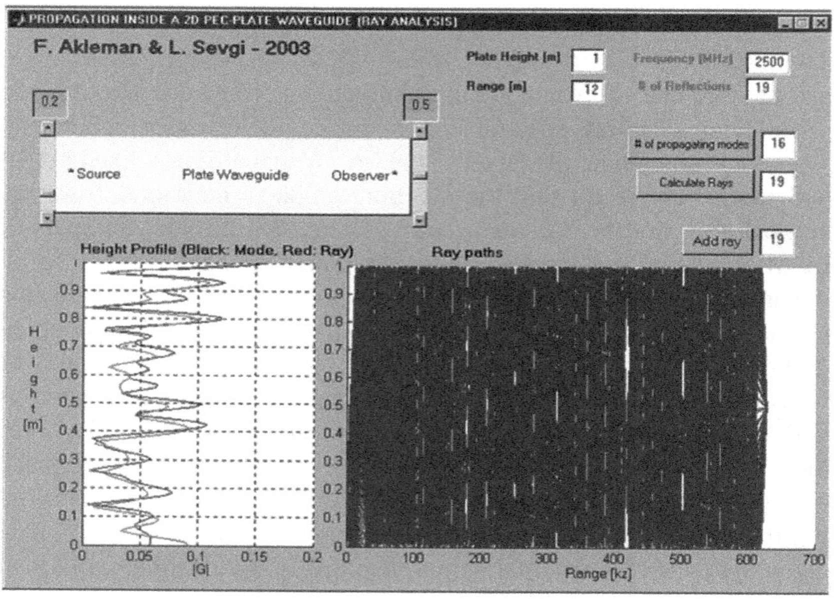

Figure 8.18. Same as in Fig. 8.14, but with $n = 19$.

8.5 EIGENVALUE EXTRACTION FROM PROPAGATION CHARACTERISTICS

A generic problem for a source-driven structure is the GF problem. The GF is the response of a linear system to a point source of unit strength. The general form of a GF problem may be postulated via a second-order nonhomogeneous differential equation with a delta function on the right-hand side, together with BC that represents the geometry of the problem. Omitting the delta function on the right-hand side (i.e., using homogeneous second-order differential equation and BC) yields the source-free (i.e., an eigenvalue) problem. An eigenvalue problem is to find out a set of complete functions that characterize the geometry of the problem (and yields eigenmodes and eigenspectral wave numbers that describe source-free wave fields which individually satisfy BC in wave-confining environments). Many of EM problems, either source-free or source driven, can be represented by a unique second-order differential equation: the Sturm–Liouville (SL) equation. Solving SL equation when λ is a fixed parameter yields nothing but the GF. GF solution may also be built from eigenfunction solutions. Alternatively, eigenfunctions may also be derived from the GF solution. It illustrates that a guided wave problem may be handled as either an eigenvalue problem or a GF problem.

8.5.1 Longitudinal Correlation Function

It is interesting to note that the eigenvalues of the problem can be extracted from the propagation characteristics as an alternative to the link between source-free and source-driven formulations [7–11]. Any transverse wave function (e.g., a transmitting antenna field profile) accumulates eigencharacteristics of the medium while longitudinally propagating. Consider

$$F(x, z_0) = \sqrt{\frac{2}{d}} \sum_{n=1}^{NM} K_n \sin\left(\frac{n\pi}{d} x\right) e^{-j\beta_n z_0} \tag{8.57}$$

as the eigenfunction expansion of the initial profile at the initial range $z_0 = 0$, with excitation the coefficients K_n specified from

$$K_n = \int_{x=x_1}^{x=x_2} F(x, z_0) U_n(x) dx, \tag{8.58}$$

and *NM* is the maximum number of modes that are required to represent the initial field profile "adequately." A longitudinal correlation function may be built while propagating this field profile through the waveguide

$$\begin{aligned} P(z) &= \int_{x=0}^{x=d} F(x, z_0) F(x, z)\, dx \\ &= \sum_{n=1}^{NM} K_n^2 e^{-j\beta_n z} \int_{x=0}^{x=d} \sqrt{\frac{2}{d}} \sin^2\left(\frac{n\pi}{d} x\right) dx = \sum_{n=1}^{NM} K_n^2 e^{-j\beta_n z}. \end{aligned} \tag{8.59}$$

Since the integral over transverse domain is equal to "1" (because of the orthonormality condition), the Fourier transformation of this correlation function yield, the line spectra

$$P(\beta) = \sum_{n=1}^{NM} K_n^2 \delta(\beta - \beta_n), \tag{8.60}$$

that is, the eigenvalues and the excitation coefficients. However, since the data are available only in a finite range interval, a suitable windowing function better be applied to reduce aliasing effects. The Fourier transform may be performed numerically via discrete Fourier transform (DFT) or fast Fourier transform (FFT); therefore, the longitudinal correlation function should be observed long enough to meet maximum longitudinal wave number ($\beta_{\max}$) and wave number resolution ($\Delta\beta$) via the relations (i.e., Nyquist sampling criteria)

$$\beta_{\max} = \frac{\pi}{2\Delta z} \quad \text{and} \quad \Delta\beta = \frac{2\pi}{z_{\max}}, \tag{8.61}$$

where Δz and $z_{\max}$ are range step size and maximum range, respectively. The three-step eigenvalue extraction procedure is as follows:

- First give any transverse source profile and find out the number of modes and their excitation coefficients using the orthonormality condition. A line source is the limiting case, and according to the completeness relation, it requires infinite number of modes.
- Use (8.57) and calculate transverse profiles analytically at a number of range points.
- Correlate them with the initial profile and build the longitudinal correlation function. Then, apply FFT to this correlation function and extract modal eigenvalues and excitation coefficients.

8.5.2 Numerical Illustrations

A simple MATLAB code is prepared for the numerical illustrations. Figure 8.19 shows the flowchart of the eigenvalue extraction procedure. A short eigenvalue extraction MATLAB script is listed in Table 8.4.

The code builds a given vertical input field profile in terms of eigenfunctions with the help of numerically computed excitation coefficients and then propagates it longitudinally. While propagating, it constructs the longitudinal correlation function as defined in (8.58). Finally, the FFT is applied to this correlation function and line spectrum is obtained. The package assumes a Gaussian initial field profile where *hh* and *tt* are vertical source position and vertical source extend, respectively. The user supplies initial parameters: width of the parallel plate (*d*), operating frequency (*fr*), vertical source position *hh*, vertical source extend *tt*, and the number of modes *mm*. The number of vertical points (*nx*) is chosen as 128 in the package. Once the user-supplied parameters

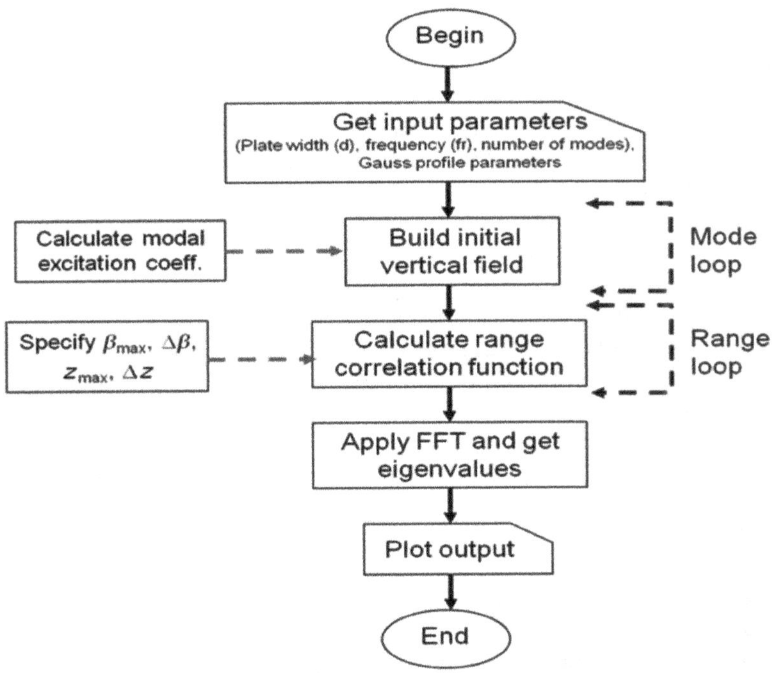

Figure 8.19. The flowchart of the MATLAB package.

Table 8.4.
Eigenvalue Extraction MATLAB Code

```
% --------------------------------------------------------
%  Program: EigenExtract.m
%  Purpose: To derive eigenvaues from longitudinal
correlation func
% --------------------------------------------------------
clear all; clc;
d = 1; %input('Plate height (e.g., d = 1) = ?');
fr = 8e8; %input('Frequency (e.g., fr = 8e8) = ?');
hh = .77; %input('Vertical source location (e.g., hh =
0.77) = ?');
tt = 0.1; %input('Vertical source extend (e.g., tt =
0.1) = ?');
mm = 20; % input('Number of Modes (e.g, mm = 20) = ?');
% ============== Calculate other parameters
==============
nx = 128; delx = d/(nx-1); c = 3.E8; k0 = (2*pi*fr)/c;
for k=1:nx    % Source profile
    xx(k)=(k-1)*delx; sgauss(k)=exp(-((xx(k)-hh)/tt)^2);
end
mp = 0;
for n = 1:mm   % propagation constants
    if k0 > ((n*pi)./d)
        beta(n) = sqrt(k0^2-(n*pi./d)^2); mp = mp+1;
    else
        beta(n) = i*sqrt((n*pi./d)^2-k0^2);
    end
end
fprintf(1,'Number of propagating modes = %4.0f \n', mp);
for m = 1:mm   % Use orthonormality for excitation
coefficients
    for n = 1:nx
        ff(n) = sgauss(n)*sqrt(2./d)*sin((m*pi*xx(n))/d);
    end
    kn(m) = trapz(ff,xx);
end
% Calculate Correlation Parameters
betamax = 1.25*k0; dbeta =
0.05*(abs(beta(mp-1))-abs(beta(mp)));
nz0 = 2.*betamax/dbeta; k = round((log(nz0)+1))+1; nz
= 2^k;
delz = pi/betamax; zmax = nz*delz; dbeta = 2.*pi/zmax;
```

Table 8.4.
(Continued)

```
for k = 1:nx    % Build initial profile
   ff(k) = 0.0;
   for n = 1:mm

ff(k)=ff(k)+kn(n)*sqrt(2./d)*sin((n*pi*xx(k))./d);
   end
end
% Build longitudinal correlation function P(z)
for kk = 1:nz
  zz(kk)=(kk-1)*delz; xbeta(kk) = -betamax +
(kk-1)*dbeta;
  for k = 1:nx
    gg(k) = 0.0;
    for n = 1:mp
       gg(k) =
gg(k)+kn(n)*sqrt(2./d)*sin(n*pi*xx(k)/d). . .
                  *exp(i*beta(n)*zz(kk));
    end
  end
  p = ff.*gg; corel(kk) = trapz(xx,p);
end
for kk = 1:nz % Apply Hanning window
   corel(kk) = 0.5*(1.-cos(2.*pi*zz(kk)/
zmax))*corel(kk);
end
% ======= Take fft of correlation function ========
Corelf0=fft(corel,nz); corelf=fftshift(corelf);
% Plot output
figure (1);
plot(sgauss,xx,abs(ff),xx,'r-','LineWidth',2,'MarkerSize',5)
xlabel('Amplitude'); ylabel('Plate height [m]');
figure (2); plot(abs(gg),xx, 'LineWidth',2)
xlabel('Amplitude'); ylabel('Plate height [m]');
figure (3);      % Correlation function vs. range
plot(zz,corel, 'LineWidth',2)
xlabel('Range [m]'); ylabel('Correlation Func');
figure (4);      % Line spectra vs. eigenvalue
bmax = betamax + betamax/5.;
plot(xbeta,abs(corelf0), 'LineWidth',2); xlim([0,bmax]);
xlabel('Long. Wavenumber (Beta)'); ylabel('Eigen. Ampl.');
% ---------------------- END -------------------------
```

are in, the package computes vertical step (*delx*) and the wave number (k_0). The 1D arrays *xx(nx)*, *sgauss(nx)*, and *beta(mm)* are reserved for the vertical points, initial Gauss field values at these points, and *mm* eigenvalues for both propagating and nonpropagating modes. The number of propagating modes (*mp*) (with real eigenvalues) is displayed by the package. Then, excitation coefficients (*kn(mm)*) are calculated by using the orthonormality property of the eigenfunctions, and the initial field profile is reproduced (and stored in the array *ff(nx)*) in terms of *mm* eigenfunctions. The first plot in the package gives initial height profiles *sgauss* and *ff* versus *xx*. The user may change *mm* and observe graphically whether or not the number of given modes is adequate to represent the given Gaussian field profile.

An example is given in Fig. 8.20 for the parameters $d = 1$ m, $fr = 800$ MHz, $h = 0.77$ m. Eigenvalues calculated analytically are also shown in the figure. As observed, eigenvalues are extracted accurately.

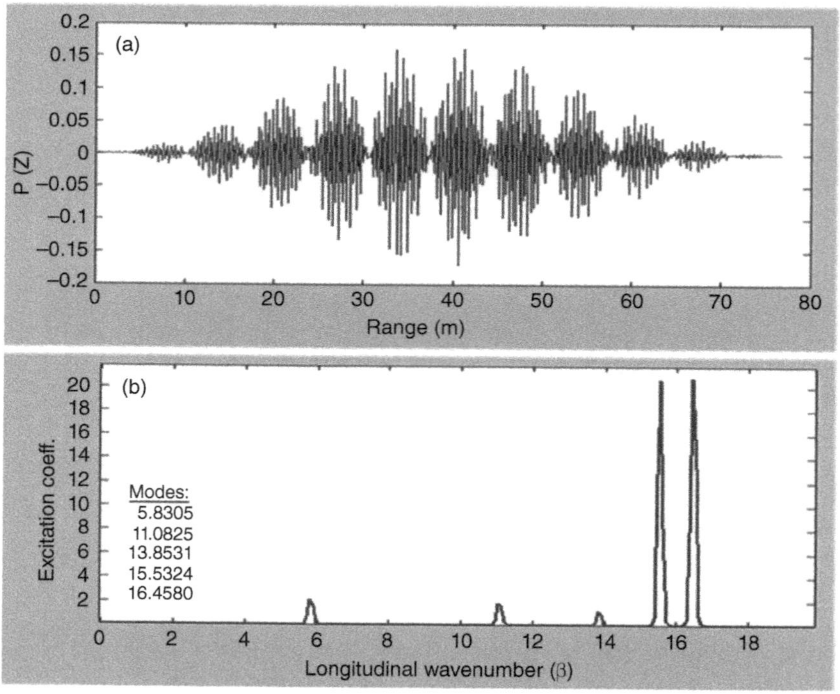

Figure 8.20. (a) Windowed longitudinal correlation function and (b) the frequency spectrum of the correlation function (eigenvalues are also listed).

8.6 TILTED BEAM EXCITATION

As discussed earlier, line source excitation necessitates infinite number of modes in mode summation model. In practice, most of antennas have finite sector beam-type radiations; therefore, Gaussian source pattern [6]

$$u(x) = U_0 e^{-ik_0 x \sin\theta_{tilt}} e^{\frac{\sin^2\left(\frac{\theta_{bw}}{2}\right)k_0^2(x-h_s)^2}{(2\ln 2)}} \tag{8.62}$$

is often used in applications. Here, U_0 is the antenna amplitude, h_s is the height of the source, θ_{bw} is the beamwidth, and θ_{tilt} is the tilt angle.

The Gaussian antenna pattern can also be defined in the vertical wave number domain as

$$U(k_x) = \exp\left[\frac{-k_x^2 \ln 2}{2k_0^2 \sin^2(\theta_{bw}/2)}\right]. \tag{8.63}$$

The tilt (or elevation) angle θ_{tilt} is introduced by shifting the antenna pattern, that is, $u(k_x) \rightarrow u(k_x - k_0 \sin\theta_{\text{tilt}})$. The vertical field in the spatial domain is then obtained by taking the inverse Fourier transform of (8.63).

Once the antenna radiation pattern is given, the modal summation is used together with the orthonormality condition and the number of modes and their excitation coefficients are derived numerically for a given error boundary. Note that modal excitation coefficients are real if the antenna pattern has no vertical tilt. These modal excitation coefficients become complex when upslope or downslope tilt is introduced. Table 8.5 lists a simple and short MATLAB code which calculates and plots three-dimensional (3D) map of field strength versus range/height under tilted beam excitation.

Figure 8.21 shows propagation inside a waveguide excited by an antenna having tilted beam-type radiation characteristics. On the top, the scenario deals with a 1-m-wide parallel PEC plate waveguide, excited by a horizontally polarized and 45°-tilted Gaussian source at 3 GHz with 3 dB beamwidth of 3° and located at the midheight of the waveguide. Here, 50 modes is superposed for the representation of this tilted beam antenna. The bottom figure shows a 30°-tilted Gaussian beam at 1 GHz with 3 dB beamwidth of 7° and located at the midheight of the 8-m-wide waveguide. Here, 50 modes are superposed for the representation of this tilted beam antenna.

Table 8.5.
A MATLAB Mode Summation Code with Tilted Beam Excitation

```
function PPlate_BeamTilt
close all; format long, clear all; clc;
% Wave Propagation between parallel plates
%% given Input Specifications
f0=1000;              % Frequency
 MHz]
tt=0.65;              % Beamwidth
 m]
rangekm=0.04;        % Range
 km]
hmax=8;               % Maximum Height                    [m]
hv=4;                 % Transmitter Height                [m]
elv=-30;              % Antenna tilt elevations
 deg]
NX=801; NZ=1001; z1=1e3*rangekm/2;   z2=1e3*rangekm;
h1=hmax/4;
h2=hmax/2; clight=3e8; lam=clight/(f0*1e6); k0=2*pi/lam;
acc=1e-8; dx=hmax/(NX-1); dz=z2/(NZ-1); z=(0:dz:z2).';
xx=(0:dx:hmax).';
%---------- Transverse Gaussian Source profile ----------
sgauss=exp(-(xx-hv).^2/(2*tt^2)).*exp(1i*k0*sin(elv/180*p
i)*xx);
nmfldz=zeros(NX,NZ); err=1; n=0;
while err>acc, n=n+1;
    if k0>((n*pi)/hmax)
        beta(n)=sqrt(k0^2-(n*pi/hmax)^2);
    else
        beta(n)=1i*sqrt((n*pi/hmax)^2-k0^2);
    end
    %---- Use orthonormality & extract excitation
coefficients ----
    kn(n)=sum(sqrt(2/hmax)*chebfun(@(t)(exp(-(t-hv).^2/
(2*tt^2)).*. . .
            (exp(1i*k0*sin(elv/180*pi)*t))).*sin(n*pi*t/
hmax),[0,hmax]));
   %---- Build initial profile in terms of given modes ----
    nm=kn(n)*sqrt(2/hmax)*sin((n*pi*xx)/hmax);
    for ii=1:NZ,
```

Table 8.5.
(Continued)

```
        nmfldz(:,ii)=nmfldz(:,ii)+nm*exp(1i*beta(n)*
(ii-1)*dz);
     end
     err=max(abs(sgauss-nmfldz(:,1)));
end
display(['Number of modes used: ',num2str(n),. . .
     ' with error: ',num2str(err)]);
display('----------------------------------------')
u=nmfldz;   x=xx;   ufield=u;   u=abs(u);u=u/
max(abs(u(:,1)));
f1=figure('Position',[10 45 874 350]);
axes1 = axes('Parent',f1,'LineWidth',2,'FontWeight','b
old',. . .
      'FontSize',16,'XTick',linspace(0,z(end),11),'YT
ick',. . .
       linspace(0,x(end),6));hold on;box on
surf(z,x,u); ylim([0 x(end)]); shading interp;
colorbar;view([0 90]),
cmax=max(u(:)); caxis([cmax/1000 cmax])
colorbar('peer',axes1,'FontWeight','bold','FontSize',16);
xlabel('z [m]','FontWeight','bold'); ylabel('x [m]','FontWeigh
t','bold');
set(gcf,'Renderer','zbuffer'); set(gcf,'PaperPositionMode','a
uto');
%------------------------- END ------------------------
```

8.7 CONCLUSIONS

A simple line-source-excited 2D parallel plate, homogeneously filled waveguide has been used in this chapter to introduce the analytical background for developing alternative GF representations by spectral wave number techniques, emphasizing the complementary roles of progressing (ray-type) and oscillatory (mode-type) wave dynamics, and culminating in the self-consistent hybrid ray-mode approach.

MATLAB visualization packages have been prepared, documented, and used to illustrate the ray, mode, and hybrid ray-mode wave dynamics through parametric numerical experiments. For the hybrid

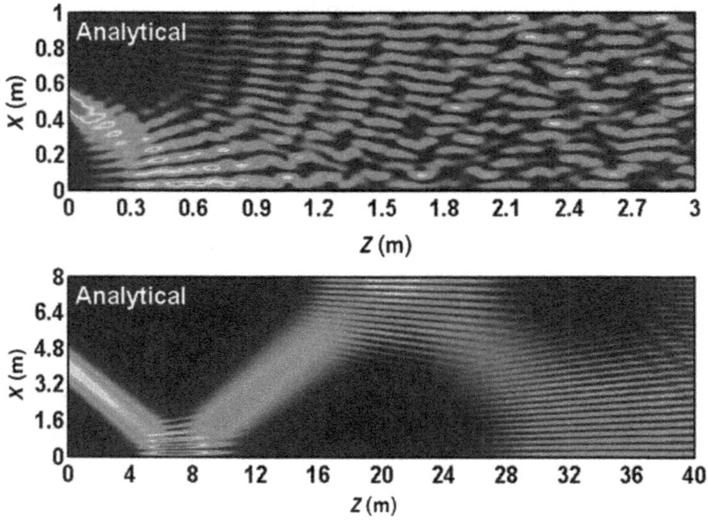

Figure 8.21. 3D color map of field versus range–height variations inside a parallel plate PEC waveguide. (Top) Waveguide height = 1 m, $f = 3$ GHz, beam width = 3°, antenna height = 0.5 m, beam tilt = −45°; (bottom) waveguide height = 8 m, $f = 1$ GHz, beam width = 7°, antenna height = 4 m, beam tilt = −30°.

formulation, it has been shown that there exists a variety of well-defined combinations of mode and ray fields that expresses the total field inside the waveguide with good accuracy. For each combination, the modal fields account for the omitted rays or vice versa; for certain parameter ranges, the addition of a remainder (in easily computable form) may be required. The modes (plus remainder term, when necessary) quantify the truncation error of an incomplete ray series, whereas the rays (plus remainder term, when necessary) quantify the truncation error of an incomplete mode series.

The two MATLAB packages designed to investigate ray, mode, and hybrid options for the canonical parallel plate waveguide problem can serve as effective educational tools toward understanding the physics-based, observable wave objects in this environment. The packages may easily be extended to include the remainder terms and to handle impedance-type boundary conditions. Moreover, based on the experience and physical insight gained here with these packages, the user may extend the algorithms to waveguiding structures with virtual boundaries (i.e., trapping inhomogeneous refractivities), where modal as well as ray solutions undergo critical transitions, such as modal cutoff, ray caustics.

REFERENCES

1. L. B. Felsen, N. Marcuvitz, *Radiation and Scattering of Waves*, Prentice-Hall, Englewood Cliffs, NJ, 1973, Classic reissue IEEE Press, 1994.
2. D. G. Dudley, *Mathematical Foundations for Electromagnetic Theory*, IEEE Press (Series on Electromagnetic Waves), New York, 1994.
3. R. E. Collin, *Field Theory of Guided Waves*, McGraw-Hill, Inc., New York, 1960.
4. L. B. Felsen, A. H. Kamel, "Hybrid Ray-Mode Formulation of Parallel Plate Waveguide Green's Functions," *IEEE Transactions on Antennas and Propagation*, **29** (4), 1981, pp. 637–649.
5. L. B. Felsen, F. Akleman, L. Sevgi, "Wave Propagation inside a Two-Dimensional Perfectly Conducting Parallel Plate Waveguide: Hybrid Ray-Mode Techniques and Their Visualizations," *IEEE Antennas and Propagation Magazine*, **46** (6), 2004, pp. 69–89.
6. G. Apaydin, L. Sevgi, "A Canonical Test Problem for Computational Electromagnetics (CEM): Propagation in a Parallel Plate Waveguide," *IEEE Antennas and Propagation Magazine*, **54** (4), 2012, pp. 290–315.
7. L. Sevgi, *Complex Electromagnetic Problems and Numerical Simulation approaches*, IEEE & John Wiley Press, New York, 2003.
8. M. D. Feit Jr., J. A. Fleck, "Computation of Mode Properties in Optical Fiber Waveguides by a Beam Propagating Method," *Applied Optics*, **19** (7), 1980, pp. 1154–1164.
9. L. Sevgi, E. Topuz, "Split step Parabolic Equation analysis of Coupled Dielectric Waveguides," *ELEKTRIK, Turkish Journal of Electrical Engineering and Computer Sciences*, **3** (2–3), 1995, pp. 85–92.
10. L. Sevgi, "Sturm-Liouville Equation: The Bridge Between Eigenvalue and Green's Function Problems," *ELEKTRIK, Turkish Journal of Electrical Engineering and Computer Sciences*, **14** (2), 2006, pp. 293–311.
11. L. Sevgi, "Numerical Fourier Transforms DFT and FFT," *IEEE Antennas and Propagation Magazine*, **49** (3), 2007, pp. 238–243.

CHAPTER 9

Wedge Waveguide with Nonpenetrable Boundaries

Many natural or man-made guiding environments are characterized by physical parameters that render the wave equation nonseparable in any of the standard coordinate systems. In particular, in the absence of transverse-longitudinal separability, it is not possible to define discrete or continuous normal modes (NMs) that individually satisfy the transverse boundary conditions (BCs) and propagate longitudinally without coupling to other modes. When transverse-longitudinal separability is only weakly perturbed, one may define local (adiabatic) modes which adapt smoothly, without intermode coupling, to the slowly changing conditions. Adiabatic modes (AMs) fail in cutoff regions and can be uniformized there by intrinsic modes (IMs), which are synthesized by a spectral continuum of AM. These concepts have been elucidated and validated previously by investigating the wave dynamics in a simple test environment: a wedge waveguide with nonpenetrable boundaries, viewed either in coordinate-separable cylindrical coordinates that yield exact field solutions in terms of NM or in nonseparable rectangular coordinates that yield approximate field solutions in terms of AM and IM [1,2]. This chapter is intended to tutorially enhance the utility and understand these analytical formulations through visualizations of the dynamical interaction between the various wave species, implemented through an educational MATLAB package. The visualizations for the

Electromagnetic Modeling and Simulation, First Edition. Levent Sevgi.

full range of ray, mode, and hybrid options, parameterized in the spectral wave number domain, has been explored by us previously for the inherently separable canonical environment of a line-source-excited parallel plate waveguide [3]. In our present investigation of the wedge waveguide, we shall not attempt to mimic the variety of options because of the substantial complications and subtleties inherent in their rectilinearly weakly nonseparable implementation. For our purposes here, a single specific option suffices to address the NM, AM, and IM phenomologies. Throughout, the intended audience is expected to be familiar with asymptotic methods for evaluation of integrals.

9.1 INTRODUCTION

Guided wave dynamics can be organized around two complementary phenomenologies—progressing and oscillatory—which are closely related to local versus global descriptions [4]. Progressing wave objects (ray fields or wave fronts) are responsible for point-to-point propagation and sample the physical environment locally along their (ray) trajectories. Oscillatory wave objects (modes or resonances) form standing waves over extended (global) portions of the physical environment.

Two-dimensional (2D) waveguiding environments with transversely and longitudinally varying properties can serve as models in many natural and man-made settings. In such environments, it is desirable to seek propagating modes that maintain their transverse shape in a suitably defined analytical reference frame because the propagation of general wave fields can then be organized systematically around these elements. The ability to define modes is intimately tied to the separability of the wave equation in the given environment.

Normal modes (NMs) are the solutions of the source-free wave equation in a coordinate system that renders the wave equation separable [5]. They satisfy transverse BCs on each transverse cross section. NM confinement may be due to fixed transverse boundaries and/or mode-dependent virtual boundaries (refractive confinement) when the ambient dielectric medium is vertically inhomogeneous. Each NM propagates in the longitudinal direction with a distinct propagation constant, each carries finite energy (can be normalized), and is independent of every other NM (orthogonality).

If no separable reference frame can be found, the concept of individual modes is ill defined, but if conditions along the (longitudinal) propagation axis change *sufficiently slowly*, approximate mode fields can be constructed by assuming *local separability*. AMs are approximate solutions of the source-free wave equation in a reference frame (coordinate system) that renders the wave equation weakly nonseparable [6,7]. AMs satisfy local transverse BCs on longitudinally changing transverse cross sections, adapt themselves to a slowly changing longitudinal environment, propagate in the longitudinal direction with distinct propagation constants, carry finite energy (can be normalized locally), and do not couple to any other AM (local orthogonality). The AM confinement may be due to fixed transverse boundaries and/or mode-dependent virtual boundaries (refractive confinement). Finally, the AMs are valid in regions away from critical transitions, such as the guiding-to-evanescent cutoff transition; they fail in such transitions.

Intrinsic modes (IMs) have been introduced to extend the concept of the AM and repair AM deficiencies near and through cutoff transitions by synthesizing the IM via a spectral continuum of AM [1,2,8–15].

In this chapter, the problem, in general, is defined and formulated in Section 9.2. The special test case of a wedge waveguide with nonpenetrable boundaries, under line source excitation (Green's function [GF] problem), is formulated and solved rigorously (in cylindrical coordinates) and approximately in terms of local rectangular mode superposition in Section 9.3. A self-contained MATLAB-based visualization package WedgeGUIDE is described in Section 9.4. Numerical tests and illustrations are described in Section 9.5. Section 9.6 is reserved for the method of moments (MoM) modeling. An appendix is included for a simplified model of the IM construction. The reader is assumed to be familiar with standard procedures for the asymptotic evaluation of integrals.

9.2 STATEMENT OF THE PROBLEM: PHYSICAL CONFIGURATION AND RAY-ASYMPTOTIC GUIDED WAVE SCHEMATIZATIONS

We consider 2D guiding regions comprising fixed boundaries or interfaces, interior guiding inhomogeneous medium profiles, or both, in

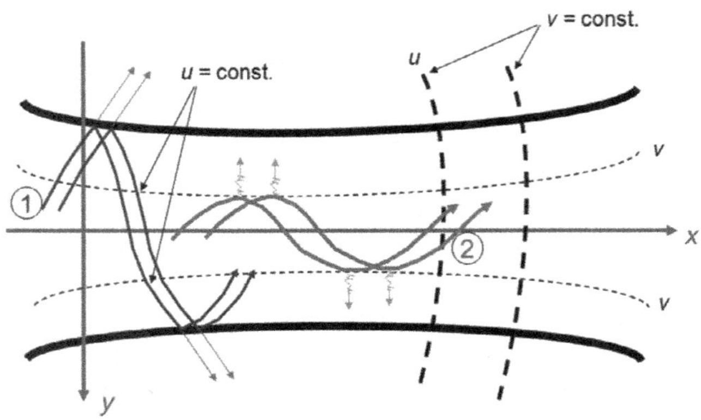

Figure 9.1. Geometry of waveguiding regions comprising fixed boundaries or interfaces, interior guiding inhomogeneous medium profiles, or both, in original (x,y) and better matched, locally separable (u,v) coordinates. Also shown are portions of modal ray congruences pertaining to modes confined by (1) physical fixed boundaries and (2) caustics. If the boundaries are penetrable, fields are refracted into the exterior.

original (x,y) and better-matched locally separable (u,v) coordinates as pictured in Fig. 9.1.

The physical wave numbers in the guiding region are denoted by $k(x,y)$ or $k(u,v)$ in the two respective coordinate systems. Schematics of modal ray congruencies pertaining to modes confined by fixed boundaries are denoted by "1" and those confined by caustics are denoted by "2." If the boundaries are penetrable, ray fields are refracted into the exterior. For the refractively trapped modes confined by interior caustics, the mode field exterior to the caustics is evanescent (wiggly arrows), but it may become leaky with energy transfer into the exterior regions if these are accessible.

9.3 SOURCE-FREE SOLUTIONS

9.3.1 Separable Coordinates: Conventional NM

The local NM problem (for mode index m) may be addressed via the wave equation

$$\left\{\nabla_{u,v}^2 + k^2\right\} \quad W_m(u,v) = 0,$$
$$\nabla_{u,v}^2 = \frac{1}{h_u h_v}\left[\frac{\partial}{\partial u}\frac{h_v}{h_u}\frac{\partial}{\partial u} + \frac{\partial}{\partial v}\frac{h_u}{h_v}\frac{\partial}{\partial v}\right], \tag{9.1}$$

subject to wave-trapping BCs on the u-coordinate and a radiation condition when $\nu \to \pm\infty$. Here, h_u, h_v are the metric coefficients along the coordinates u and v, respectively. If the problem is separable in the (u,v) coordinate frame, wave fields can be obtained by applying the separation-of-variables method,

$$W_m(u,v) = A_m(v)U_m(u). \tag{9.2}$$

The resulting reduced one-dimensional (1D) equations in the spectral wave number domain become

$$\left\{\frac{1}{h_u h_v}\frac{\partial}{\partial v}\frac{h_u}{h_v}\frac{\partial}{\partial v} + \beta_m^2\right\} A_m(v) = 0;\ \text{radiation condition when}\ v \to \pm\infty \tag{9.3}$$

$$\left\{\frac{1}{h_u h_v}\frac{\partial}{\partial u}\frac{hv}{h_u}\frac{\partial}{\partial u} + k^2(u,v) - \beta_m^2\right\} U_m(u) = 0;\quad \text{transverse BC,} \tag{9.4}$$

and can be used to construct the NM solution together with the BCs on u. In (9.3) and (9.4), the "separation constant" β_m is the spectral wave number.

9.3.2 Weakly Nonseparable Coordinates: AM

If the problem is weakly nonseparable along the *v-coordinate* in the (u,v) coordinate frame, *local separability* can be invoked by introducing a highly oscillatory phase function $\bar{U}_m(u,v)$ and a relatively slowly varying amplitude function $\bar{A}_m(v)$ as

$$\bar{W}_m(u,v) = \bar{A}_m(v)\bar{U}_m(u,v), \tag{9.5}$$

where the overbar "–" denotes the AM regime. Local separability on v-dependent cross sections *u(v)* yields

$$\left\{\frac{1}{h_u h_v}\frac{\partial}{\partial v}\frac{h_u}{h_v}\frac{\partial}{\partial v} + \bar{\beta}_m^2(v)\right\} \bar{A}_m(v) = 0;\ \text{radiation condition when}\ \nu \to \pm\infty \tag{9.6a}$$

$$\left\{\frac{1}{h_u h_v}\frac{\partial}{\partial u}\frac{hv}{h_u}\frac{\partial}{\partial u} + k^2(u,v) - \bar{\beta}_m^2(v)\right\} \bar{U}_m(u,v) = 0;\quad \text{transverse BC,} \tag{9.6b}$$

with $\bar{\beta}_m(v)$ now ν-dependent through the specified inhomogeneity profile. Within the Wentzel–Kramers–Brillouin (WKB) high frequency asymptotic (HFA) approximation, the amplitude function $\bar{A}_m(v)$ and the phase function $\bar{\beta}_m(v)$ are given by (Ref. 4, p. 337)

$$\overline{A}_m^{\pm}(v) \approx \frac{e^{\pm j\overline{\psi}_m(v)}}{\sqrt{\overline{\beta}_m(v)}}, \tag{9.7a}$$

$$\overline{\psi}_m(v) = \int_{v_0}^{v} \overline{\beta}_m(v)dv. \tag{9.7b}$$

9.3.3 Uniformizing the AM Near Caustics: IM

The AM solution in (9.7) becomes invalid when

$$\overline{\beta}_m(v) \to 0 \tag{9.8}$$

(i.e., near and within the AM modal propagating-to-evanescent caustic region).

The IM solution can be constructed by postulating a wave number-spectral-integral superposition of the AM functions [8] (for a tutorial simplified derivation, see Appendix 9.A),

$$\begin{aligned}\widehat{W}_m(u, v) &= \int_{C_\beta} \hat{f}_m(\beta)e^{j\hat{P}_m(\beta)}d\beta \\ &= \int_{C_\beta} \overline{a}_m(\beta)\overline{U}_m(u; \overline{v}_m(\beta))e^{-j[\Omega_m(\beta)+\beta(v-\overline{v}_m(\beta))]}d\beta,\end{aligned} \tag{9.9}$$

where the caret "^" denotes the IM regime. In (9.9)

$$\hat{P}_m(\beta) = \Omega_m(\beta) - \beta_m(x - \overline{x}(\beta)) \text{ (see (A.8))}, \tag{9.10a}$$

$$\overline{v}_m\left[\beta_{(m)}(v)\right] = v, \tag{9.10b}$$

$$a_m(\beta) = \left\{\frac{1}{\beta}\frac{d}{d\beta}\overline{v}_m(\beta)\right\}^{1/2}, \tag{9.10c}$$

$$\Omega_m(\beta) = \int_0^{\beta} \beta' \frac{d}{d\beta'}\overline{v}_m(\beta')d\beta' = \int_0^{\overline{v}_m} \beta_{(m)}(v)dv, \tag{9.10d}$$

$$\overline{A}_m(\beta) = A_m\left[\overline{v}_m(\beta')\right]\frac{d}{d\beta}\overline{v}_m(\beta') = a_m(\beta)e^{j\Omega_m(\beta)}. \tag{9.10e}$$

The IM integral in (9.9) together with (9.10), taken along a suitably specified integration path C_β in the complex-β plane, uniformizes the AM field in a transitional guiding region in the same sense that a generalized ray (plane wave spectral) integral uniformizes the spectrally deficient ray field in a transitional region of the *geometrical theory of diffraction* (GTD). Thus, by isolated stationary phase evaluation of (9.9) at the saddle points (SPs) $\beta_m(\nu)$ defined by $\overline{\nu}_m\left[\beta_{(m)}(\nu)\right]=\nu$ in (9.10b), which is valid in *regular* regions, the IM integral should be reducible to the AM field in (9.7). To within these leading order asymptotic considerations that reduce the IM integral to the AM which generate it, the IM construction from the AM is unique. Asymptotic evaluation in transition regions, where the AMs fail due to the confluence of stationary points and (or) singularities, can be performed by reducing the IM integral to the appropriate canonical forms.

More specifically, the IM integral in (9.9) has been constructed from the AM solution in such a manner that lowest order SP evaluation along the local steepest descent path(s) (LSDP)—when the two real SPs $\overline{\beta}_m^{(1)}(\nu)$ and $\overline{\beta}_m^{(2)}(\nu)$ are isolated—reduces the IM to the AM in (9.7) (details are given in the appendix). When the AM travels in the direction of narrowing waveguide cross section (i.e., "upslope propagation"), the two SPs which correspond to waves propagating along $+\nu$ and $-\nu$, respectively, approach one another and coalesce at the cutoff (turning) point determined by $\overline{\beta}_m(\nu)=0$. The cutoff region converts the incoming AM into a reflected AM on the propagating side and into an evanescent field (complex β_m) behind the caustic. The integration path C_β in (9.9) can be chosen so as to traverse one or both SPs, depending on which side of the caustic the observation point resides. The corresponding *uniform asymptotics* near and at the cutoff transition involves Airy functions and their derivatives (Ref 4, p. 410),

$$\widehat{W}_m^{\mathrm{IM}}(u,v)\approx\frac{1}{2}e^{\pm j\Delta}\left[\hat{\Lambda}_1+\hat{\Lambda}_2\right]\mathrm{Ai}(s)+\frac{1}{2\sqrt{s}}e^{\pm j\Delta}\left[\hat{\Lambda}_1-\hat{\Lambda}_2\right]\mathrm{Ai}'(s), \tag{9.11a}$$

$$\hat{\Lambda}_{1,2}=f_m(\beta_{1,2})h_{1,2},\ h_{1,2}=\left[\frac{2\sqrt{s}}{\hat{P}''(\nu,\beta_{1,2})}\right]^{1/2}, \tag{9.11b}$$

$$\hat{\Delta}=\frac{1}{2}\left\{P(\nu,\overline{\beta}_m^{(1)})+P(\nu,\overline{\beta}_m^{(2)})\right\},\ s=-\left\{\frac{3}{4}\left[\hat{P}''(\nu,\beta_1)-\hat{P}''(\nu,\beta_2)\right]\right\}^{2/3}, \tag{9.11c}$$

$$\hat{f}_m(\beta)=\overline{a}_m(\beta)\overline{U}_m(u;\overline{\nu}_m(\beta)). \tag{9.11d}$$

Here, Ai(.) is the Airy function and the prime denotes the derivative with respect to the argument. This uniform asymptotic representation given in (9.11) is valid in *all* regions where the two SPs are either isolated or coincide. A simpler "less uniform" version valid *in the vicinity* of the caustic is given by (Ref. 4, p. 591)

$$\hat{W}_m^{\rm IM}(u,v) \approx 2\pi\left[\frac{2}{P'''(\nu,\beta_0)}\right]^{1/3} f(\beta_0)\mathrm{Ai}(\tau)e^{j\Delta}, \quad \beta_0=0 \quad (9.12a)$$

$$\tau \approx -\left[\frac{1}{2}P'''(\nu,\beta_0)\right]^{2/3}\left\{\frac{\beta_m^{(1)}-\beta_m^{(2)}}{2}\right\}^2 \quad (9.12b)$$

$$\hat{\Delta} = \frac{1}{2}\left\{\hat{P}(\nu,\bar{\beta}_m^{(1)})+\hat{P}(\nu,\bar{\beta}_m^{(2)})\right\}, \quad (9.12c)$$

and *at* the caustic by

$$\hat{W}_m^{\rm IM}(u,v) \approx 2\pi\left[\frac{2}{\hat{P}'''(\nu,\beta_0)}\right]^{1/3} f(\beta_0)\mathrm{Ai}(0). \quad (9.12d)$$

9.4 TEST PROBLEM: THE 2D LINE-SOURCE-EXCITED NONPENETRABLE WEDGE WAVEGUIDE

9.4.1 Exact Solution in Cylindrical Coordinate

9.4.1.1 The Reduced Field Equations

Referring to the waveguide configuration and coordinates pictured in Fig. 9.2, the 2D z-directed line-source-excited vector Maxwell equations in cylindrical coordinates (with suppressed $\exp(j\omega t)$ time dependence and $\partial/\partial z \equiv 0$) contain only three nonvanishing scalar field components: E_z, H_ρ, H_φ [4]. E_z is determined from the source-excited 2D wave equation,

$$\left[\nabla_{\rho\varphi}^2+k^2\right]E_z = j\omega\mu I_0\frac{\delta(\rho-\rho')\delta(\varphi-\varphi')}{\rho}, k=\omega\sqrt{\varepsilon\mu}, \quad (9.13a)$$

$$\nabla_{\rho\varphi}^2 = \frac{1}{\rho}\left[\frac{\partial}{\partial\rho}\left(\rho\frac{\partial}{\partial\rho}\right)+\frac{1}{\rho^2}\frac{\partial^2}{\partial\varphi^2}\right], \quad (9.13b)$$

subject to the BCs:

$$E_z \text{ finite as } \rho\to 0, \text{ radiation condition at } \rho\to\infty, \quad (9.14)$$

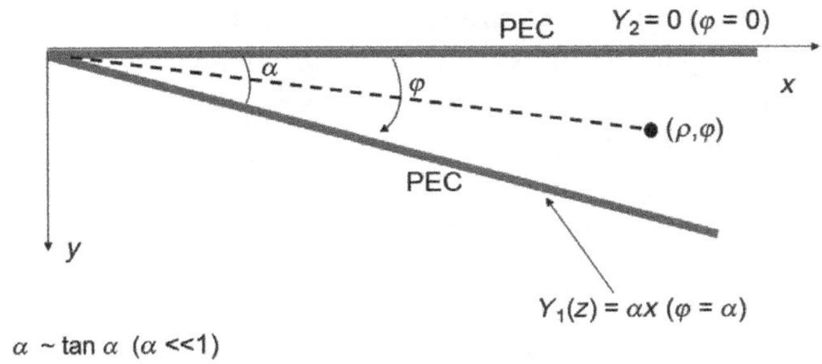

Figure 9.2. Physical configuration and coordinate designation for homogeneously filled waveguide with weakly wedge-tapered PEC boundaries (weakly nonseparable in rectangular coordinates: $[u,v] \rightarrow [x,y]$. Strictly separable in cylindrical coordinates: $[u,v] \rightarrow [\rho,\varphi]$). Note that the "height" coordinate y increases in the downward direction.

$$E_z = 0 \text{ at } \varphi = 0, \alpha,$$

with I_0 representing the line current amplitude. The remaining field components are then obtained from

$$H_\rho = -\frac{1}{j\omega\mu\rho}\frac{\partial}{\partial\varphi}E_z, \tag{9.15}$$

$$H_\varphi = \frac{1}{j\omega\mu}\frac{\partial}{\partial\rho}E_z. \tag{9.16}$$

Using the normalization

$$G = -\frac{E_z}{j\omega\mu I_0} \tag{9.17}$$

in (9.13a) yields the 2D GF problem for the wedge-type waveguide with apex angle α,

$$\left\{\frac{1}{\rho}\frac{\partial}{\partial\rho}\left(\rho\frac{\partial}{\partial\rho}\right) + \frac{1}{\rho^2}\frac{\partial^2}{\partial\varphi^2} + k^2\right\}G(\rho,\varphi;\rho',\varphi') = -\frac{\delta(\rho-\rho')}{\rho}\delta(\varphi-\varphi'), \tag{9.18}$$

subject to the BCs in (9.14),

$$G(\rho,\phi;\rho',\phi') = 0 \text{ at } \phi = 0, \alpha,$$
$$G\begin{cases} \text{bounded at } \rho \rightarrow 0, \\ \text{radiation cond. at } \rho \rightarrow \infty \end{cases}. \tag{9.19}$$

9.4.1.2 GF Solution: Alternative Field Representations

In coordinate-separable physical configurations, the method of characteristic (spectral) GFs (set in the complex spectral wave number plane) furnishes a concise framework for generating alternative representations of the field solutions governed by (9.13a,b) with (9.14) (Ref. 4, Chapter 3) [5]. Here, the one which best suits the problem is chosen.

GF Solution:

$$G(\rho,\varphi;\rho',\varphi') = \sum_{m=1}^{\infty} f_m(\varphi) f_m(\varphi') g_\rho(\rho,\rho';k^2-\eta_m^2). \qquad (9.20)$$

In (9.20), $f_m(\varphi)$ denotes a normalized eigenfunction in the φ-domain with spectral eigenvalue η_m and g_ρ is the 1D characteristic GF in the ρ-domain.

1D Problem in ρ-Domain: Characteristic GF Problem. This is

$$\left\{\frac{1}{\rho}\frac{d}{d\rho}\left(\rho\frac{d}{d\rho}\right) + k^2\rho - \frac{\eta^2}{\rho}\right\} g_\rho(\rho,\rho';\eta) = -\frac{\delta(\rho-\rho')}{\rho}, \qquad (9.21a)$$

with g_ρ bounded at $\rho \to 0$,

$$\text{radiation condition at } \rho \to \infty, \qquad (9.21b)$$

and η_m specified in (9.26a).

The solution of (9.21) is given by (Ref. 4, chapter 3)

$$g_\rho(\rho;\rho';\eta) = \frac{\overleftarrow{\psi}(\rho_<,\eta)\overrightarrow{\psi}(\rho_>,\eta)}{W(\overleftarrow{\psi},\overrightarrow{\psi})}, \qquad (9.22a)$$

$$W(\overleftarrow{\psi},\overrightarrow{\psi}) \equiv \overleftarrow{\psi}\frac{d\overrightarrow{\psi}}{d\rho} - \overrightarrow{\psi}\frac{d\overleftarrow{\psi}}{d\rho}, \qquad (9.22b)$$

with

$$\overleftarrow{\psi}(\rho_<,\eta_m) = J_{\eta_m}(k\rho), \qquad (9.23a)$$

$$\overrightarrow{\psi}(\rho_>,\eta_m) = H^{(2)}_{\eta_m}(k\rho). \qquad (9.23b)$$

Here, W is the Wronskian, while $\overleftarrow{\psi}$, $\overrightarrow{\psi}$ are source-free solutions satisfying the BCs at $\rho = 0$ and $\rho \to \infty$, respectively. The result is

$$g_\rho(\rho,\rho';\eta) = \frac{-j\pi}{2} J_\eta(k\rho_<) H^{(2)}_\eta(k\rho_>), \qquad (9.24a)$$

where

$$\rho_< = \begin{cases} \rho & \text{for} \quad \rho < \rho' \\ \rho' & \text{for} \quad \rho > \rho' \end{cases}, \quad \rho_> = \begin{cases} \rho & \text{for} \quad \rho > \rho' \\ \rho' & \text{for} \quad \rho < \rho' \end{cases}. \tag{9.24b}$$

1D Problem in φ-Domain: Eigenvalue Problem. The φ-domain eigenvalue problem is defined by

$$\left(\frac{d^2}{d\varphi^2} + \eta^2\right) f_m(\varphi) = 0, \tag{9.25a}$$

with

$$f_m(\varphi) = 0 \text{ at } \varphi = 0, \alpha. \tag{9.25b}$$

The corresponding completeness relation for the discrete set of eigenfunctions $f_m(\varphi)$ with spectral wave numbers (eigenvalues) η_m is given by

$$\delta(\varphi - \varphi') = \sum_{m=1}^{\infty} f_m(\varphi') f_m(\varphi), \tag{9.26a}$$

$$f_m(\phi) = \sin(\eta_m \phi), \quad \eta_m = \frac{m\pi}{\alpha}. \tag{9.26b}$$

2D Problem in (ρ,φ) Domain. Combining (9.24) and (9.26) yields

$$G(\rho, \rho', \phi, \phi') = \frac{-j\pi}{\alpha} \sum_{m=1}^{\infty} f_m(\phi) f_m(\phi') J_{\eta_m}(k\rho_<) H^{(2)}_{\eta_m}(k\rho_>). \tag{9.27}$$

This is the exact solution for line-source-excited propagation inside a wedge-type waveguide having an arbitrary apex angle α and PEC boundaries. It should be noted that for upslope propagation toward the apex, the mode fields in (9.27) undergo a cutoff transition because the Bessel function $J_{\eta_m}(k\rho)$ decays when $k\rho \leq \eta_m$. Using the terminology of ray theory, this implies the existence of a modal caustic at [16]

$$\rho_{cm} = \frac{\eta_m}{k} = \frac{m\pi}{k\alpha}. \tag{9.28}$$

The geometrical illustration of caustic formation inside the wedge waveguide is given in Fig. 9.3. A few numerical examples are presented at this stage to illustrate the behavior of the eigenfunctions for different choices of the physical parameters. Consider a 3° wedge waveguide

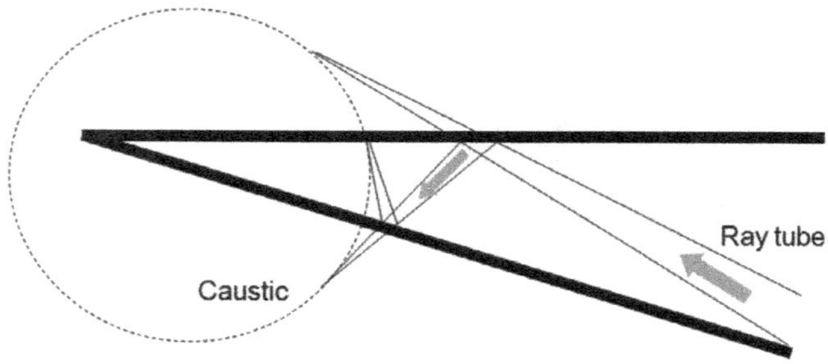

Figure 9.3. Wedge waveguide and modal caustic formation, showing track of incident-reflected modal ray tube.

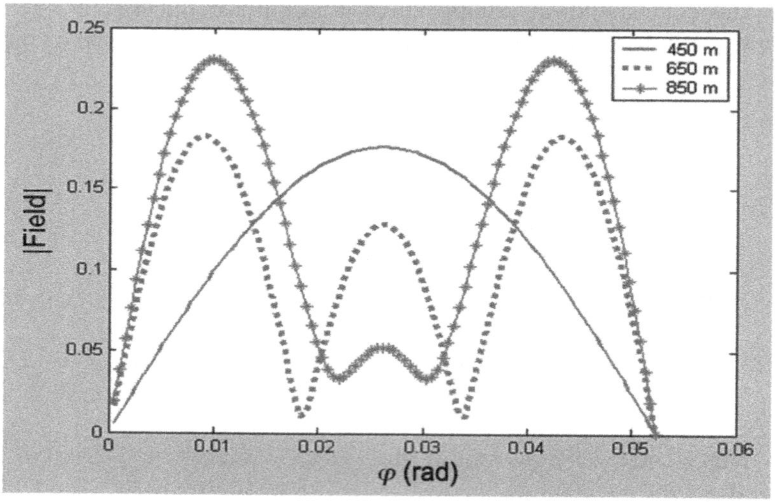

Figure 9.4. Transverse mode fields at different ranges.

and a frequency of 15 MHz. The source is located 1 km away from the tip ($\rho_k = 1$ km) at $\varphi' = \alpha/4$. For upslope propagation, with these parameters, five propagating modes exist at the source location. Since this location coincides with modal nulls of the even-numbered modes, only the $m = 1,3,5$ are excited. The modal caustic locations for these modes are calculated from (30) to be 191, 572, and 954 m, respectively. Figure 9.4 illustrates the angular field profile at three observation ranges; 850, 650, and 450 m (calculated from (9.29)). As seen in the figure, the narrowing waveguide acts like a mode filter, with the first

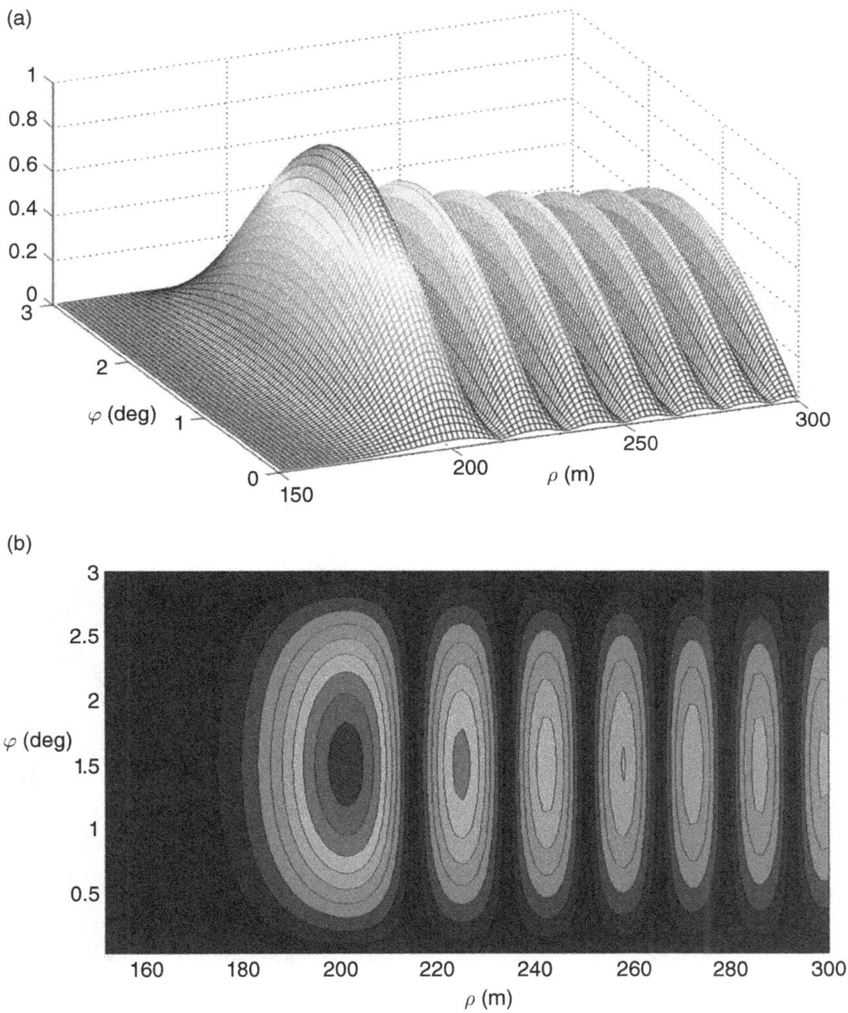

Figure 9.5. (a) Longitudinal variation of the dominant $m = 1$ mode showing the propagation–cutoff transition: surfer plot. (b) Longitudinal variation of the dominant $m = 1$ mode showing the propagation–cutoff transition: contour plot.

and third modes interfering at ranges of 850 and 650 m, but only the first mode surviving at 450 m.

Three-dimensional (3D) plots for the first and second modes are displayed in Fig. 9.5 and Fig. 9.6. In Fig. 9.5, the source is located at $\rho' = 300$ m, $\varphi' = \alpha/2$ so that only the $m = 1$ mode is excited. The

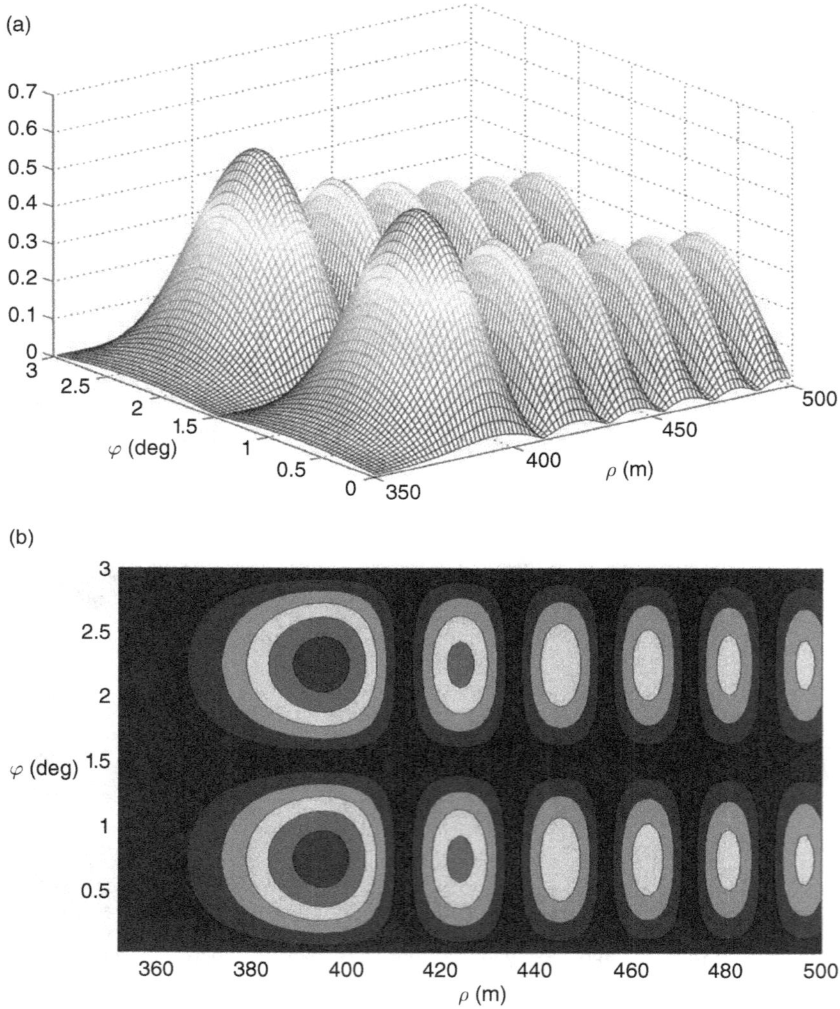

Figure 9.6. (a) As in Fig. 9.5a, but for $m = 2$. (b) As in Fig. 9.5b, but for $m = 2$.

surfer plot in Fig. 9.5a and contour plot in Fig. 9.5b clearly reveal the propagating—evanescent cutoff transition and the upslope propagation interference with the caustic—reflected (i.e., downslope propagating) mode. In Fig. 9.6, the source is located at $\rho' = 500$ m, $\varphi' = \alpha/3$, and only the $m = 3$ mode is plotted.

9.4.2 Approximate Solutions in Rectangular Coordinates

9.4.2.1 Problem Formulation

Propagation inside the wedge waveguide may also be formulated in rectangular coordinates (see Fig. 9.2). In this case, the upper boundary is parallel to the x-axis and independent of the y-axis. On the other hand, the lower boundary is (x,y)-dependent, making the wave equation (which will be derived from Maxwell's equations later) nonseparable. When the lower boundary depends weakly on the longitudinal coordinate (x-axis), local separability may be applied. The modes derived under this assumption are the AMs. Referring to the waveguide configuration and coordinates pictured in Fig. 9.2, the 2D z-directed line-source-excited vector Maxwell equations in cylindrical coordinates (with suppressed $\exp(j\omega t)$ time dependence, and $\partial/\partial z \equiv 0$) contain only three nonvanishing scalar field components: E_z, H_x, H_y [5]. E_z is determined from the source-excited 2D wave equation,

$$\frac{\partial^2 E_z}{\partial x^2}+\frac{\partial^2 E_z}{\partial y^2}+k^2 E_z = j\omega\mu I_0 \delta(x-x')\delta(y-y'), \tag{9.29}$$

subject to the BCs

$$\text{radiation condition at } x \to \infty, \text{ finiteness at } x \to 0 \tag{9.30a}$$

$$E_z = 0 \quad \text{at} \quad y = 0, Y_1, \quad Y_1 = x\tan\alpha. \tag{9.30b}$$

In (9.29), I_0 is the line current amplitude. The remaining field components are then obtained from

$$H_x = \frac{j}{\omega\mu}\frac{\partial}{\partial y}E_z, \quad H_y = \frac{-j}{\omega\mu}\frac{\partial}{\partial x}E_z, \tag{9.31}$$

where Y_1 is the local width of the waveguide. Whereas the conventional radiation condition applies to the expanding waveguide cross sections as $x \to \infty$ it requires introduction of a reflected wave that makes the total modal field undergo a propagating-to-evanescent cutoff transition $x \to 0$ as will be implemented later. Accordingly, the normalized function,

$$\bar{G}(x,y;x',y') = \frac{-E_z}{j\omega\mu I_0}, \tag{9.32}$$

defines the AM GF $\bar{G}$ in rectangular coordinates as

$$\{\nabla^2 + k^2\}\bar{G}(x, y; x', y') = -\delta(x - x')\delta(y - y'), \tag{9.33}$$

subject to BCs

$$\bar{G}(x, y; x', y') = 0 \quad \text{at} \quad y = 0, Y_1, \quad Y_1 = x \tan\alpha, \tag{9.34a}$$

$$\text{application of (9.32a) in the } x\text{-domain.} \tag{9.34b}$$

9.4.2.2 GF Solution: (cf. (9.27)])
The GF can be given as

$$\bar{G}(x, y; x', y') = \sum_{m=1}^{\infty} \bar{f}_m(y)\bar{f}_m(y')\bar{g}_{xm}(x, x'; k_{xm}). \tag{9.35}$$

In (9.35), $\bar{f}_m(y)$ denotes a normalized eigenfunction and $\bar{g}_{xm}$ is the 1D modal (spectral) GF in x (see later).

1D Problem in x-Domain: Characteristic GF Problem:
The problem can be postulated via

$$\left(\frac{d^2}{dx^2} + k_{xm}^2\right)\bar{g}_{xm}(x, x'; \beta(x)) = -\delta(x - x'), k_{xm} = \beta(x) = \sqrt{k^2 - k_{ym}^2}, \tag{9.36a}$$

$$\text{subject to a radiation condition at } x \to \infty, \tag{9.36b}$$

$$\text{incident-reflected wave condition as } x \to 0. \tag{9.36c}$$

The HFA WKB solution of (9.36) is given by (see (9.7)),

$$\bar{g}_x(x, x'; \beta(x)) = \frac{\exp\left\{-j\int_{x_<}^{x_>} \beta(\eta)d\eta\right\}}{2j\sqrt{\beta(x)\beta(x')}}. \tag{9.37}$$

1D Problem in y-domain: Eigenvalue Problem:
The problem ın this domain can be postulated as

$$\left(\frac{d^2}{dy^2} + k_{ym}^2\right)f_m(y) = 0, \text{BCs as in (9.36a).} \tag{9.38}$$

The solution is given by

$$\bar{f}_m(y) = \sqrt{\frac{2}{x \tan\alpha}} \sin\left(\bar{k}_{ym} y\right), \quad \bar{k}_{ym} = \frac{\bar{\eta}_m}{x}, \quad \bar{\eta}_m = \frac{m\pi}{\tan\alpha}, \quad m = 1, 2, 3, \ldots \tag{9.39}$$

2D Problem in *x*-*y*-Domain: As in (9.20), the relevant AM GF $\bar{G}$ is given by

$$\bar{G}(x, y; x', y') = \sum_{m=1}^{\infty} \bar{f}_m(x, y) \bar{f}_m(x', y') \bar{g}_{xm}(x, x'; \beta(x)). \tag{9.40}$$

Referring to the HFA explicit solutions in (9.37), in conjunction with the BCs in (9.36a,b), one obtains

$$\bar{G}(x, y; x', y') \approx \frac{j}{\sqrt{xx'} \tan\alpha} \sum_{m=1}^{\infty} \sin\left(\frac{m\pi y}{x \tan\alpha}\right) \sin\left(\frac{m\pi y'}{x' \tan\alpha}\right)$$
$$\times \left\{ \frac{\exp\left\{-j \int_{x_<}^{x_>} \beta_m(\eta) d\eta\right\} - j \exp\left\{-j \left[\int_{x_m}^{x_<} + \int_{x_m}^{x_>}\right] \beta_m(\eta) d\eta\right\}}{\sqrt[4]{k^2 - \left(\frac{m\pi}{x \tan\alpha}\right)^2} \sqrt[4]{k^2 - \left(\frac{m\pi}{x' \tan\alpha}\right)^2}} \right\}. \tag{9.41}$$

With respect to the phase terms in the exponential function, the first term inside the brackets describes a wave propagating directly from the source to the observer. The first and second parts of the exponentials in the second term inside the brackets correspond, respectively, to waves propagating from the source to the mth modal caustic and from the mth modal caustic to the observer. Expression (9.41) is valid for all observation ranges except in the vicinity of modal caustics (i.e., $x \approx X_{cm}$). The higher the mode index (m), the farther removed is the modal caustic from the apex. Therefore, the number of contributing modes decreases as the observer moves in the upslope direction.

Figure 9.7 displays longitudinal variations of the dominant ($m = 1$) NM and AM, calculated from (9.27) and (9.41), respectively. As observed, the AM fails to represent the modal field variation around the modal cutoff range.

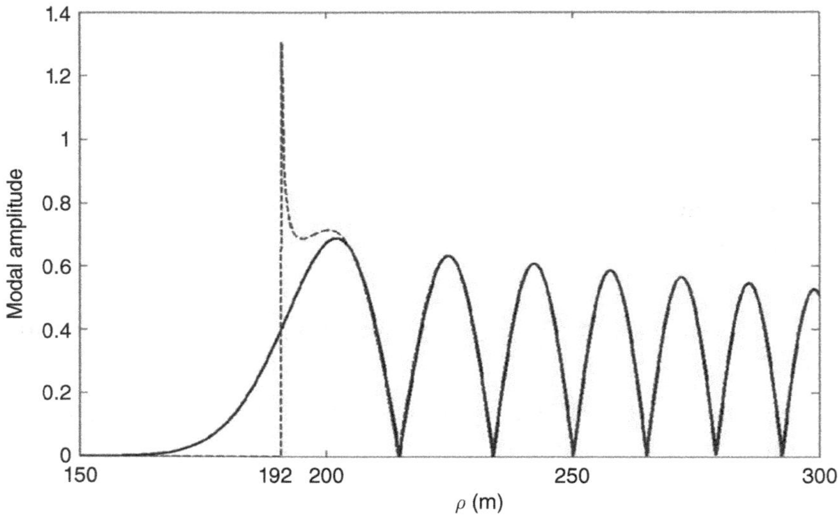

Figure 9.7. Dominant mode versus range ($f = 15$ MHz, $\alpha = 3°$, $\varphi_s = 1.5°$, $\varphi_o = 1.5°$). Solid: NM, dashed: AM (modal cutoff [caustic] range is 192 m).

9.4.3 IM Spectral Representation

As noted earlier, the IMs are constructed via a continuous spectral domain superposition of the AM so as to overcome the deficiency of the AM solution in critical guiding-to-evanescent cutoff transition regions. Referring to (9.9) and (9.10), with (9.41), one obtains, for the IM solution $\hat{W}(\cdot)$

$$\hat{W}_m(\overline{x}(\beta), y) \approx \frac{1}{2\pi}\sqrt{\frac{k}{\tan\alpha}} \int_{C_\beta} \frac{\sin\left\{\sqrt{k^2-\beta^2}\,y\right\}}{\sqrt{k^2-\beta^2}} e^{-j\left[\beta x - \overline{\eta}_m \sin^{-1}\left(\frac{\beta}{k}\right)\right]} d\beta, \quad (9.42)$$

since

$$\overline{x}_m(\beta) = \frac{\overline{\eta}_m}{\sqrt{k^2-\beta^2}}, \quad (9.43a)$$

$$a_m(\beta) = \frac{\overline{\eta}_m}{\left(k^2-\beta^2\right)^{3/4}}, \quad (9.43b)$$

$$\Omega_m(\beta) = \overline{\eta}_m \left\{\frac{\beta}{\sqrt{k^2-\beta^2}} - \sin^{-1}\left(\frac{\beta}{k}\right)\right\}, \quad (9.43c)$$

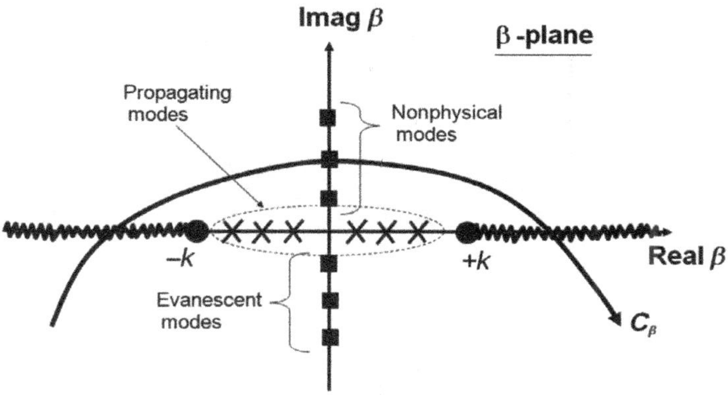

Figure 9.8. Integration path C_β and singularities in the complex-β plane: Branch cuts. Poles associated with propagating, evanescent, and nonphysical modes.

$$U_m(y;\bar{x}_m(\beta)) = \sqrt{\frac{2}{\bar{\eta}_m \tan\alpha}}\left(k^2-\beta^2\right)^{1/4}\sin\left(\sqrt{k^2-\beta^2}\,y\right). \quad (9.43\text{d})$$

As detailed in Section 9.3.3, the integration contour must be a permissible contour which ensures the existence of the integral representation everywhere inside the wedge, that is, for $y \le x\tan\alpha$. Moreover, the contour in the complex β-plane should be deformable into the LSDPs through the relevant SP(s) in such a way that the radiation condition is satisfied for $x \to \infty$ (see Fig. 9.8), thereby ensuring that (9.42) will reduce asymptotically to the (x,y)-parameterized mth AM. Using the mapping $\beta = k\sin\theta$, one obtains in the complex θ-plane,

$$\hat{W}_m(x,y) = \frac{1}{2\pi}\sqrt{\frac{k}{\tan\alpha}}\int_{C_\theta}\sin(ky\cos\theta)\,e^{-j[kx\sin\theta-\bar{\eta}_m\theta]}d\theta, \quad (9.44)$$

with SPs at $\theta_{ms} = \cos^{-1}(\bar{\eta}_m / k_x)$.

The LSDPs associated with the propagating mth AM are depicted in Fig. 9.9. Since the contour must terminate in one of the valley regions, it is concluded that there are two possible contours which are designated by C_θ and C'_θ in Fig. 9.9. Note that C_θ is deformable into both LSDPs and hence results in standing wave-type modal fields, whereas C'_θ can only be deformed into the LSDP through the SP $\theta_{ms} = (0, \pi/2)$ and results in the traveling wave-type mode field which

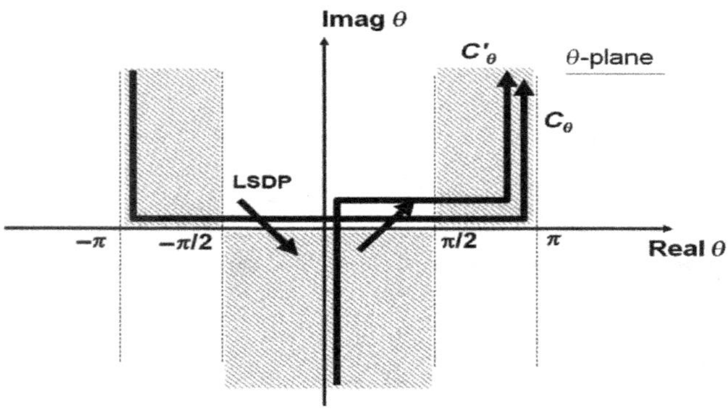

Figure 9.9. Integration paths C_θ and C'_θ in the complex angular spectrum θ-plane $\beta = k\sin\theta$. Relevant local steepest descent paths (LSDPs) through the SP(s). Permissible paths should terminate in the shaded valley regions.

properly accounts for the radiation condition. After introducing a change of variables to polar coordinates $y = \rho\sin\varphi$, $x = \rho\cos\varphi$, $\rho > 0$, $\varphi < \pi/4$, and decomposing the sine function into exponential constituents, one can write the integral in (9.44) as

$$I_1 - I_2 = \frac{1}{2j}\int_{C_\theta} e^{-j[k\rho\sin(\theta+\phi)-\overline{\eta}_m\theta]}d\theta - \frac{1}{2j}\int_{C_\theta} e^{-j[k\rho\sin(\theta-\phi)-\overline{\eta}_m\theta]}d\theta. \tag{9.45}$$

Invoking the final mapping $\zeta = \theta + \phi$ in I_1, $\zeta = \theta - \phi$ in I_2 permits explicit evaluation of (9.42), by recognizing that I_1 and I_2 are both deformable into the original contour C_ξ in Fig. 9.10, when for $\varphi \to 0$, one obtains

$$I_1 - I_2 = \sin(\overline{\nu}_m\phi)\int_{C_\zeta} e^{-j[k\rho\sin\zeta-\overline{\nu}_m\zeta]}d\zeta, \tag{9.46}$$

where C_ζ in the ζ-plane is identical with C_θ (or C'_θ) in the θ-plane, the integrals can be recognized as representations of $J_{\overline{\eta}_m}(k\rho)$ and $H^{(2)}_{\overline{\eta}_m}(k\rho)$ for $C_\zeta = C_\theta$ and $C_\zeta = C'_\theta$, respectively [16]. Substituting these results back into (9.44), one can express the (x,y) parameterized mth IM of the wedge waveguide via the alternatives

$$\hat{W}_m(\rho_<, \phi_<, \xi) = \sqrt{\frac{\xi}{\tan\alpha}}\hat{f}_m(\phi_<)J_{\overline{\eta}_m}(\xi\rho_<), \tag{9.47a}$$

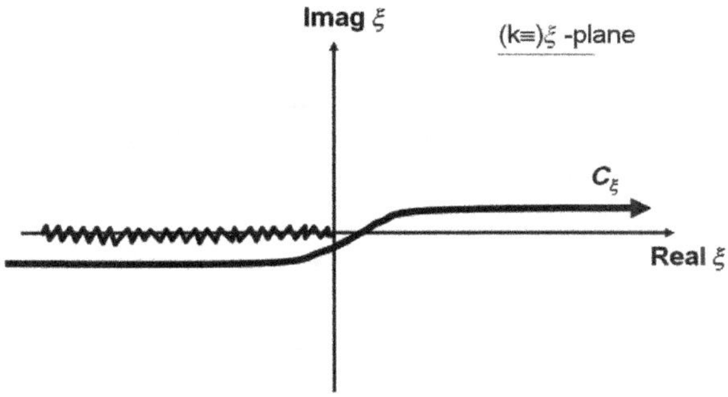

Figure 9.10. Integration path in the complex ξ-plane.

$$\hat{W}_m(\rho_>, \phi_>, \xi) = \sqrt{\frac{\xi}{\tan\alpha}} \hat{f}_m(\phi_>) H^{(2)}_{\overline{\eta}_m}(\xi\rho_>), \tag{9.47b}$$

with

$$\rho = \sqrt{x^2 + y^2}, \varphi = \tan^{-1}\left(\frac{y}{x}\right), \tag{9.48}$$

where $f_m(\varphi)$ is obtained from (9.26) upon replacing η_m with $\overline{\eta}_m$. The corresponding asymptotic solution for the IM GF becomes

$$\hat{G}(\rho, \rho', \phi, \phi') = \frac{-j\pi}{\tan\alpha} \sum_{m=1}^{\infty} \hat{f}_m(\phi) \hat{f}_m(\phi') J_{\overline{\eta}_m}(k\rho_<) H^{(2)}_{\overline{\eta}_m}(k\rho_>). \tag{9.49}$$

It is interesting to note that the approximate GF representation in (9.49) obtained in Cartesian coordinates parameterized via the spectral IM integral is in identical functional form with the exact solution of the problem given in (9.27) and can be obtained therefrom by simply replacing α with $\tan\alpha$.

9.5 THE MATLAB PACKAGE "WEDGEGUIDE"

A MATLAB package—WedgeGUIDE—has been prepared for the exploration of wave propagation inside a 2D nonpenetrable, homogeneously filled wedge waveguide. It is designed to investigate line-source-excited wave fields in terms of NM, AM, and IM solutions. The

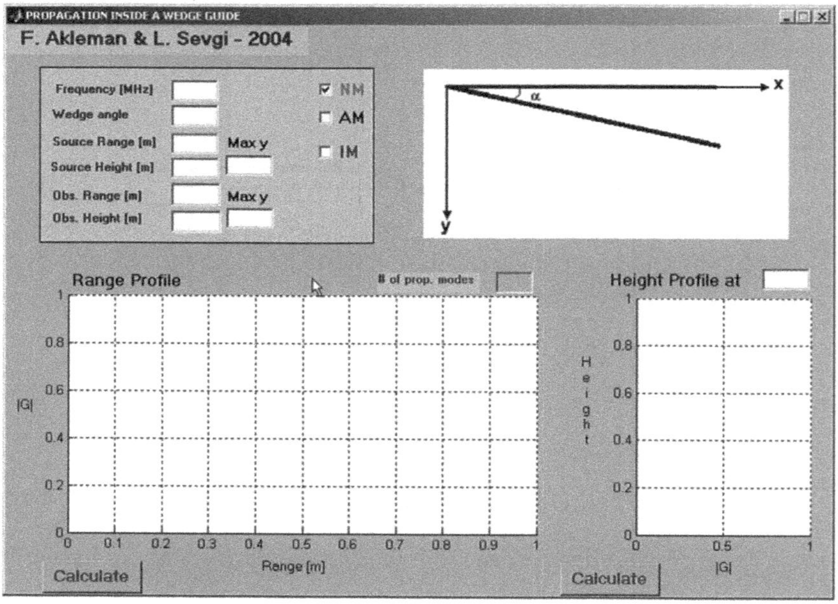

Figure 9.11. The run-time window of the WedgeGUIDE MATLAB package.

run-time window is shown in Fig. 9.11, depicting the wedge waveguide, wave field versus range, and wave field versus height plot windows. The geometrical parameters that should be supplied by the user are the wedge angle (α) and the source and observer locations; the operational parameter is the frequency in megahertz. Once the user supplies the source (observer) range, the package automatically calculates and displays the wedge width at the source (observer) distance. The user then selects the height coordinate of the source (observer) which may be any value between zero and the displayed number. There are two operational buttons. The first button is for the calculation of the range profile. When it is clicked, the package first calculates the number of propagating modes (i.e., the modes with positive real modal propagation constants) and displays it. Then it plots wave fields versus range. The second button is for the calculation of the transverse wave field profile inside the wedge at the observation range. The user may select to include NM only calculations, as well as NM plus AM, NM plus IM, AM plus IM, and NM plus AM plus IM.

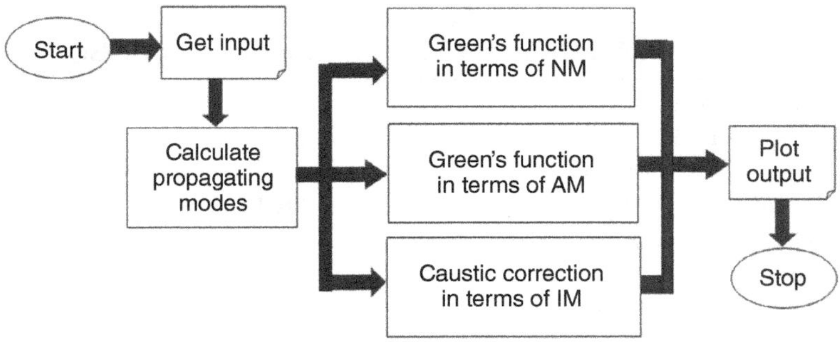

Figure 9.12. The flowchart of WedgeGUIDE algorithm.

The flowchart of the WedgeGUIDE package is given in Fig. 9.12. Here, the user first supplies required input parameters. Once they are supplied, the number of propagating modes is calculated automatically. Then, the package calls three subroutines for NM, AM, and AM-IM calculations and plots the output as wave field versus range and/or height. As an example, a simple MATLAB program code *WedgeNMHBC.m* is given in Table 9.1. The code computes NM fields versus range at a specified height and NM fields versus height at a specified range when other parameters are supplied. The modules may easily be modified to calculate AM and IM equations, as well as to account for nonpropagating mode contributions.

9.6 NUMERICAL TESTS AND ILLUSTRATIONS

We have run the WedgeGUIDE package for characteristic parameter sets and present the outputs in this section. The geometry, with "height" measured downward, is shown in Fig. 9.13 for a typical scenario pertaining to a 3° wedge waveguide and a frequency of 15 MHz. The source is located 1 km away from the tip at a height of 40 m. Upslope propagation is investigated first. For the specified parameters, the number of propagating modes at the source range is found to be 5. The modal caustics for these five modes are calculated from (30) and are found to be 191, 382, 572, 763, and 954 m for modes $m = 1$–5, respectively. The outcome for this example is shown in Fig. 9.14a, with plots of wave field versus range for the NM (reference) and IM solutions.

Table 9.1.
A MATLAB Code for NM Summation with HBC

```
function WedgeNMHBC
clc, clear all, close all, format long
%% Input Parameters
% frequency [MHz], % wedge angle [Deg], % source
position [m,m],
% observer [m,m]: TE (HBC)
fr = 15;  alfa = 15; xs = 1000; ys = 15; xo = 150; yo
= 2;
% ---------------------------------------------------
if ys>xs*tand(alfa), error(['ys must be smaller than
',num2str(xs*tand(alfa))]),end
if yo>xo*tand(alfa), error(['yo must be smaller than
',num2str(xo*tand(alfa))]),end
xx=linspace(xo,xs,501); yy=linspace(0,xo*tand(alfa),501);
k=2*pi*fr/300;
Uv=NaN(size(yy)); Uvmom=Uv; Uh=NaN(size(xx)); Uhmom=Uh;
[Uv,Uh]=wedgeanalysiss(k,alfa,xs,ys,xo,yo,xx,yy); %
Green's function
plotfigure(xx,yy,Uv,Uh);
% ---------------------------------------------------
function [Uver,Uhor]=wedgeanalysiss(k,alfa,xs,ys,xo,yo,xx
,yy)
r0 = sqrt(xs^2+ys^2); angle0 = atan(ys/xs); rver =
sqrt(xo^2+yy.^2);    Arver = atan(yy/xo); rhor =
sqrt(xx.^2+yo^2); Arhor = atan(yo./xx);  alfa =
alfa*pi/180;
% Exact Series Solution
Uver = zeros(1,length(Arver));
for m = 1:length(Arver)
     Uver(m) = Exact_Series(k,alfa,r0,angle0,rver(m),Arve
r(m));
end
Uhor = zeros(1,length(Arhor));
for m = 1:length(Arhor)
     Uhor(m) = Exact_Series(k,alfa,r0,angle0,rhor(m),Arho
r(m));
end
% ---------------------------------------------------
function U=Exact_Series(k,alfa,r0,phi0,r,phi)
```

Table 9.1.
(Continued)

```
error=1e-12; kr=k*r; kr0=k*r0;
vn1=0*pi/alfa; vn2=1*pi/alfa; n=1;
if r<=r0,
  U=0.5*besselj(vn1,kr)*besselh(vn1,2,kr0)*cos(vn1*phi0)*
cos(vn1*phi);
  U2=besselj(vn2,kr)*besselh(vn2,2,kr0)*cos(vn2*phi0)*cos
(vn2*phi);
  while abs(U2/U)>error
      U=U+U2; n=n+1; vn=n*pi/alfa;
      U2=besselj(vn,kr)*besselh(vn,2,kr0)*cos(vn*phi0)*co
s(vn*phi);
  end
else
  U=0.5*besselj(vn1,kr0)*besselh(vn1,2,kr)*cos(vn1*phi0)*
cos(vn1*phi);
  U2=besselj(vn2,kr0)*besselh(vn2,2,kr)*cos(vn2*phi0)*cos
(vn2*phi);
  while abs(U2/U)>error
      U=U+U2; n=n+1; vn=n*pi/alfa;
      U2=besselj(vn,kr0)*besselh(vn,2,kr)*cos(vn*phi0)*co
s(vn*phi);
  end
end
U=-1i*pi/alfa*U;
% ---------------------------------------------------
function plotfigure(xx,yy,Uver,Uhor)
scrsz = get(0,'ScreenSize');
figure('Position',round([scrsz(3)/50 scrsz(4)/5 scrsz(3)/4
scrsz(4)/2.5]),'Color',[1 1 1]);
    p1=plot(abs(Uver),yy,'-');hold on,set(p1,'linewidth',2);
    ylabel('Height [m]','FontWeight','bold');xlabel('Field','Fo
ntWeight','bold')
figure('Position',round([scrsz(3)/50 scrsz(4)/5 scrsz(3)/3
scrsz(4)/2.7]),'Color',[1 1 1]);
    p1=plot(xx,abs(Uhor),'-');hold on,set(p1,'linewidth',2);
    xlabel('Range [m]','FontWeight','bold');ylabel('Field','Fon
tWeight','bold')
%   -------------------- END ------------------
```

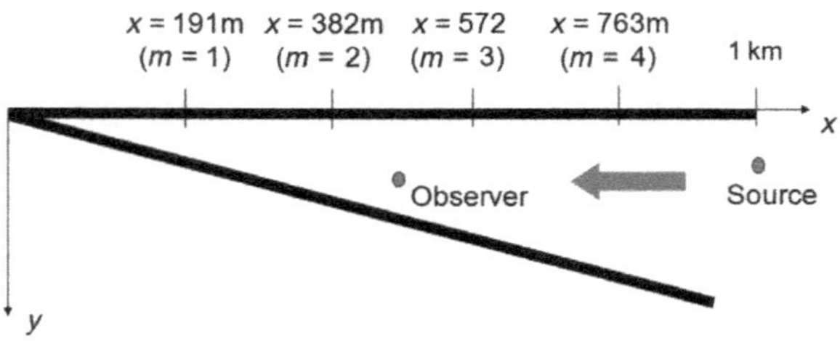

Figure 9.13. Test 1; upslope propagation. Source: $x = 1000$ m, $y = 40$ m, observation along $y = 20$ m, $f = 15$ MHz, $\alpha = 3°$. Number of propagating modes $m = 5$. Modal cutoff ranges: $x_{c1} = 191$ m, $x_{c2} = 382$ m, $x_{c3} = 572$ m, $x_{c4} = 763$ m, $x_{c5} = 954$ m.

The range in which all five modes undergo their cutoff transitions extends from the source to 100 m. The height of the source is chosen to be $y = 15$ m. The observation line is parallel to the x-axis at a height of $y = 2$ m. The wave field height profile at $x = 300$ m is also shown. Excellent agreement between the (ρ, φ)-parameterized NM and the (x, y)-parameterized IM solutions is clearly observed in the figure. For the same physical parameters, Fig. 9.14b compares the NM, AM, and IM solutions. The various options agree very well, except at modal caustics where AM fails, as indicated by spikes at ranges surrounding strongly excited AM.

In Fig. 9.15, GF range and height profiles are presented at longer ranges extending from 2000 to 2500 m, where the number of propagating modes increases to 13. The plots show results for the NM, AM, and IM. The caustics of the 13th and 11th modes are clearly visible in the figure, but the 12th modal caustic is not clear. This is because of the range samples used in the package. The program is adjusted to use 1000 range points (i.e., the range step size is equal to (max range) – (min range) divided by 1000). Changing source/observer ranges (and/or readjusting the number of range points) may cause a range point to coincide with a modal caustic, where the AM fails in the GF representation. If a range point is very close to a modal caustic, the AM solution blowup can be identified in the AM field profile. If none of the range samples lie on, or come close to, a caustic, then there are no visible irregularities in the AM curve. Again, very good agreement is obtained among the NM, AM, and IM representations.

(a)

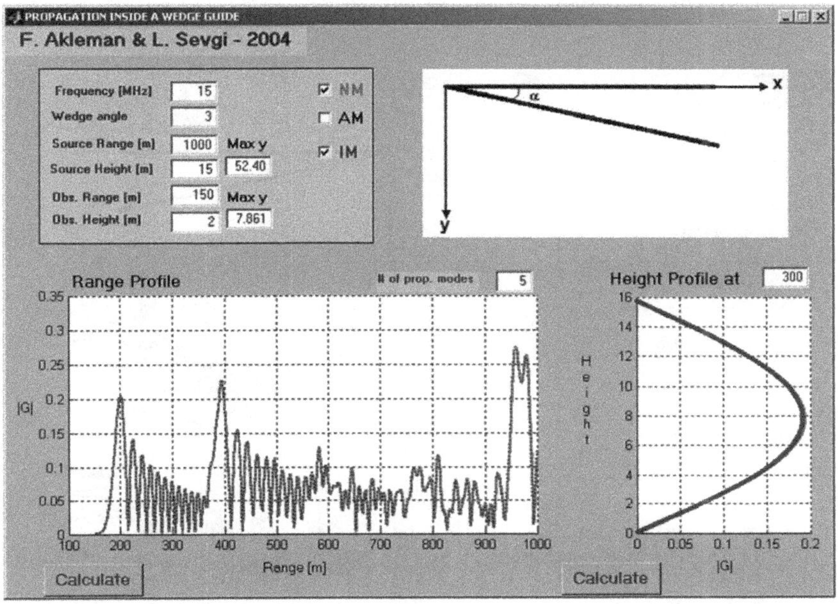

(b)

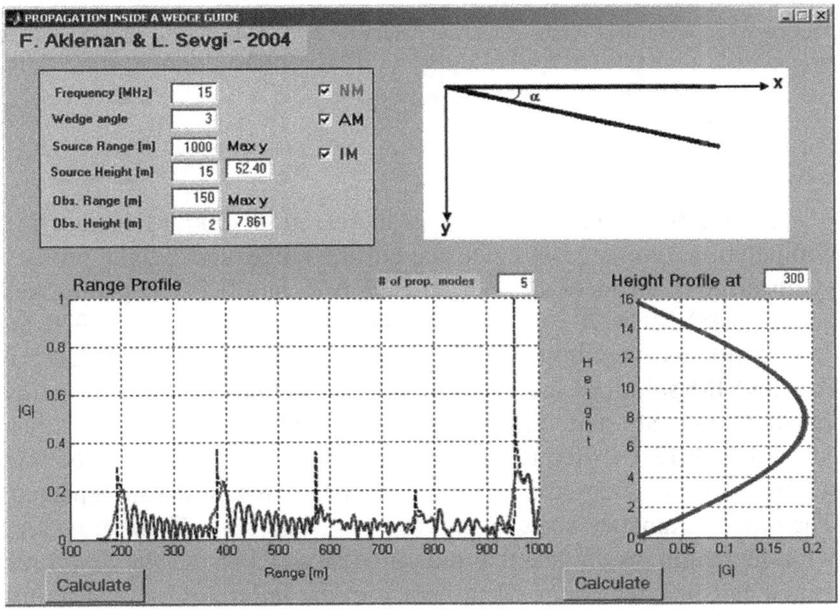

Figure 9.14. (a) Wave field versus range for the line source ($x_s = 1000$ m) in terms of NM (9.27) and IM summations (9.41). The range extends from the source to 100 m. The height of the source is $y = 15$ m. The observation track is parallel to the top x-axis at a height $y = 2$ m ($f = 15$ MHz, $\alpha = 3°$). The wave field versus height profile at $x = 300$ m is plotted as well. (b) Same as Fig. 9.14a with NM, AM, and IM (9.49) summations.

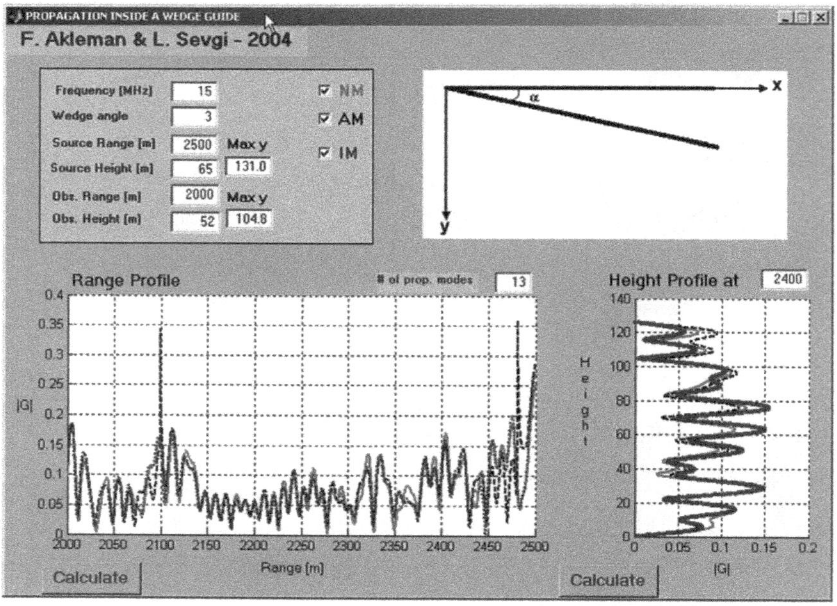

Figure 9.15. Wave field versus range and height for NM, AM, and IM formulations at longer ranges. The parameters are $f = 15$ MHz, $\alpha = 3°$, $x_s = 2500$ m, $y_s = 65$ m, $x_o = 2000$ m, $y_o = 52$ m ($x = 2400$ m).

In Fig. 9.16, the frequency is increased to 100 MHz and the computations are repeated. Displayed are range and height profiles synthesized in terms of all relevant NM, AM, and IM. The parameters are $f = 100$ MHz, $\alpha = 3°$, $x_s = 1000$ m, $y_s = 26$ m, $x_o = 970$ m, $y_o = 25$ m, and the number of propagating modes is 34. Although the AM and IM solutions are in close agreement (except near the modal caustics), they are quite different from the NM solution.

In Fig. 9.17, the frequency is lowered again to $f = 15$ MHz, but the wedge angle is increased to $\alpha = 15°$, and the computations are repeated. The other parameters are $x_s = 250$ m, $y_s = 33$ m, $x_o = 20$ m, $y_o = 3$ m, with the height profile taken at the range of $x = 140$ m. The first three modal caustics, which are identified from the failure of the AM representation, are clearly observed in the figure. In Fig. 9.18, the calculations for the same wedge angle and operating frequency are repeated for longer ranges. Here the parameters are $x_s = 1000$ m, $y_s = 135$ m,

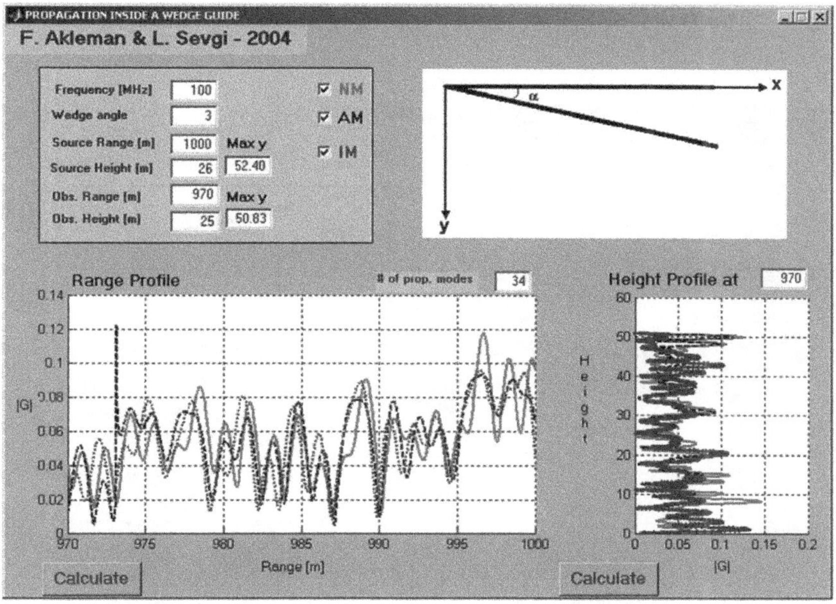

Figure 9.16. Wave field versus range and height for NM, AM, and IM formulations with the following parameters: $f = 100$ MHz, $\alpha = 3°$, $x_s = 1000$ m, $y_s = 26$ m, $x_o = 970$ m, $y_o = 25$ m. The number of propagating modes is 34 ($x = 970$ m).

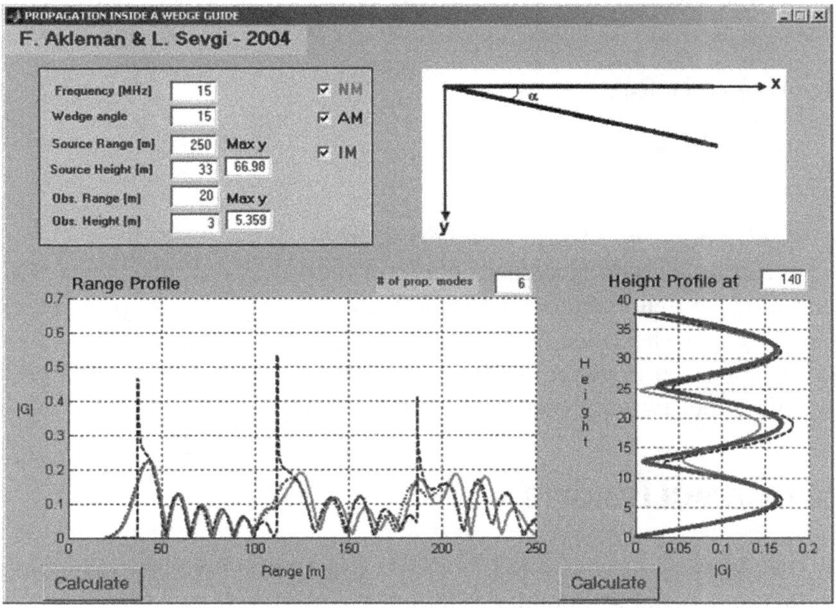

Figure 9.17. Wave field versus range and height computed from NM, AM, and IM formulations with the following parameters: $f = 15$ MHz, $\alpha = 15°$, $x_s = 250$ m, $y_s = 33$ m, $x_o = 20$ m, $y_o = 3$ m. The number of propagating modes is 6. The wave field height profile is taken at $x = 140$ m. The AM/IM spikes near modal caustics are explained in the text.

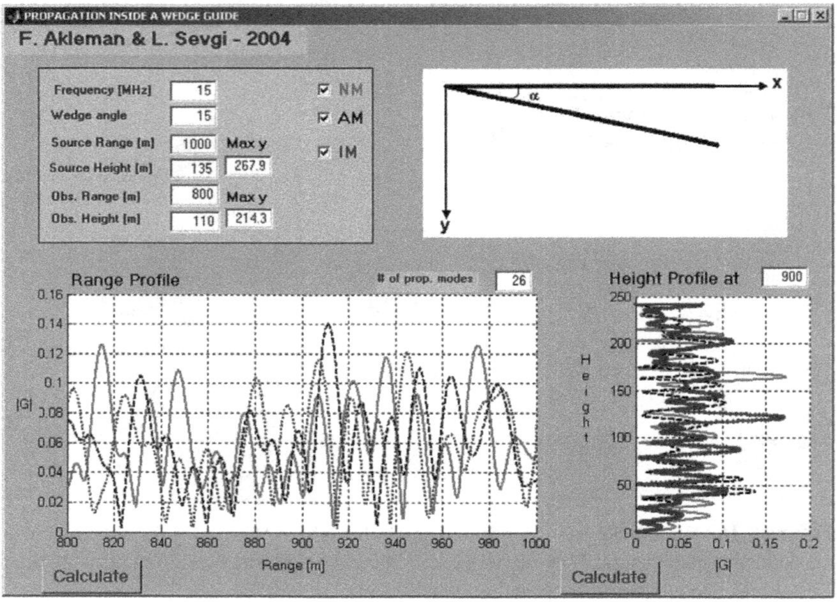

Figure 9.18. Wave field versus range and height computed from NM, AM, and IM formulations. The parameters are $f = 15$ MHz, $\alpha = 15°$, $x_s = 1000$ m, $y_s = 135$ m, $x_o = 800$ m, $y_o = 110$ m. The number of propagating modes is 26. The height profile is taken at $x = 900$ m (TM Pol/SBC).

$x_o = 800$ m, $y_o = 110$ m. The height profile is taken at the range of $x = 900$ m. The disagreement between the (ρ, φ)-parameterized NM and the (x, y)-parameterized AM and IM solutions is seen to increase. These results show that as the number of propagating modes increases (either because of range or because of frequency increases), the NM and the other two representations tend to disagree.

9.7 CONCLUSIONS

A simple line-source-excited, homogeneously filled wedge waveguide with nonpenetrable boundaries has been used here to introduce the analytical background for developing GF representations in terms of NMs [1,2], AMs [6,7], and IMs [8–15,17,18].

The intent has been tutorial, targeting graduate and upper level undergraduate students. MATLAB visualization packages have been prepared, documented, and used to illustrate the effects of various modal approaches through parametric numerical experiments.

The MATLAB package WedgeGUIDE [19] designed to investigate modal options for the canonical wedge-waveguide problem can serve as effective educational tools toward understanding the physics-based, observable wave objects in this environment. The packages may easily be extended to handle impedance-type or penetrable BCs. Moreover, based on the experience and physical insight gained here with these packages, the user may extend the algorithms to waveguiding structures with virtual boundaries (i.e., trapping inhomogeneous refractivities).

APPENDIX 9A: FORMATION OF THE SPECTRAL IM INTEGRAL IN SECTION 9.3.3

A.1 The Spectral Integral

Tutorially, the derivation of the IM formalism leading to (9.9) may be better understood by considering the simpler 1D wave equation (see Ref. 12 for details),

$$\left[\frac{d}{dx}+\gamma_m^2(x)\right]\hat{V}_m(x)=0, \tag{A.1}$$

whose asymptotic solution is traditionally given by the WKB method (Ref. 4, p. 337), which is not uniform in the vicinity of the turning points (caustics) where $\gamma_m^2(x)$ vanishes. One solution method is to postulate a uniform representation in terms of a spectral integral,

$$\hat{V}_m(x)=\int_{C_\beta} a_m(\beta)e^{j\Omega_m(\beta)}e^{-j\beta(x-\overline{x}_m(\beta))}d\beta. \tag{A.2}$$

Here, $\overline{x}_m(\beta)$ is the function that assigns a point in configuration space to a particular local wave number $\beta_m(x)$. It is the continuation, into the complex β-plane, of the inverse of the function $\gamma_m(x)$, that is,

$$p=\gamma_m(q) \quad \Rightarrow \quad q=\overline{x}_m(p). \tag{A.3}$$

The amplitude and phase functions $a_m(\beta)$ and $\Omega_m(\beta)$, respectively, and the integration contour C_β are yet to be specified. Note that (A.2) expresses $\hat{V}_m(z)$ as a Fourier integral with the Fourier transform given in terms of $a_m(\beta)\exp\{j\Omega_m(\beta)\}\exp\{j\beta\bar{x}_m(\beta)\}$.

Further insight is gained by formally changing the variable of integration in (A.2) to $\bar{x}$

$$\hat{V}_m(x) = \int_{\bar{x}} a_m\left[\beta(\bar{x})\right]\frac{d\beta}{d\bar{x}} e^{j\Omega_m[\beta(\bar{x})]} e^{-j\beta(\bar{x})(x-\bar{x}_m)} d\bar{x}, \tag{A.4}$$

from where it is seen that each point $\bar{x}$ along the contour of the integration path contributes to the field a plane wave with a wave number $\beta(\bar{x})$. The plane wave is normalized to unity at $x = \bar{x}$ and is weighted by a factor

$$\hat{V}_m(x) = a_m\left[\beta(\bar{x})\right]\frac{d\beta}{d\bar{x}} e^{j\Omega_m[\beta(\bar{x})]}. \tag{A.5}$$

Where the WKB solution of (A.2) exists, it corresponds physically to plane waves propagating with local wave numbers $\beta = \pm\gamma_m(x)$, and it reconstructs the uniformizing spectral integral from the WKB solution.

To proceed, if (A.2) is substituted into (A.1), it is implied that

$$\int_{C_\beta} \left(\gamma_m^2 - \beta^2\right) a_m(\beta) e^{j\Omega_m(\beta)} e^{-j\beta(x-\bar{x}_m(\beta))} d\beta = 0. \tag{A.6}$$

In general, it is not possible to satisfy (A.6) exactly. However, if the integral is evaluated asymptotically, then it is possible to choose $a_m(\beta)$ and $\Omega_m(\beta)$ in such a way that the integrand in (A.6) is small near the SPs $\beta = \pm\beta_s$, from where most of the contribution to the integral would normally.

The first step is to choose $\Omega_m(\beta)$, such that at $\beta = \pm\beta_s$,

$$\gamma_m^2 - \beta_s^2 = 0 \quad \text{or} \quad x = \bar{x}(\beta_s). \tag{A.7}$$

Denoting the total phase of the integral by

$$\hat{P}_m(x, \beta) = \Omega_m(\beta) - \beta_m(x - \bar{x}(\beta)), \tag{A.8}$$

the SPs are defined by

$$\frac{d}{d\beta}P_m(x, \beta) = \frac{d}{d\beta}\Omega_m(\beta) - (x - \bar{x}(\beta)) + \frac{d}{d\beta}\bar{x}(\beta) = 0, \tag{A.9}$$

whence the appropriate phase function is of the form

$$\Omega_m(\beta)=\int_0^{\beta}\hat{\beta}\frac{d}{d\hat{\beta}}\overline{x}\left(\hat{\beta}\right)d\hat{\beta} \tag{A.10}$$

Via (A.10), the equality in (A.6) is satisfied to the leading asymptotic order at the SPs, and the resulting phase term reads

$$\hat{P}_m(x,\beta)=\int_0^{\beta}\left(x-\overline{x}(\hat{\beta})\right)d\hat{\beta}, \tag{A.11}$$

In a similar way, $a_m(\beta)$ may be chosen to enforce vanishing of (A.6) to the next asymptotic order. To proceed, (A.6) is integrated by parts along the steepest descent path (SDP) to obtain

$$\int_{C_\beta}\frac{d}{d\beta}\left\{\frac{\left(\gamma_m^2-\beta^2\right)a_m(\beta)}{j\dfrac{d}{d\beta}P_m(x,\beta)}\right\}e^{-jP_m(x,\beta)}d\beta\approx 0, \tag{A.12}$$

which requires that

$$\lim_{\beta\to\beta_s}\frac{d}{d\beta}\left\{\frac{\left(\gamma_m^2-\beta^2\right)a_m(\beta)}{j\dfrac{d}{d\beta}P_m(x,\beta)}\right\}=0. \tag{A.13}$$

In order to solve (A.13) for $a_m(\beta)$, we define auxiliary functions $q(\beta)$ and $f(x,\beta)$,

$$q(\beta)=\beta-\beta_s,\ f(x,\beta)=\frac{\beta^2-\beta_s^2}{\overline{x}(\beta)-x}, \tag{A.14}$$

Using Taylor series expansion of $\overline{x}(\beta)$ around β_s gives

$$\begin{aligned}\overline{x}(\beta)=\overline{x}(\beta_s)+\frac{d}{d\beta_s}\overline{x}(\beta_s)(\beta-\beta_s)+\frac{d^2}{2d\beta_s^2}\overline{x}(\beta_s)(\beta-\beta_s)^2\\+\frac{d^3}{3d\beta_s^3}\overline{x}(\beta_s)(\beta-\beta_s)^3+\ldots.\end{aligned} \tag{A.15}$$

$$\begin{aligned}\overline{x}(\beta)-x &= q(\beta)\frac{d}{d\beta_s}\overline{x}(\beta_s)+\frac{1}{2}q^2(\beta)\frac{d^2}{d\beta_s^2}\overline{x}(\beta_s)\\ &+\frac{1}{3}q^3(\beta)\frac{d^3}{d\beta_s^3}\overline{x}(\beta_s)+\ldots,\end{aligned} \tag{A.16}$$

where $d()/d\beta_s$ is used to denote $d()/d\beta$ at $\beta=\beta_s$. By combining (A.14) and (A.16), $f(x, \beta)$ can be rewritten as

$$f(x,\beta)=\frac{q(\beta)+2\beta_s}{\frac{d}{d\beta_s}\overline{x}(\beta)+\frac{1}{2}q(\beta_s)\frac{d^2}{d\beta_s^2}\overline{x}(\beta)+\ldots}. \tag{A.17}$$

Therefore, at the SP $q(\beta_s)=0$,

$$f(x,\beta_s)=2\beta_s\left\{\frac{d}{d\beta_s}\overline{x}(\beta)\right\}^{-1}, \tag{A.18}$$

$$\frac{\partial}{\partial\beta_s}f(x,\beta_s)=\left\{\frac{d}{d\beta_s}\overline{x}(\beta)\right\}^{-1}-\beta_s\left\{\frac{d}{d\beta_s}\overline{x}(\beta)\right\}^{-2}\frac{d^2}{d\beta_s^2}\overline{x}(\beta). \tag{A.19}$$

Using (A.17) and (A.18) in (A.13) gives

$$\begin{aligned}&\left[\left\{\frac{d}{d\beta_s}\overline{x}(\beta)\right\}^{-1}-\beta_s\left\{\frac{d}{d\beta_s}\overline{x}(\beta)\right\}^{-2}\frac{d^2}{d\beta_s^2}\overline{x}(\beta)\right]a(\beta_s)\\ &+2\beta_s\left\{\frac{d}{d\beta_s}\overline{x}(\beta)\right\}^{-1}\frac{d}{d\beta_s}a(\beta_s)=0.\end{aligned} \tag{A.20}$$

Integration of (A.20) yields

$$a(\beta_s)=\left\{\frac{1}{\beta_s}\frac{d}{d\beta_s}\overline{x}(\beta)\right\}^{1/2}. \tag{A.21}$$

Since the SP β_s is arbitrary, (A.21) is valid for all β, that is,

$$a_m(\beta)=\left\{\frac{1}{\beta}\frac{d}{d\beta}\overline{x}_m(\beta)\right\}^{1/2}. \tag{A.22}$$

A.2 The Steepest Descent Paths

If Taylor series expansion is applied to the phase term in (A.10) inside the IM integral in (A.2) around the SP β_s, one obtains

$$-jP(x,\beta) \approx -jP(x,\beta_s) - \frac{j}{2}\frac{d^2P(x,\beta_s)}{d\beta_s^2}(\beta-\beta_s)^2. \tag{A.23}$$

Equation (A.23) may also be written as

$$jP(x,\beta) \approx jP(x,\beta_s) + \left|\frac{1}{2}\frac{d^2P(x,\beta_s)}{d\beta_s^2}(\beta-\beta_s)^2\right| e^{j2\psi}, \tag{A.24}$$

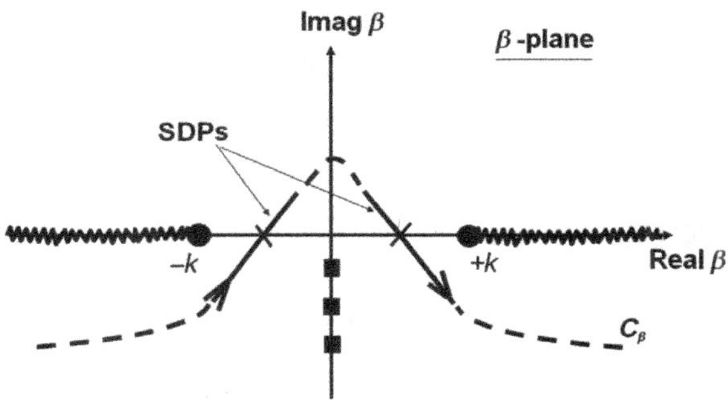

Figure 9A.1. Branch cuts and SDPs in the complex-β plane for propagating modes.

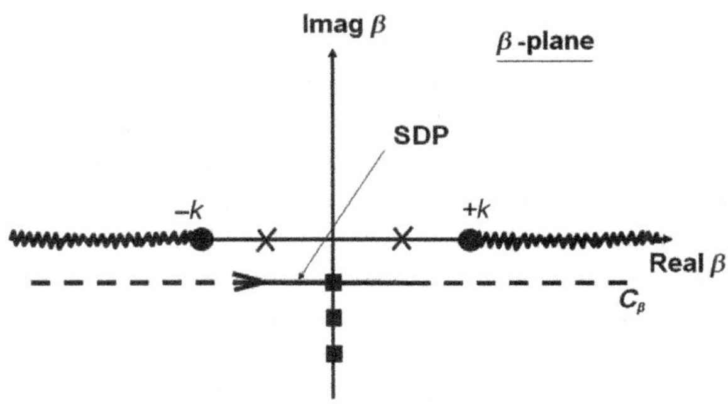

Figure 9A.2. Branch cuts and SDP in the complex-β plane for evanescent-cutoff modes.

where

$$\psi = \frac{\pi}{4} + \arg(\beta - \beta_s) + \frac{1}{2}\arg\left\{\frac{d^2P(x,\beta_s)}{d\beta_s^2}\right\}. \qquad \text{(A.25)}$$

Since the last term in (A.24) is constant, ψ changes only with $(\beta - \beta_s)$. The SDP is that integration contour whereon the phase term behaves like $\exp\{-|P(x,\beta)|\}$. Accordingly, as it is well known (see Ref. 4), the SDPs paths of the propagating modes (with real β_s) trace local contour along which

$$\arg(\beta - \beta_s) = \begin{cases} -\dfrac{\pi}{4} \quad \ldots\ \text{at} \quad \beta = +|\beta_s| \\ \dfrac{\pi}{4} \quad \ldots\ \text{at} \quad \beta = -|\beta_s| \end{cases}, \qquad \text{(A.26)}$$

as shown in Fig. 9A.1. Similar procedures yield the SDP for nonpropagating modes as shown in Fig. 9A.2.

REFERENCES

1. L. B. Felsen, L. Sevgi, "Adiabatic and Intrinsic Modes for Wave Propagation in Guiding Environments with Longitudinal and Transverse Variations: Formulation and Canonical Test," *IEEE Transactions on Antennas and Propagation*, **39** (8), 1991, pp. 1130–1136.
2. L. B. Felsen, L. Sevgi, "Adiabatic and Intrinsic Modes for Wave Propagation in Guiding Environments with Longitudinal and Transverse Variations: Continuously Refracting Media," *IEEE Transactions on Antennas and Propagation*, **39** (8), 1991, pp. 1137–1143.
3. L. B. Felsen, F. Akleman, L. Sevgi, "Wave Propagation Inside a Two-Dimensional Perfectly Conducting Parallel-Plate Waveguide: Hybrid Ray-Mode Techniques and Their Visualizations," *IEEE Antennas and Propagation Magazine*, **46** (6), 2004, pp. 69–89.
4. L. B. Felsen, N. Marcuvitz, *Radiation and Scattering of Waves*, Prentice-Hall, Englewood Cliffs, NJ, 1973. classic reissue by IEEE Press, Piscataway, New Jersey, 1994.
5. D. G. Dudley, *Mathematical Foundations for Electromagnetic Theory*, IEEE Press (Series on Electromagnetic Waves), New York, 1994.
6. A. D. Pierce, "Extension of the Method of Normal Modes to Sound Propagation in almost Stratified Medium," *The Journal of the Acoustical Society of America*, **37**, 1965, pp. 19–27.

7. A. D. Pierce, "Guided Mode Disappearance during Upslope Propagation in Variable Depth Shallow Water Overlying a Fluid Bottom," *The Journal of the Acoustical Society of America*, **72**, 1982, pp. 523–631.
8. J. M. Arnold, L. B. Felsen, "Intrinsic Modes in a Wedge Shaped Ocean," *The Journal of the Acoustical Society of America*, **76**, 1984, pp. 850–860.
9. E. Topuz, L. B. Felsen, "Intrinsic Modes: Numerical Implementation in a Wedge-Shaped Ocean," *The Journal of the Acoustical Society of America*, **88**, 1985, pp. 1735–1745.
10. J. M. Arnold, A. Bewlgharof, A. Dendane, "Intrinsic Mode Theory of Tapered Optical Waveguides," *IEE Proceedings J*, **6**, 1985, pp. 314–318.
11. J. M. Arnold, L. B. Felsen, "Local Intrinsic Modes: Layer with Non-Planar Interface," *Wave Motion*, **8**, 1986, pp. 1–14.
12. J. M. Arnold, A. Dendane, "Intrinsic Mode Theory of Corrugated Horns," *IEE Proceedings H*, 1988, pp. 511–518.
13. M. Cada, F. Xiang, L. B. Felsen, "Intrinsic Modes in Tapered Optical Waveguides," *IEEE Journal of Quantum Electronics*, **24**, 1988, pp. 758–765.
14. M. Cada, F. Xiang, L. B. Felsen, "A Substantially Improved Treatment of Intrinsic Modes in Tapered Optical Waveguides," *IEEE Journal of Quantum Electronics*, **25**, 1989, pp. 933–938.
15. J. M. Arnold, L. B. Felsen, "Spectral Reconstruction of Uniformized Wavefields from Non-Uniform Ray or Adiabatic Mode Forms for Acoustic Propagation and Diffraction," *The Journal of the Acoustical Society of America*, **87**, 1990, pp. 587–600.
16. M. Abramowitz, I. A. Stegun, *Handbook of Mathematical Functions*, Dover, New York, 1969.
17. E. Topuz, L. Sevgi, "Intrinsic Mode Representations of the Green's Functions for a Benchmark Problem: The Non-Penetrable Wedge," *JEMWA Journal of Electromagnetic Waves and Applications*, **9** (7/8), 1995, pp. 1065–1082.
18. L. Sevgi, "Local Intrinsic Modes: A New Method to Solve Non-Separable Wave Problems," Ph.D. Dissertation, ITU Institute of Science, Istanbul, 1990.
19. L. Sevgi, F. Akleman, L.B. Felsen, "Visualizations of Wave Dynamics in a Wedge-Waveguide with Non-penetrable Boundaries: Normal, Adiabatic, and Intrinsic Mode Representations," *IEEE Antennas and Propagation Magazine*, **49** (3), June 2007, pp. 76–94.

CHAPTER 10

High Frequency Asymptotics: The 2D Wedge Diffraction Problem

Electromagnetic (EM) waves interact with objects and scatter. The word *scattering* includes *reflection*, *refraction*, and *diffraction*. Early analytical studies on EM wave scattering are usually named high frequency asymptotics (HFA) (see Refs. 1–4, for recent tutorials and MATLAB packages). HFA methods include geometrical optics (GO), geometrical theory of diffraction (GTD), its uniform extension uniform theory of diffraction (UTD) [5–11], physical optics (PO), physical theory of diffraction (PTD), elementary edge waves (EEWs), and parabolic equation (PE) methods [12–21].The GO, GTD, and UTD are *ray-based* diffraction methods; PO and PTD are *induced-source-based* methods. The GO describes a wave field in the limiting case when a wavelength tends to zero. It deals with reflection and refraction on the illuminated side of objects and cannot take into account the diffracted field. According to GO, the field in the shadow region behind the opaque objects equals zero; including the points on the object surface. On the illuminated side of the object, the GO field is the sum of the incident and reflected waves.

HFA techniques are reviewed in this chapter through a classical, canonical problem: EM wave scattering from a wedge shaped object with perfectly electrical conductor (PEC) boundaries. A MATLAB-based virtual diffraction tool WedgeGUI, using analytical exact as well

Electromagnetic Modeling and Simulation, First Edition. Levent Sevgi.

as well-known HFA techniques, is also introduced. Both line source (LS) and plane wave (PW) illuminations may be analyzed with this tool. Effects of various parameters on the reflection, refraction, and diffraction can be investigated. Comparisons among GO, UTD, PO, PTD, and PE models through many scenarios and illustrations are possible.

10.1 INTRODUCTION

Electromagnetic (EM) wave scattering has been substantially investigated for many decades and just reviewed with a tutorial in Ref. 1 on the canonical 2D wedge problem. Numerical difficulties of various analytical models based on complex integration as well as series summation are also discussed in details. Comparisons against numerical techniques such as the finite-difference time-domain (FDTD) method are given in Refs. 3 and 4. In all these studies, wave pieces such as *reflection*, *refraction*, and *diffraction* which are the components of scattering are revisited through analytical exact as well as HFA models.

This wedge diffraction problem plays a fundamental role in the construction of HFA techniques as well as for numerical tests. The exact solution to this scattering problem was first obtained by Sommerfeld [22] in the particular case of a half-plane. For a wedge with an arbitrary angle between its faces, the solution was obtained by Macdonald [23] and later on by Sommerfeld, who has developed the method of branched wave functions [24].

The 2D wedge scattering scenario is pictured in Fig. 10.1. The semi-infinite wedge with PEC boundaries is located in free space. The polar coordinates r, φ, z are used here. The z-axis is aligned along the edge of the wedge. The angle φ is measured from the top face of the wedge. The exterior angle of the wedge equals α. The wedge is illuminated by an LS at a distance r_0 from a direction $\varphi = \varphi_0$. In other words, source and observation points are given by (r_0,φ_0) and (r,φ), respectively.

The scenario for the LS-1 ($0 < \varphi_0 < \pi$) belongs to the single side illumination (SSI) where *the top face is always illuminated*. In this case, the 2D scattering plane around the wedge may be divided into three regions in terms of critical wave phenomena occurring there. In region-I ($0 \leq \varphi < \pi - \varphi_0$), all the field components—incident field, reflected

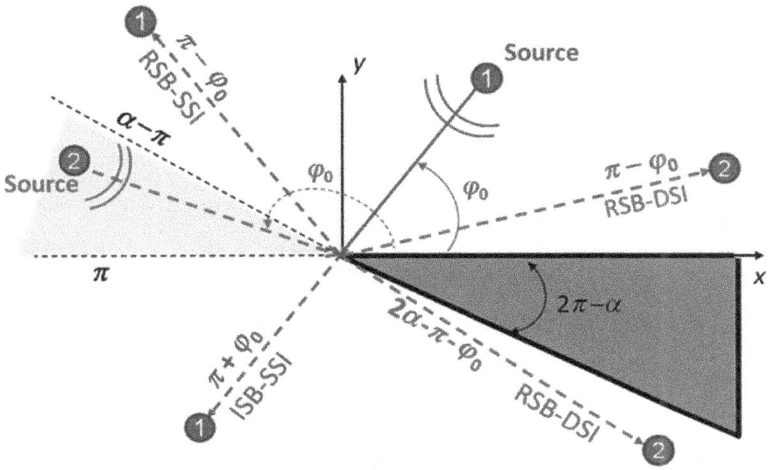

Figure 10.1. Geometry of the wedge scattering problem. SSI, single side illumination; DSI, double side illumination.

field, and diffracted field—exist. The angle $\varphi = \pi - \varphi_0$ is the limiting boundary of reflected fields and region I (RSB). In region-II $(\pi - \varphi_0 \leq \varphi < \pi + \varphi_0)$, only incident and diffracted fields exist. The angle $\varphi = \pi + \varphi_0$ is the limiting boundary of the incident field and region-II (*incident shadow boundary* [ISB]). In region-III (i.e., in the shadow region, $\pi + \varphi_0 \leq \varphi \leq \alpha$), only diffracted fields exist.

The scenario for the LS-2 $(\alpha - \pi < \varphi < \pi)$ belongs to the double side illumination (DSI) where *both faces are always illuminated.* In this case, the 2D scattering plane around the wedge may also be divided into three regions. In region-I $(0 \leq \varphi < \pi - \varphi_0)$, all the field components—incident field, reflected fields from the top face, and diffracted field—exist. The angle $\varphi = \pi - \varphi_0$ is the limiting boundary and called *reflection shadow boundary* (RSB). Similarly, in region-III $(2\alpha - \pi - \varphi_0 \leq \varphi < \alpha)$, all the field components—incident field, reflected fields from the bottom face, and diffracted field—exist. The angle $\varphi = 2\alpha - \pi - \varphi_0$ is also an *RSB*. The region between these two (i.e., region-II; $\pi - \varphi_0 \leq \varphi < 2\alpha - \pi - \varphi_0$) contains no reflected fields and only incident and diffracted fields exist. In summary, both incident and diffracted fields exist everywhere, but reflected fields exist only in region-1 and region-3.

Except for the UTD formulations, the time dependence $\exp(-i\omega t)$ is accepted here. The field outside the wedge satisfies the wave equation

$$\left(\frac{\partial^2}{\partial r^2}+\frac{1}{r}\frac{\partial}{\partial r}+\frac{1}{r^2}\frac{\partial^2}{\partial\varphi^2}+k^2\right)u=\frac{I_0}{r}\delta(r-r_0)\delta(\varphi-\varphi_0), \quad (10.1)$$

and the boundary conditions (BCs)

$$u_s=0 \quad \text{or} \quad \partial u_h/\partial n=0 \text{ on } \varphi=0,\alpha, \quad (10.2a)$$

and the Sommerfeld's radiation condition (SRC) at infinity,

$$\lim_{r\to\infty}\sqrt{kr}\left(\frac{du}{dr}-iku\right)=0. \quad (10.2b)$$

In the case of EM waves, these BCs are appropriate for the PEC wedge and function u_s represents the z-component of electric field intensity E_z (TM), while function u_h is the z-component of magnetic field intensity H_z (TE). In the case of acoustic waves, these conditions refer to acoustically soft (TM → SBC) and hard (TE → HBC) wedges, respectively.

10.2 PLANE WAVE ILLUMINATION AND HFA MODELS

Mathematical equations of different models are included here for both LS and PW illuminations. The PW models presented in this section are exact by series summation, PO, PTD, UTD, and PE models.

10.2.1 Exact Solution by Series Summation

The total field solutions of the Helmholtz equation with SBC and HBC for both SSI and DSI are [1]

$$u_{s,h}=u_0\cdot[u(kr,\varphi-\varphi_0)\mp u(kr,\varphi+\varphi_0)], \quad (10.3)$$

where

$$u(kr,\psi)=\frac{2\pi}{\alpha}\sum_{l=0}^{\infty}\varepsilon_l e^{-i(\pi/2)\nu_l}J_{\nu_l}(kr)\cos(\nu_l\psi), \quad (10.4a)$$

$$u_0 = \frac{1}{4i} I_0 H_0^{(1)}(kr_0). \tag{10.4b}$$

Here, $\nu_l = l\pi/\alpha$ and $\varepsilon_0 = 1/2$, $\varepsilon_1 = \varepsilon_2 = \varepsilon_3 = \ldots = 1$. The diffracted fields $u_s^{d,\text{exact}}$ and $u_h^{d,\text{exact}}$ can be calculated by subtracting suitable components (i.e., incident and/or reflected fields) from (10.3) in different regions as

$$u_{s,h}^{d,\text{exact}} / u_0 = u_{s,h}^{t,\text{exact}} / u_0 - u_{s,h}^{\text{GO}}. \tag{10.5}$$

where, for SSI ($0 < \varphi_0 < \alpha - \pi$),

$$u_{s,h}^{\text{GO}} = \begin{cases} e^{-ikr\cos(\varphi-\varphi_0)} \mp e^{-ikr\cos(\varphi+\varphi_0)}, & 0 \le \varphi < \pi - \varphi_0 \\ e^{-ikr\cos(\varphi-\varphi_0)}, & \pi - \varphi_0 \le \varphi < \pi + \varphi_0, \\ 0, & \pi + \varphi_0 \le \varphi \le \alpha \end{cases} \tag{10.6a}$$

and for DSI ($\alpha - \pi < \varphi_0 < \pi$),

$$u_{s,h}^{\text{GO}} = \begin{cases} e^{-ikr\cos(\varphi-\varphi_0)} \mp e^{-ikr\cos(\varphi+\varphi_0)}, & 0 \le \varphi < \pi - \varphi_0 \\ e^{-ikr\cos(\varphi-\varphi_0)}, & \pi - \varphi_0 \le \varphi < 2\alpha - \pi - \varphi_0. \\ e^{-ikr\cos(\varphi-\varphi_0)} \mp e^{-ikr\cos(2\alpha-\varphi-\varphi_0)}, & 2\alpha - \pi - \varphi_0 \le \varphi \le \alpha \end{cases} \tag{10.6b}$$

This model is based on the series summation in (10.4a) and the critical issue is the specification of the number of terms included which increases with frequency and/or distance.

Figure 10.2 presents total field versus angle around the wedge computed with both exact mode summation and the GO model. The interference of the incident and reflected fields dominates in region I. On this background, the contribution of the diffracted field is not appreciable away from the boundary of the reflected wave, where it is of the order of $1/\sqrt{kr}$. However, approaching this boundary, the diffracted field becomes comparable with the reflected field: exactly at this boundary it asymptotically equals one half of the reflected field. In region II, the diffracted field clearly reveals itself in the interference with the incident wave. Only diffracted fields contribute in region III.

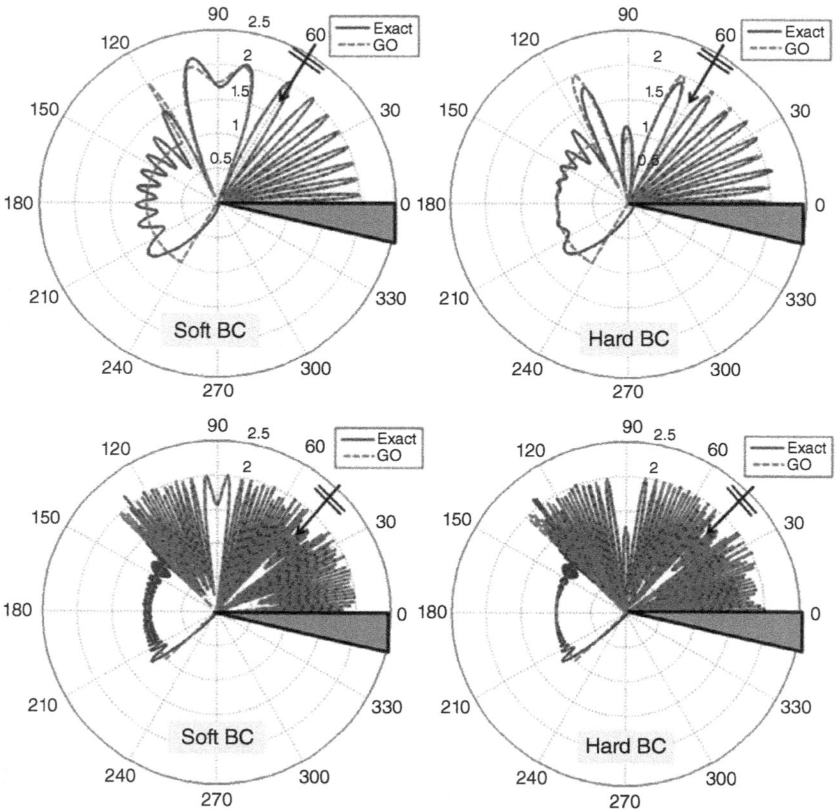

Figure 10.2. Exact versus GO solutions for the total fields. (Top) $\varphi_0 = 60°$, $r = 50$ m, $kr = 31.4159$, $N = 200$; (bottom) $\varphi_0 = 45°$, $r = 500$ m, $kr = 314.1593$, $N = 800$ ($\alpha = 350°$, $f = 30$ MHz).

10.2.2 The Physical Optics (PO) Solution

Physics optics (PO) takes uniform surface currents into account. The diffracted fields for the PO model are given as [1]

$$u_{s,h}^{d,\mathrm{PO}} / u_0 = v_{s,h}^{(0)}(kr, \varphi, \varphi_0) \tag{10.7a}$$

for SSI ($0 < \varphi_0 < \alpha - \pi$) and

$$u_{s,h}^{d,\mathrm{PO}} / u_0 = v_{s,h}^{(0)}(kr, \varphi, \varphi_0) + v_{s,h}^{(0)}(kr, \alpha - \varphi, \alpha - \varphi_0) \tag{10.7b}$$

for DSI ($\alpha - \pi < \varphi_0 < \pi$), where

$$v_s^{(0)}(kr,\varphi,\varphi_0) = \frac{1}{\pi\sqrt{2}} e^{i\left(kr+\frac{\pi}{4}\right)} \int_{-\infty}^{\infty} e^{-krs^2} \left[\frac{1}{2}\cot\frac{\varsigma+\psi_1-\pi}{2} - \frac{1}{2}\csc\frac{\varsigma+\psi_1-\pi}{2} \right.$$
$$\left. - \frac{1}{2}\cot\frac{\varsigma+\psi_2-\pi}{2} + \frac{1}{2}\csc\frac{\varsigma+\psi_2-\pi}{2}\right] \frac{ds}{\cos\frac{\varsigma}{2}}$$
$$- e^{-ikr\cos(\psi_1)} \frac{e^{-i\frac{\pi}{4}}}{\sqrt{\pi}} \int_{\sqrt{2kr}\cos\frac{\psi_1}{2}}^{\infty\cos\frac{\psi_1}{2}} e^{ix^2} dx$$
$$+ e^{-ikr\cos(\psi_2)} \frac{e^{-i\frac{\pi}{4}}}{\sqrt{\pi}} \int_{\sqrt{2kr}\cos\frac{\psi_2}{2}}^{\infty\cos\frac{\psi_2}{2}} e^{ix^2} dx \tag{10.8a}$$

$$v_h^{(0)}(kr,\varphi,\varphi_0) = \frac{1}{\pi\sqrt{2}} e^{i\left(kr+\frac{\pi}{4}\right)} \int_{-\infty}^{\infty} e^{-krs^2} \left[\frac{1}{2}\cot\frac{\varsigma+\psi_1-\pi}{2} - \frac{1}{2}\csc\frac{\varsigma+\psi_1-\pi}{2} \right.$$
$$\left. + \frac{1}{2}\cot\frac{\varsigma+\psi_2-\pi}{2} - \frac{1}{2}\csc\frac{\varsigma+\psi_2-\pi}{2}\right] \frac{ds}{\cos\frac{\varsigma}{2}}$$
$$- e^{-ikr\cos(\psi_1)} \frac{e^{-i\frac{\pi}{4}}}{\sqrt{\pi}} \int_{\sqrt{2kr}\cos\frac{\psi_1}{2}}^{\infty\cos\frac{\psi_1}{2}} e^{ix^2} dx$$
$$- e^{-ikr\cos(\psi_2)} \frac{e^{-i\frac{\pi}{4}}}{\sqrt{\pi}} \int_{\sqrt{2kr}\cos\frac{\psi_2}{2}}^{\infty\cos\frac{\psi_2}{2}} e^{ix^2} dx. \tag{10.8b}$$

Here, $\psi_1 = \varphi - \varphi_0$ and $\psi_2 = \varphi + \varphi_0$. The total fields, in these cases, can be obtained via

$$u_{s,h}^{t,\mathrm{PO}} / u_0 = u_{s,h}^{d,\mathrm{PO}} / u_0 + u_{s,h}^{\mathrm{GO}}, \tag{10.9}$$

with the same GO solutions given in (10.6). This model is based on the numerical integral of (8) [1] and needs special care.

10.2.3 The PTD Solution

The PO proceeds from the hypothesis that a surface current on the illuminated side of a scattering object is determined by GO. This hypothesis is acceptable in the case of smooth objects whose radii of curvature are large compared with the wavelength; however, it neither satisfies the *reciprocity principle* nor the *BCs*. The PO also fails to predict depolarization of the diffracted field. The PTD improves PO by taking into account nonuniform surface currents and concentrates near edges. The PTD model is given as [11]

$$u_{s,h}^{d,\mathrm{PTD}} / u_0 = v_{s,h}^{(0)} + \frac{e^{i\left(kr+\frac{\pi}{4}\right)}}{\sqrt{2\pi kr}} f_{s,h}^{(1)}(\varphi, \varphi_0, \alpha), \tag{10.10}$$

where

$$f_s^{(1)}(\varphi, \varphi_0, \alpha) = f_s(\varphi, \varphi_0, \alpha) - f_s^{(0)}(\varphi, \varphi_0, \alpha), \tag{10.11a}$$

$$f_h^{(1)}(\varphi, \varphi_0, \alpha) = f_h(\varphi, \varphi_0, \alpha) - f_h^{(0)}(\varphi, \varphi_0, \alpha), \tag{10.11b}$$

$$\begin{aligned} f_s(\varphi, \varphi_0, \alpha) = \frac{1}{2n}\Bigg[& \cot\frac{\pi-\varphi-\varphi_0}{2n} + \cot\frac{\pi+\varphi+\varphi_0}{2n} - \cot\frac{\pi-\varphi+\varphi_0}{2n} \\ & - \cot\frac{\pi+\varphi-\varphi_0}{2n}\Bigg], \end{aligned} \tag{10.12a}$$

$$\begin{aligned} f_h(\varphi, \varphi_0, \alpha) = -\frac{1}{2n}\Bigg[& \cot\frac{\pi-\varphi-\varphi_0}{2n} + \cot\frac{\pi+\varphi+\varphi_0}{2n} \\ & + \cot\frac{\pi-\varphi+\varphi_0}{2n} + \cot\frac{\pi+\varphi-\varphi_0}{2n}\Bigg], \end{aligned} \tag{10.12b}$$

$$f_s^{(0)}(\varphi, \varphi_0, \alpha) = \frac{\sin\varphi_0}{\cos\varphi_0 + \cos\varphi} + \varepsilon(\alpha - \varphi_0)\frac{\sin(\alpha-\varphi_0)}{\cos(\alpha-\varphi)+\cos(\alpha-\varphi_0)}, \tag{10.12c}$$

$$f_h^{(0)}(\varphi, \varphi_0, \alpha) = \frac{-\sin\varphi}{\cos\varphi_0 + \cos\varphi} + \varepsilon(\alpha - \varphi_0)\frac{-\sin(\alpha-\varphi)}{\cos(\alpha-\varphi)+\cos(\alpha-\varphi_0)}, \tag{10.12d}$$

and

$$\varepsilon(\varphi_0)=\begin{cases}0 & \text{if} \quad 0<\varphi_0<\alpha-\pi \\ 1 & \text{if} \quad \alpha-\pi<\varphi_0<\pi\end{cases}. \tag{10.13}$$

Again, the total fields, in these cases, can be obtained via

$$u_{s,h}^{t,\mathrm{PTD}} / u_0 = u_{s,h}^{d,\mathrm{PTD}} / u_0 + u_{s,h}^{\mathrm{GO}}, \tag{10.14}$$

with the same GO solutions given in (10.6). The PTD model requires the computation of the PO model first, then its modification with the extra term in (10.10) as explained in Ref. 1.

10.2.4 The UTD Solution

The GTD is the extension of GO to account for the diffraction. Unfortunately, GTD fails to express diffracted fields around ISB and RSB as well as near caustics. The UTD was introduced to overcome GTD's deficiency around ISB and RSB [6].

The UTD diffracted fields are in the form of

$$u_{s,h}^{d,\mathrm{UTD}} = u_0 d_{s,h}^{\mathrm{UTD}} \frac{e^{-jkr}}{\sqrt{r}}, \tag{10.15}$$

where $u_0 = H_0^{(2)}(kr_0)$ and the time dependence is $\exp(j\omega t)$. According to the UTD, the diffraction coefficients for SBC and HBCs with LS are as follows [6]:

$$d_s^{\mathrm{UTD}}(r,\varphi,\varphi_0,k) = \frac{-e^{-j\pi/4}}{2n\sqrt{2\pi k}} \begin{bmatrix} \cot\left(\frac{\pi-\xi^-}{2n}\right)F\left(kLg^+(\xi^-)\right)+\cot\left(\frac{\pi+\xi^-}{2n}\right)F(kLg^-(\xi^-)) \\ -\cot\left(\frac{\pi-\xi^+}{2n}\right)F\left(kLg^+(\xi^+)\right)-\cot\left(\frac{\pi+\xi^+}{2n}\right)F\left(kLg^-(\xi^+)\right) \end{bmatrix}, \tag{10.16a}$$

$$d_h^{\mathrm{UTD}}(r,\varphi,\varphi_0,k) = \frac{-e^{-j\pi/4}}{2n\sqrt{2\pi k}} \begin{bmatrix} \cot\left(\frac{\pi-\xi^-}{2n}\right)F\left(kLg^+(\xi^-)\right)+\cot\left(\frac{\pi+\xi^-}{2n}\right)F\left(kLg^-(\xi^-)\right) \\ +\cot\left(\frac{\pi-\xi^+}{2n}\right)F\left(kLg^+(\xi^+)\right)+\cot\left(\frac{\pi+\xi^+}{2n}\right)F\left(kLg^-(\xi^+)\right) \end{bmatrix}, \tag{10.16b}$$

where $\zeta^+ = \varphi + \varphi_0$, $\zeta^- = \varphi - \varphi_0$, and $F(X)$ is the Fresnel function given as

$$F(X)=2j\sqrt{X}e^{jX}\int_{\sqrt{X}}^{\infty} e^{-j\tau^2}d\tau \tag{10.17}$$

and L, $g^{\pm}$ are determined as given in [6] by

$$L=r,\ g^{\pm}(\xi)=2\cos^2\left(\frac{2n\pi N^{\pm}-\xi}{2}\right),\ N^{\pm}=\frac{\pm\pi+\xi}{2n\pi}. \tag{10.18}$$

Here, $N^{\pm}$ are the integers which most nearly satisfy the last equation given in (10.18). Note that the cotangent functions in (10.16a,b) become singular at ISB and RSB and are replaced as [6]

$$\cot\left(\frac{\pi\pm\beta}{2n}\right)F\left(kLg^{\pm}(\xi)\right)\simeq n\left[\sqrt{2\pi kL}\,\mathrm{sgn}(\varepsilon)-2kL\varepsilon e^{-j\frac{\pi}{4}}\right]e^{-j\frac{\pi}{4}} \tag{10.19}$$

for small ε. This term is finite but discontinuous at ISB and RSB and compensates for the discontinuities in the incident or reflected fields at these boundaries. The UTD-based total fields are then obtained by adding the GO fields appropriately.

Again, the total fields, in these cases, can be obtained via

$$u_{s,h}^{t,\mathrm{UTD}}/u_0=u_{s,h}^{d,\mathrm{UTD}}/u_0+u_{s,h}^{\mathrm{GO}} \tag{10.20}$$

with the GO solutions given for SSI ($0 < \varphi_0 < \alpha - \pi$) as

$$u_{s,h}^{\mathrm{GO}}=\begin{cases} e^{jkr\cos(\varphi-\varphi_0)}\pm e^{jkr\cos(\varphi+\varphi_0)}, & 0\le\varphi<\pi-\varphi_0\\ e^{jkr\cos(\varphi-\varphi_0)}, & \pi-\varphi_0\le\varphi<\pi+\varphi_0\\ 0, & \pi+\varphi_0\le\varphi\le\alpha \end{cases} \tag{10.21a}$$

and for DSI ($\alpha - \pi < \varphi_0 < \pi$) as

$$u_{s,h}^{\mathrm{GO}}=\begin{cases} e^{jkr\cos(\varphi-\varphi_0)}\pm e^{jkr\cos(\varphi+\varphi_0)}, & 0\le\varphi<\pi-\varphi_0\\ e^{-jkr\cos(\varphi-\varphi_0)}, & \pi-\varphi_0\le\varphi<2\alpha-\pi-\varphi_0.\\ e^{jkr\cos(\varphi-\varphi_0)}\pm e^{jkr\cos(2\alpha-\varphi-\varphi_0)}, & 2\alpha-\pi-\varphi_0\le\varphi\le\alpha \end{cases} \tag{10.21b}$$

Note that the UTD diffraction formulas are exactly the same for the LS excitation, except $L = r$ in (10.17) is replaced with $L = rr_0/(r + r_0)$. The critical issue of the UTD model is the numerical computation of the Fresnel integral (10.17) [1,6].

10.2.5 The Parabolic Equation (PE) Solution

The method of PE provides a correct first-order approximation to the diffracted field in the case when $kr >> 1$ and $kr_0 >> 1$ [12],

$$u_{s,h}^{d,\mathrm{PE}} = u_0 e^{ikr} W_{s,h}(kr, \varphi, \varphi_0), \tag{10.22}$$

where

$$W_{s,h}(kr, \varphi, \varphi_0) = \frac{e^{i\frac{\pi}{4}}}{\pi\sqrt{2}}[w(kr, \psi_1) \pm w(kr, \psi_2)], \tag{10.23a}$$

$$w(kr, \psi) = \int_{-\infty}^{\infty} \frac{\frac{1}{n}\sin\left(\frac{\pi}{n}\right) e^{-krs^2}}{\cos\left(\frac{\pi}{n}\right) - \cos\left(\frac{\eta + \psi}{n}\right)} ds, \tag{10.23b}$$

with $\psi_{1,2} = \varphi \mp \varphi_0$, $n = \alpha/\pi$, and $\eta = \sqrt{2}e^{-\frac{i\pi}{4}}s$. Again, total fields, in these cases, can be obtained via

$$u_{s,h}^{t,\mathrm{PE}} / u_0 = u_{s,h}^{d,\mathrm{PE}} / u_0 + u_{s,h}^{\mathrm{GO}}, \tag{10.24}$$

with the same GO solutions given in (10.6). The critical issue of the PE model is the numerical computation of the integral given in (10.23b) [1,12].

10.3 HFA MODELS UNDER LINE SOURCE (LS) EXCITATIONS

Mathematical equations of the models exact by series, exact by integral, the UTD, and PE under LS illumination are presented in this section. Note that numerical results of these models match with the PW models given in the previous section when the source-to-tip distance is large as compared with the wavelength.

10.3.1 Exact Solution by Series Summation

The total field solutions of the Helmholtz equation with SBC and HBC for both SSI and DSI are [1]

$$u_s^{t,\text{Exact}} = \begin{cases} \dfrac{4\pi}{\alpha}\displaystyle\sum_{l=1}^{\infty} J_{\upsilon_l}(kr)H_{\upsilon_l}^{(1)}(kr_0)\sin(\upsilon_l\varphi_0)\sin(\upsilon_l\varphi)\ldots & r \le r_0 \\ \dfrac{4\pi}{\alpha}\displaystyle\sum_{l=1}^{\infty} J_{\upsilon_l}(kr_0)H_{\upsilon_l}^{(1)}(kr)\sin(\upsilon_l\varphi_0)\sin(\upsilon_l\varphi)\ldots & r \ge r_0 \end{cases}, \tag{10.25a}$$

$$u_h^{t,\text{Exact}} = \begin{cases} \dfrac{4\pi}{\alpha}\displaystyle\sum_{l=0}^{\infty} \varepsilon_l J_{\upsilon_l}(kr)H_{\upsilon_l}^{(1)}(kr_0)\cos(\upsilon_l\varphi_0)\cos(\upsilon_l\varphi)\ldots & r \le r_0 \\ \dfrac{4\pi}{\alpha}\displaystyle\sum_{l=0}^{\infty} \varepsilon_l J_{\upsilon_l}(kr_0)H_{\upsilon_l}^{(1)}(kr)\cos(\upsilon_l\varphi_0)\cos(\upsilon_l\varphi)\ldots & r \ge r_0 \end{cases}, \tag{10.25b}$$

with the same u_0 given in (10.4b). Normalized diffracted fields $u_{s,h}^{d,\text{Exact}}$ can be obtained by subtracting GO fields,

$$u_{s,h}^{d,\text{Exact}} / u_0 = \left(u_{s,h}^{t,\text{Exact}} - \frac{I_0}{4i} u_{s,h}^{\text{GO}} \right) \Big/ u_0. \tag{10.26}$$

The GO field that under LS excitation is given as follows:

$$u_{s,h}^{\text{GO}} = \begin{cases} H_0^{(1)}(kR_1) \pm H_0^{(1)}(kR_2), & 0 \le \varphi < \pi - \varphi_0 \\ H_0^{(1)}(kR_1), & \pi - \varphi_0 \le \varphi < \pi + \varphi_0 \\ 0, & \pi + \varphi_0 \le \varphi \le \alpha \end{cases} \tag{10.27a}$$

for SSI and

$$u_{s,h}^{\text{GO}} = \begin{cases} H_0^{(1)}(kR_1) \pm H_0^{(1)}(kR_2), & 0 \le \varphi < \pi - \varphi_0 \\ H_0^{(1)}(kR_1), & \pi - \varphi_0 \le \varphi < 2\alpha - \pi - \varphi_0 \\ H_0^{(1)}(kR_1) \pm H_0^{(1)}(kR_3), & 2\alpha - \pi - \varphi_0 \le \varphi \le \alpha \end{cases} \tag{10.27b}$$

for DSI, where (−) and (+) are for SBC and HBCs, respectively, and

$$R_1 = \sqrt{r^2 + r_0^2 - 2rr_0\cos(\varphi - \varphi_0)}, \tag{10.28a}$$

$$R_2 = \sqrt{r^2 + r_0^2 - 2rr_0 \cos(\varphi + \varphi_0)}, \tag{10.28b}$$

$$R_3 = \sqrt{r^2 + r_0^2 - 2rr_0 \cos(2\alpha - \varphi - \varphi_0)}. \tag{10.28c}$$

This model is based on the series summation in (10.25) and the critical issue is the specification of the number of terms included which increases with frequency and/or distance. One also needs to pay attention to the source/receiver distances [1].

10.3.2 Exact Solution by Integral

Solutions for LS illumination are also well known and presented by Bowman and Senior in [8] as

$$\begin{aligned} u_s^{d,\mathrm{BS}} &= \{V_d(-\pi - \varphi + \varphi_0) - V_d(\pi - \varphi + \varphi_0)\} \\ &\quad - \{V_d(-\pi - \varphi - \varphi_0) - V_d(\pi - \varphi - \varphi_0)\}, \end{aligned} \tag{10.29a}$$

$$\begin{aligned} u_h^{d,\mathrm{BS}} &= \{V_d(-\pi - \varphi + \varphi_0) - V_d(\pi - \varphi + \varphi_0)\} \\ &\quad + \{V_d(-\pi - \varphi - \varphi_0) - V_d(\pi - \varphi - \varphi_0)\}, \end{aligned} \tag{10.29b}$$

where

$$V_d(\beta) = \frac{1}{2\pi n} \int_0^{\infty} H_0^{(1)}[kR(it)] \frac{\sin(\beta / n)}{\cosh(t/n) - \cos(\beta/n)} dt, \tag{10.30}$$

$n = \alpha/\pi$, and $R(\eta) = \sqrt{r^2 + r_0^2 + 2rr_0 \cos(\eta)}$. In this case, total fields are obtained by adding the GO fields:

$$u_{s,h}^{t,\mathrm{BM}} / u_0 = \left(u_{s,h}^{d,\mathrm{BM}} + u_{s,h}^{\mathrm{GO}}\right) / u_0, \tag{10.31}$$

with the same GO solutions given in (10.27). The critical issue of the Bowman–Senior model is the numerical computation of the integral given in (10.30) [1].

10.3.3 The Parabolic Equation (PE) Solution

The PE solution under a LS illumination when $kr >> 1$ and $kr_0 >> 1$ can be given as [12]

$$u_{s,h}^{d,\mathrm{PE}} / u_0 = e^{ikr} \sqrt{\frac{r_0}{r + r_0}} W_{s,h}(k\rho, \varphi, \varphi_0), \ \rho = \frac{rr_0}{r + r_0}, \tag{10.32}$$

where

$$W_{s,h}(k\rho,\varphi,\varphi_0)=\frac{e^{i\frac{\pi}{4}}}{\pi\sqrt{2}}[w(k\rho,\psi_1)\pm w(k\rho,\psi_2)], \tag{10.33a}$$

$$w(k\rho,\psi)=\int_{-\infty}^{\infty}\frac{\frac{1}{n}\sin\left(\frac{\pi}{n}\right)e^{-k\rho s^2}}{\cos\left(\frac{\pi}{n}\right)-\cos\left(\frac{\eta+\psi}{n}\right)}ds, \tag{10.33b}$$

with $\psi_{1,2}=\varphi\mp\varphi_0$, $n=\alpha/\pi$, and $\eta=\sqrt{2}e^{-\frac{i\pi}{4}}s$. Total fields, in these cases, can be obtained via

$$u_{s,h}^{t,\mathrm{PE}}/u_0=\left(u_{s,h}^{d,\mathrm{PE}}+\frac{I_0}{4i}u_{s,h}^{\mathrm{GO}}\right)\Big/u_0, \tag{10.34}$$

with the same GO solutions given in (10.27). This model also needs the numerical computation of the integral given in (10.33b) [1,4].

Note that analytical exact solutions of the total fields are continuous; therefore, their numerical implementations only require to take accuracy into account. This is achieved by controlling contributions of the higher-order terms in the series summation formulation and by dynamically changing integration steps in the integral formulation. The computations may be carried out for any angles including critical angles of the incident and shadow boundaries. On the other hand, diffracted fields are discontinuous at these two critical angles. However, at these angles, normalized modulus of the diffracted fields asymptotically tends to 1/2 and have the discontinuity of the order of $1/\sqrt{kr}$. Hence, for large $\sqrt{kr}$, these fields are graphically continuous.

10.4 BASIC MATLAB SCRIPTS

Different diffraction models listed earlier have different mathematical representations. The exact solution for either PW illumination (10.3)–(10.6) or for LS excitation (10.25)–(10.28) is based on mode summation approach. Although, mathematically exact, mode summation includes infinite number of terms; therefore, its accurate numerical computation is a challenge. An accuracy criterion has to be set for the termination of the summation after a finite number of terms. Table 10.1 lists a sample MATLAB code for the computation of total and

Table 10.1.
A Sample MATLAB Code for the Exact Mode Summation (10.3)–(10.6).

```
%-------------------------------------------------------
% Program: WedgeExact.m
%-------------------------------------------------------
clear all;clc; close all; format long
alfamax = 350; %input('wedge angle [Deg] = ');
I0 = 1; %input('strength of line source  = ');
r = 30; %input('distance of the observer to wedge  =
');
r0 = 25; %input('distance of the source to wedge  = ');
angle0 = 45; %input('incident angle [Deg] = ');
freq = 2e7; %input('frequency = ');
N = 200; %input('Number of Terms in Exact Solution (N)
= ');
c = 3e8; % speed of light
alfa = alfamax*pi/180; % change wedge angle degree to
radians
angle0 = angle0*pi/180; % change incident angle degree
to radians
k = 2*pi*freq/c; kr = k*r; kr0 = k*r0;
coeff1 = 2*pi/alfa; coeff2 = exp(1i*(kr+pi/4))/
sqrt(2*pi*kr);
coeff3 = exp(-1i*pi/4)/sqrt(pi);
U0 = I0*besselh(0,1,kr0)/(4*1i); % amplitude of the
incident field
A = 0.01:.5:alfamax; [dd nA]=size(A); Ar=A*pi/180;  %
Obs angles
Rsoft=-1; % Reflection coefficient for soft boundary
conditions
Rhard=1;  % Reflection coefficient for hard boundary
conditions
source=1; % Source=1 Plane wave, Source=0 Line source
for m=1:nA
 fprintf(1, 'Calculating Angle %5.1f :  \n',A(m)); angle
= Ar(m);
 if (angle >= 0) & (angle < pi-angle0)   % Region I
   Us_old=1e6; Us=0; err=1e6; n=0; vn=0;
   while N > n
```

(*Continued*)

Table 10.1.
(Continued)

```
    vn = n*pi/alfa;  if n == 0 eps = 0.5; else eps =
1; end
    cc = eps*besselj(vn,kr)*exp(-1i*pi*vn/2);
    Term1=cc*cos(vn*(angle-angle0)); Term2=cc*cos(vn*(an
gle+angle0));
    Us=Us+U0*coeff1*(Term1-Term2); err=Us_old-Us;Us_
old=Us; n=n+1;
   End
   Us1=Us-U0*(exp(-1i*kr*cos(angle-angle0))
         +Rsoft*exp(-1i*kr*cos(angle+angle0)));
   Uh_old=1e6; Uh=0; err=1e6; n=0; vn=0;
   while N > n
     vn = n*pi/alfa;  if n == 0 eps = 0.5; else eps
= 1; end
     cc = eps*besselj(vn,kr)*exp(-1i*pi*vn/2);
     Term1=cc*cos(vn*(angle-angle0)); Term2=cc*cos(vn*(
angle+angle0));
     Uh=Uh+U0*coeff1*(Term1+Term2); err=Uh_old-Uh; Uh_
old=Uh; n=n+1;
   end
   Uh1 = Uh-U0*(exp(-1i*kr*cos(angle-angle0))
           +Rhard*exp(-1i*kr*cos(angle+angle0)));
   Us1_exact_diff(m)=Us1/U0; Uh1_exact_diff(m)=Uh1/U0;
% exact DF
   Us1_exact_total(m)=Us/U0; Uh1_exact_total(m)=Uh/U0;
% exact TF
  elseif  (angle >= pi-angle0) & (angle < pi+angle0)
% Region II
   Us_old=1e6; Us=0; err=1e6; n=0; vn=0;
   while N > n
     vn = n*pi/alfa; if n == 0 eps = 0.5; else eps =
1; end
     cc = eps*besselj(vn,kr)*exp(-1i*pi*vn/2);
     Term1=cc*cos(vn*(angle-angle0)); Term2=cc*cos(vn*(
angle+angle0));
     Us=Us+U0*coeff1*(Term1-Term2); err=Us_old-Us;Us_
old=Us; n=n+1;
   end
   Us1 = Us-U0*(exp(-1i*kr*cos(angle-angle0)));
```

Table 10.1.
(Continued)

```
    Uh_old=1e6; Uh=0; err=1e6; n=0; vn=0;
    while N > n
      vn = n*pi/alfa;  if n == 0 eps = 0.5; else eps
= 1; end
      cc = eps*besselj(vn,kr)*exp(-1i*pi*vn/2);
      Term1=cc*cos(vn*(angle-angle0)); Term2=cc*cos(vn*(
angle+angle0));
      Uh=Uh+U0*coeff1*(Term1+Term2); err=Uh_old-Uh; Uh_
old=Uh; n=n+1;
    end
    Uh1 = Uh-U0*(exp(-1i*kr*cos(angle-angle0)));
    Us1_exact_diff(m)=Us1/U0; Uh1_exact_diff(m)=Uh1/U0;
% exact DF
    Us1_exact_total(m)=Us/U0; Uh1_exact_total(m)=Uh/U0;
% exact TF
   elseif (angle >= pi+angle0) & (angle <= alfa)
% Region III
    Us_old=1e6; Us=0; err=1e6; n=0; vn=0;
    while N > n
      vn = n*pi/alfa;  if n == 0 eps = 0.5; else eps
= 1; end
      cc = eps*besselj(vn,kr)*exp(-1i*pi*vn/2);
      Term1=cc*cos(vn*(angle-angle0)); Term2=cc*cos(vn*(
angle+angle0));
      Us=Us+U0*coeff1*(Term1-Term2); err=Us_old-Us;Us_
old=Us; n=n+1;
    end
    Us1 = Us;  Uh_old=1e6; Uh=0; err=1e6; n=0; vn=0;
    while N > n
      vn = n*pi/alfa;  if n == 0 eps = 0.5; else eps
= 1; end
      cc = eps*besselj(vn,kr)*exp(-1i*pi*vn/2);
      Term1=cc*cos(vn*(angle-angle0)); Term2=cc*cos(vn*(
angle+angle0));
      Uh=Uh+U0*coeff1*(Term1+Term2); err=Uh_old-Uh; Uh_
old=Uh; n=n+1;
    end
    Uh1 = Uh;
```

(Continued)

Table 10.1.
(Continued)

```
   Us1_exact_diff(m)=Us1/U0; Uh1_exact_diff(m)=Uh1/U0;
% exact DF
   Us1_exact_total(m)=Us/U0; Uh1_exact_total(m)=Uh/U0;
% exact TF end
end
figure(1)    % Diffracted fields for Soft and Hard BC
polar(Ar,abs(Us1_exact_diff)); hold on;
polar(Ar,abs(Uh1_exact_diff),'r-');
title('Diffracted Fields vs. Angle'); legend('SBC','HBC');
figure(2)   % Total fields for Soft and Hard BC
polar(Ar,abs(Us1_exact_total)); hold on;
polar(Ar,abs(Uh1_exact_total),'r-');
title('Total Fields vs. Angle'); legend('SBC','HBC');
%-------------------- END ----------------------
```

diffracted fields under PW illumination for both SBC and HBC. Note that the number of terms in this series summation increases drastically (1) as the frequency increases and (2) as the observation point moves away from the tip of the wedge.

Table 10.2 lists a short MATLAB script for the calculation of PO and PTD fields using (10.7)–(10.14) for both polarizations under a LS illumination. The script first computes PO fields and requires a complex integration (10.8). This integral has a highly oscillatory term; therefore, contour deformation is used to speed up numerical integration [1]. A sample code for this integration is listed in Table 10.3.

Table 10.4 lists a short MATLAB code for the UTD formulations given with (10.15)–(10.22) which can be used for both PW and LS illuminations. Note that the UTD code listed in Table 10.4 uses the Fresnel integral in (10.17). A short code for the numerical evaluation of this integral using MATLAB command ***trapz()*** is listed in Table 10.5.

The final MATLAB script listed in Table 10.6 belongs to the numerical computation of PE fields (10.32)–(10.34). The PE representation also includes a highly oscillatory complex integral (10.33b); therefore, attention must be paid its numerical evaluation. A simple code for its numerical evaluation is listed in Table 10.7.

Table 10.2.
A Sample MATLAB Code for the PO-PTD Formulations

```
%-----------------------------------------------------
% Program : WedgePOPTD.m
%-----------------------------------------------------
clear all;clc; close all; format long
alfamax=350; I0=1; r=100; r0=300; angle0=150;
angle0=150; freq=3e7; c=3e8; k=2*pi*freq/c;
alfa=alfamax*pi/180; angle0=angle0*pi/180; kr=k*r;
kr0=k*r0; error=1e-12;
A = 0:1.13:alfamax; [dd nA]=size(A); Ar=A*pi/180;
phi0 = -angle0; n = (alfa)/pi; coeff = exp(-1i*pi/4)/
(2*n*sqrt(2*pi*k));
B = -Ar; coeff1 = 2*pi/alfa; coeff2 = exp(i*(kr+pi/4))/
sqrt(2*pi*kr);
coeff3 = exp(-1i*pi/4)/sqrt(pi); U0 =
I0*besselh(0,1,kr0)/(4*1i);
Rsoft=-1; Rhard=1;
for m=1:nA
 angle = Ar(m);
 fprintf(1, 'Calculating PO+PTD Model for LS = > Angle
%5.1f :  \n',A(m));
%% *******PO and PTD TOTAL and DIFFRACTED FIELDS
***************
psi1 = angle-angle0; psi2 = angle+angle0;
t0_1 = sqrt(2*kr)*cos(psi1/2);  t0_2 =
sqrt(2*kr)*cos(psi2/2);
I1 = VInt(t0_1);  I2 = VInt(t0_2);v_plus =
-sin(psi1/2)*exp(-1i*kr*cos(psi1))*I1;
v_minus = sin(psi2/2)*exp(-1i*kr*cos(psi2))*I2;
vs_0 = coeff3*(v_plus+v_minus);  vh_0 =
coeff3*(v_plus-v_minus);
Us_PO_diff(m) = vs_0; % PO diffracted field for soft BC
Uh_PO_diff(m) = vh_0; % PO diffracted field for hard BC
%% ***************** Calculate f1 and g1
********************
angle1 = (pi-psi2)/(2*n);   angle2 = (pi-psi1)/(2*n);
angle3 = (pi+psi2)/(2*n);   angle4 = (pi+psi1)/(2*n);
```

(*Continued*)

Table 10.2.
(Continued)

```
if (psi1 == pi)
   f = (1/(2*n))*(cot(angle1)+cot(angle3)-cot(angle4));
   g = -(1/(2*n))*(cot(angle1)+cot(angle3)+cot(angle4));
   f0 = 0.5*(tan(psi2/2));     g0 = -0.5*(tan(psi2/2));
   f1 = f-f0;   g1 = g-g0;
elseif (psi2 == pi)
   f = (1/(2*n))*(-cot(angle2)+cot(angle3)-cot(angle4));
   g = -(1/(2*n))*(cot(angle2)+cot(angle3)+cot(angle4));
   f0 = -0.5*(tan(psi1/2));   g0 = -0.5*(tan(psi1/2));
   f1 = f-f0;   g1 = g-g0;
else
   f = (1/
(2*n))*(cot(angle1)-cot(angle2)+cot(angle3)-cot(angle4));
   g = -(1/(2*n))*(cot(angle1)+cot(angle2)+cot(angle3)+c
ot(angle4));
   f0 = 0.5*(tan(psi2/2)-tan(psi1/2));
   g0 = -0.5*(tan(psi2/2)+tan(psi1/2));
   f1 = f-f0;   g1 = g-g0;
end
vs_1(m) = f1*coeff2; vh_1(m) = g1*coeff2;
Us_PTD_diff(m) = Us_PO_diff(m)+vs_1(m); % PTD diffracted
field for SBC
Uh_PTD_diff(m) = Uh_PO_diff(m)+vh_1(m); % PTD diffracted
field for HBC
%% ************ Compute total PO and PTD fields
****************
 if (angle >= 0) & (angle < pi-angle0)
    Us_PO_total(m) = Us_PO_diff(m)+exp(-1i*kr . . .
          *cos(angle-angle0))+Rsoft*exp(-1i*kr*cos(angle+
angle0));
    Uh_PO_total(m) = Uh_PO_diff(m)+exp(-1i*kr . . .
          *cos(angle-angle0))+Rhard*exp(-1i*kr*cos(angle+
angle0));
    Us_PO_total(m) =
```

Table 10.2.
(Continued)

```
   Us_PTD_total(m) = Us_PTD_diff(m)+exp(-1i*kr . . .
        *cos(angle-angle0))+Rsoft*exp(-1i*kr*cos(angle+
angle0));
   Uh_PTD_total(m) = Uh_PTD_diff(m)+exp(-1i*kr . . .
        *cos(angle-angle0))+Rhard*exp(-1i*kr*cos(angle+
angle0));
 elseif  (angle >= pi-angle0) & (angle < pi+angle0)

Us_PO_diff(m)+exp(-1i*kr*cos(angle-angle0));
   Uh_PO_total(m) =
Uh_PO_diff(m)+exp(-1i*kr*cos(angle-angle0));
   Us_PTD_total(m) =
Us_PTD_diff(m)+exp(-1i*kr*cos(angle-angle0));
   Uh_PTD_total(m) =
Uh_PTD_diff(m)+exp(-1i*kr*cos(angle-angle0));
 elseif (angle >= pi+angle0) & (angle <= alfa)
   Us_PO_total(m) = Us_PO_diff(m); Uh_PO_total(m) =
Uh_PO_diff(m);

Us_PTD_total(m)=Us_PTD_diff(m); Uh_PTD_total(m)=Uh_PTD_
diff(m);
 end
end
%% ------------ POLAR PLOTS ------------
polar(Ar,abs(Us_PO_diff),'-r'); hold on;
polar(Ar,abs(Us_PTD_diff),'-.k');
legend('u^d^,^P^O / u_0','u^d^,^P^T^D / u_0');
figure(2)
polar(Ar,abs(Us_PO_total),'-r'); hold on;
polar(Ar,abs(Us_PTD_total),'-.k');
legend('u^t^o^t^,^P^O / u_0','u^t^o^t^,^P^T^D / u_0');
% --------------------- END ---------------------
```

Table 10.3.
Numerical Calculation of the Integrals in (10.8)

```
function result = VInt(t0)
% ------------------------------------------------
% Numerical integration of (8)
% t0 is the upper integral limit
% ------------------------------------------------
  M=20; Ml=0.005;Mr=80;
  if abs(t0) <= 1e-12; a0=t0; t0=0.0; end
  if (t0 == 0)
    if (a0 < 0.0)
      result = -sqrt(pi)/2*exp(1i*pi/4);
    elseif (a0 >= 0.0)
      result = sqrt(pi)/2*exp(1i*pi/4);
    end
  elseif (t0 > 0) && (t0 < Ml)
    dq=t0/100; q = 0.0:dq:t0; y = exp(1i*q.*q);
    result = -trapz(q,y)+sqrt(pi)/2*exp(1i*pi/4);
  elseif (t0 >= Ml) && (t0 < Mr)
    dq=0.0001; qr = t0:dq:t0+M; qi = 0:dq:M;
    q=complex(qr,qi); y=exp(1i*q.*q);
result=trapz(q,y);
  elseif(t0 >= Mr)
    result = -exp(1i*t0^2)/(2*1i*t0);
  elseif (t0 < 0) && (t0 > -Ml)
    dq=abs(t0)/100; q = 0.0:-dq:t0; y = exp(1i*q.*q);
    result = -trapz(q,y)-sqrt(pi)/2*exp(1i*pi/4);
  elseif (t0 <= -Ml) && (t0 > -Mr)
    dq=0.0001; qr = t0:-dq:t0-M; qi = 0:-dq:-M;
    q=complex(qr,qi); y=exp(1i*q.*q);
result=trapz(q,y);
  elseif(t0 <= -Mr)
    result = -exp(1i*t0^2)/(2*1i*t0);
  end
end
% ------------------- END --------------------
```

Table 10.4.
A Short MATLAB Code for the UTD Calculations

```
%------------------------------------------------------------
% Program : WedgeUTD.m
%------------------------------------------------------------
clear all;clc; close all; format long
alfamax=350; I0=1; r=100; r0=300; angle0=150; freq=3e7;
c=3e8; k=2*pi*freq/c; alfa=alfamax*pi/180;
angle0=angle0*pi/180; kr=k*r; kr0=k*r0; error=1e-12;
A = 0:1.13:alfamax; [dd nA]=size(A); Ar=A*pi/180;
phi0 = -angle0; n = (alfa)/pi; U0 =  besselh(0,2,kr0);
coeff = exp(-1i*pi/4)/(2*n*sqrt(2*pi*k)); B = -Ar;
for m=1:nA
  angle = Ar(m);
  fprintf(1, 'Calculating UTD (Line Source)=> Angle
%5.1f: \n',A(m));
  phi = B(m);  ksi_p = phi+phi0; ksi_m = phi-phi0;
  Npp=(pi+ksi_p)/(2*pi*n); % N+ for ksi+ & ksi-
  Npm=(pi+ksi_m)/(2*pi*n); % N+ for ksi+ & ksi-
  Nmp=(-pi+ksi_p)/(2*pi*n); % N- for ksi+ & ksi-
  Nmm=(-pi+ksi_m)/(2*pi*n); % N- for ksi+ & ksi-
% -------------- N+ values-------------------------
  if (Npp<1) || (Npm<1)
    Npp = 0; Npm = 0;
  elseif (Npp>=1) || (Npm>=1)
    Npp = 1; Npm = 1;-
  end
% ------------- N- values--------------------------
  if (Nmp<0) || (Nmm<0)
    Nmp = -1; Nmm = -1;
  elseif (0<=Nmp<1) || (0<=Nmm<1)
    Nmp = 0; Nmm = 0;
  elseif (Nmp>=1) || (Nmm>=1)
    Nmp = 1; Nmm = 1;
  end
  gpp = 1+cos(ksi_p-2*n*pi*Npp);  gpm =
1+cos(ksi_m-2*n*pi*Npm);
  gmp = 1+cos(ksi_p-2*n*pi*Nmp); gmm =
1+cos(ksi_m-2*n*pi*Nmm);
```

(Continued)

Table 10.4.
(Continued)

```
  x1 = k*r*r0*gpm/(r+r0); x2 = k*r*r0*gmm/(r+r0);   x3
= k*r*r0*gpp/(r+r0); x4 = k*r*r0*gmp/(r+r0);
  F1 = Fresnel_Int(x1); F2 = Fresnel_Int(x2); % Fresnel
integral
  F3 = Fresnel_Int(x3); F4 = Fresnel_Int(x4); % Fresnel
integral
  UsDIFF_UTD(m) = -coeff*(cot((pi+ksi_m)/(2*n))*F1+
. . .
      cot((pi-ksi_m)/(2*n))*F2-cot((pi+ksi_p)/(2*n))*F3
. . .
     -cot((pi-ksi_p)/(2*n))*F4); % coef for SBC
  UhDIFF_UTD(m) = -coeff*(cot((pi+ksi_m)/(2*n))*F1
. . .
     +cot((pi-ksi_m)/(2*n))*F2+cot((pi+ksi_p)/(2*n))*F3
. . .
     +cot((pi-ksi_p)/(2*n))*F4); % coef for HBC
  UsDIFF_UTD(m) =UsDIFF_UTD(m)*exp(1i*kr)/sqrt(r);
  UhDIFF_UTD(m) =UhDIFF_UTD(m)*exp(1i*kr)/sqrt(r);
% ----------- Total Field for UTD -------------
  if (angle >= 0) && (angle < pi-angle0)   % Region I
    R1=sqrt(r*r+r0*r0-2*r*r0*cos(phi-phi0));
    R2=sqrt(r*r+r0*r0-2*r*r0*cos(phi+phi0));
    Uxs=besselh(0,2,k*R1)-besselh(0,2,k*R2);
    Uxh=besselh(0,2,k*R1)+besselh(0,2,k*R2);
    UsTOTAL_UTD(m)=UsDIFF_UTD(m)+Uxs/U0; UhTOTAL_
UTD(m)=UhDIFF_UTD(m)+Uxh/U0;
  elseif  (angle >= pi-angle0) && (angle < pi+angle0)
% Region II
    R1=sqrt(r*r+r0*r0-2*r*r0*cos(phi-phi0));
    Ux=besselh(0,2,k*R1);
    UsTOTAL_UTD(m)=UsDIFF_UTD(m)+Ux/U0;
    UhTOTAL_UTD(m)=UhDIFF_UTD(m)+Ux/U0;
  elseif (angle >= pi+angle0) && (angle <= alfa)  %
Region III

UsTOTAL_UTD(m)=UsDIFF_UTD(m); UhTOTAL_UTD(m)=UhDIFF_
UTD(m);
  end
end
polar(Ar,abs(UsDIFF_UTD),'r'); hold on; legend('UTD');
```

Table 10.4.
(Continued)

```
ttle=['Soft BC / Freq= ', num2str(freq/1e6),' MHz /', '
kr= ', num2str(kr),' / Alpha= ', num2str(alfamax),' Deg
','/ Incident field= ', num2str(angle0*180/pi),' Deg'];
title(ttle);
% -------------------- END ------------------
```

Table 10.5.
Numerical Calculation of Fresnel Integral

```
function result = Fresnel_Int(x)
% ----------------------------------------------------
% Numerical integration of exp(-iq*q) from sqrt(x) to
infinity
% ----------------------------------------------------
  Ml = 0.3;  Mu = 5.5; coeff = 2*1i*sqrt(x)*exp(1i*x);
  if (x < Ml)
    result=exp(1i*(pi/4+x))*(sqrt(p
i*x)-2*x*exp(1i*pi/4)-..
            (2/3)*x*x*exp(-1i*pi/4));
  elseif (x > Mu)
    result=1+1i/(2*x)-3/(4*x^2)-15*1i/(8*x^3)+75/
(16*x^4);
  else
  eps = 1e-12; Mmax = 5000;
  d = 1e-3; y_old = 1e6; ii = 0;
  for M=sqrt(x)+1:Mmax
    ii=ii+1;  t1=sqrt(x):d:M; t2=0:d:M-sqrt(x); t =
t1-1i*t2;
    fun=exp(-1i*t.*t); y(ii)=trapz(t,fun);
    error=abs(y(ii)-y_old);
    y_old = y(ii);
    if (error<eps)
      result = y(ii)*coeff; break;
    else
      continue
    end
  end
  result = y(ii)*coeff;
end
%------------------------ END ---------------------
```

Table 10.6.
Numerical Calculation of PE Fields

```
%---------------------------------------------------
% Program : WedgePOPTD.m
%---------------------------------------------------
clear all;clc; close all; format long
alfamax=350; I0=1; r=100; r0=300; angle0=150;
angle0=150;
freq=3e7; c=3e8; k=2*pi*freq/c; alfa=alfamax*pi/180; n
= (alfa)/pi;
angle0=angle0*pi/180; kr=k*r; kr0=k*r0; error=1e-12;
A=0:1.13:alfamax; [dd nA]=size(A); Ar=A*pi/180;
Rsoft=-1; Rhard=1; rho=r*r0/(r+r0); krho=k*rho;
U0=I0*besselh(0,1,kr0);
for m = 1:nA
   fprintf(1, 'Calculating PE Fields for Angle %5.1f :
\n',A(m));
   angle = Ar(m); psi = angle-angle0;
   w1 = IntPE(alfa,krho,psi)*exp(1i*(krho+pi/4))/
(pi*sqrt(2));
   psi = angle+angle0;
   w2 = IntPE(alfa,krho,psi)*exp(1i*(krho+pi/4))/
(pi*sqrt(2));
   Ws=(w1-w2)*exp(-1i*krho); Wh=(w1+w2)*exp(-1i*krho);
   Us_PE(m)=Ws*sqrt(r0/(r0+r))*exp(1i*kr); % norm PO
dif fields for SBC
   Uh_PE(m)=Wh*sqrt(r0/(r0+r))*exp(1i*kr); % norm PO
dif fields for HBC
   if (angle >= 0) && (angle < pi-angle0)    % Region
I
      R1=sqrt(r*r+r0*r0-2*r*r0*cos(angle-angle0));
      R2=sqrt(r*r+r0*r0-2*r*r0*cos(angle+angle0));
      Uxs=besselh(0,1,k*R1)-besselh(0,1,k*R2);
      Uxh=besselh(0,1,k*R1)+besselh(0,1,k*R2);
      UsTOTAL_PE(m) = Us_PE(m)+Uxs/U0; UhTOTAL_PE(m) =
Uh_PE(m)+Uxh/U0;
   elseif (angle >= pi-angle0) && (angle < pi+angle0)
% Region II
      R1=sqrt(r*r+r0*r0-2*r*r0*cos(angle-angle0));
      Ux=besselh(0,1,k*R1);
      UsTOTAL_PE(m) = Us_PE(m)+Ux/U0; UhTOTAL_PE(m) =
Uh_PE(m)+Ux/U0;
```

Table 10.6.
(Continued)

```
    elseif (angle >= pi+angle0) && (angle <= alfa)   %
Region III
      UsTOTAL_PE(m) = Us_PE(m); UhTOTAL_PE(m) =
Uh_PE(m);
    end
end
figure(1); polar(Ar,abs(Us_PE)); figure(2);
polar(Ar,abs(UsTOTAL_PE));
% --------------------- END ----------------------
```

Table 10.7.
A MATLAB Script for the Numerical Integration of (10.33b)

```
function result = IntPE(alfa,kr,psi)
  eps = 1e-12; n = alfa/pi; y_old = 1e6;
  for M=1:100
    s = -M:0.0001:M; eta=sqrt(2).*s.*exp(-1i*pi/4);
    F2 = (1/n).*sin(pi/n)./(cos(pi/n)-cos((eta+psi)/n));
    fun = exp(-kr*s.*s).*F2;
    y(M) = trapz(s,fun);
    error = abs(y(M)-y_old);
    y_old = y(M);
    if (error<eps) result = y(M);
       N = M; break;
    else continue
    end
  end
% ---------------------- END --------------------
```

10.5 THE WEDGEGUI VIRTUAL TOOL AND SOME EXAMPLES

The MATLAB-based simulation package WedgeGUI with the front panel displayed in Fig. 10.3 is organized for the investigation of wedge diffraction in 2D with various exact and HFA models. The panel is divided into three parts. The top block is reserved for the structure. The wedge figure is shown on the top right. The wedge exterior angle and

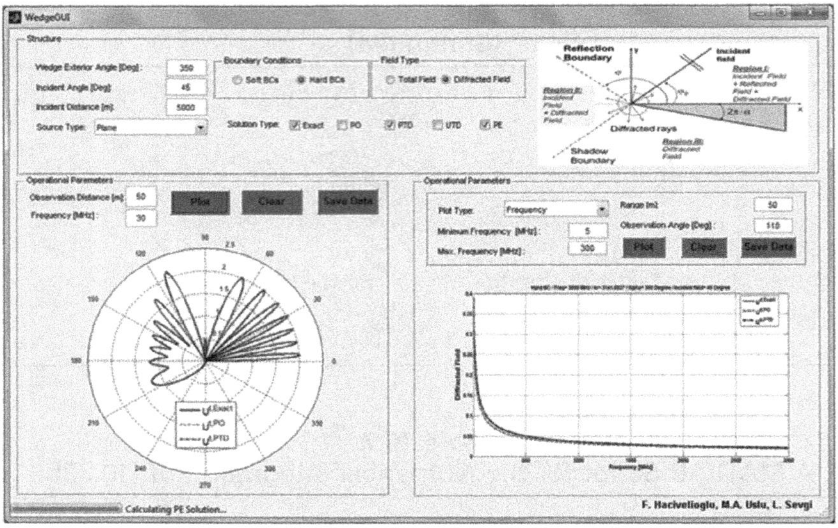

Figure 10.3. Front panel of WedgeGUI.

incident distance/angle are supplied on the top left. The user also selects either of the soft and hard BCs: total and diffracted fields in this block. A pop-up menu allows the user to choose the type of the source: PW excitation or cylindrical wave excitation. For each source type, the methods used in simulations are given with tick boxes. Multiple selection is possible. The flowchart of the virtual tool is given in Fig. 10.4.

The area below the structure block is divided into two parts. On the left, the results are presented in terms of polar plots (i.e., fields vs. angle), on the right in terms of Cartesian plots (i.e., field vs. radial range or field vs. frequency). For the polar plots on the left block, one needs to specify frequency and radial distance from the tip. The rest is handled automatically once the Plot button is pressed and fields are calculated at N observation points equidistant from the tip located $\Delta\varphi = \alpha/N$ apart. The progress bar on the bottom left shows the status of the calculations. Figure 10.5, Figure 10.6, and Figure 10.7 show example scenarios and simulation results.

The user may edit/modify the plot in another figure window. This is possible by clicking on the polar plot. The angle of the illumination is also shown in these figures. Similarly, one can plot field versus range

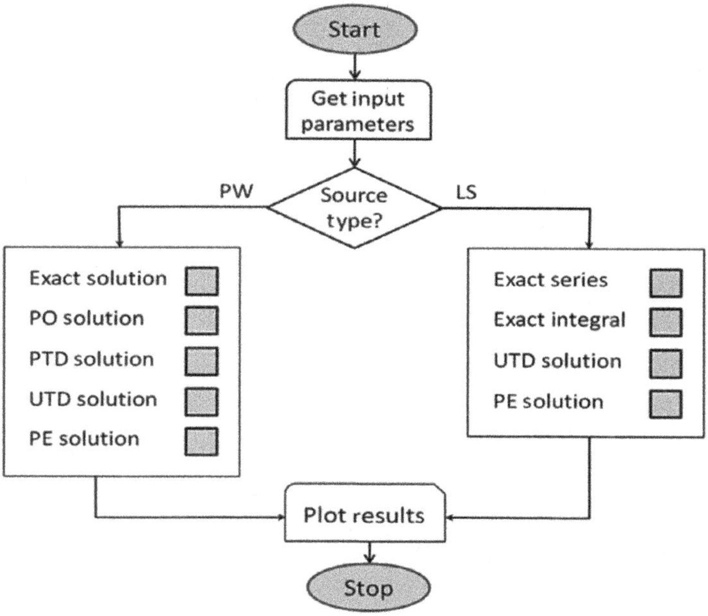

Figure 10.4. The basic flowchart of WedgeGUI.

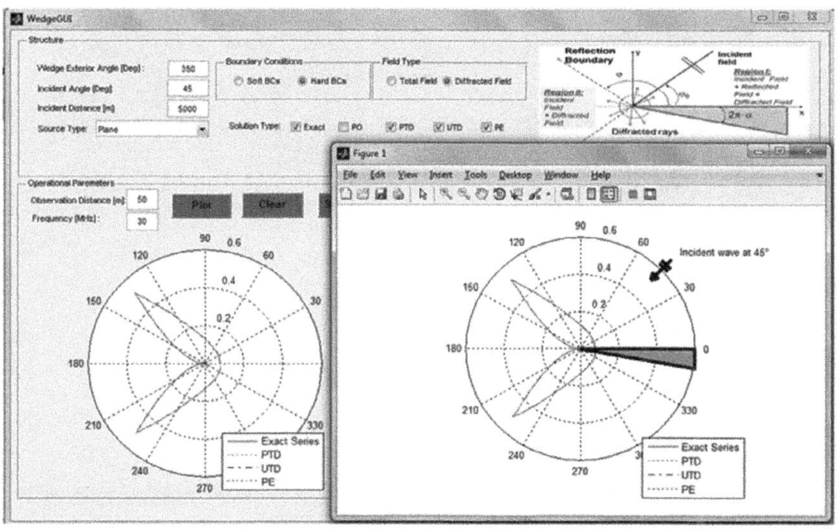

Figure 10.5. WedgeGUI: Diffracted fields versus angle for HBC ($\alpha = 350°$, $\varphi_0 = 45°$, $f = 30$ MHz, $r = 50$ m, $kr = 31.4159$).

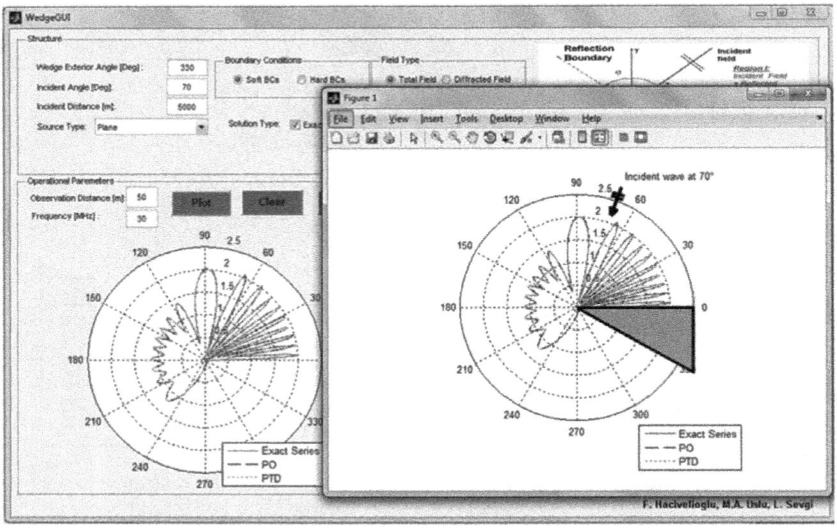

Figure 10.6. WedgeGUI: Total fields vs. angle for SBC ($\alpha = 330°$, $\varphi_0 = 70°$, $f = 30$ MHz, $r = 50$ m, $kr = 31.4159$).

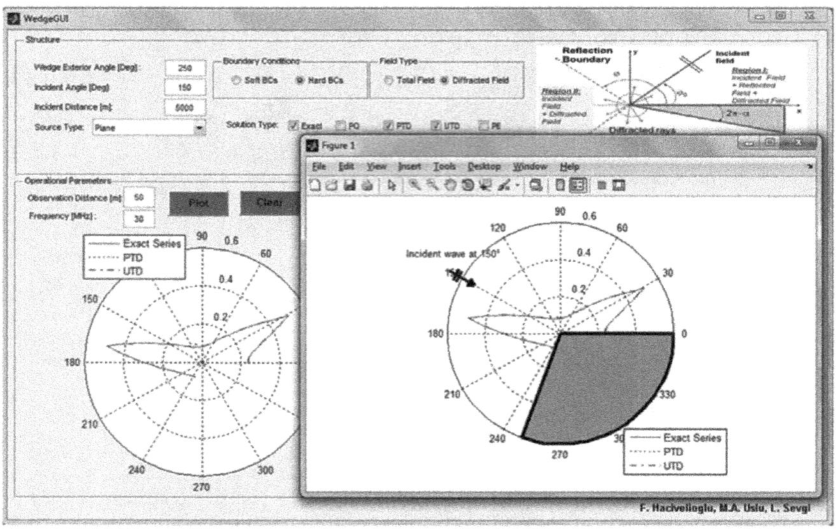

Figure 10.7. WedgeGUI: Diffracted fields versus angle for HBC ($\alpha = 250°$, $\varphi_0 = 150°$, $f = 30$ MHz, $r = 50$ m, $kr = 31.4159$).

or field versus frequency on the right block. Clear button clears the plot for the next simulation and Save Data button records the data in a text file with the name given by the user. Table 10.8 lists a sample recorded data which belongs to the simulations presented in Fig. 10.7. Data recording is important because WedgeGUI does not run simulations for both SBC and HBC, or both total and diffracted fields at the same time. One needs to run WedgeGUI twice for the examples presented in Fig. 10.8, Fig. 10.9, and Fig. 10.10.

Physical optics (PO) does not satisfy rigorously the reciprocity principle and the BCs but it describes some features of the scattered field within approximations acceptable in the engineering practice. In particular, PO correctly determines the total scattering cross section of large opaque objects and predicts the field behavior in the vicinity of

Table 10.8.
Sample Recorded Data for Angle versus Diffracted Fields

Amplitude vs. Angle, Created by WEDGE GUI Version 1.0			
Angle (deg)	Exact Series	PTD	UTD
$0.0000 \times 10^{+000}$	$0.0000 \times 10^{+000}$	$0.0000 \times 10^{+000}$	$0.0000 \times 10^{+000}$
$1.3300 \times 10^{+000}$	2.0847×10^{-003}	2.0904×10^{-003}	2.0889×10^{-003}
$2.6600 \times 10^{+000}$	4.1726×10^{-003}	4.1850×10^{-003}	4.1823×10^{-003}
$3.9900 \times 10^{+000}$	6.2679×10^{-003}	6.2882×10^{-003}	6.2845×10^{-003}
$5.3200 \times 10^{+000}$	8.3748×10^{-003}	8.4045×10^{-003}	8.4001×10^{-003}
$6.6500 \times 10^{+000}$	1.0497×10^{-002}	1.0538×10^{-002}	1.0533×10^{-002}
$7.9800 \times 10^{+000}$	1.2641×10^{-002}	1.2694×10^{-002}	1.2689×10^{-002}
$9.3100 \times 10^{+000}$	1.4810×10^{-002}	1.4878×10^{-002}	1.4873×10^{-002}
$1.0640 \times 10^{+001}$	1.7008×10^{-002}	1.7094×10^{-002}	1.7089×10^{-002}
$1.5960 \times 10^{+001}$	2.6197×10^{-002}	2.6393×10^{-002}	2.6388×10^{-002}
$1.7290 \times 10^{+000}$	2.8619×10^{-002}	2.8856×10^{-002}	2.8852×10^{-002}
$1.8620 \times 10^{+000}$	3.1102×10^{-002}	3.1389×10^{-002}	3.1385×10^{-002}
$1.9950 \times 10^{+001}$	3.3652×10^{-002}	3.3998×10^{-002}	3.3995×10^{-002}
$2.1280 \times 10^{+001}$	3.6275×10^{-002}	3.6693×10^{-002}	3.6690×10^{-002}
$2.2610 \times 10^{+001}$	3.8978×10^{-002}	3.9482×10^{-002}	3.9480×10^{-002}
$2.7930 \times 10^{+001}$	5.0729×10^{-002}	5.1789×10^{-002}	5.1791×10^{-002}
$3.1920 \times 10^{+001}$	6.0741×10^{-002}	6.2595×10^{-002}	6.2600×10^{-002}
$3.3250 \times 10^{+000}$	6.4352×10^{-002}	6.6584×10^{-002}	6.6590×10^{-002}
$3.4580 \times 10^{+001}$	6.8116×10^{-002}	7.0801×10^{-002}	7.0808×10^{-002}

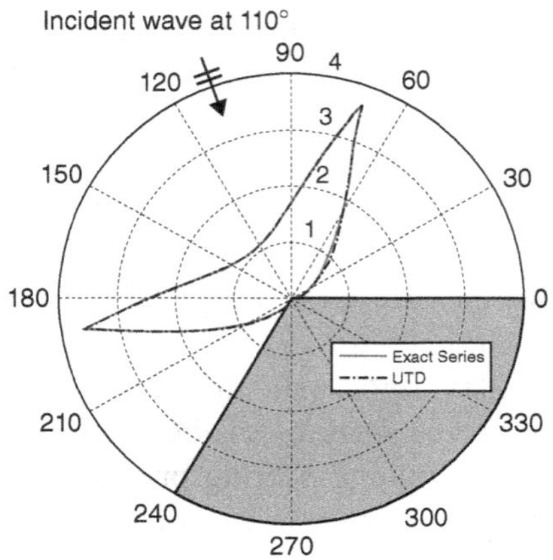

Figure 10.8. Diffracted fields versus angle for (left) SBC and (right) HBC computed with exact and UTD models ($\alpha = 240°$, $\varphi_0 = 110°$, $f = 30$ MHz, $r = 50$ m, $kr = 31.4159$).

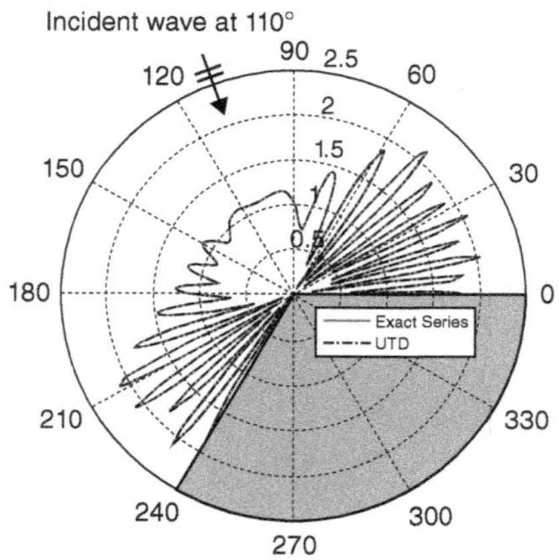

Figure 10.9. Total fields versus angle for (left) SBC and (right) HBC computed with exact and UTD models ($\alpha = 240°$, $\varphi_0 = 110°$, $f = 30$ MHz, $r = 50$ m, $kr = 31.4159$).

the GO boundaries. PO also takes into account diffracted fields up to a certain extent (diffracted fields of the uniform currents), but this is not sufficient to describe the diffraction effects fully because it neglects the contributions of nonuniform currents. Finally, PO diffracted fields can be analytically separated into the reflected part and the shadow radiation.

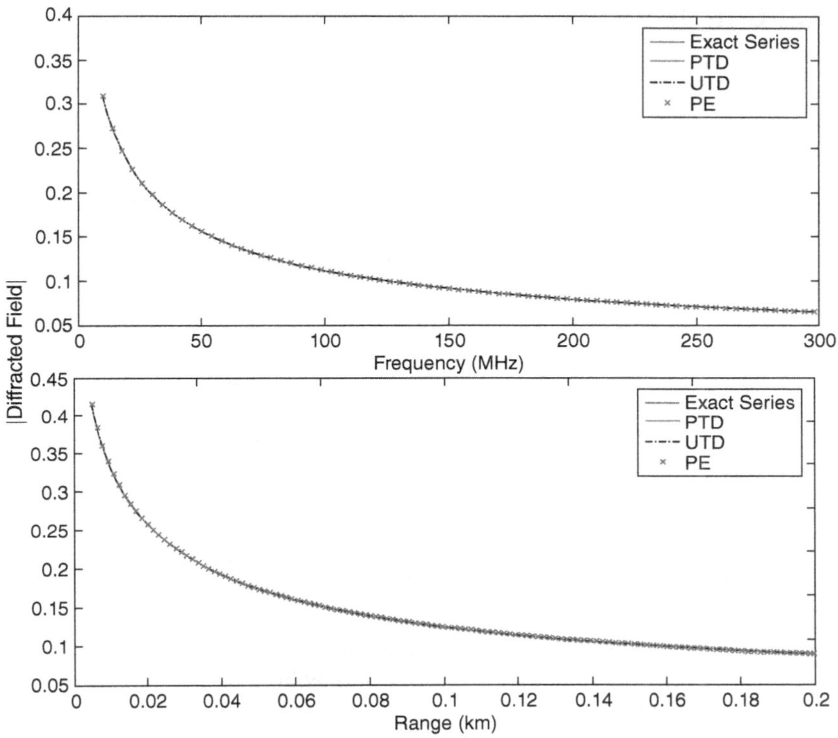

Figure 10.10. Diffraction coefficients versus (top) frequency and (bottom) range for the PW illumination and SBC ($\alpha = 350°$, $\varphi_0 = 60°$, $f = 30$ MHz).

10.6 CONCLUSIONS

The problem of EM wave scattering from a nonpenetrable wedge is posed and solutions existing in the literature have been surveyed in this chapter. Then a MATLAB-based virtual diffraction tool WedgeGUI has been introduced. The WedgeGUI presents results of EM wave scattering from a wedge shaped object under different structural as well as operational parameters. Various models under both LS and PW illuminations are included. Comparisons among UTC, PO, PTD, PE models through many scenarios and illustrations are possible. The WedgeGUI virtual tool can be used in many graduate level courses such as advanced EM theory, high frequency asymptotic methods in EM, diffraction theory, and so on.

REFERENCES

1. F. Hacıvelioğlu, L. Sevgi, P. Ya Ufimtsev, "Electromagnetic Wave Scattering from a Wedge with Perfectly Reflecting Boundaries: Analysis of Asymptotic Techniques," *IEEE Antennas and Propagation Magazine*, **53** (3), 2011, pp. 232–253.
2. F. Hacıvelioğlu, M. A. Uslu, L. Sevgi, "A MATLAB-Based Simulator for the Electromagnetic Wave Scattering from a Wedge with Perfectly Reflecting Boundaries," *IEEE Antennas and Propagation Magazine*, **53** (6), 2011, pp. 234–243.
3. G. Çakır, L. Sevgi, P. Y. Ufimtsev, "FDTD Modeling of Electromagnetic Wave Scattering from a Wedge with Perfectly Reflecting Boundaries: Comparisons against Analytical Models and Calibration," *IEEE Transactions on Antennas and Propagation*, **60** (7), 2012, pp. 3336–3342.
4. M. A. Uslu, L. Sevgi, "MATLAB-Based Virtual Wedge Scattering Tool for the Comparison of High Frequency Asymptotics and FDTD Method," *ACES, International Journal on Applied Computational Electromagnetics* **27** (9), 2012, pp. 697–705.
5. J. B. Keller, "The Geometric Optics Theory of Diffraction," presented at the 1953 McGill Symposium of Microwave Optics, *A. F. Cambridge Res. Cen. Rep. TR-59-118* (II), 1959, pp. 207–210.
6. R. G. Kouyoumjian, P. H. Pathak, "A Uniform Geometrical Theory of Diffraction for an Edge in a Perfectly Conducting Surface," *Proceedings of IEEE*, **62**, 1974, pp. 1448–1461.
7. C. Hansen, ed., *Geometrical Theory of Diffraction*, IEEE Press, New York, 1981.
8. J. J. Bowman, T. B. A. Senior, P. L. E. Uslenghi, eds., *Electromagnetic and Acoustic Scattering by Simple Shapes*, Hemisphere Publishing Corporation, New York, 1987.
9. Yu. A. Kravtsov, N. Y. Zhu, *Theory of Diffraction: Heuristic Approaches*, Alpha Science Series on Wave Phenomena, Oxford, 2010.
10. G. R. Kirchhoff, "Zur Theorie der Lichtstrahlen," *Annalen der Physik und Chemie*, **18** (4), 1883, pp. 663–695.
11. H. M. Macdonald, "The Effect Produced by an Obstacle on a Train of Electric waves," *Philosophical Transactions of the Royal Society of London. Series A: Mathematical and Physical Sciences*, **212**, 1912, pp. 299–337.
12. P. Ya. Ufimtsev, *Theory of Edge Diffraction in Electromagnetics: Origination & Validation of PTD*, SciTech Publishing, Inc., Secaucus, NJ, 2009.
13. P. Ya. Ufimtsev, *Fundamentals of the Physical Theory of Diffraction*, Wiley & Sons, Inc., Hoboken, NJ, 2007.
14. P. Ya. Ufimtsev, "Backscatter," in *Wiley Encyclopedia of Electrical and Electronics Engineering*, John Wiley & Sons, Inc., New York, 1999.
15. P. Ya. Ufimtsev, "New Insight into the Classical Macdonald Physical Optics Approximation," *IEEE Antennas and Propagation Magazine*, **50** (3), 2008,

pp. 11–20. Corrections: *IEEE Antennas & Propagation Magazine*, **50** (4), Aug 2008, p. 65.

16. P. Ya. Ufimtsev, "On Polarization Coupling in the PO and PTD Approximations," *IEEE Transactions on Antennas and Propagation*, **56** (12), 2008, pp. 3883–3885.
17. P. Ya. Ufimtsev, *Metod Kraevykh Voln v Fizicheskoi Teorii Diffraktsii [Method of Edge Waves in the Physical Theory of Diffraction]*, Sovetskoe Radio, Moscow, 1962. Translated into English by the U.S. Air Force, Foreign Technology Division (National Air Intelligence Center), Wright-Patterson AFB, OH, 1971. Technical Report AD 733203, Defense Technical Information Center of USA, Cameron Station, Alexandria, VA, 22304-6145, USA.
18. P. Ya. Ufimtsev, "Diffraktsiya na Kline i Lente" [Diffraction from Wedge and Strip], *Zurnal Technicheskoi Fiziki*, **27** (8), 1957, pp. 1840–1849. [English translation published by *Soviet Physics-Technical Physics*].
19. P. Ya. Ufimtsev, "Fast Convergent Integrals for Nonuniform Currents on Wedge Faces," *Electromagnetics*, **18** (3), 1998, pp. 289–313. Corrections in *Electromagnetics*, 19(5), 1999, p.473.
20. P. Ya. Ufimtsev, "Transverse Diffusion for Diffraction by a Wedge," *Radio Engineering and Electronic Physics*, **10** (6), 1965, pp. 866–875.
21. P. Ya. Ufimtsev, "Comments on Diffraction Principles and Limitations of RCS Reduction Techniques," *Proceedings of IEEE*, **84** (12), 1996, pp. 1828–1851.
22. A. Sommerfeld, "Mathematische Theorie der Diffraction," *Mathematische Annalen*, **16**, 1896, pp. 317–374.
23. H. M. Macdonald, *Electric Waves*, The University Press, Cambridge, England, 1902, pp. 186–198.
24. A. Sommerfeld, Theorie der Beugung, Chapter 20 P. Frank, R. M. Friedr, eds., *Die Differential- und Integralgleichungen der Mechanik und der Physik*, Vieweg & Sohn, Braunschweig, 1935. The second (Physical) part. American publication by Mary S. Rosenberg, New York, 1943.

CHAPTER 11

Antennas: Isotropic Radiators and Beam Forming/Beam Steering

A simple MATLAB package (ARRAY), accommodated with beam steering capabilities, is designed and introduced which plots two-dimensional (2D) and three-dimensional (3D) radiation patterns of a number of selected and user-located isotropic radiators [1]. The reader is advised to refresh their memory on antennas and arrays. A few of many useful books are References 2–7.

11.1 INTRODUCTION

Antenna is an electromagnetic (EM) radiator that emits radio frequency power. In other words, it is a transducer that converts voltage to the electric field (or vice versa). The Institute of Electrical and Electronics Engineering (IEEE) definition of an antenna is given as "that part of a transmitting or receiving system that is designed to radiate or receive EM waves" [8]. Most antennas are reciprocal devices and behave the same on transmit and on receive.

A receiver (transmitter) *antenna factor* (AF) is the ratio of electric field (voltage) to the voltage (electric field) at the receiver (transmitter).

Electromagnetic Modeling and Simulation, First Edition. Levent Sevgi.

Angular variation of radiation around the antenna is called *radiation pattern*. The simplest antenna is a point radiator that radiates *isotropically*. Antennas can be arranged to radiate along some chosen direction. *Directivity* is the ratio of power density in the direction of the pattern maximum to the average power density at the same distance from the antenna. Antenna *efficiency* is a parameter between 0 and 1, which is defined as the power radiated by the antenna divided by the power delivered to the antenna. Efficiency is 1 (100%) for ideal (loss-free) antennas. Directivity is reduced by the losses on the antenna. (It should be noted that directivity and *gain* often misused among antenna engineers. Directivity can be calculated and/or measured, while gain of the antenna can only be measured, since it is related to all, cable material, connector, etc., losses.)

Directive antennas direct majority of its power to a desired direction. This is called *main lobe* (beam). Unwanted lobes at the sides and back are called side lobes and back lobes, respectively. The ratio of the power in the main lobe to the back lobe is the front-to-back ratio (FBR) of the antenna. Half-power beamwidth (HPBW) is the angular distance between half-power (3 dB) point on each side of the main beam maximum. First-null beamwidth (FNBW) is the angle between first null points of the main lobe. When receive, total power incident on the receive antenna is the sum of power density over the "area" of the antenna, which is called *effective aperture*. It is the ratio of the available power at the terminals of the antenna to the power flux density of a plane wave incident upon the antenna, which is polarization matched to the antenna. *Polarization* is the figure traced out with time by the instantaneous electric field vector associated with the radiation from an antenna (e.g., linear, circular, or elliptical polarization).

Several antennas can be arranged in free space and interconnected to produce a directional radiation pattern. Such a configuration or an arrangement of multiple radiators is referred to as an *antenna array*. Antenna arrays have the capability of *beam forming*. The direction of the directional radiation pattern can also be changed, which is called beam forming or *beam steering*. Beam steering can be achieved by either mechanical scanning or electronic (phase) scanning. The latter is then called *phased arrays*.

It is called a *broadside antenna* if its radiation pattern with main beam maximum in a direction normal to the plane containing itself. An *endfire antenna* is the one for which the main beam is in the plane

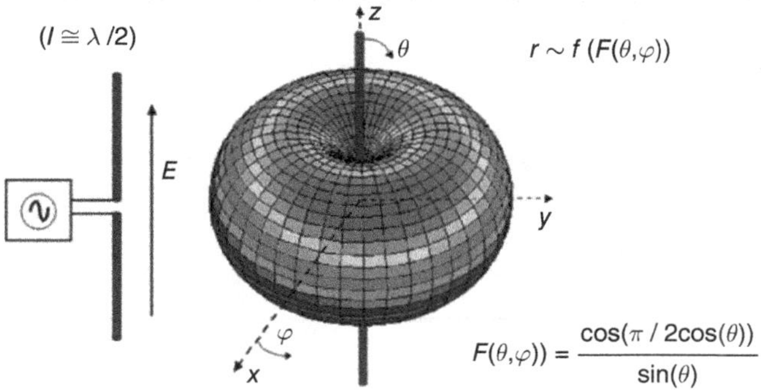

Figure 11.1. A vertical wire antenna and its 3D radiation pattern at resonance.

containing the antenna. Figure 11.1 illustrates a wire antenna located vertically (along z-axis) and its 3D radiation pattern at a frequency that corresponds to a length equal to half-wavelength ($l \cong \lambda/2$). It is customary to use vertical and horizontal radiation patterns in antenna engineering. The vertical pattern gives θ-variation at a constant φ-angle. For example, the radiation pattern on xz-plane corresponds to θ-variation when $\varphi = 0°$. Similarly, $\varphi = 90°$ corresponds to the pattern on yz-plane. Horizontal (xy-plane) radiation pattern corresponds to φ-variation when $\theta = 90°$. It is easy to visualize isotropic horizontal pattern for the half-wavelength dipole and radiation null along z-axis in Fig. 11.1.

The radiation pattern of a group of radiators (i.e., array) is determined by the type of individual elements, their location, spacing, orientation, excitation amplitudes, and phases. In practice, radiation pattern of an array is divided into two: element pattern and array factor. *Array factor* is the radiation pattern of an array, where all array elements are isotropic radiators.

11.2 ARRAYS OF ISOTROPIC RADIATORS

An isotropic radiator is a hypothetical, lossless antenna occupying a point in space and, when transmitting, radiates uniformly in all directions. Figure 11.2 shows geometrical parameters associated with this problem. In general, a radiation source is located at (x_1, y_1, z_1) with respect to a phase reference at the origin (0, 0, 0).

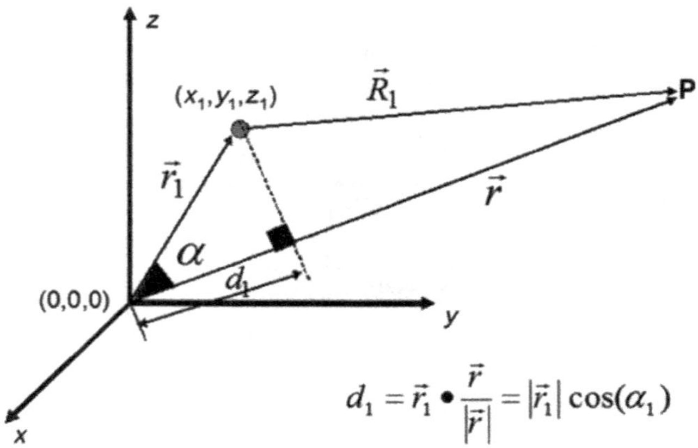

Figure 11.2. An isotropic radiator located at (x_1, y_1, z_1). The ith radiator's location vector and the observation vector are $\vec{r}_i$ and $\vec{r}$, respectively.

When it is an array of N-elements, the coordinates of each radiator with respect to the phase reference are (x_i, y_i, z_i) $(i = 1, \ldots, N)$ and the total electric field calculated at a far field point P is

$$E(\theta, \varphi) = \sum_{i=1}^{N} I_i e^{j\psi_i(\theta,\varphi)}, \tag{11.1}$$

where I_i is the complex current for the ith element and ψ_i is phase contribution at the far field point from the ith radiator with respect to the origin, which can be expressed as

$$\psi_i = k\left(\vec{r}_i \cdot \frac{\vec{r}}{|\vec{r}|}\right). \tag{11.2}$$

Here, $j = \sqrt{-1}$, $k = 2\pi/\lambda$ is the wave number (λ is the wavelength), $\vec{r}_i = x_i\vec{e}_x + y_i\vec{e}_y + z_i\vec{e}_z$ is the vector from the origin to the ith element (radiator vector), $\vec{r} = x\vec{e}_x + y\vec{e}_y + z\vec{e}_z$ is the vector from the origin to the observation point P (radiation direction vector), and $(\vec{e}_x, \vec{e}_y, \vec{e}_z)$ are the unit vectors along x, y, and z, respectively. Using the spherical coordinates, with $x = r\sin\theta\cos\varphi$, $y = r\sin\theta\sin\varphi$, $z = r\cos\theta$, the phase of the ith radiator may be obtained as

$$\psi_i = k\left(x_i \sin\theta\cos\varphi + y_i \sin\theta\sin\varphi + z_i \cos\theta\right). \tag{11.3}$$

For planar arrays located on the xy-plane, the radiators' coordinates are $x_i = r_i \cos\varphi_i$, $y_i = r_i \sin\varphi_i$, and $z_i = 0$, where $r_i = \sqrt{x_i^2 + y_i^2}$ and $\varphi_i = a\tan^{-1}(y_i/x_i)$. For the planar arrays (i.e., $\theta_i = 90°$), the total electric field, which is known as the array factor, can be expressed as

$$E(\theta,\varphi) = \sum_{i=1}^{N} I_i \exp\left(jkr_i \sin\theta \cos(\varphi_i - \varphi)\right). \tag{11.4}$$

This expression corresponds to the pattern of arbitrarily located, isotropic, equiamplitude, nonphased radiators when $I_i = 1.0$. Arrays may have special geometrical configurations, such as shown in Fig. 11.3, for example, arrays of arbitrarily located radiators, linear arrays, planar arrays, circular arrays, and so on.

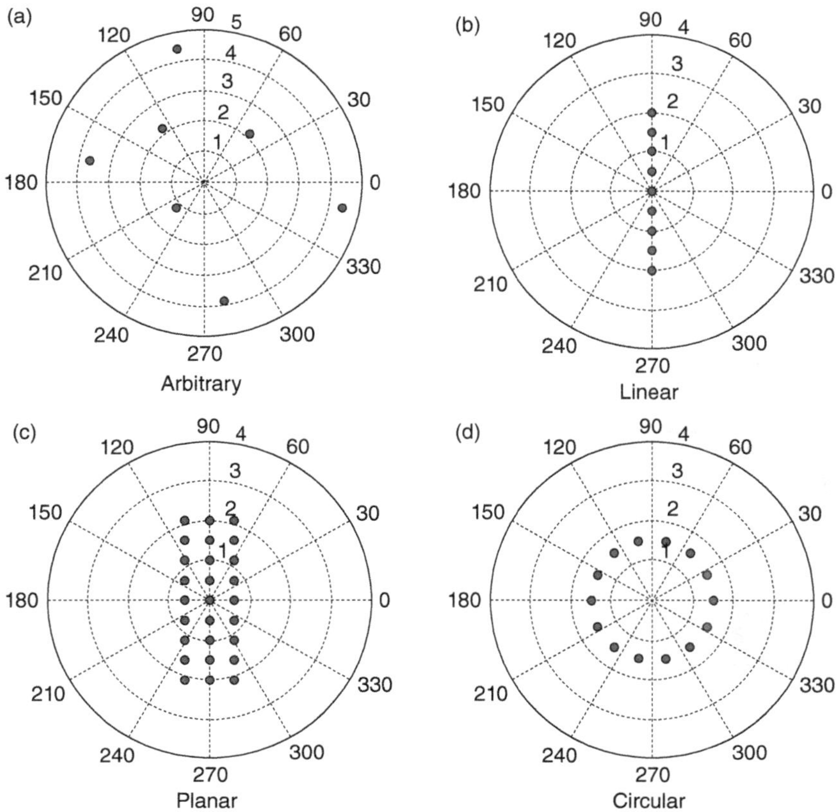

Figure 11.3. Geometrical configurations of planar arrays of isotropic radiators (a) arbitrary, (b) linear, (c) planar, and (d) circular.

11.3 THE ARRAY PACKAGE

The MATLAB-based ARRAY, as displayed in Fig. 11.4, has been prepared for the visualization of radiation characteristics of planar arrays of isotropic radiators based on (11.1), (11.2), (11.3), and (11.4). This package allows the user to choose from a popup menu (a) arbitrary, (b) linear, (c) planar, or (d) circular arrays with a number of user-selected and located isotropic radiators. The default array type is *arbitrary*. One of the exciting features of ARRAY is to let the user locate the radiators arbitrarily. Once the number of radiators (N), maximum radius (in meters) of the polar region where the radiators will be located, and the

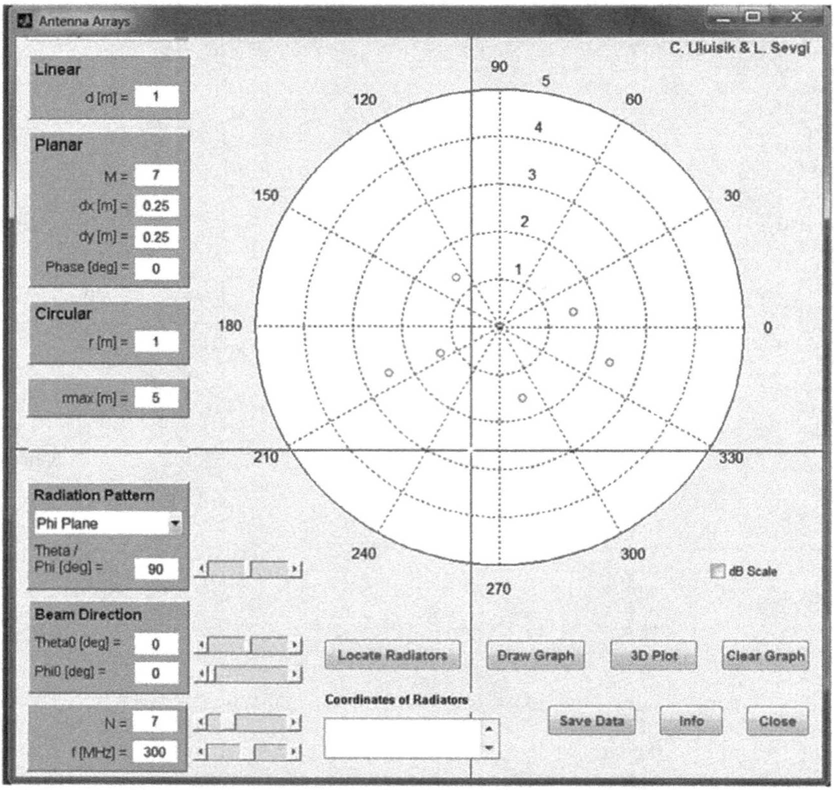

Figure 11.4. Front panel of ARRAY package and radiator locating.

operating frequency (in megahertz) are specified, the user may proceed to locate the radiators one by one after pressing the "Locate Radiators" button. As shown in Fig. 11.4, five of the seven radiators have already been located and the sixth one is being located by the cursor. The coordinates of the located radiators are also displayed in the list box at the bottom.

By this procedure (i.e., by using the mouse), the user may design any kind of an array, but the package also locates the radiators automatically when one of the other types of the arrays is selected and required parameters are supplied. If the array type is set to *linear*, N-elements are located on the y-axis symmetrically, with an interelement distance d. If it is planar, $N \times M$ elements are located on xy-plane. In this case, N and M are the numbers of the elements along y- and x-directions, respectively. The distances between the radiators along x- and y-directions are dx and dy, respectively. If the circular array type is selected, N-elements are located symmetrically on a circle whose center is at the origin and radius is r.

After locating the radiators, the user may visualize a 2D radiation pattern at the selected plane by pressing the "Draw Graph" button. The 2D radiation pattern plane to be displayed is specified by the *Theta/Phi Plane* popup menu and *Theta/Phi value* box. If the *θ-plane* is selected (vertical pattern), the angle θ varies in $0° \leq \theta \leq 360°$ at a fixed φ value (specified by the *Theta/Phi* box). If the *φ-plane* is selected, the angle φ varies between $0° \leq \varphi \leq 360°$ at a fixed θ-value specified by the user; the radiation pattern in this case is called "horizontal." All the 2D radiation patterns are plotted with an angular resolution of $\Delta\theta = \Delta\varphi = 1°$.

The main beam of an array can be steered electronically by varying the phases of the currents applied to the array elements. Equation (11.4) takes on the form of

$$E(\theta,\varphi)=\sum_{i=1}^{N} I_i \exp\{jkr_i[\sin\theta\cos(\varphi_i-\varphi)-\sin\theta_0\cos(\varphi_i-\varphi_0)]\} \tag{11.5}$$

if the beam angle is specified as (θ_0, φ_0). Figure 11.5 shows the radiation pattern of arbitrarily located seven-element array. The user selects the radiation pattern plane and the beam direction from the front panel

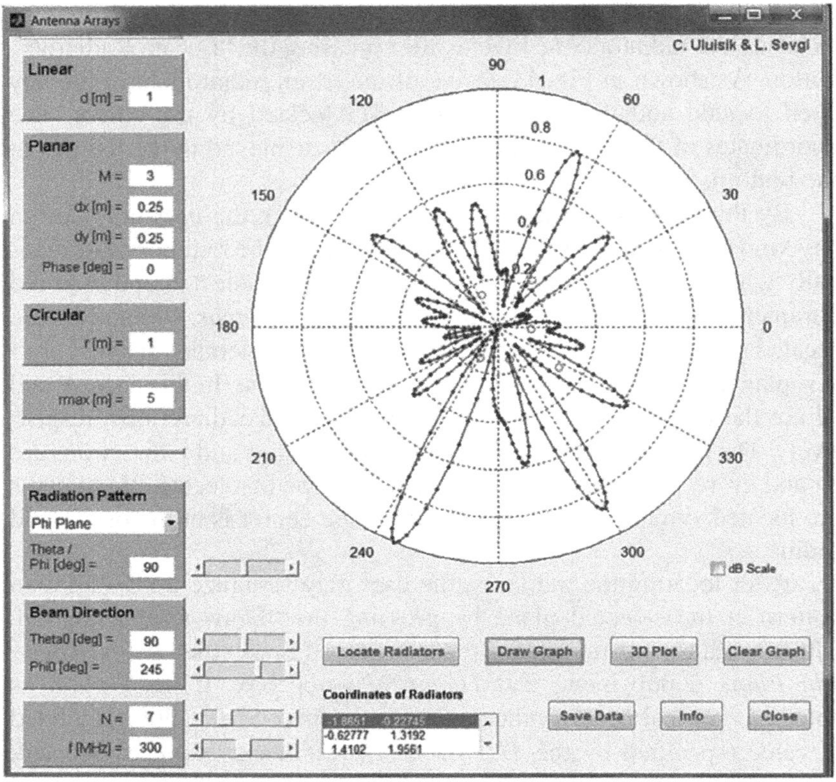

Figure 11.5. Horizontal radiation pattern of arbitrarily located seven-element array. The beam angles are selected to be $\theta = 0°$ and $\varphi = 245°$.

(on the bottom left). In this example, the pattern belongs to φ-plane (i.e., on the xy-plane) since $\theta = 90°$. The beam angles are selected to be $\theta = 90°$ and $\varphi = 245°$.

It should be noted that (11.5) is used in the ARRAY package. This means that each element is phased accordingly to direct the beam of the array along (θ_0, φ_0) direction. Equation (11.5) reduces to (11.4) when $\theta_0 = 0°$; therefore, it represents "no-phasing." The default θ_0 value is zero in ARRAY and should be kept as it is if one does not want to phase elements.

A 3D radiation pattern plotting capability has also been included in the ARRAY package. The capability is based on normalization of the radius of the unit sphere for each observation angle (θ_i, φ_i) in 3D Cartesian coordinates according to the radiation intensity calculated from (11.5). The same approach (in decibel scale with a 30 dB margin along each axis) has been followed here and similar codes have been embedded in the ARRAY package. The angular resolution for the 3D plots is set to be $\Delta\theta = \Delta\varphi = 2.5°$, which means the calculations are repeated at $151 \times 151 = 22{,}801$ vertices.

Another exciting feature of the ARRAY package is that the user can use sliding bars to change $N, f, \theta_0, \varphi_0$. For instance, by pressing the sliding bar of the frequency continuously, the user can increase or decrease the frequency, see the corresponding radiation pattern immediately, and may easily understand the effect of frequency change on the radiation pattern. The package also allows the user to save the radiation pattern data (field intensity vs. angle) to a file named *Arraypattern.dat*. The first column of this file corresponds to 361 observation angle values (in radians), and corresponding array factors are in the second column.

Extra line phasing capability has been included in the ARRAY package to change FBR characteristics in the planar array option. The user may change the phase angle $\Delta\beta$ of the lined-up radiators and observe how FBR improves or gets worse. In this case, complex currents I_i in (11.5) is replaced with I_{in} as

$$I_i \to I_{\text{in}} = \sum_{n=1}^{M} I_i \exp\{j(n-1)\Delta\beta\}. \tag{11.6}$$

For an $N \times M$ planar array, if, for example, a phase of $\Delta\beta = 30°$ is selected, then the phases of the second row ($M = 2$) lag 30° the phases of the first row ($M = 1$); the phases of the third row ($M = 3$) lags 30° the phases of the second row ($M = 2$) and lags 60° the phases of the first row ($M = 1$), etc.

Pressing the "Info" button opens the MATLAB Help window that includes explanations on how to use the package. Typing "help antenna_arrays" at the MATLAB command line also displays the same explanations. The "Clear Graph" button clears the graph and also deletes the coordinates of the radiators. The "Close" button exits the package. Finally, a small check box is reserved for the radiation patterns in decibel scale. If this is selected, the patterns are plotted in decibel scale with a 30 dB margin.

11.4 BEAM FORMING/STEERING EXAMPLES

The ARRAY package is used to plot 2D and 3D radiation patterns of various types of antenna arrays, and the examples are given in Fig. 11.6, Fig. 11.7, Fig. 11.8, Fig. 11.9, Fig. 11.10, Fig. 11.11, Fig. 11.12, Fig. 11.13, and Fig. 11.14. First, a three-element linear array is taken into account. Figure 11.6 and Figure 11.7 display this array's 2D and 3D radiation patterns, respectively. In Fig. 11.6, the following are the parameters: frequency is $f = 300$ MHz, interelement distance is $d = 0.5$ m ($d = \lambda/2$), beam angles are $\theta_0 = 90°$ and $\varphi_0 = 0°$, and the

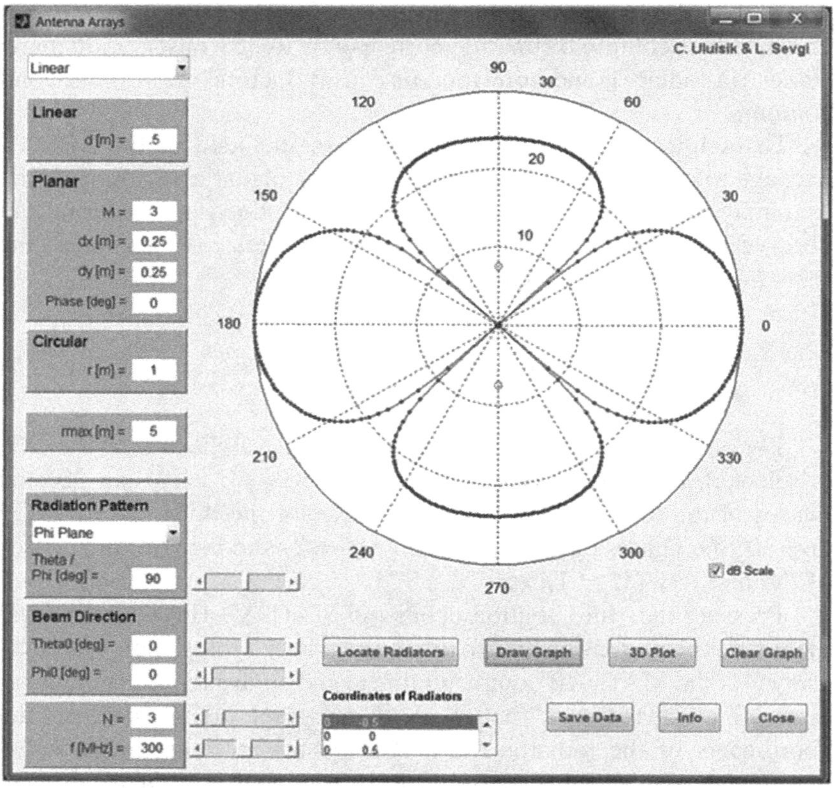

Figure 11.6. Radiation pattern of a three-element linear array on the horizontal plane ($f = 300$ MHz, $d = 0.5$ m [$d = \lambda/2$]; beam angles are $\theta_0 = 90°$ and $\varphi_0 = 0°$).

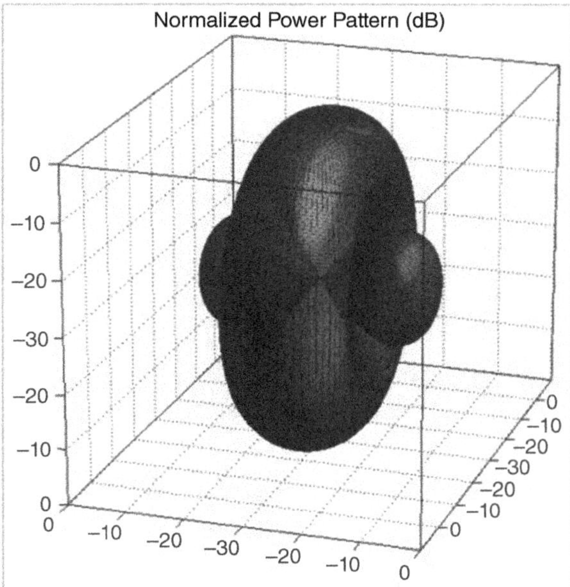

Figure 11.7. The 3D radiation pattern of the same three-element array (f = 300 MHz, d = 0.5 m [= $\lambda/2$]; beam angles are $\theta_0 = 90°$ and $\varphi_0 = 0°$).

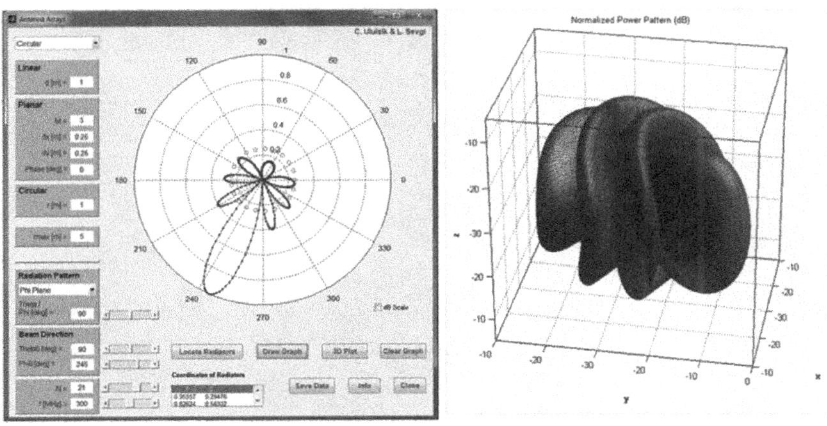

Figure 11.8. A 21-element circular array (left) its horizontal plane (right) 3D patterns (f = 300 MHz, r = 0.5 m [$\lambda/2$]; beam angles are $\theta_0 = 90°$ and $\varphi_0 = 245°$).

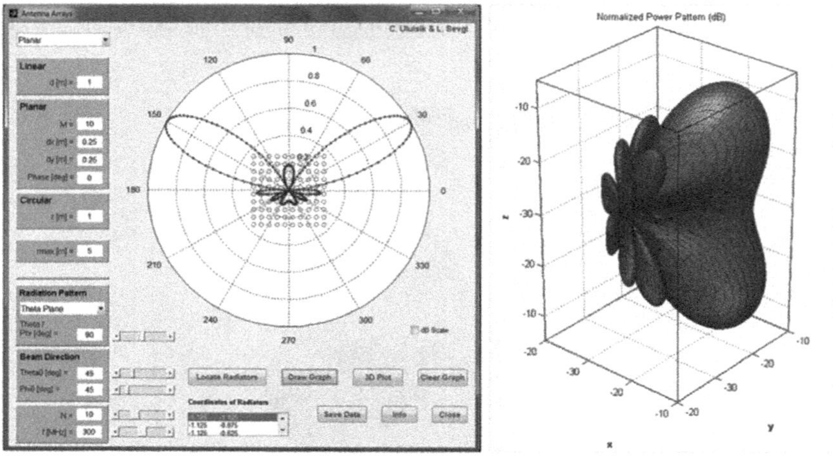

Figure 11.9. A 10 × 10 planar array and its (left) horizontal plane, (right) 3D patterns ($f = 300$ MHz, $dx = dy = 0.25$ m; beam angles are $\theta_0 = 45°$ and $\varphi_0 = 45°$).

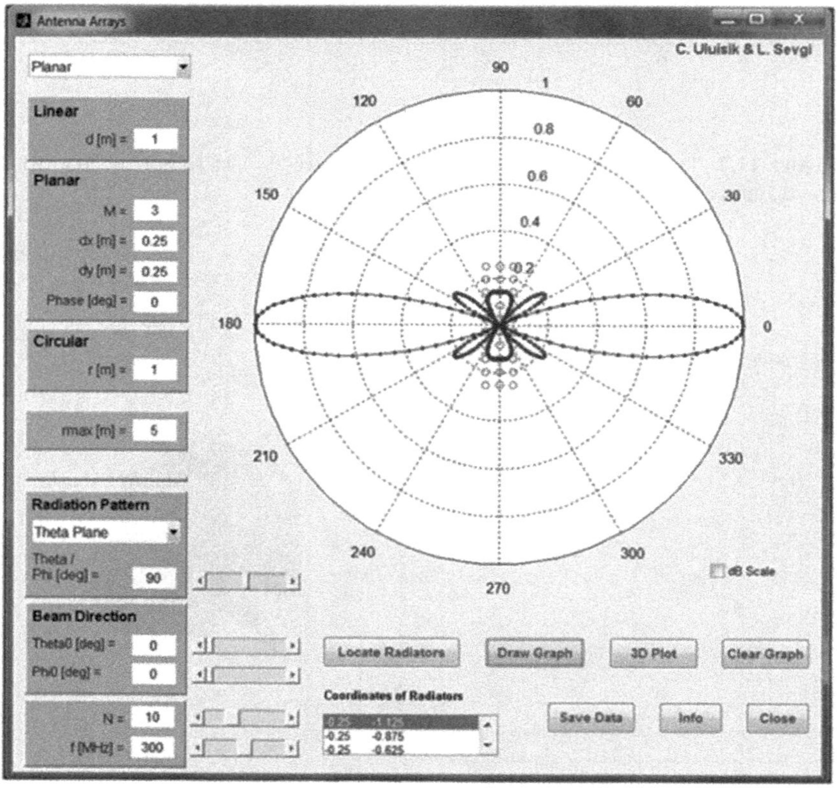

Figure 11.10. A 3 × 10 planar array and its vertical radiation for $\theta_0 = \phi_0 = 0°$.

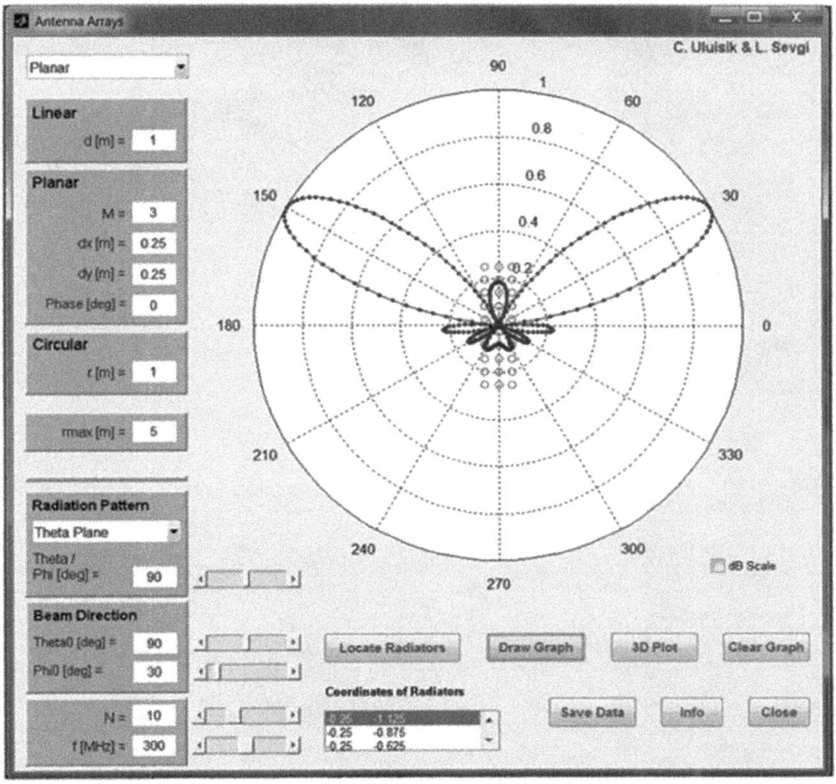

Figure 11.11. A 3 × 10 planar array and its vertical radiation for $\theta_0 = 90°$, $\phi_0 = 30°$.

pattern belongs to *xy*-plane. The 3D radiation pattern of this three-element linear array is given in Fig. 11.7.

The next example belongs to a 21-element circular array, whose radius is 1.0 m. The rest of the parameters are as follows: frequency $f = 300$ MHz, beam angles $\theta_0 = 90°$, and $\varphi_0 = 245°$. The 2D pattern, which belongs to *xy*-plane, and its 3D radiation pattern are given in Fig. 11.8.

The third example is a 10 × 10 planar array, with interelement distances of 0.25 m in both *x*- and *y*-directions. This corresponds to $d = \lambda/4$ *since the frequency is* $f = 300$ *MHz*. Figure 11.9 shows the radiation pattern on the horizontal (*xy*) plane. The beam angles are

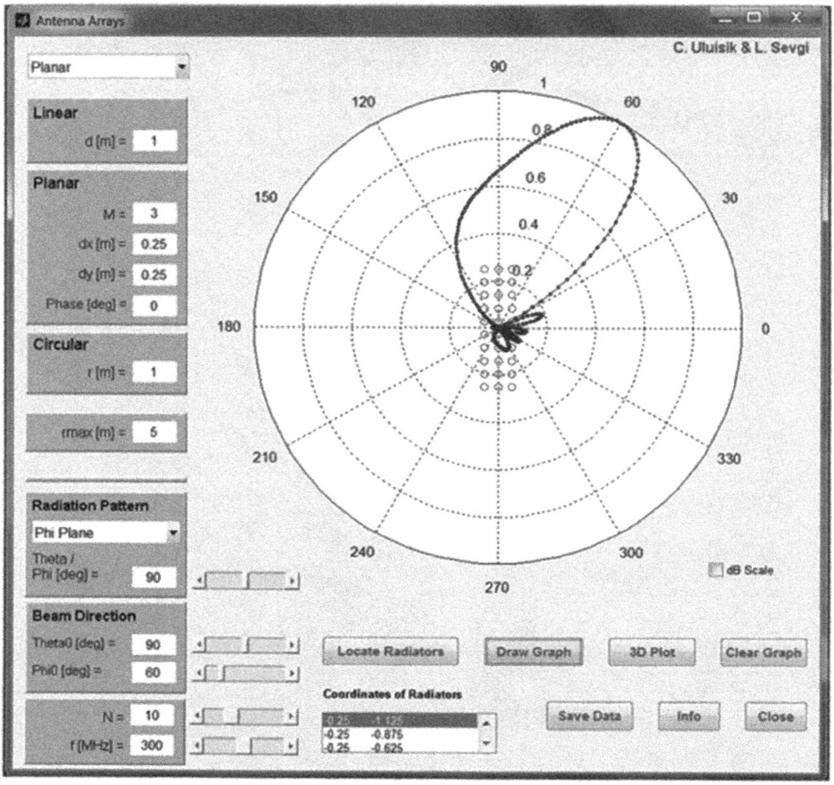

Figure 11.12. A 3 × 10 planar array and its horizontal radiation for $\theta_0 = 90°$, $\phi_0 = 60°$.

$\theta_0 = 45°$ and $\varphi_0 = 45°$. The 3D radiation pattern is also given in Fig. 11.9.

The fourth example takes a 3 × 10 element planar array into account, and beam steering capability of this array is illustrated in Fig. 11.10. Interelement distances are $dx = dy = 0.25$ m and the frequency is 300 MHz. Figure 11.10, Figure 11.11, and Figure 11.12 correspond to 2D radiation patterns with beam pointing angles of ($\theta_0 = 0°$, $\varphi_0 = 0°$), ($\theta_0 = 90°$, $\varphi_0 = 30°$), and ($\theta_0 = 90°$, $\varphi_0 = 60°$), respectively. As observed from the figures, although there is a significant beam deformation, beam forming capability still works for $\varphi_0 = 60°$.

The last example belongs to line phasing of a $M \times N$ planar array. Examples are given in Fig. 11.13 and Fig. 11.14. In Fig. 11.13, the

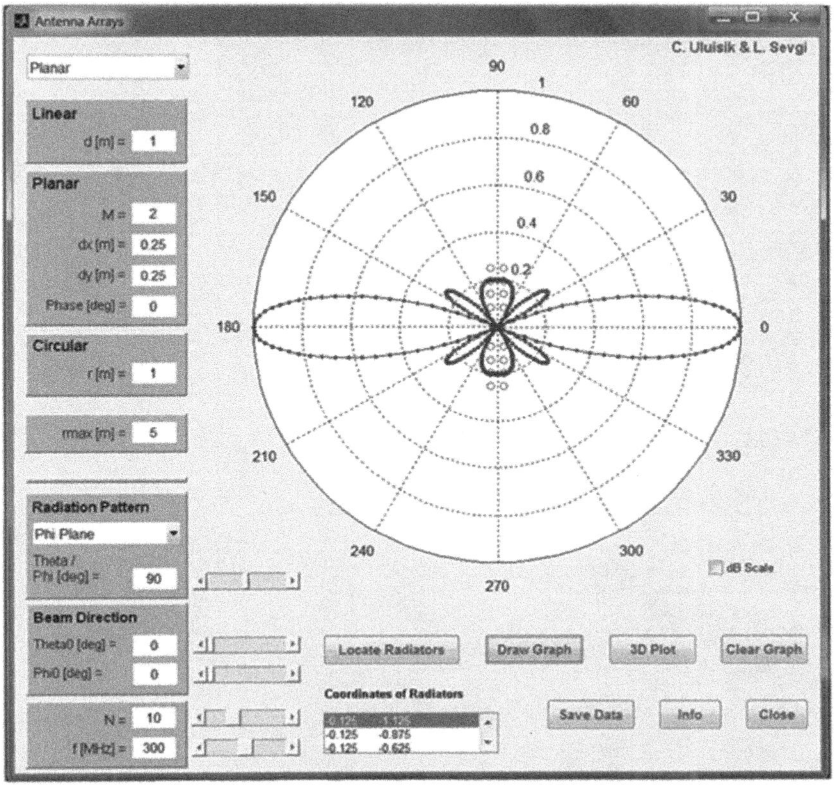

Figure 11.13. A 2 × 5 planar array and its horizontal radiation patterns without line phasing ($dx = 0.25$ m, $dy = 0.40$ m, $f = 300$ MHz) (FBR is zero).

horizontal radiation pattern of a 2 × 5 planar array with interelement distance of $dy = 0.40$ m and interarray distance of $dx = 0.25$ m is plotted without phasing the radiators. The radiation of this array is broadside and the FBR is zero. In many applications (e.g., a shore-based high-frequency ocean surveillance radar), the pattern of the array is desired to extend in only one direction (e.g., toward ocean), therefore a large FBR is required. This may be achieved by phasing the second row radiators (and the others if there are more than two rows) with respect to the first row. This is illustrated in Fig. 11.14, where a 90° phase difference is given for the second row radiators. As observed in the figure, a very good FBR is achieved.

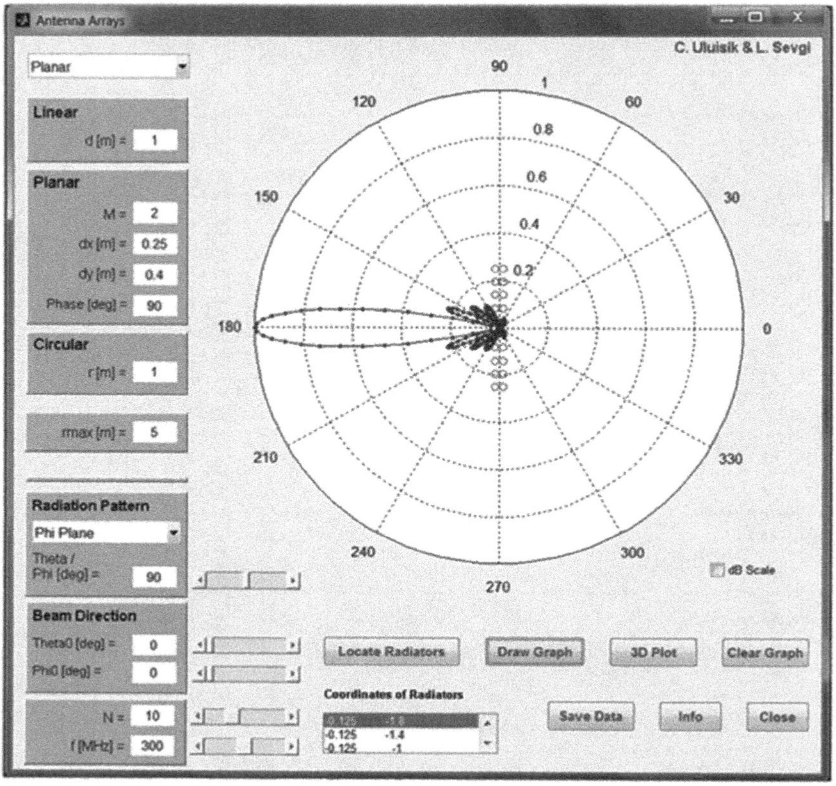

Figure 11.14. A 2 × 5 planar array and its horizontal radiation patterns with 90° line phasing ($dx = 0.25$ m, $dy = 0.40$ m, $f = 300$ MHz) (more than 30 dB FBR is achieved).

Note that isotropic radiators are fed with equal amplitudes. The ARRAY package in its current form may be used to investigate array characteristics. In other words, array analysis may be done. The reader may easily modify ARRAY.m code and introduce amplitude weighing; that is, may excite array elements with different amplitudes. With this modification, the ARRAY package may also be used for array synthesis (i.e., beam shaping).

It should be noted that the running code of ARRAY is simple. For example, the MATLAB script listed in Table 11.1 is prepared for the circular arrays. The user supplies the number of radiators, the radius of the circle that the radiators will be located, the operating frequency, and

Table 11.1.
A MATLAB Module for Circular Array Investigations

```
% ----------------------------------------------------
%  Program: Circular.m
%  Purpose: Radiation pattern of a circular array of
isotropic radiators
% ----------------------------------------------------
clc; clear all;
N = input('Number of radiators = ?');
r = input('Radius [m]= ?');
fr = input('Frequency [MHz] = ?');
phi0 = input('Beam direction [degree]= ?');
c=3e8; fr=fr*1e6; k=2*pi*fr/c; dtr=pi/180; rtd=180/pi;
phi0r=phi0*dtr;
for i=1:N % Locate radiators on a circular path
  phi=2*pi*(i-1)/N; phin(i)=phi;
  xn(i)=r*cos(phi); yn(i)=r*sin(phi);
end
i = 0;
for phi = 0:1:359 % Angle loop
  i = i+1; phir(i) = phi*dtr; Af(i) = 0;
  for n = 1:N % Radiation pattern calculation
     Af(i)=Af(i)+exp(1i*k*r*(cos(phi
r(i)-phin(n))-cos(phi0r-phin(n))));
  end
  Af(i)=abs(Af(i));
end
Array=Af/max(Af);
polar(phir,Array); hold on;
plot(xn*0.25/max(xn),yn*0.25/max(xn),'ro')
% ---------------------- End ------------------------
```

the beam direction. The script then forms the circular array of isotropic radiators, calculates required element phases to point the main beam along the user-specified direction, and then plots the radiation pattern.

11.5 CONCLUSIONS

A simple antenna array MATLAB package is introduced in this chapter. Any 2D array (i.e., linear, planar, circular, etc.) may be designed by the reader and its 2D and 3D radiation characteristics can be investigated (it should be noted that addition of a module for the formation of 3D arrays, i.e., radiators in a volume, is straightforward, but is not included here for simplicity). Beam forming capabilities for different locations, number of radiators, as well as for operating frequencies can be visualized. The package may be used as an educational tool in many undergraduate antenna lectures. It may also be used to validate and verify the finite-difference time-domain (FDTD) and method of moments (MoM) packages in public domain or, for example, the ones supplied in Reference 9. Moreover, the reader may improve package and add novel features by using the supplied source codes.

REFERENCES

1. L. Sevgi, Ç. Uluışık, "A MATLAB-based Visualization Package for Planar Arrays of Isotropic Radiators," *IEEE Antennas and Propagation Magazine*, **47** (1), 2005, pp. 156–163.
2. E. Jordan, K. G. Balmain, *Electromagnetic Waves and Radiating Systems*, Prentice-Hall, Englewood Cliffs, NJ, 1950.
3. J. D. Kraus, *Antennas*, 2nd ed., MacGraw-Hill Book Co., New York, 1988.
4. C. A. Balanis, *Antenna Theory—Analysis & Design*, 2nd ed., John Wiley and Sons, Inc., New York, 1997.
5. R. E. Collins, *Antennas and Radiowave Propagation*, MacGraw-Hill Book Co., New York, 1985.
6. W. L. Stutzman, G. A. Thiele, *Antenna Theory and Design*, 2nd ed., John Wiley and Sons, Inc., New York, 1998.
7. B. R. Mahafza, A. Z. Elsherbeni, *MATLAB Simulations for Radar Systems Design*, Chapman & Hall/CRC Press, Boca Raton, FL, 2003.
8. IEEE Standard Definitions of Terms for Antennas, IEEE Standards, Std 145-1983.
9. L. Sevgi, *Complex Electromagnetic Problems and Numerical Simulation Approaches*, IEEE Press–John Wiley and Sons, New York, 2003.

CHAPTER 12

Simple Propagation Models and Ray Solutions

The topic of electromagnetic (EM) wave propagation through complex environments has been and will continue to be of great interest for communication and radar systems. Propagation engineers, service planners, site surveyors, and others in their office desire to have access in *real time* to the propagation characteristics on a digital map on a computer when they select the locations of a transmitter/receiver pair. The modelers are still away from satisfying this requirement directly in three dimensions (3D), and the field strength prediction can be simulated for a two-dimensional (2D) path (projected from the ground profile between those two selected points—see Fig. 12.1). Even propagation along these 2D projected propagation paths poses problems requiring further investigation.

Simple propagation models are introduced in this chapter. First, a simple ray-shooting MATLAB package is designed and discussed [1]. The package may be used as an educational tool in various EM lectures (e.g., EM wave theory, antennas and propagation, wireless propagation, etc.) as well as a research tool to predict ducting and/or antiducting conditions in the presence of buildings and under various atmospheric refractivity conditions. The user may improve the MATLAB code to include more obstacles as well as subroutines to accommodate ray fields. Then, early analytical ray-based models are discussed. Propagation over

Electromagnetic Modeling and Simulation, First Edition. Levent Sevgi.

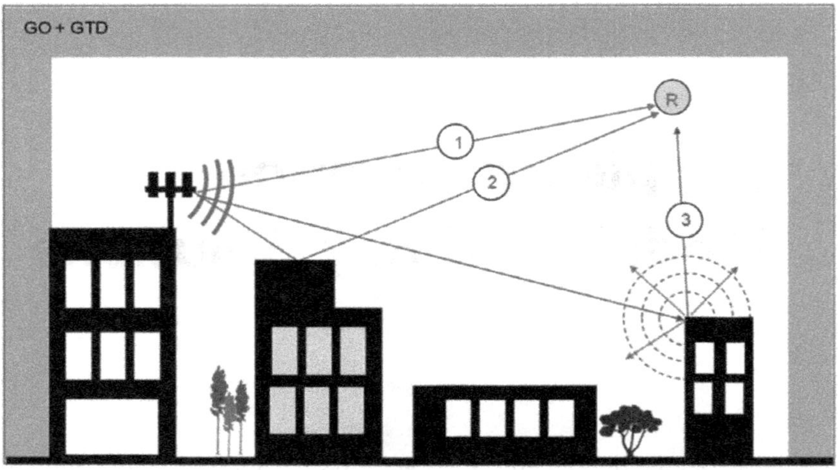

Figure 12.1. A 2D projection of an urban propagation environment with buildings along the propagation path and GO/GTD components: (1) direct ray between transmitter and receiver, (2) reflected rays (from the top of buildings as well as from the bottom surface), and (3) tip-diffracted rays.

nonpenetrable flat earth with the simple two ray (2Ray) model and single knife-edge problem with the four ray (4Ray) model are summarized with basic examples and illustrations.

12.1 INTRODUCTION

The matched coordinates for the problem of 3D wave propagation over a spherical earth with a nonflat terrain profile plus impedance boundary conditions, and a radially dependent atmosphere, excited by a vertical electric dipole located near the earth's surface is the spherical system (the earth's radius is $r = a$; finite conductivity of the earth's surface at $r = a$ is described by the surface impedance). For the relevant source (r') and observation (r) heights, which satisfy the inequalities $((r' - a)/a) << 1$ and $((r - a)/a) << 1$, respectively, the 3D problem in spherical coordinates can be approximated in "earth-flattening" nondimensional 2D rectangular (x, z) coordinates, and the earth's curvature and the standard atmosphere condition can be included by using $n = n_0 + x/a_e$, where n_0 is the refractivity value at the surface and

$a_e = 4a/3 = 8504$ km is the effective earth's radius (a: earth's radius). Extra terms are also added to model various super- or subrefraction propagation cases. Because of the closeness of n to unity, it is customary to use the refractivity N defined as

$$N = (n-1)\times 10^6. \tag{12.1}$$

N is dimensionless, but is measured in "N units" for convenience. N depends on the pressure P (mbar), the absolute temperature T (K), and the partial pressure of water vapor e (mbar) as [2]

$$N = 77.6\frac{P}{T} + 3.73\times 10^5 \frac{e}{T^2}, \tag{12.2}$$

which is valid in earth-troposphere waveguides and can be used in ground wave propagation modeling. If the refractive index were constant, radio waves would propagate in straight lines. Since n decreases with height, radio waves are bent downward toward the earth, so that the radio horizon lies further away than the optical horizon. Note that the radio horizon effect is taken into account either by using N with a_e or by introducing a fictitious medium where N is replaced by the modified refractivity M,

$$M = N + \frac{x}{a}\times 10^6 = N + 157x, \tag{12.3}$$

with the height x given in kilometers. In (12.2),

$$a = 6378 \text{ km}, \quad \frac{10^6}{a} = 157, \quad \text{and} \quad \frac{\partial M}{\partial x} = \frac{\partial N}{\partial x} + 157. \tag{12.4}$$

For the standard atmosphere (i.e., for a vertical linearly decreasing refractive index), N decreases by about 40 N unit/km, while M increases by about 117 N unit/km. Subrefraction (super-refraction) occurs when the rate of change in N with respect to height (i.e., $\partial N/\partial x$) is less (more) than 40 N unit/km.

12.2 RAY-TRACING APPROACHES

Ray-tracing models have gained special attention for the last decade parallel to the introduction of personal communication systems [3–9] (propagation models based on powerful frequency time and frequency domain techniques are beyond the scope of this chapter and will be treated in details in the following chapters).

The most difficult part of propagation simulation in terms of ray summation is to predict ray paths between the transmitter and receiver. The reflected and refracted ray paths are constructed by either *eigenray search* (i.e., by finding out characteristic rays between the transmitter and receiver through a specified propagation path) or by *brute-force* (shooting and bouncing rays [SBRs], i.e., by considering a bundle of transmitted rays that may or may not reach the receiver) methods. The challenge to a ray-tracing propagation model is to find a computationally fast way to specify the dominant ray paths that account for the field strength prediction.

An EM field contribution of a ray at a distance $R = R(x,y,z)$ for a point source (x',y',z') (in 3D environment) (with the assumption of $\exp(j\omega t)$ time dependence) is

$$E(R) = \frac{e^{-jkR}}{R} \tag{12.5}$$

and, at a distance $\rho = \rho(x,z)$ for a line source (x',z') (i.e., in 2D environment), is

$$E(\rho) = \frac{e^{-jk\rho}}{\sqrt{\rho}}, \tag{12.6}$$

where $R = \sqrt{(x-x')^2 + (y-y')^2 + (z-z')^2}$, $\rho = \sqrt{(x-x')^2 + (z-z')^2}$, $k = \omega/\nu$ is the wave number and $\omega = 2\pi f$ is the radial frequency. As mentioned earlier, once a ray is shot (i.e., leaves the transmitter), there are three kinds of mechanisms that change the direction of propagation of a ray: reflection from hard boundaries, refraction because of the refractivity variations of the propagation environment, and diffraction from the edges and tips of obstacles. Reflections are accounted for by using the reflection coefficient for an impedance boundary,

$$R = \frac{\sin\phi' - \sin\theta}{\sin\phi' + \sin\theta}, \sin\theta = \begin{cases} 1/\eta & \text{for soft boundary} \\ \mu & \text{for hard boundary} \end{cases}, \tag{12.7}$$

where η is normalized surface impedance and ϕ' is the angle between the surface and the ray incidence direction (depending on the polarization, $R = \pm 1$ for a perfectly electrical conducting [PEC] boundary). Different kinds of diffraction may be accounted, from introducing a simple knife-edge diffraction coefficient to double-wedge diffraction, and so on (any time a ray is incident on an edge, a cone of diffracted rays, any time a ray is incident on a tip, a circle of tip diffracted rays is created). These are not included here.

12.3 A RAY-SHOOTING MATLAB PACKAGE

A short and very simple MATLAB package has been prepared for visualization purposes by using a ray-shooting technique. The SNELL package shoots a number of rays, whose angles of departures are specified by the user, through a propagation medium characterized by various linear vertical refractivity profiles. Range-independent refractivity is assumed here, but its extension to range-dependent cases is straightforward. The user selects maximum range (*zlast*), maximum height (*xlast*), horizontal layer thickness (Δx), source height (*xsource*), refractivity parameters, and may choose to locate one or two PEC obstacles along the propagation path. Figure 12.2 schematizes the core of the package, which is based on direct application of the Snell's law between adjacent horizontal layers whose refractive indexes are assumed constant. The ray angle is measured from the vertical (x)-axis.

As long as the ray coordinates are less than maximum range and height, the package shoots rays emanating from the source one by one and stores the points on the ray path as (x,z) pairs. Five different procedures are included in the package. The first procedure belongs to consecutive application of Snell's law through a multi-horizontal-layered propagation medium. The height of each layer (Δx) is user specified and constant. The initial layer is the layer that corresponds to the user-supplied source height. The first ray emanates the source with the first ray angle and Snell's law,

$$n_i \sin\phi_i = n_{i+1} \sin\phi_{i+1}, \tag{12.8}$$

gives the refracted ray angle in the next layer (here i is the index of the source layer). The horizontal distance Δz and the ray segment Δs of the next ($i + 1$) layer are then calculated by using the trigonometric relations

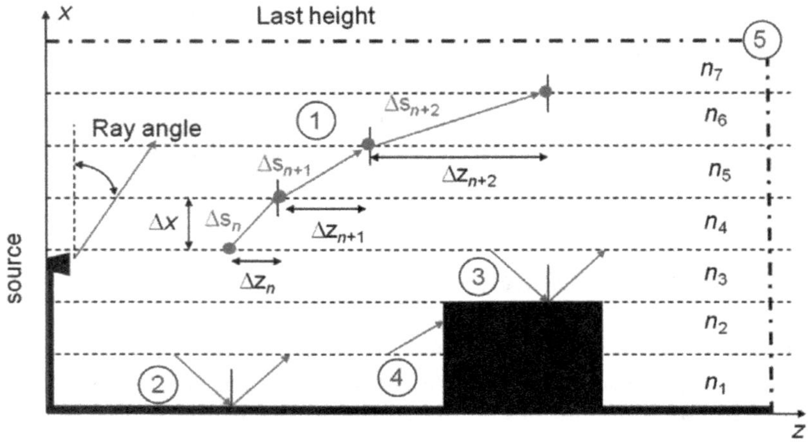

Figure 12.2. 2D propagation environment with a PEC obstacle along the path. The medium is assumed multiplanar layers with constant refractive indexes. The height (Δx) of each layer is user selected and constant. The ray path segments (Δs) and range segments (Δz) are variable and determined by the Snell's law. (1) Consecutive applications of Snell's law and calculations of Δs and Δz. (2) Reflection from the PEC bottom surface. (3) Reflection from the top of the obstacle. (4) Ray end at the left of the PEC obstacle (the package may be improved also to handle ray reflection at walls of the obstacles). (5) Maximum chosen height.

$$\Delta z_{i+1} = |\Delta x| \sin \phi_{i+1}, \tag{12.9}$$

$$\Delta s_{i+1} = \sqrt{(\Delta x)^2 + (\Delta z_{i+1})^2}. \tag{12.10}$$

Ray propagates through the layer via the same way until it reaches maximum range or maximum height; ray segments' coordinates

$$z(i+1) = z(i) + \Delta z_{i+1}, \tag{12.11}$$

$$x(i+1) = x(i) \pm \Delta z_{i+1} \tag{12.12}$$

(where $\pm$ in (12.12) corresponds to upward and downward propagating rays, respectively) are stored, and the ray loci are then plotted. The total ray path, when ray fields are going to be calculated, is given (as the superposition of the ray segments) as

$$R_i = \sum_{n=1}^{M} \Delta s_n, \tag{12.13}$$

where i and M are ray index and number of ray segments, respectively.

The second procedure belongs to the full reflection from the PEC bottom surface, and this is realized by just bending the ray upward with the same angle it hits the bottom surface. In MATLAB, this is achieved by just replacing Δx with $-\Delta x$. Procedure 3 is same with procedure 2, except the reflection occurs on top of the user-specified obstacles. Procedure 4 is reserved for a case when a ray hits the left wall of an obstacle; in this case, ray ends there and the package starts shooting the next ray. (It should be noted that the package presented here is designed only for visualization purposes and forward propagating rays, but it is quite straightforward to add a procedure to visualize backward propagating rays. This is especially essential to account for rays and fields between nearby buildings, where multiforward and backward reflections—*resonances*—occur.) An important procedure for rays that reach almost 90° ray angles is the treatment of the ray *caustics*. In terms of programming, this is very similar with the procedure 2. Rays that reach 90° ray angle and propagating upward are bend downward by replacing Δx with $-\Delta x$ once again (the same applies for rays that reach −270° ray angle and propagating downward which are bend upward by replacing $-\Delta x$ with Δx). The flowchart explaining all these procedures is given in Fig. 12.3. Here, the thick-arrow loop is for an individual ray trace. The thin-arrow loop shows the complete bundle of ray loop. Table 12.1 lists a short ray-shooting MATLAB script.

The trilinear vertical refractivity of the environment is specified via

$$n(x) = \begin{cases} 1 - a_0 x & x \le X_{D1} \\ 1 - a_0(2X_{D1} - x) & X_{D1} < x \le X_{D2}, \\ 1 - 2a_0(X_{D1} - X_{D2}) + a_0 x & X_{D2} < x \end{cases} \quad (12.14)$$

where a_0 is the slope of the refractivity (not the earth's radius), and X_{D1} and X_{D2} are ducting and antiducting heights, respectively (all of which are user specified).

Figure 12.4 shows an example for the set of parameters listed in Table 12.1.

It should be noted that the package presented here is designed to draw ray paths through a complex environment with a variable refractivity profile along a path which may or may not include obstacles. The

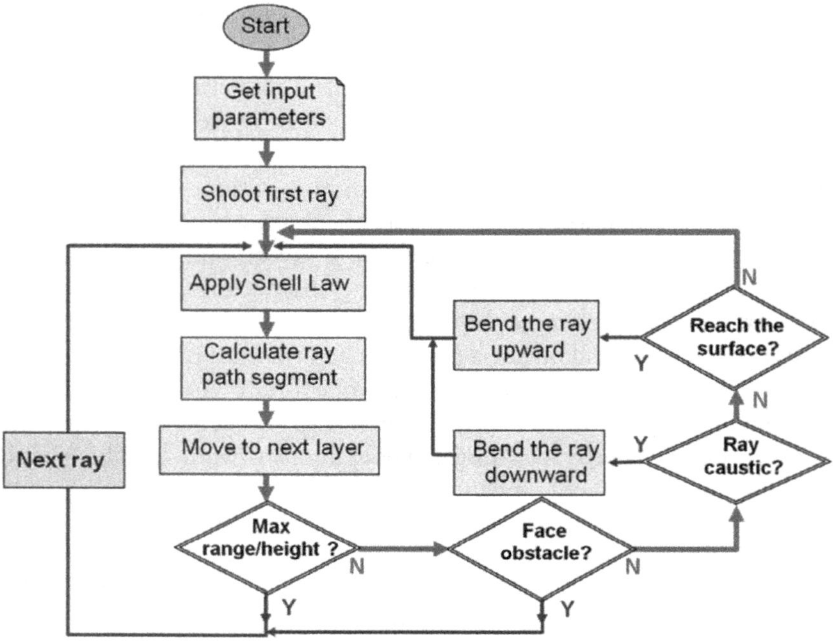

Figure 12.3. Flowchart of SNELL (input parameters: maximum range/height, layer height, source location, first/last ray angles measured from surface normal, ray increment, trilinear refractivity parameters, and obstacle heights and locations).

EM field contributions of rays at a chosen observation range or height (i.e., ray fields vs. height at a specified range, or vs. range at a constant height) may be added by using (12.5), (12.6), and (12.7), and accumulating rays' contributions that pass the same observation point:

- calculate the ray path of each ray by using (12.13);
- store number of reflections (*RN*) from PEC surfaces (either boundaries or obstacles);
- store number of turning points (*CN*) (i.e., ray caustics);
- and calculate the EM field contribution from

$$E(\mathrm{Ray}_i) = \frac{e^{-jkR_i}}{R_i} \prod_{k=0}^{RN} e^{-j\pi} \prod_{l=0}^{CN} e^{-j\frac{\pi}{2}}. \tag{12.15}$$

Table 12.1.
A MATLAB Ray-Shooting Script

```
%------------------------------------------------
% Program: RayShoot.m
%------------------------------------------------
clc; clear all;
xbas=0; xson=300; zson=2500; tbas=45; tson=120;
tart=0.23;
xart=0.2; xd1=80; xd2=150; a0=100*1e-5;
acison=.99999999999999999999;
zbas=0.0; eps=pi/20000; smax=10000; nx=xson/xart;
for n=1:nx  % build refractive index array
    yx(n)=(n-1)*xart;
   if yx(n) <= xd1
      ref(n)=1-a0*yx(n);
   elseif yx(n) > xd1 && yx(n) <= xd2
      ref(n)=1-a0*(2*xd1-yx(n));
   else
      ref(n)=1-2*a0*(xd1-xd2)-a0*yx(n);
   end
end
nteta=(tson-tbas)/tart;
for n=1:nteta+1    % ----- Ray loop -----
    for m=1:smax
       x(m)=0; z(m)=0;
    end
    teta=tbas+(n-1)*abs(tart); xart=abs(xart);
    if teta > 90
       teta=180-teta; xart=-xart;
    end
    teta=teta*pi/180.; sayac=1; x(sayac)=xbas;
z(sayac)=zbas;
    while z(sayac) < zson
      if x(sayac) >= xson      % exit when reach last
height
          break
        end
        if x(sayac) <= xd1      % refractive index
           nf=1-a0*x(sayac);
        elseif x(sayac) > xd1 && x(sayac) <= xd2
           nf=1-a0*(2*xd1-x(sayac));
```

(*Continued*)

Table 12.1.
(Continued)

```
        else
            nf=1-2*a0*(xd1-xd2)-a0*x(sayac);
        end
        if abs(pi-teta) > eps
           zart=abs(xart)*tan(teta);
        else
           zart=zart;
        end
        sayac=sayac+1;
        if sayac > smax     % stop when iterations exceeds
             break;
        end
        x(sayac)=x(sayac-1)+xart;    % calculate next ray
        z(sayac)=z(sayac-1)+zart;
        if x(sayac) <= 0    % reflection from the bottom
            xart=-xart;
        end
        if x(sayac) <= xd1    % refractive index
            ns=1-a0*x(sayac);
        elseif x(sayac) > xd1 && x(sayac) <= xd2
           ns=1-a0*(2*xd1-x(sayac));
        else
           ns=1-2*a0*(xd1-xd2)-a0*x(sayac);
        end
        s=nf*(sin(teta)/ns);
        if s >= acison      % reflection from the ray caustic
           xart=-xart;
        end
        teta=asin(s);
    end % of while
    for k=1:sayac-1
        xx(k)=x(k); zz(k)=z(k);
    end
    plot(zz,xx,'k-'); axis([0 zson 0 xson]); hold on;
    xlabel('Range [m]'); ylabel('Height [m]');
    title(['Ray Shooting Through Non-homogeneous
Atmosphere']);
end % of Ray Loop
%------------------------ END ------------------------
```

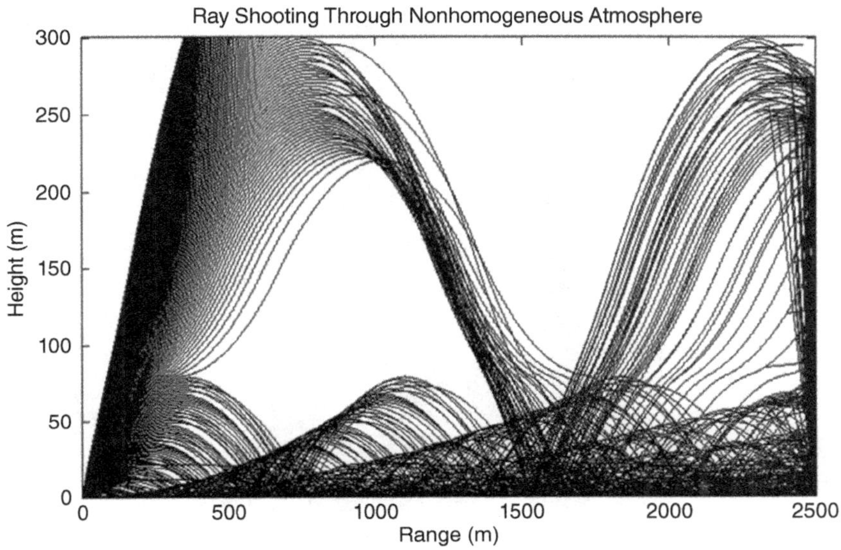

Figure 12.4. Ray paths through trilinear atmosphere.

12.4 CHARACTERISTIC EXAMPLES

The front panel of the SNELL package with the first propagation scenario is given in Fig. 12.5. Here, the left plot is reserved for vertical refractivity profile $n = n(x)$. The ray paths are plotted in the right plot. Ray shooting through an environment with the homogeneous atmosphere (i.e., $n = n_0 = 1.0$) and with no obstacle where rays are straight lines is shown in this figure. The plot shows a total of 150 rays emanating from the source at 50 m height. First and last ray angles are 45° and 120°, respectively, and the ray increment is 0.5°. Maximum range and height are 2500 and 300 m, respectively.

The scenario is changed a little and two PEC obstacles are added, and ray paths are plotted with the same parameters. The ray paths are plotted in Fig. 12.6. Reflections from tops of the buildings and ray ending at left sides of the obstacles are clearly observed in the figure.

Another scenario is given, and ray paths are plotted in Fig. 12.7. Here, the parameters are source height of 50 m, maximum height of 300 m, layer height of 0.2 m, maximum range of 2500 m, first ray angle

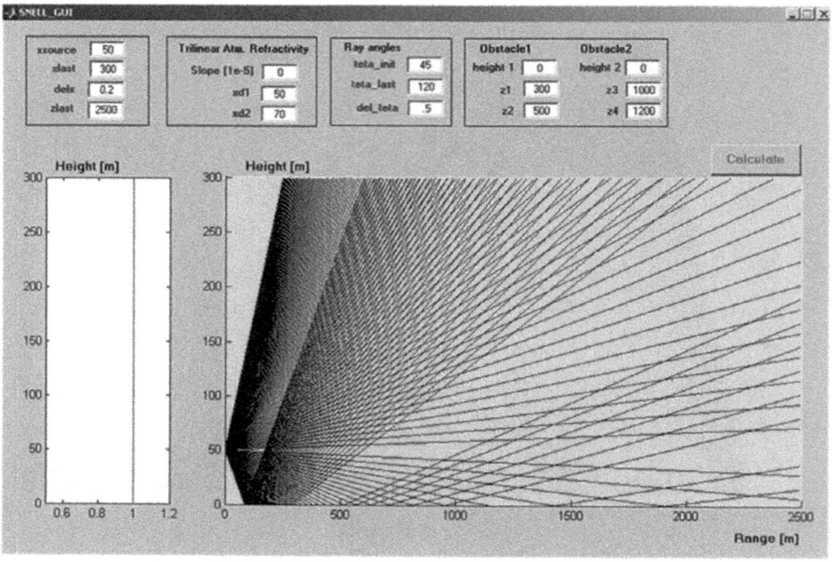

Figure 12.5. Front panel of the SNELL package. The left plot is reserved for vertical refractivity profile $n = n(x)$. The right plot is for ray paths.

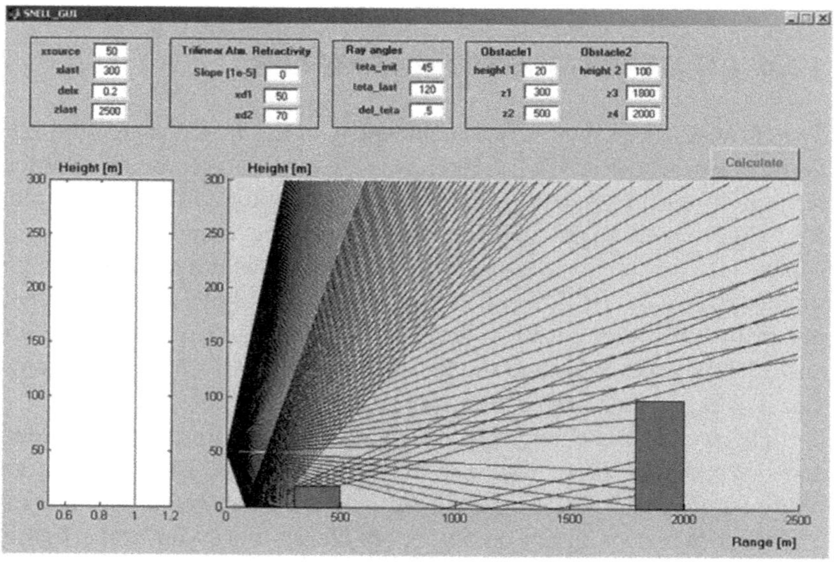

Figure 12.6. Ray shooting through an environment with the homogeneous atmosphere (i.e., $n = n_0 = 1.0$) and with two obstacles where rays are straight lines (first obstacle: height = 10 m, base length = 200 m, range = 300 m; second obstacle: height = 100 m, base length = 200 m, range = 1800 m).

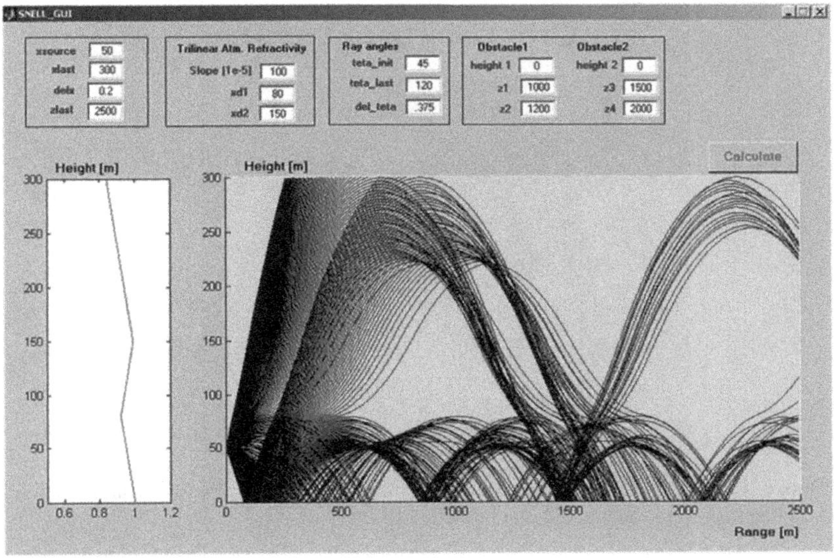

Figure 12.7. Ray shooting through an atmosphere with a typical trilinear refractivity profile and with no obstacles. Ray reflection, refraction, bending, and caustics are clearly observed.

of 45°, last ray angle of 120°, ray increment of 0.375°, slope of refractivity of 0.001, and the heights of ducting and antiducting are 80 and 150 m, respectively. A trilinear vertical refractivity profile is as specified. Heights between the surface and 80 m correspond to *surface ducting*, while heights between 80 and 150 m correspond to *antiducting*. There are a certain number of surface-trapped rays and a certain number of other rays that are trapped inside an elevated duct.

The next scenario belongs to a bilinear vertical refractivity profile that causes an elevated-ducted propagation (see Fig. 12.8). Here, the height of the boundary between antiducting and ducting is 150 m, and the propagation path does not include any obstacle. Other parameters are source height of 50 m, maximum height of 300 m, layer height of 0.2 m, maximum range of 2500 m, first ray angle of 45°, last ray angle of 120°, ray increment of 0.375°, and the slope of refractivity of 0.001. The number of total rays is again 150. The elevated duct may be controlled by changing the slope of the refractivity and height of the source.

Figure 12.9 belongs to another trilinear vertical refractivity profile, where surface ducting extends up to 80 m in height and elevated duct occurs around 150 m in height. Plotting ray paths as explained here

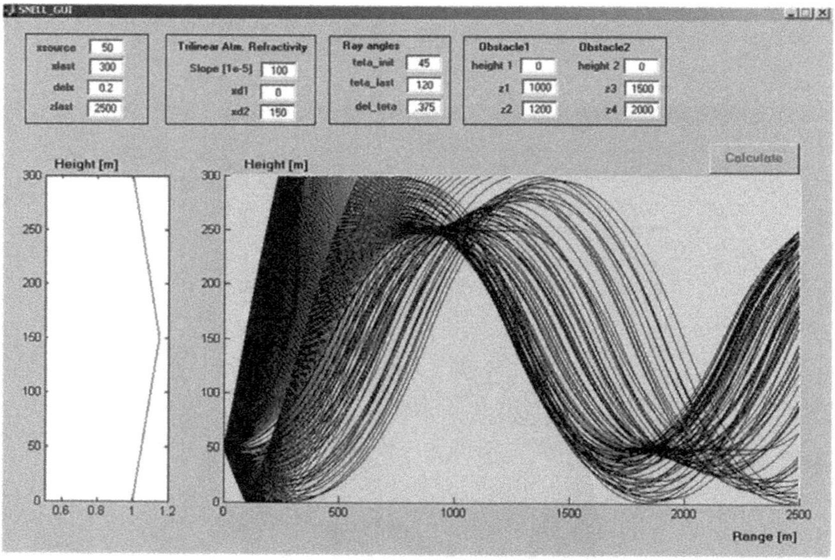

Figure 12.8. Ray shooting through an atmosphere with a typical bilinear refractivity (elevated ducting) profile and with no obstacles.

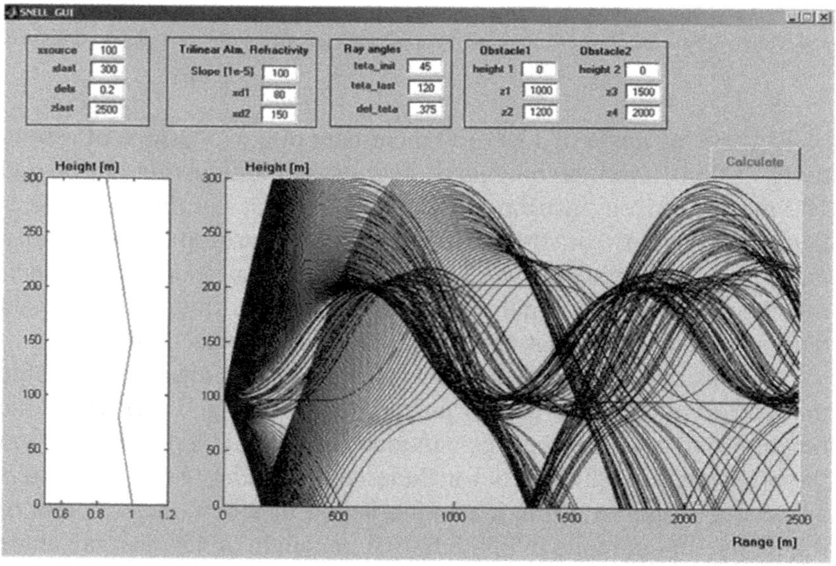

Figure 12.9. Ray shooting with the same trilinear refractivity profile in Fig. 12.6, but with different source locations (no obstacles, source height = 100 m, maximum height = 300 m, layer height = 0.2 m, maximum range = 2500 m, first ray angle = 45°, last ray angle = 120°, ray increment = 0.375°, the slope of refractivity = 0.001, and the heights of ducting and antiducting are 80 and 150 m, respectively).

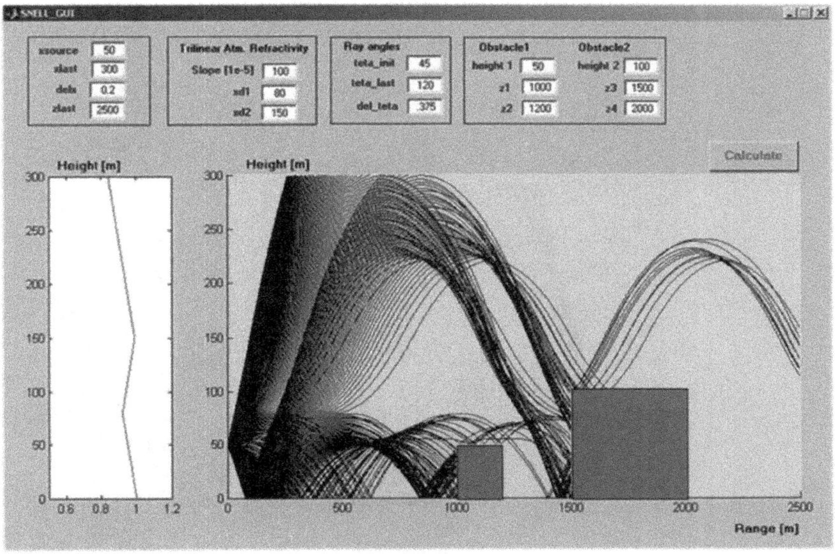

Figure 12.10. Ray shooting with the same refractivity in Fig. 12.6, with two obstacles (source height = 50 m, maximum height = 300 m, layer height = 0.2 m, maximum range = 2500 m, first ray angle = 45°, last ray angle = 120°, and ray increment = 0.375°).

certainly helps to predict strong and/or weak EM signal regions. For example, the signal is extremely weak at ranges between 500 and 800 m up to heights of 50 m (it should be noted that not all rays are plotted in this figure, only finite number of rays between 45° and 120°, 0.5° apart from each other are plotted). Regions with multirays (dark regions in the figure) may either be field maximum or minimum depending on constructive or destructive interference of rays.

Finally, Fig. 12.10 and Fig. 12.11 correspond to two different trilinear refractivity profiles and two obstacle cases, where illuminated and shadowed regions can be distinguished easily.

12.5 FLAT-EARTH PROBLEM AND 2RAY MODEL

One of the simple, canonical propagation scenario taught in many undergraduate lectures is the nonpenetrable flat earth. This scenario is pictured in Fig. 12.12. A line source is located above the flat earth at

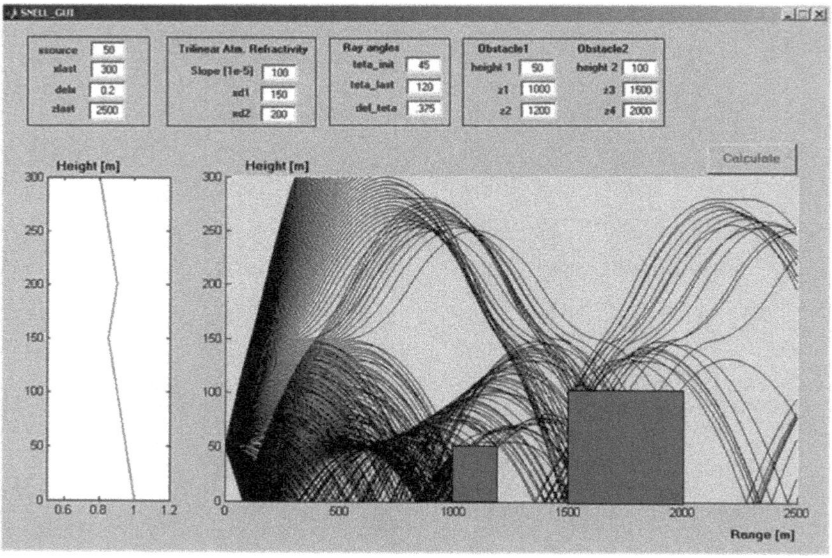

Figure 12.11. Ray shooting with a trilinear refractivity profile and two obstacles (source height = 50 m, maximum height = 300 m, layer height = 0.2 m, maximum range = 2500 m, first ray angle = 45°, last ray angle = 120°, ray increment = 0.375°, the slope of refractivity = 0.001, and the heights of ducting and antiducting are 150 and 200 m, respectively).

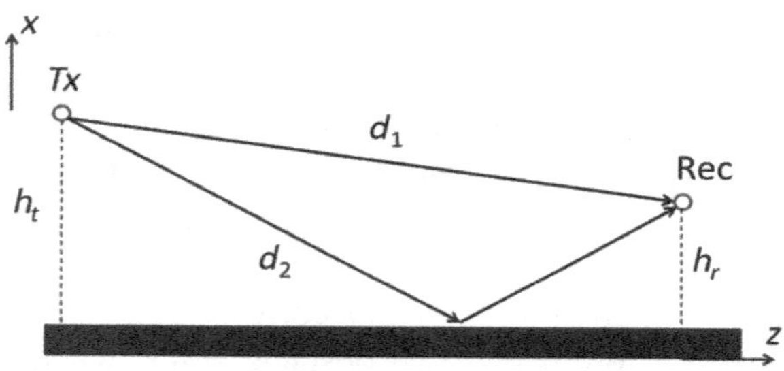

Figure 12.12. Propagation over PEC flat-earth scenario.

an altitude of h_t. Cylindrical waves emanating from the source reaches the receiver at a distance d and altitude h_r via direct and ground-reflected paths. Two possible cases are the horizontal (electric field is perpendicular to the 2D propagation plane) and vertical (magnetic field is perpendicular to the 2D propagation plane) polarizations. The ground-reflection coefficients are $\Gamma = -1$ and $\Gamma = +1$ for the horizontal (TE_z) and perpendicular (TM_z) polarizations, respectively.

The total field at the receiver can directly be given as

$$E_t = E_{d_1} + E_{d_2} = E_0 \frac{e^{jkd_1}}{d_1} + \Gamma E_0 \frac{e^{jkd_2}}{d_2}. \tag{12.16}$$

The path difference between the direct and ground-reflected waves Δ is

$$\Delta = d_2 - d_1 = \sqrt{(h_t + h_r)^2 + d^2} - \sqrt{(h_t - h_r)^2 + d^2}, \tag{12.17}$$

and if the transmitter–receiver distance is much greater than their heights, the path difference can be approximated as $\Delta \approx 2h_t h_r / d$.

The received power for the 2Ray scenario is given as

$$P_r = \frac{P_t G_t G_r \lambda^2}{(4\pi d)^2} \left| 1 - e^{j2\pi\Delta/\lambda} \right|, \tag{12.18}$$

where P_t, G_t, G_r, and λ refer to the transmit power, transmit antenna gain, receive antenna gain, and the wavelength, respectively. The received power for the long-range approximation is given as

$$P_r = \frac{P_t G_t G_r h_t^2 h_r^2}{d^4}. \tag{12.19}$$

Finally, the received power for the free space is

$$P_r = \frac{P_t G_t G_r \lambda^2}{(4\pi d)^2}. \tag{12.20}$$

Table 12.2 lists a short MATLAB code for the calculation of receiver power versus range using (12.16). A typical simulation result is given in Fig. 12.13. Here, the frequency is 300 MHz; source is located at 50 m; receiver height is 5 m. The reflection coefficient is 1, which represents vertical polarization. As observed, two rays interact and form an interference pattern. Maxima are 6 dB which show that when two rays are in phase, the amplitude of the total field becomes 2 (rays are excited with unit amplitude); therefore, propagation factor (PF) is 6 dB. One can easily replace the range loop in the code with

Table 12.2.
A Short MATLAB Code for the 2Ray Model

```
%-------------------------------------------------------
% Program: TwoRay.m
% Purpose: To compute path loss over 2D flat PEC ground
%      by using direct and ground-reflected rays
%      x: height, z: range, k0: wavenumber
%      REF: Reflection coefficient (1:Ver Pol, -1:Hor Pol)
%      R1: direct ray, R2: ground reflected ray
%      fgan(): propagation factor in dB (20Log10(|E/
E0|))
%-------------------------------------------------------
clear all; clc;
E0=1.0; freq=1e8; k0=2*pi*freq/3e8; lambda=3e8/freq;
REF=1; xsource=50; xobs=5; zlast=150;
delz=lambda/10; Nz=zlast/delz;
% ----------- Range loop ------------
for n=1:Nz
  zobs(n)=(2*n-1)*delz/2;
  R1=sqrt(zobs(n)^2+(xsource-xobs)^2);
  R2=sqrt(zobs(n)^2+(xsource+xobs)^2);
  gfree(n)=E0/(120*pi)*exp(-1i*k0*R1)/R1;
  gref(n)=REF*E0/(120*pi)*exp(-1i*k0*R2)/R2;
  gtot(n)=gfree(n)+gref(n); gpf(n)=gtot(n)/gfree(n);
end
plot(zobs,20*log10(abs(gpf)));
xlabel('Distance [m]','Fontweight','bold');
ylabel('Propagation Factor [dB]','Fontweight','bold');
title('Propagation over flat-Earth: 2Ray','Fontweight',
'bold');
% ------------------------- END -------------------------
```

the height loop and simulate field versus height at a given range. This is illustrated in Fig. 12.14 with parameters are frequency of 100 MHz, source height of 20 m, receiver height of 5 m, and reflection coefficient of 1.

The code in Table 12.2 can be modified easily by coding (12.18), (12.19), and (12.20) which make the comparison of all three cases possible. A typical example computed with the three formulas given in

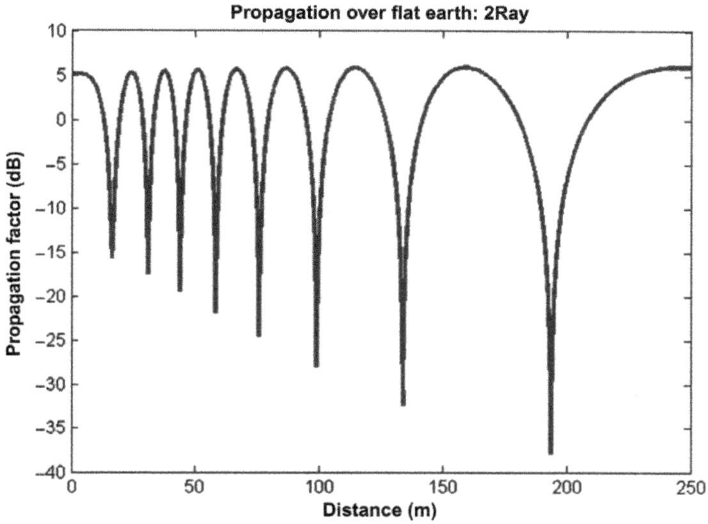

Figure 12.13. Field versus range showing interaction of direct and ground-reflected rays.

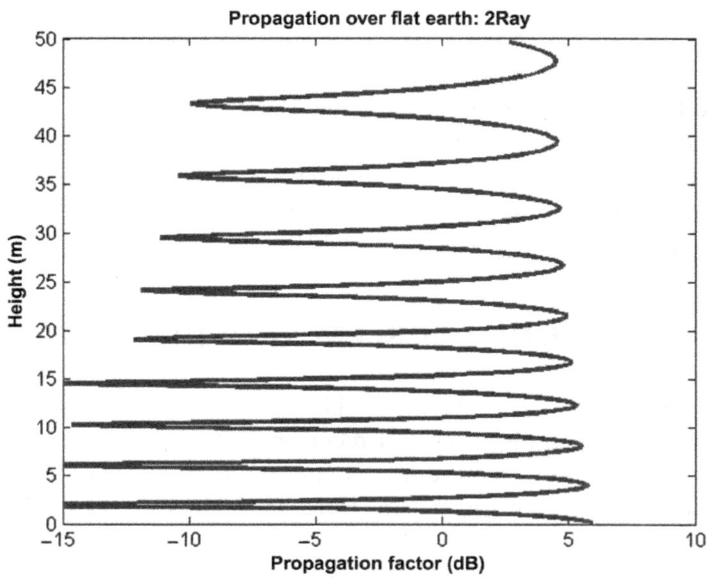

Figure 12.14. Field versus height showing interaction of direct and ground-reflected rays.

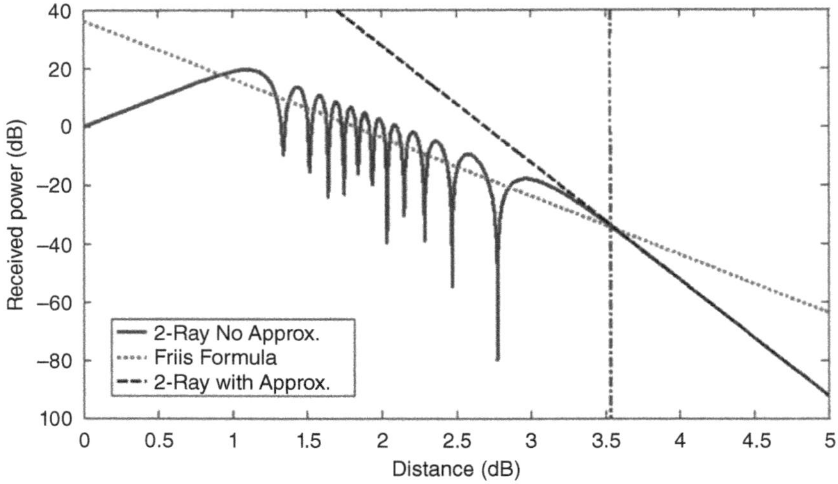

Figure 12.15. Received power versus range.

(12.18), (12.19), and (12.20) is plotted in Fig. 12.15. Here, the parameters are frequency of 900 MHz, transmitter height of 50 m, receiver height of 2 m, antenna gains, $G_t = G_r = 1$, and total distance of 100 km. The vertical line at around 35 km is the critical distance. As observed, direct and ground-reflected waves interfere and maxima/minima occur at certain ranges. The mean variation of the received power agrees well with the free-space value. Beyond the critical range $dc = 2\pi h_t h_r / \lambda$, there is no interference and (12.18) and (12.19) agree well.

12.6 KNIFE-EDGE PROBLEM AND 4RAY MODEL

The single-knife-edge problem [10,11] models one dominating sharp obstacle between transmitter and receiver. In this problem, the 4Ray and Fresnel integral representation are used. The scenario of this problem is pictured in Fig. 12.16. Here, h_t, h_r, and h_w are the heights of the transmitter, receiver, and the knife-edge obstacle; d_1 and d_2 are the distances from source to obstacle and from obstacle to receiver, respectively.

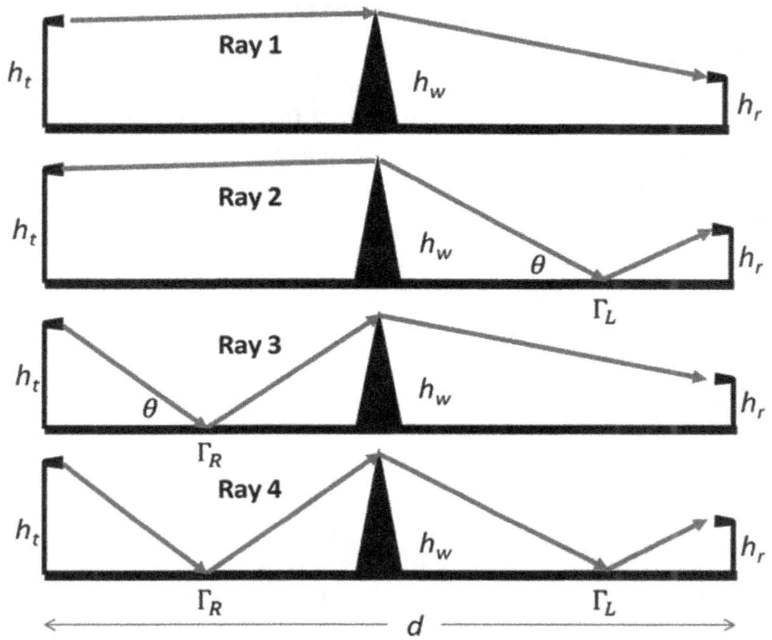

Figure 12.16. Single knife-edge problem and possible four rays.

Possible four rays are as follows.

- Ray 1 is the direct path between the transmitter and the receiver.
- Ray 2 is considered as the ray from the transmitter reflected from the right side of the knife-edge obstacle. This ray reaches the receiver directly or tip diffraction may occur.
- Ray 3 is considered as the ray from the transmitter reflected from the left side of the knife-edge obstacle. Same as before, this ray also reaches the receiver directly or tip diffraction will occur.
- Ray 4 is considered as the ray from the transmitter reflected from both left side and the right side of the knife-edge obstacle. Note that this model uses both reflections and the diffraction from edge.

The parameters of the Fresnel integrals are derived by using the image source/receiver for the reflected waves. The Fresnel clearance, the height of the knife edge above the line of sight (LOS), may be positive or negative [11]. When the direct ray between the transmitter and receiver intersects obstacle, h_0 is taken negative. The Fresnel integrals $C(\nu)$ and $S(\nu)$ are evaluated, where $v = h_0\sqrt{2}$ with h_0 equal to the ray clearance over the knife edge. The pattern PF is equal to

$$F = \frac{E}{E_0} = \sum_{q=1}^{4} E_q \exp(i\psi_q), \tag{12.21}$$

where $Eq = A(\nu q)\Gamma q$, $\Gamma_1 = 1$, $\Gamma_2 = \Gamma_R$, $\Gamma_3 = \Gamma_L$, $\Gamma_4 = \Gamma_L\Gamma_R$, and $\psi_q = \beta_q + (R_q - R_1)k_0$. Here,

$$A(v_q) = \sqrt{\frac{(C(v_q)+0.5)^2 + (S(v_q)+0.5)^2}{2}} \tag{12.22}$$

for $C(v_q) = \int_0^v \cos(v_q^2)dv$, $S(v_q) = \int_0^v \sin(v_q^2)dv$; the distances of four rays are

$$R_q = \begin{Bmatrix} \sqrt{d^2 + (h_t - h_r)^2} & \text{for} \quad q = 1, q = 3 \\ \sqrt{d^2 + (h_t + h_r)^2} & \text{for} \quad q = 2, q = 4 \end{Bmatrix}, \tag{12.23}$$

$$\beta_q = \begin{Bmatrix} \arctan\left(\dfrac{S(v_q)+0.5}{C(v_q)+0.5}\right) - \dfrac{\pi}{4} & \text{if} \quad C(v_q) \geq -0.5 \\ \arctan\left(\dfrac{S(v_q)+0.5}{C(v_q)+0.5}\right) + \dfrac{3\pi}{4} & \text{if} \quad C(v_q) < -0.5 \end{Bmatrix}, \tag{12.24}$$

and the complex reflection coefficients, for the horizontal and vertical polarizations, respectively, are

$$\Gamma = \frac{\sin(\theta) - \sqrt{\varepsilon - \cos^2(\theta)}}{\sin(\theta) + \sqrt{\varepsilon - \cos^2(\theta)}}, \quad \Gamma = \frac{\varepsilon\sin(\theta) - \sqrt{\varepsilon - \cos^2(\theta)}}{\varepsilon\sin(\theta) + \sqrt{\varepsilon - \cos^2(\theta)}}, \tag{12.25}$$

where $\varepsilon = \varepsilon_r - i60\sigma_g\lambda$, σ_g is the conductivity, ε_r is the relative permittivity of ground, and θ is the angle of incidence in radians. The ray clearances for the four rays are

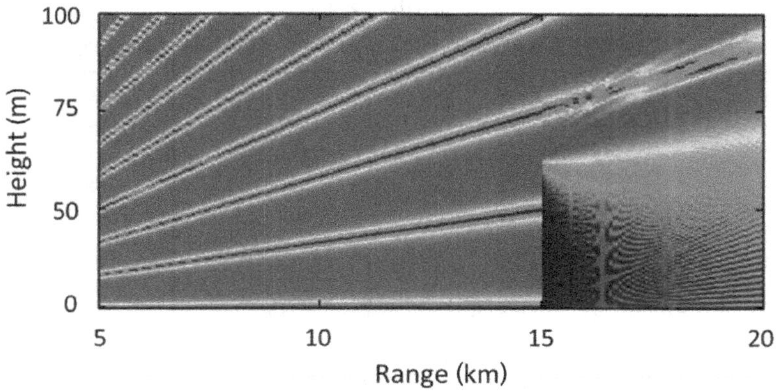

Figure 12.17. PF versus range/height for horizontal polarization over PEC ground with a 75-m-high knife edge at 15-km range (source height = 20 m, f = 3 GHz).

$$h_{01} = \sqrt{\frac{2d}{\lambda d_1 d_2}}\left(-h_t + \frac{(h_r - h_t)d_1}{d} - h_w\right), \quad (12.26a)$$

$$h_{02} = \sqrt{\frac{2d}{\lambda d_1 d_2}}\left(-h_t + \frac{(h_r + h_t)d_1}{d} - h_w\right), \quad (12.26b)$$

$$h_{03} = \sqrt{\frac{2d}{\lambda d_1 d_2}}\left(h_t + \frac{(-h_r - h_t)d_1}{d} - h_w\right), \quad (12.26c)$$

$$h_{04} = \sqrt{\frac{2d}{\lambda d_1 d_2}}\left(-h_t + \frac{(h_t - h_r)d_1}{d} - h_w\right). \quad (12.26d)$$

A typical example computed with the 4Ray model is shown in Fig. 12.17. The scenario belongs to one-way propagation for horizontal polarization over PEC ground with a 75-m-high knife edge at 15-km range. The line source is at 15-m height at $z = 0$.

Table 12.3 and Table 12.4 list short MATLAB scripts for the same 4Ray model which plot height and range profiles, respectively. An example produced with this code is presented in Fig. 12.18. Here, a 75-m-tall single knife edge is located 3 km away from the source. The source height is 50 m. For the field versus height profile, the receiver is located at 7 km and receiver height is changed from zero to 150 m. PF versus height at 1-GHz frequency is plotted for horizontal polarization.

Table 12.3.
A Short MATLAB Code for 4Ray Model (Height Profile)

```
function 4RayHeight
% ---------------------------------------------------
%  Purpose : To compute path loss over 2D flat PEC
ground
%       by using four rays
%       x: height, z: range, k0: wavenumber
%       R1: direct ray, R2: ground reflected ray
%       fgan(): propagation factor in dB (i.e.,
20Log10(|E/E0|))
% ---------------------------------------------------
clc, clear all, close all;
fr=1;       %fr=input('Frequency [GHz] = ? ');
ht=50;      %ht=input('Transmitter height [m] = ? ');
hmax=150; %hmax=input('Maximum receiver height [m] = ?
');
nx=751;     %nx=input('Number of height points [] = ? ');
hw=75;      %hw=input('Obstacle height [m] = ? ');
d1=3;       %d1=input('Source - Obstacle distance [km] =
? ');
d=7;        %d=input('Source - Receiver distance [km] =
? ');
pol=2;%pol=input('Polarization ([1] Vertical, [2]
Horizontal) = ? ');
d=d*1e3; d1=d1*1e3; lamda=3e8/(fr*1e9); k0=2*pi/lamda;
d2=d-d1; delx=hmax/(nx-1);
% ========== Height loop begins ============
xg=0:delx:hmax; PF_free=zeros(size(xg));
PL_diff=zeros(size(xg));
for k=1:nx
 hr=xg(k);   ref=-1;% PEC Hor Pol.
 rd=sqrt(d^2+(ht-hr)^2); rr=sqrt(d^2+(ht+hr)^2);
 f0=exp(-1i*k0*rd)/sqrt(rd); f1=ref*exp(-1i*k0*rr)/
sqrt(rr);
 PF_free(k)=20.*log10(abs((f0+f1)/f0));
 h0=sqrt((lamda*d1*d2)/d); sq2=sqrt(2);
 % Calculate r1, r2, r3, and r4
 r1=sqrt(d^2+(hr-ht)^2); r2=sqrt(d^2+(-hr-ht)^2);
r3=r2; r4=r1;
 % Calculate v1, v2, v3 and v4
 delh=(hr-ht)*d1/d;   v1=sq2/h0*(ht+delh-hw);
 delh=(hr+ht)*d1/d;   v2=sq2/h0*(-ht+delh-hw);
```

Table 12.3.
(Continued)

```
delh=(-ht-hr)*d1/d; v3=sq2/h0*(ht+delh-hw);
delh=(ht-hr)*d1/d;  v4=sq2/h0*(-ht+delh-hw);
% Calculate grazing angle at left
if v2 >= 0;psil=atan((ht+hr)/
d);else;psil=atan((ht+delh)/d1);end
% Calculate grazing angle at right
if v3 >=
0;psir=atan((ht+hr)/d);else;psir=atan((hr+delh)/d);end
% Calculate complex reflection coefficients for left and
right
refl=-1;%refl=cref(epsr,sigma,lamda,psil,pol);
refr=-1;%refr=cref(epsr,sigma,lamda,psir,pol);
% Calculate diffaction loss contribution for RAY1
[ic,is]=determineintegral(v1); a=1/sq2*sqrt((ic+0.5)^2+
(is+0.5)^2);
if (ic+0.5) >= 0
  b=atan((is+0.5)/(ic+0.5))-pi/4;
else
  b=pi+atan((is+0.5)/(ic+0.5))-pi/4;
end
ce1=a*(cos(b)+1i*sin(b));
% Calculate diffaction loss contribution for RAY2
[ic,is]=determineintegral(v2); a=1/sq2*sqrt((ic+0.5)^2+
(is+0.5)^2);
if (ic+0.5) >= 0
  b=atan((is+0.5)/(ic+0.5))-pi/4;
else
  b=pi+atan((is+0.5)/(ic+0.5))-pi/4;
end
b=b+(2*pi)/lamda*(r2-r1); ce2=a*(cos(b)+1i*sin(b))*refl;
% Calculate diffaction loss contribution for RAY3
[ic,is]=determineintegral(v3); a=1/sq2*sqrt((ic+0.5)^2+
(is+0.5)^2);
if (ic+0.5) >= 0
  b=atan((is+0.5)/(ic+0.5))-pi/4;
else
  b=pi+atan((is+0.5)/(ic+0.5))-pi/4;
end
b=b+(2*pi)/lamda*(r3-r1);
ce3=a*(cos(b)+1i*sin(b))*refr;
% Calculate diffaction loss contribution for RAY4
```

Table 12.3.
(Continued)

```
 [ic,is]=determineintegral(v4); a=1/sq2*sqrt((ic+0.5)^2+
(is+0.5)^2);
 if (ic+0.5) >= 0
     b=atan((is+0.5)/(ic+0.5))-pi/4;
 else
     b=pi+atan((is+0.5)/(ic+0.5))-pi/4;
 end
 ce4=a*(cos(b)+1i*sin(b))*refl*refr;
 % Total diffraction loss
 PL_diff(k)=20.*log10(abs(ce1+ce2+ce3+ce4));
 PL_knife(k)=20.*log10(abs(ce1));
end
plot(PL_diff,xg); hold on; plot(PL_knife,xg, 'b-');
xlim([-60 10])
ylabel('Height [m]');xlabel('Diffraction Loss [dB]')
legend('4Ray','Knife-edge');
title('Field behind a knife-edge
obstacle','Fontweight','bold');
function ref=cref(epsr,sigma,lamda,ang,pol)
cim=-60*lamda*sigma; cak=complex(epsr,cim);
if pol == 1
   c1=cak*sin(ang)-sqrt(cak-cos(ang)^2);
   c2=cak*sin(ang)+sqrt(cak-cos(ang)^2);
else
   c1=sin(ang)-sqrt(cak-cos(ang)^2);
   c2=sin(ang)+sqrt(cak-cos(ang)^2);
end
ref=c1/c2;
function [ic,is]=determineintegral(v)
if v>0,
   ic= sum(chebfun(@(t) cos(pi/2*t.^2),[0,v]));
   is= sum(chebfun(@(t) sin(pi/2*t.^2),[0,v]));
elseif v<0,
   ic=-sum(chebfun(@(t) cos(pi/2*t.^2),[v,0]));
   is=-sum(chebfun(@(t) sin(pi/2*t.^2),[v,0]));
else
   ic=0; is=0;
end
% ------------------------- END -------------------------
```

Table 12.4.
A Short MATLAB Code for 4Ray Model (Range Profile)

```
function 4RayRange
% ------------------------------------------------
%  Purpose : To compute path loss over 2D flat PEC
ground
%      by using four rays
%      x: height, z: range, k0: wavenumber
%      R1: direct ray, R2: ground reflected ray
%      fgan(): propagation factor in dB (i.e.,
20Log10(|E/E0|))
% ------------------------------------------------
clc, clear all, close all;
fr=.3; ht=50; Xg=100; nz=751; hw=75; d1=3; Zmax=7;
pol=2;
Zmax=Zmax*1e3; d1=d1*1e3; delz=Zmax/(nz-1);
Zg=0:delz:Zmax;
lamda=3e8/(fr*1e9); k0=2*pi/lamda;
%----------------------- Range loop begins --------------------
PF_free=zeros(size(Zg)); PL_diff=zeros(size(Zg));
for k=1:nz
 d=Zg(k); d2=d-d1;
 hr=Xg;   ref=-1;% PEC Hor Pol.
 %----------------------- 2Ray model-----------------------
 R1=sqrt(d^2+(ht-hr)^2); R2=sqrt(d^2+(ht+hr)^2);
 F2Ray(k)=exp(-1i*k0*R1)/sqrt(k0*R1)+ref*exp(-1i*k0*R2)/
sqrt(k0*R2);
 PF2Ray(k)=F2Ray(k)/(exp(-1i*k0*R1)/sqrt(k0*R1)); %
Calculate PF=E/E0
 % -----------------------------------------------------
 rd=sqrt(d^2+(ht-hr)^2); rr=sqrt(d^2+(ht+hr)^2);
 f0=exp(-1i*k0*rd)/sqrt(rd); f1=ref*exp(-1i*k0*rr)/
sqrt(rr);
 PF_free(k)=20.*log10(abs((f0+f1)/f0));
 h0=sqrt((lamda*d1*d2)/d); sq2=sqrt(2);
 % Calculate r1, r2, r3, and r4
 r1=sqrt(d^2+(hr-ht)^2); r2=sqrt(d^2+(-hr-ht)^2);
r3=r2; r4=r1;
 % Calculate v1, v2, v3 and v4
 delh=(hr-ht)*d1/d;   v1=sq2/h0*(ht+delh-hw);
 delh=(hr+ht)*d1/d;    v2=sq2/h0*(-ht+delh-hw);
 delh=(-ht-hr)*d1/d;  v3=sq2/h0*(ht+delh-hw);
```

(Continued)

Table 12.4.
(Continued)

```
 delh=(ht-hr)*d1/d;  v4=sq2/h0*(-ht+delh-hw);
 % Calculate grazing angle at left
 if v2>=0;psil=atan((ht+hr)/
d);else;psil=atan((ht+delh)/d1);end
 % Calculate grazing angle at right
 if
v3>=0;psir=atan((ht+hr)/d);else;psir=atan((hr+delh)/
d);end
 % Calculate complex reflection coefficients for left and
right
 refl=-1;refr=-1;%refr=cref(epsr,sigma,lamda,psir,pol);
 % Calculate diffaction loss contribution for RAY1
 [ic,is]=determineintegral(v1); a=1/sq2*sqrt((ic+0.5)^2+
(is+0.5)^2);
 if (ic+0.5) >= 0
     b=atan((is+0.5)/(ic+0.5))-pi/4;
 else
     b=pi+atan((is+0.5)/(ic+0.5))-pi/4;
 end
 ce1=a*(cos(b)+1i*sin(b));
 % Calculate diffaction loss contribution for RAY2
 [ic,is]=determineintegral(v2); a=1/sq2*sqrt((ic+0.5)^2+
(is+0.5)^2);
 if (ic+0.5) >= 0
     b=atan((is+0.5)/(ic+0.5))-pi/4;
 else
     b=pi+atan((is+0.5)/(ic+0.5))-pi/4;
 end
 b=b+(2*pi)/lamda*(r2-r1); ce2=a*(cos(b)+1i*sin(b))*refl;
 % Calculate diffaction loss contribution for RAY3
 [ic,is]=determineintegral(v3); a=1/sq2*sqrt((ic+0.5)^2+
(is+0.5)^2);
 if (ic+0.5) >= 0
     b=atan((is+0.5)/(ic+0.5))-pi/4;
 else
     b=pi+atan((is+0.5)/(ic+0.5))-pi/4;
 end
 b=b+(2*pi)/lamda*(r3-r1);
ce3=a*(cos(b)+1i*sin(b))*refr;
 % Calculate diffaction loss contribution for RAY4
```

Table 12.4.
(Continued)

```
 [ic,is]=determineintegral(v4); a=1/sq2*sqrt((ic+0.5)^2+
(is+0.5)^2);
 if (ic+0.5) >= 0
     b=atan((is+0.5)/(ic+0.5))-pi/4;
 else
     b=pi+atan((is+0.5)/(ic+0.5))-pi/4;
 end
 ce4=a*(cos(b)+1i*sin(b))*refl*refr;
 % Total diffraction loss
 PL_diff(k)=20.*log10(abs(ce1+ce2+ce3+ce4));
 PL_knife(k)=20.*log10(abs(ce1));
end
plot(Zg/1e3,PL_diff,'Linewidth',2); hold on;
plot(Zg/1e3,PL_knife,'k:','Linewidth',2); hold on;
plot(Zg/1e3,20.*log10(PF2Ray),'r-','Linewidth',2);
legend('4Ray', 'Knife-Edge','2Ray');
xlim([0 Zmax/1e3]); ylim([-25 10]);
xlabel('Range [km]');ylabel('Diffraction Loss [dB]')
title('Field behind a knife-edge
obstacle','Fontweight','bold');
function ref=cref(epsr,sigma,lamda,ang,pol)
cim=-60*lamda*sigma; cak=complex(epsr,cim);
if pol == 1
   c1=cak*sin(ang)-sqrt(cak-cos(ang)^2);
   c2=cak*sin(ang)+sqrt(cak-cos(ang)^2);
else
   c1=sin(ang)-sqrt(cak-cos(ang)^2);
   c2=sin(ang)+sqrt(cak-cos(ang)^2);
end
ref=c1/c2;
function [ic,is]=determineintegral(v)
if v>0,
   ic= sum(chebfun(@(t) cos(pi/2*t.^2),[0,v]));
   is= sum(chebfun(@(t) sin(pi/2*t.^2),[0,v]));
elseif v<0,
   ic=-sum(chebfun(@(t) cos(pi/2*t.^2),[v,0]));
   is=-sum(chebfun(@(t) sin(pi/2*t.^2),[v,0]));
else
   ic=0; is=0;
end % ---------------------- END ----------------------
```

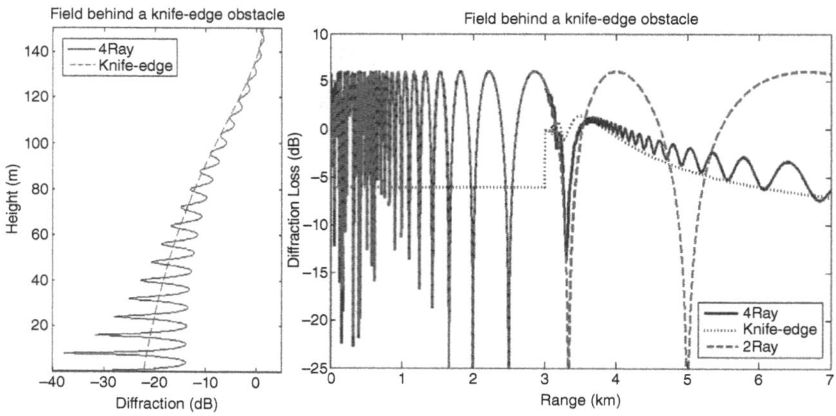

Figure 12.18. PF versus (left) height and (right) range computed with the 4Ray model.

The ground is assumed PEC. The dashed line corresponds to simple knife-edge model (i.e., Ray-1 contribution only). The solid line corresponds to the 4Ray model. For the range profile, a 75-m knife-edge obstacle is located at 100 m fixed height, 3 km from the source. The frequency is 300 MHz. The transmitter is at 50-m height. The range is changed from source to 7 km. The 2Ray model result is also plotted. The 2Ray model belongs to the case without the knife-edge obstacle between source and receiver. The interference between direct and ground-reflected rays is clearly observed. When they have the same phase, they add up constructively to 6 dB and deep nulls occur at heights where the two rays have opposite phases. As observed, only diffracted fields contribute beyond the obstacle in the shadow region. Above the LOS where transmitter and receiver see each other, the interference is clearly observed.

12.7 RAY PLUS DIFFRACTION MODELS

Electromagnetic wave propagation over the earth's surface is affected by several scattering phenomena, such as reflection, refraction, and diffraction. Developing a rigorous method for understanding the effects of varying conditions on ground wave propagation is usually a challenging task because of the vast variability of the medium parameters, and also the complexity of the surfaces and obstacles that redirect the propagating energy. Initially, the prediction of wave propagation above earth's surface problem has been achieved by means of some *analytical exact* or *analytical asymptotic* techniques [12], which require the geometry to be represented as a member of a set of some canonical geometries. The exact techniques express the solution in the form of infinite series or integral and may serve as reference, but might be poorly convergent at high frequencies when the dimensions of the objects are much larger than the wavelength. Asymptotic techniques have been widely used for decades to solve problems at high (usually optical) frequencies. These high frequency asymptotics (HFA) approaches [12] (see also Chapter 10) employ simplified models of EM wave reflection, refraction, and diffraction such as GO, PO, GTD, UTD, and PTD. The GO model describes only the incident, reflected, and refracted waves on the illuminated side of the scatterers. The GTD model is complementary to the GO model, in the sense that it overcomes some limitations of GO by including the diffraction mechanism. However, the GTD model exhibits singularities along the incident and reflection shadow boundaries (ISB and RSB). The UTD model achieves smooth wave behavior along these boundaries. This family (GO, GTD, UTD) is also known as ray-tracing techniques in the literature. The PO and PTD models, on the other hand, are based on induced currents on the surfaces of the scatterers. The PTD model supplements PO by introducing nonuniform ("fringe") edge currents to incorporate the effect of diffraction into the solution. In spite of simplicity, the asymptotic techniques suffer from two major drawbacks: (a) solutions are available only for certain canonical objects, such as wedge (or knife edge), cylinder; (b) (unless ray shooting is not used) vertically varying refractivity profiles in the troposphere cannot be handled.

One scenario that can be handled by ray plus one of variety of diffraction models [12] is pictured in Fig. 12.19. If the edges are located on a conducting surface, the image theory can be employed to account for the multiple reflections from the ground or edges. Multiple bouncing of the rays can also be handled by a successive implementation of GO and diffraction algorithms. The total field is then obtained by the sum of the direct ray, the reflected rays emanating from image sources, and the diffracted rays from the tip of the edges, and also by checking LOS conditions between the source and the observation point. The beauty of this model is that it yields physical insight, and contributions

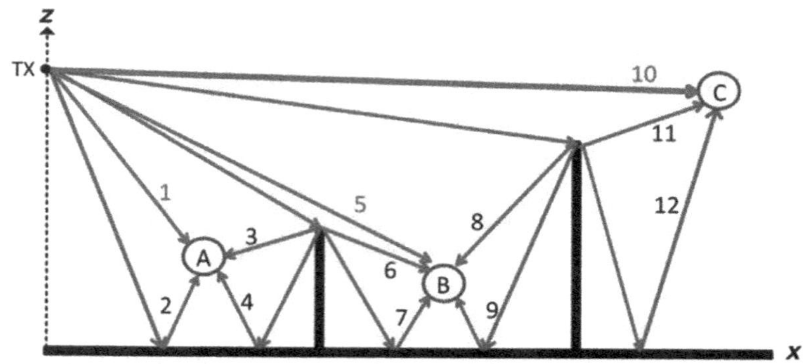

Figure 12.19. Two knife edges and possible ray (either reflected or diffracted) paths at three receiver locations between source and receiver which contribute the total EM field at the receiver.

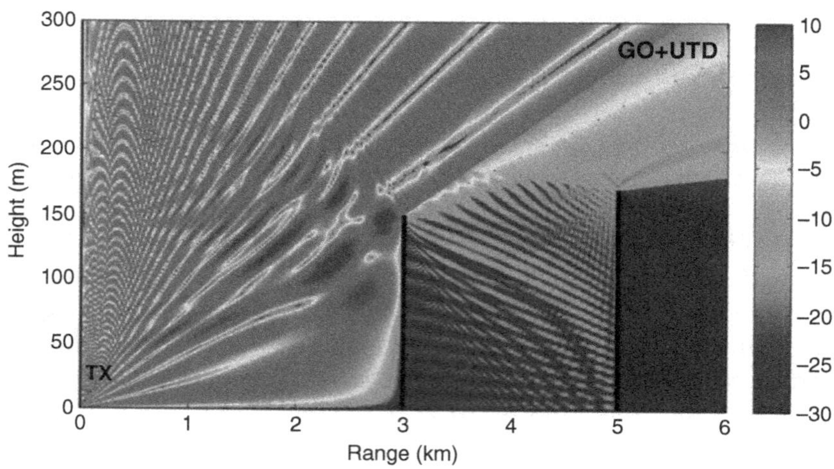

Figure 12.20. 3D PF maps for double knife edges above flat earth at 3 and 5 km ranges illuminated by a line source at 25 m height (Hor Pol, GO + UTD, $f = 30$ MHz).

of every ray item can be observed separately. Note that both forward and backward scattered (reflected and/or diffracted) waves are included in this model.

An example is presented in Fig. 12.20 for double knife edges above flat earth at 3- and 5-km ranges illuminated by a line source at 25-m height at 30-MHz frequency. The polarization is horizontal. Here, GO and edge-diffracted fields computed by the UTD model discussed in Chapter 10 are used in numerical simulations.

12.8 CONCLUSIONS

Simple 2D propagation models are presented in this chapter. An interesting ray-shooting MATLAB package is designed and discussed for the illustration of variations of ray paths because of atmospheric changes. The package may be used as an educational tool in various EM lectures (e.g., EM wave theory, antennas and propagation, wireless propagation, etc.) as well as a research tool to predict ducting and/or antiducting conditions in the presence of buildings and under various atmospheric refractivity conditions. The user may improve the MATLAB code to include more obstacles as well as subroutines to accommodate ray fields. Also, early analytical propagation models such as 2Ray model, knife-edge and 4Ray model, and others are presented.

REFERENCES

1. L. Sevgi, "A Ray Shooting Visualisation MATLAB Package for 2D Groundwave Propagation Simulations," *IEEE Antennas and Propagation Magazine*, **46** (4), 2004, pp. 140–145.
2. M. P. M. Hall, L. W. Barclay, M. T. Hewitt, eds., *Propagation of Radiowaves*, IEE Press, London, 1996.
3. L. Sevgi, F. Akleman, L. B. Felsen, "Ground Wave Propagation Modeling: Problem-Matched Analytical Formulations and Direct Numerical Techniques," *IEEE Antennas and Propagation Magazine*, **44** (1), 2002, pp. 55–75.
4. K. Rizk, J. F. Wagen, F. Gardiol, "Two-Dimensional Ray-Tracing Modeling for Propagation Prediction in Microcellular Environments," *IEEE Transactions on Vehicular Technology*, **46** (2), 1997, pp. 508–518.
5. J. H. Tarng, W. R. Chang, B. J. Hsu, "Three-Dimensional Modeling of 900-MHz and 2.44-GHz Radio Propagation in Corridors," *IEEE Transactions on Vehicular Technology*, **46** (2), 2002, pp. 519–527.
6. G. Liang, H. L. Bertoni, "A New Approach to 3D Ray Tracing for Propagation Prediction in Cities," *IEEE Transactions on Antennas and Propagation*, **46** (6), 1998, pp. 853–863.
7. G. Durgin, N. Patwari, T. S. Rappaport, "Improved 3D Ray Launching Method for Wireless Propagation Prediction," *Electronics Letters*, **33** (16), 1997, pp. 1412–1413.

8. D. Erricolo, P. L. E. Uslenghi, "Two-Dimensional Simulator for Propagation in Urban Environments," *IEEE Transactions on Vehicular Technology*, **50** (4), 2001, pp. 1158–1168.
9. L. Sevgi, *Complex Electromagnetic Problems and Numerical Simulation Approaches*, IEEE Press & John Wiley and Sons Inc., Piscataway, NJ, 2003.
10. M. L. Meeks, *Radar Propagation at Low Altitudes*, Artech House, Dedham, MA, 1982.
11. G. Apaydın, L. Sevgi, "Validation, Verification and Calibration in Applied Computational Electromagnetics," *ACES, International Journal on Applied Computational Electromagnetics*, **25** (12), 2010, pp. 1026–1035.
12. O. Ozgun, L. Sevgi, "Comparative Study of Analytical and Numerical Techniques in Modeling Electromagnetic Scattering from Single and Double Knife-Edge in 2D Ground Wave Propagation Problems," *ACES, International Journal on Applied Computational Electromagnetics*, **27** (5), 2012, pp. 376–388.

CHAPTER 13

Method of Moments

Method of moments (MoM) is a numerical technique used to approximately solve linear operator equations such as differential equations or integral equations. The unknown function is approximated by a finite series of known expansion functions with unknown expansion coefficients. The approximate function is substituted into the original operator equation and the resulting approximate equation is tested so that the weighted residual is zero. This results into a number of simultaneous algebraic equations for the unknown coefficients. These equations are then solved using matrix calculus. MoM has been used to solve vast number of electromagnetic (EM) problems during the last five decades.

13.1 INTRODUCTION

Method of moments is a general procedure for solving linear equations. Many problems that cannot be solved *exactly* can be solved *approximately* by this method.

For example, consider the simple problem of the parallel plate capacitor. The approximate analytical formula for the capacitance is $C_0 = \varepsilon A/d$, where A is the area of each plate and d is the distance between them. This formula neglects the fringing fields and is inaccurate except for very small d. In a later section, we use MoM to compute more accurate capacitance (including the fringing effects) for

Electromagnetic Modeling and Simulation, First Edition. Levent Sevgi.

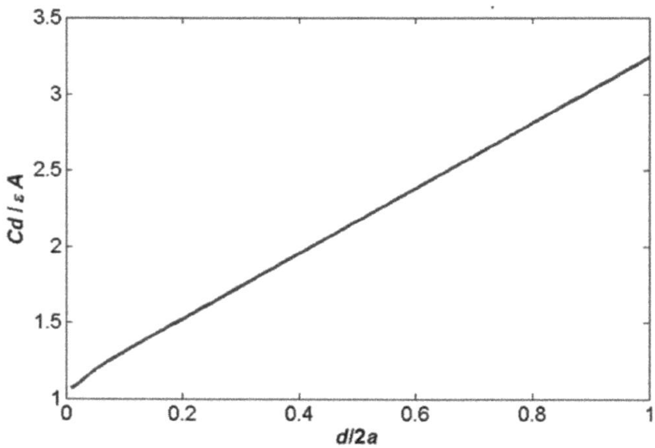

Figure 13.1. Normalized capacitance of the parallel plate capacitor versus plate distance, simulated with MoM (plates are identical and square with width = length = $2a$).

arbitrary d. Figure 13.1 shows the computed capacitance C normalized to C_0 given earlier. The figure shows the limitation of C_0 even for quite small values of d.

The MoM owes its name to the process of taking moments by multiplying with appropriate *weighing* functions and integrating. It has been applied to a broad range of EM problems since the publication of the book by Harrington [1]. A comprehensive bibliography is too vast to be given here. A selected short list includes References 2–9. A quick Internet search yields a very long list of publications on MoM.

To reinforce the concept of minimizing error or residual, the Fourier series (FS) expansion of a continuous function in a specified finite region (presented in Chapter 4) is revisited here. Then, the basics of MoM are presented. After giving some simple examples on the application of MoM, more advanced work using MoM are reviewed.

13.2 APPROXIMATING A PERIODIC FUNCTION BY OTHER FUNCTIONS: FOURIER SERIES REPRESENTATION

A known periodic function $f(x)$ can be represented by a FS as in (4.7) with the expansion (or basis) *sine* and *cosine* functions, and the expansion coefficients A_0, A_n, and B_n given by (4.8) in Chapter 4. To explain the reasoning behind the choice of expansion coefficients (4.8), let the series on the right side of (4.7) be denoted by $S(x)$ and the difference between the known function $f(x)$ and the series $S(x)$ be the *error* $e(x) = f(x) - S(x)$. Ideally, we would like to have the error itself to be zero identically. However, we will force the *weighted error* to be zero. In other words, for a certain set of weighting functions w_m, we require

$$\langle w_m(x), e(x)\rangle = \int_{-P/2}^{P/2} w_m(x)e(x)dx = 0. \tag{13.1}$$

Equation (13.1) can be written as

$$\langle w_m(x), f(x)\rangle = \langle w_m(x), S(x)\rangle. \tag{13.2}$$

If we let $w_0(x) = 1$ in (13.2), we obtain (4.8a) of Chapter 4. Note that $w_0(x)$, and the other sine and cosine functions in (4.7) of Chapter 4 are orthogonal. Hence, $w_n(x) = \cos(2n\pi x/P)$ and $w_n(x) = \sin(2n\pi x/P)$ in (13.2) yield (4.8a) and (4.8b), respectively. Summarizing, the FS is an expansion of a known periodic function $f(x)$ in terms of sine and cosine functions. The expansion coefficients are determined so that the weighted error is zero.

Depending on the function $f(x)$, the FS $S(x)$ converges to in some sense. For example, if $f(x) = \cos^2 x$ with $P = \pi$, then only A_0 and A_1 in (4.8) are nonzero, and the series $S(x) = [1 + \cos(2x)]/2$, hence the error

$$e(x) = \cos^2 x - \left\{\frac{1}{2} + \frac{1}{2}\cos 2x\right\}. \tag{13.3}$$

is identically zero. That is, the series equals the function $f(x)$ exactly. On the other hand, for the periodic rectangular function shown in Fig. 13.2 with $P = 2\pi$, we have

$$f(x) \sim \frac{4}{\pi}\left\{\sin x + \frac{\sin 3x}{3} + \frac{\sin 5x}{5} + \cdots\right\}. \tag{13.4}$$

Let S_N be the series with N terms, so that $S_1 = 4\sin x/\pi$, $S_2 = S_1 + 4\sin 3x/3\pi$, and $S_3 = S_2 + 4\sin 5x/5\pi$. Figure 13.2 shows that as N increases, the FS representation resembles the function $f(x)$ better; however, it is never equals to $f(x)$.

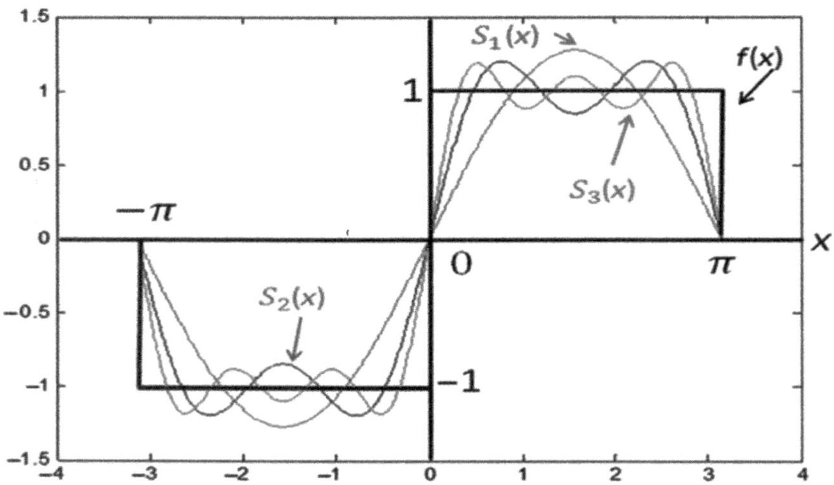

Figure 13.2. Periodic rectangular function with $P = 2\pi$ and its FS representation.

Nonperiodic functions may also be approximated by FS inside a finite region by assuming the finite region as the period of that function. In this case, it should be remembered that the FS representation no longer represents the function outside the region (see the example in Fig. 13.5). Table 4.4 of Chapter 4 lists a simple MATLAB code for the FS representation of several functions [10]. A few examples plotted with this MATLAB code are given in Fig. 13.3, Fig. 13.4, and Fig. 13.5.

The smooth Gauss function used in the first example is $f(x) = \exp(-(x - 2.5)^2/0.32)/2\pi$ and the interval is $0 \leq x \leq 10$. On top of Fig. 13.3, the function and the first term in (1) is shown. Here, A_0, A_1, A_3, A_5, . . . (odd numbered A coefficients) and B_2, B_4, B_6, . . . (even numbered B coefficients) are zero. The solid and dashed lines in the figures correspond to the function and its FS approximation, respectively. On the bottom, the function and its FS representation with the first 14 terms are shown. The coefficients A_n and B_n are also listed in the figure. As observed, 14 terms in the FS is enough to represent the function.

Next, the smooth Gauss function is modulated with a sine function and is compared with its FS representation ($f(x) = 100\sin(8\pi x)\exp(-(x - 2.5)^2/0.32)/2\pi$) in Fig. 13.4. The interval is the same. On

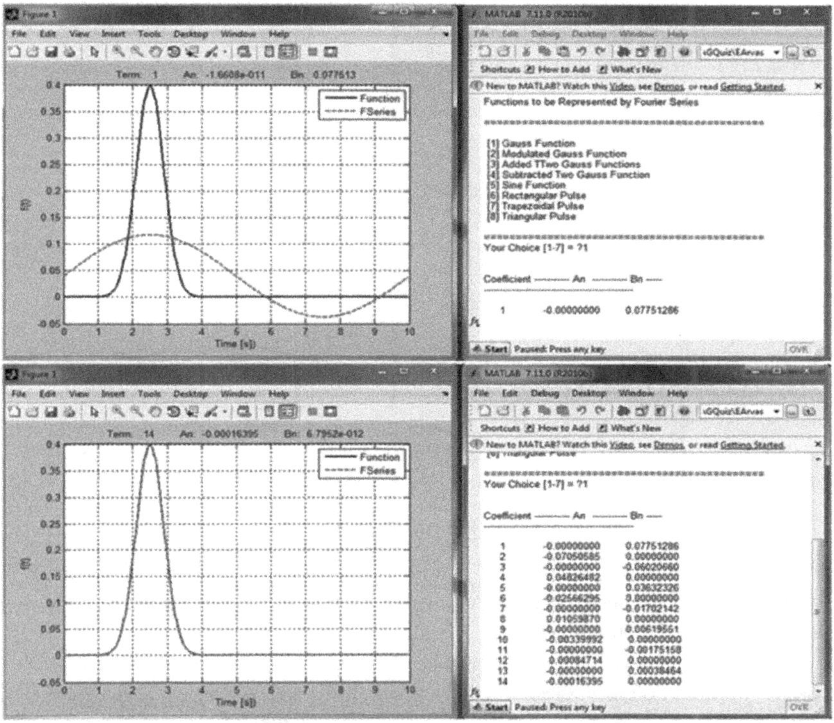

Figure 13.3. A smooth Gauss function and its FS approximations with (top) first term and (bottom) 14 terms.

top of Fig. 13.4, the function and FS representation with the first 20 terms are plotted. As observed, the first 20 terms are far from representing this highly oscillatory function. On the bottom, the function and FS representation with the first 64 terms are plotted. As observed, the FS representation with the first 64 terms agrees very well with the function.

The last example in this section is shown in Fig. 13.5. Here, the function $f(x) = x^2\sin(3x)/\exp(x/2)$ is nonperiodic and is represented in terms of its FS in the interval of $0 \leq x \leq 10$. The two curves in the figure correspond the function itself (solid curve) and its FS representation with the first 32 terms (dashed curve). The horizontal axis extends up to $x = 20$ (i.e., beyond the validity range of FS representation). As observed, 32-term FS represents the function very well for $0 \leq x \leq 10$.

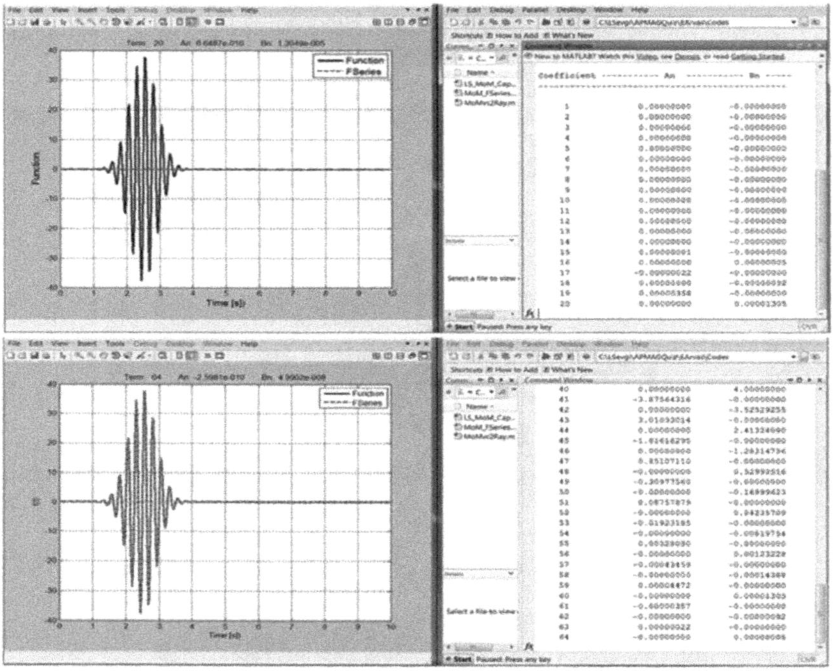

Figure 13.4. A sine-modulated smooth Gauss function and its FS approximations with (top) 20 terms and (bottom) 64 terms.

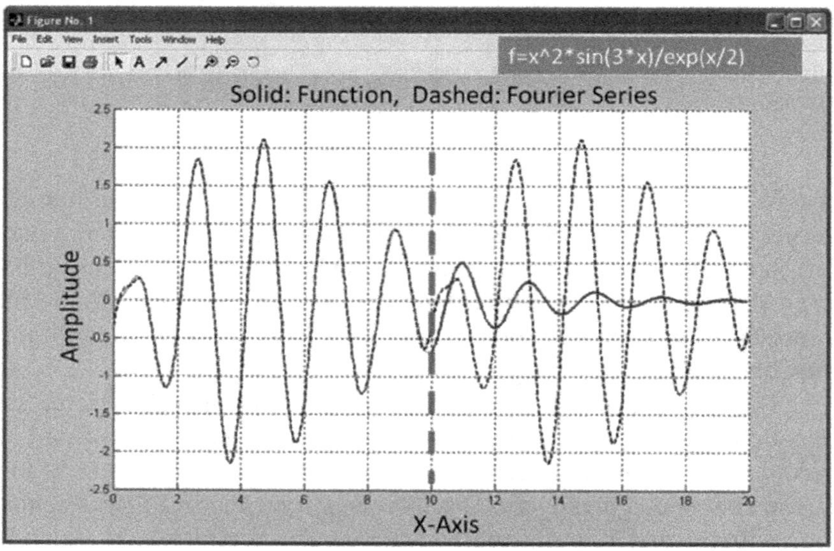

Figure 13.5. A function and its FS representation in $0 \leq x \leq 20$. The FS expansion is used for $0 \leq x \leq 10$ and perfect agreement is obtained with the first 32 terms. The function and its FS are totally different for $10 \leq x \leq 20$.

On the other hand, they are totally different in the range of $10 \leq x \leq 20$. This clearly demonstrates that nonperiodic function and its FS representation only agree in the expansion interval (i.e., in an assumed period). Their behaviors beyond this range may totally be different.

13.3 INTRODUCTION TO THE MoM

Consider a linear operator equation,

$$L(f(x)) = g(x), \tag{13.5}$$

where L is a *linear* operator, $g(x)$ is a known function (usually excitation in a linear system), and $f(x)$ is the unknown function (usually response) to be found. Note that a linear operator must satisfy

$$L(\alpha f_1(x) + \beta f_2(x)) = \alpha L(f_1(x)) + \beta L(f_2(x)), \tag{13.6}$$

where f_1 and f_2 are two functions, and α and β are constants. The following is a set of examples of linear operators:

$$L(f(x)) = 5f(x), \tag{13.7}$$

$$L(f(x)) = \frac{df(x)}{dx}, \tag{13.8}$$

$$L(f(x)) = 5\frac{d^2 f(x)}{dx^2} + 7\frac{df(x)}{dx} + 3f(x), \tag{13.9}$$

$$L(f(x)) = \int_0^x f(x')\sin(x - x')dx', \tag{13.10}$$

$$L(f(x)) = \nabla^2 f(x), \tag{13.11}$$

$$L\vec{x} = [A]\vec{x}. \tag{13.12}$$

In (13.11), ∇^2 is the Laplacian operator, and in (13.12) $[A]$ is a matrix and $\vec{x}$ is a column vector. The following is a simple example of a linear operator equation:

$$-\frac{d^2 f(x)}{dx^2} = 1 + 4x^2, \tag{13.13}$$

with $f(0) = f(1) = 0$. Equation (13.13) is in the form of (13.9) with $L = -d^2/dx^2$ and $g(x) = 1 + x^2$.

We now summarize the common steps in MoM. Given the linear operator (13.5), the MoM starts by expanding the unknown function f into a series of known *expansion functions* $h_1, h_2, \ldots, h_N$ in the domain of L as follows:

$$f(x) \approx f_\alpha(x) = \sum_{n=1}^{N} \alpha_n h_n(x). \tag{13.14}$$

Here, the constants α_n are called the unknown *expansion coefficients* to be found. Note that the function $f_\alpha(x)$ represented by the series in (13.14) is an approximation to the exact unknown function $f(x)$. Since the exact function is not known, the error function defined by $e(x) = f(x) - f_\alpha(x)$ is also unknown. Substituting $f_\alpha(x)$ into (13.5) yields

$$\begin{aligned} L(f_\alpha(x)) &= \sum_{n=1}^{N} \alpha_n L(h_n(x)) \\ &= \alpha_1 L(h_1(x)) + \alpha_2 L(h_2(x)) + \cdots + \alpha_N L(h_N(x)) \approx g(x). \end{aligned} \tag{13.15}$$

For a given $f_\alpha(x)$, the residual is defined by

$$r(x) = L(f(x)) - L(f_\alpha(x)) = g(x) - L(f_\alpha(x)). \tag{13.16}$$

Although we do not know the exact function, it is clear that if the approximate function $f_\alpha(x)$ is equal to the exact function $f(x)$, then the residual will be identically zero. Hence our purpose should be to minimize the residual. In MoM, instead of minimizing the residual itself, we force the *weighted residual* to be zero.

Defining a set of *weighting* or *testing* functions $w_1, w_2, \ldots, w_N$ and taking the inner product of (13.16) with each w_m give (for $m = 1,2, \ldots, N$)

$$\begin{aligned} \left\langle w_m(x), \sum_{n=1}^{N} \alpha_n L(h_n(x)) \right\rangle &= \alpha_1 \langle w_m(x), L(h_1(x)) \rangle + \alpha_2 \langle w_m(x), L(h_2(x)) \rangle \\ &\quad + \cdots + \alpha_N \langle w_m(x), L(h_N(x)) \rangle \\ &\approx \langle w_m(x), g(x) \rangle. \end{aligned} \tag{13.17}$$

Equations in (13.17) can be written in matrix form,

$$\begin{bmatrix} Z_{11} & Z_{12} & .. & Z_{1N} \\ Z_{21} & Z_{22} & .. & Z_{2N} \\ \vdots & \vdots & .. & \vdots \\ Z_{N1} & Z_{N2} & .. & Z_{NN} \end{bmatrix} \begin{bmatrix} \alpha_1 \\ \alpha_2 \\ \vdots \\ \alpha_N \end{bmatrix} = \begin{bmatrix} \langle w_1(x), g(x) \rangle \\ \langle w_2(x), g(x) \rangle \\ \vdots \\ \langle w_N(x), g(x) \rangle \end{bmatrix}, \qquad (13.18)$$

where

$$Z_{mn} = \langle w_m(x), L(h_n(x)) \rangle. \qquad (13.19)$$

The particular choice of $w_n = h_n$ in MoM is known as *Galerkin's* method [1].

A suitable inner product $\langle f, g \rangle$ of two functions f and g may be to multiply the functions and integrate in the domain of definition as

$$\langle f(x), g(x) \rangle = \int_a^b f(x) g(x) dx; \quad a \leq x \leq b, \qquad (13.20)$$

where a and/or b can go to infinity. The solution of the system in (13.18) yields the set of unknown expansion coefficients α_n, which completes the MoM procedure. The choice of expansion and testing functions affects the existence and accuracy of the solution to (13.18). Different choices of expansion and testing functions affect the existence and accuracy of the solution.

13.4 SIMPLE APPLICATIONS OF MoM

Three simple problems are included in this section.

13.4.1 An Ordinary Differential Equation

Use MoM to find $f(x)$ satisfying [1]

$$-\frac{d^2 f(x)}{dx^2} = 1 + 4x^2, \quad f(0) = f(1) = 0. \qquad (13.21)$$

Here, the linear operator and the excitation functions are $L = -d^2/dx^2$ and $g(x) = 1 + x^2$, respectively. The first step is to expand the unknown function in terms of known expansion functions. The expansion functions must be in the domain of the operator. In this example, twice differentiable expansion functions $h_n = x - x^{2n}$ are chosen. Note that they satisfy the boundary conditions at $x = 0$ and $x = 1$. For the weighting functions, we choose $w_m(x) = \delta(x - x_m)$, where $x_m = m/(N + 1)$. The inner product defined in (13.20) will be used with $a = 0$ and $b = 1$. By choosing the weighting functions as delta functions located at points x_m, we force $L(f_\alpha(x))$ to be equal to $g(x)$ at discrete points x_m. That is, we *match* the right and left hand sides of (13.15) at discrete points. This method makes residual equal to zero at these points and is called *point matching*.

Choosing $N = 1$; $f(x) \approx f_\alpha(x) = \alpha_1 h_1(x) = \alpha_1(x - x^2)$ and inserting it into (13.21) yields

$$-\frac{d^2 f_\alpha(x)}{dx^2} = -\frac{\alpha_1 d^2(x - x^2)}{dx^2} = 2\alpha_1. \tag{13.22}$$

Since $f(x)$ is not known, α_1 is determined to minimize the residual function defined as

$$r(x) = L(f) - L(f_\alpha) = g(x) - L(f_\alpha). \tag{13.23}$$

For this example, the residual function is $r(x) = [1 + 4x^2] - 2\alpha_1$. The residual function may be set to zero anywhere in $0 \leq x \leq 1$. Choosing the midpoint (i.e., $x = 0.5$) yields $\alpha_1 = 1$. Hence, $f(x) = x - x^2$ and $r(x) = 4x^2 - 1$. The variation of the residual function in the domain is plotted in Fig. 13.6. Note that point matching at the point $x = 0.5$ is the same as equating the inner product of the residual in (13.23) with the delta function $\delta(x - 0.5)$ to zero [1].

For a better solution let, $N = 2$, then the function can be approximated as $f_\alpha(x) = \alpha_1(x - x^2) + \alpha_2(x - x^3)$, which yields $L(f_\alpha) = 2\alpha_1 + 6\alpha_2 x$ and $r(x) = (1 + 4x^2) - 2\alpha_1 - 6\alpha_2 x$. Equating the inner product of this residual with $w_1 = \delta(x - 1/3)$ and to zero (i.e., equating $r(x) = 0$ at two points $x = 1/3$ and $x = 2/3$) yields $\alpha_1 = 1/18$ and, $\alpha_2 = 12/18$. Replotting Fig. 13.6 for this case would show that the computed function is a better approximation now. Finally, choosing $N = 3$, that is, $f_\alpha(x) = \alpha_1(x - x^2) + \alpha_2(x - x^3) + \alpha_3(x - x^4)$ and equating $r = 0$ at three points, $x = 1/4, x = 2/4$, and $x = 3/4$ yields $\alpha_1 = 1/2$, $\alpha_2 = 0$, and $\alpha_3 = 1/3$. The residual function in this case is $r \equiv 0$ and the approximate function $f_\alpha(x) = 5x/6 - x^2/2 - x^4/3$ is equal to the exact function.

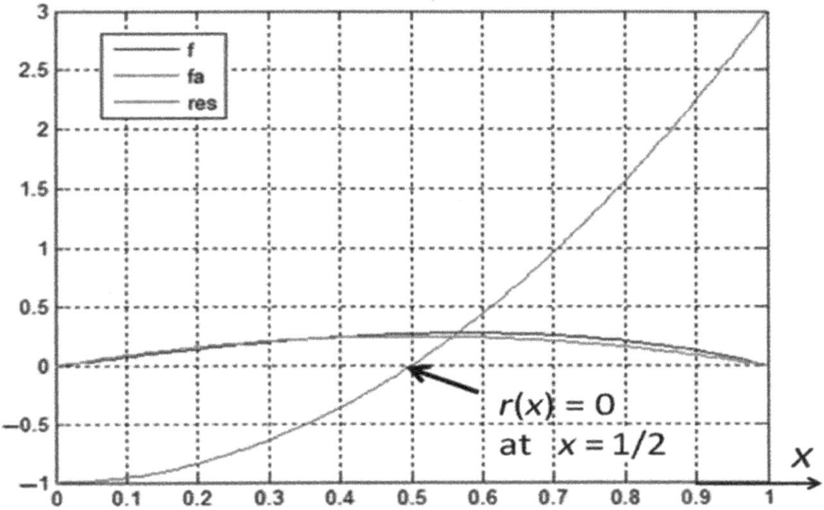

Figure 13.6. Residue function versus x.

The problem in (13.21) may be solved by Galerkin's method, where $h_n(x) = \sin(n\pi x)$, $w_m(x) = \sin(m\pi x)$. For $N = 1$ case, the moment equation (13.18) is simply $4.9348\alpha_1 = 1.3938$, yielding $\alpha_1 = 0.28245$. When $N = 2$, the moment equation is as follows:

$$\begin{bmatrix} 4.9348 & 0 \\ 0 & 19.7392 \end{bmatrix} \begin{bmatrix} \alpha_1 \\ \alpha_2 \end{bmatrix} = \begin{bmatrix} 1.3938 \\ -0.636 \end{bmatrix}.$$

The solution of this equation gives $\alpha_1 = 0.28245$, $\alpha_2 = -0.03225$. Figure 13.7 shows the computed results of this Galerkin's method for different N values.

Note that unlike the point matching case considered earlier, α_1 for $N = 2$ here is the same as α_1 for $N = 1$ case. This is the result of the fact that $L(f_\alpha(x))$ and w_m are orthogonal. This property is similar to the case found in the FS, whereas N is increased, only the expansion coefficients A_n and B_n with $n > N$ need to be computed, as those with $n < N$ remain the same. Note also that here the computed results for $N = 3$ or more are not the same as the exact result. However, for the earlier case where $h_n = x - x^{2n}$ and point matching was used, the computed results were the same as the exact function for $N \geq 3$. This shows how

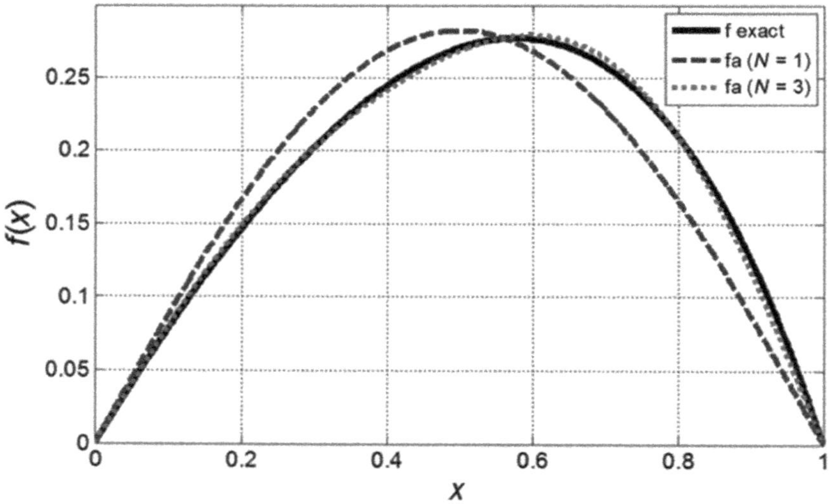

Figure 13.7. Exact and computed functions.

the choice of expansion and weighting functions affects the computed solution in MoM.

13.4.2 The Parallel Plate Capacitor

Consider the parallel plate capacitor shown in Fig. 13.8 [1]. To find the capacitance, we assume that the plate at $z = d$ is at potential V and the plate at $z = 0$ is at zero potential. Then, the capacitance $C = Q/V$, where Q is the charge on the upper plate. This charge is the surface integral of the charge density on the surface of the upper plate. Let $\rho(x', y')$ denote the charge density on both plates. Then, the electrostatic potential at an arbitrary point (x, y, z) produced by this charge distribution is given by

$$\varphi(x, y, z) = \iint\limits_{S_T + S_B} \frac{\rho(x', y')}{4\pi\varepsilon_0 R} dx'dy', \tag{13.24}$$

where $R = \sqrt{(x - x')^2 + (y - y')^2 + (z - z')^2}$ and S_T and S_B, respectively, denote the surfaces of the top and bottom plates. The variable $z' = d$ on the upper plate and is equal to zero on the bottom plate. When

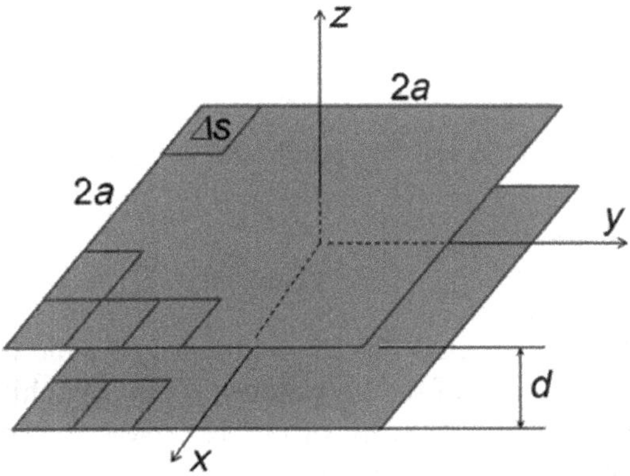

Figure 13.8. Parallel plate capacitor having identical PEC plates with length = width = $2a$, the patch area is $\Delta s = 4b^2$, and the number of patches on both plates is N.

the observation point (x, y, z) is on the top (bottom) plate, (13.24) equals V (zero). In other words, the unknown charge density satisfies the following integral equation:

$$\iint\limits_{S_T+S_B} \frac{\rho(x', y')}{4\pi\varepsilon_0 R} dx'dy' = \begin{cases} V & (x, y, z) \in S_T \\ 0 & (x, y, z) \in S_B \end{cases}. \tag{13.25}$$

Equation (13.25) is a linear operator equation in the form of (13.5), where

$$L(\rho) = \iint\limits_{S_T+S_B} \frac{\rho(x', y')}{4\pi\varepsilon_0 R} dx'dy' \tag{13.26}$$

and

$$g = \begin{cases} V & (x, y, z) \in S_T \\ 0 & (x, y, z) \in S_B \end{cases}. \tag{13.27}$$

To solve (13.25), we divide each plate into N square patches with area $\Delta s = 4b^2$ and assume that $\rho(x', y')$ is equal to a constant α_n on the nth patch. In other words, we expand $\rho(x', y')$ into pulse expansion functions as follows:

$$\rho(x', y') \approx \sum_{n=1}^{N} \alpha_n f_n, \tag{13.28}$$

where f_n is equal to 1 on the nth patch and is zero elsewhere. That is,

$$f_n = \begin{cases} 1 & \text{on } \Delta S_n \\ 0 & \text{elsewhere} \end{cases}. \tag{13.29}$$

Substituting (13.28) into (13.25) and satisfying the resulting equation at the center of each of the $2N$ patches (point matching), we obtain the moment equation in the form of (13.18). Now the moment matrix is $2N \times 2N$, and a typical element Z_{mn} is the electrostatic potential at the center of the mth patch. It is produced by a charge with density of 1 C/m^2 residing on the nth patch. When m is different from n, this charge is approximated by a point charge $q_n = 1 \times \Delta S_n$ at the center of the nth patch. When $m = n$ ([1], equation (2.31)),

$$Z_{nn} = \frac{2b}{\pi\varepsilon_0}(0.8814). \tag{13.30}$$

A typical element on the right hand side of (13.18) is equal to V if the mth patch is on the top plate and is zero otherwise. Once the expansion functions are computed, the total charge on the upper plate can easily be found. Then the capacitance is obtained from $C = Q/V$.

Figure 13.1 shows the computed normalized capacitance versus d. Note that the charge density on the lower plate is negative of the one on the top plate. Therefore, one could use this symmetry information to solve an $N \times N$ equation instead of $2N \times 2N$ that we considered earlier. The details are given in Reference 1.

Table 13.1 lists the short MATLAB script prepared for the MoM computation of this parallel plate capacitor. The program computes capacitor values for different plate separations and plots the results as the normalized capacitor versus plate distance.

13.4.3 Propagation over PEC Flat Earth

One of the simple, canonical propagation scenario taught in many undergraduate lectures is the perfectly electrical conducting (PEC) flat earth. The formulation and solution of this problem are given in Chapter

Table 13.1.

A Short MATLAB-MoM Code for the Plate Capacitor

```
% ---------------------------------------------------------
%  Program: Capacitor
% ---------------------------------------------------------
clear all; clc; nseg=100; % number of patches
nside=sqrt(nseg); % total number of patches
a=1; % half of plate width/height
b=a/nside; % half of segment width/height
A=4*a^2; % plate area
epsilon=8.8541e-12; % epsilon
for i=1:nside   % Coefficent Matrix
  for j=1:nside
     k=nside*(i-1)+j; x(k)=(2*i-1)*b; y(k)=(2*j-1)*b;
  end
end
% MoM Simulation
q=0.01:0.01:1;
for g=1:numel(q)
  d=2*a*q(g);
  for ii=1:nseg
    for jj=1:nseg
      if ii~=jj
        Itt(ii,jj)=b^2/pi/epsilon/
sqrt((x(ii)-x(jj))^2. . .
                   +(y(ii)-y(jj))^2);
        Itb(ii,jj)=b^2/pi/epsilon/
sqrt((x(ii)-x(jj))^2. . .
                   +(y(ii)-y(jj))^2+d^2);
      else
        Itt(ii,jj)=0.8814*2*b/pi/epsilon;
        Itb(ii,jj)=0.282/epsilon*2*b*(sqrt(1+pi/4*((d/b
)^2)). . .
                  -sqrt(pi)*d/2/b);
      end
    end
  end
  tmp=inv(Itt-Itb); total=sum(sum(tmp));
  capacitance(g)=2*b^2*total/(epsilon*A/d);
end
plot(q,capacitance); xlabel('d/2a','Fontweight','bold');

ylabel('Cd / \epsilon A','Fontweight','bold');
title('MoM Parallel Capacitance Simulation','Fontweight','bo
ld');
% ------------------------ END ------------------------
```

12 (see Section 12.5; the scenario is pictured in Fig. 12.11). A line source is located above the flat earth at an altitude of h_t. Cylindrical waves emanating from the source reach the receiver at a distance d and altitude h_r via direct and ground-reflected paths. Two possible cases are the horizontal (electric field is perpendicular to the two dimensional [2D] propagation plane) and vertical (magnetic field is perpendicular to the 2D propagation plane) polarizations. The ground reflection coefficients are $\Gamma = -1$ and $\Gamma = +1$ for the horizontal (transverse electric, TE_z) and perpendicular (transverse magnetic, TM_z) polarizations, respectively. The total field at the receiver can directly be given by (12.16) with the path difference given in (12.17) in Chapter 12. The received power for the 2Ray scenario is also given in (12.18), (12.19), and (12.20) there. Table 12.1 of Chapter 12 lists a short MATLAB code for the 2Ray model.

The same problem can be solved with MoM. The MoM-based propagation model requires first analytical derivation of the 2D Green's function. Then, the flat earth from transmitter to receiver (i.e., the propagation path is discretized and replaced with a number of neighboring segments. The segment lengths are specified according to the wave frequency. As a rough criterion, the length of each segment should be equal to or less than one-tenth wavelength ($\lambda/10$).

Assuming that (a) the incident field is of finite extent in space and illuminates only the propagation path portion between the transmitter and receiver, and (b) currents induced on constant-length segments when illuminated by the source are constant, one can apply the point-matching MoM technique and obtain the closed form matrix equation

$$[V]=[Z][I], \tag{13.31}$$

where $[I]$ contains the unknown segment currents I_m ($m = 1,2, \ldots N$), $[V]$ denotes the incident field ($V_n = -(E_0/2\pi)e^{-ik_0 d_n}/d_n$) evaluated at the matching points, and $[Z]$ is the $N \times N$ impedance matrix. Necessary equations for both polarizations are [11] as follows.

TE_z case:

$$V_n = -E_y^{\text{inc}}(\boldsymbol{\rho}_n) = -E_0 \frac{e^{-jkd_n}}{\sqrt{k_0 d_n}}, \tag{13.32}$$

$$d_n = \sqrt{[x(\boldsymbol{\rho}_n) - x']^2 + [z(\boldsymbol{\rho}_n) - z')]^2}, \tag{13.33}$$

$$Z_{nm} \cong \begin{cases} -\dfrac{k_0\eta_0\Delta z}{4} H_0^{(2)}(k_0|\boldsymbol{\rho}_n - \boldsymbol{\rho}_m|) & \dots m \neq n \\ -\dfrac{k_0\eta_0\Delta z}{4}\left[1 - j\dfrac{2}{\pi}\log\left(\dfrac{\gamma k_0 \Delta z}{4e}\right)\right] & \dots m = n \end{cases}, \quad (13.34)$$

$$E_y^{\text{scat}}(\boldsymbol{\rho}_n) \cong -\frac{k_0 Z_0 \Delta z}{4}\sum_{m=1}^{N} I_m H_0^{(2)}(k_0|\boldsymbol{\rho}_n - \boldsymbol{\rho}_m|), \quad (13.35)$$

$$E_y^{\text{tot}} = E_y^{\text{scat}} + E_y^{\text{inc}}. \quad (13.36)$$

TM_z case:

$$V_n = -H_y^{\text{inc}}(\boldsymbol{\rho}_n) = -\frac{E_0}{\eta_0}\frac{e^{-jkd_n}}{\sqrt{k_0 d_n}}, \quad (13.37)$$

$$d_n = \sqrt{[x(\boldsymbol{\rho}_n) - x']^2 + [z(\boldsymbol{\rho}_n) - z')]^2}, \quad (13.38)$$

$$Z_{nm} \cong \begin{cases} j\dfrac{k_0\Delta z}{4} H_1^{(2)}(k_0|\boldsymbol{\rho}_n - \boldsymbol{\rho}_m|)(\hat{\mathbf{n}}_m \cdot \hat{\boldsymbol{\rho}}_{nm}) & \dots m \neq n \\ 0.5 & \dots m = n \end{cases}, \quad (13.39)$$

$$H_y^{\text{scat}}(\boldsymbol{\rho}_n) \cong j\frac{k_0 \Delta z}{4}\sum_{m=1}^{N} I_m H_1^{(2)}(k_0|\boldsymbol{\rho}_n - \boldsymbol{\rho}_m|)(\hat{\mathbf{n}}_m \cdot \hat{\boldsymbol{\rho}}_{nm}), \quad (13.40)$$

$$H_y^{\text{tot}} = H_y^{\text{scat}} + H_y^{\text{inc}}, \quad (13.41)$$

where Δz is the segment length, $\eta_0 \approx 120\pi$ is the intrinsic impedance of free space, $H_0^{(2)}$ and $H_1^{(2)}$ are the second kind Hankel functions with order zero and one, respectively, $\gamma \approx 1.781$ is the exponential of the Euler constant, $\hat{\mathrm{n}}_m$ denotes the unit normal vector of the plate at ρ_m, and $\hat{\rho}_{nm}$ is the unit vector in the direction from source ρ_m to the receiving element ρ_n.

The unknown segment currents are then obtained by solving the system in (13.31). Finally, direct wave and scattered wave from each segment caused by induced segment current are added total wave at the observer is obtained. Table 13.2 lists a short MATLAB code for the MoM computation of this problem.

Figure 13.9 shows the comparison of MoM result with the 2Ray model. Here, propagation factor (PF) defined as $\text{PF(dB)} = 20\log_{10}|E/E_0|$ versus range is shown. Source and observer heights are 30 and 5 m, respectively. Maximum distance is 100 m, $\lambda = 2\pi$, and the segment

Table 13.2.
A Short MoM Code for the Flat-Earth Problem

```
% -----------------------------------------------------
%  Program : MoMvs2Ray.m
% -----------------------------------------------------
clear all; clc;
E0=1.0; freq=1e8; k0=2*pi*freq/3e8; lambda=3e8/freq;
REF=1;
xsource=50; xobs=5; zlast=150; delz=lambda/10;
Nz=zlast/delz;
for n=1:Nz % 2Ray Model - Range calculations
    zobs(n)=(2*n-1)*delz/2;
R1=sqrt(zobs(n)^2+(xsource-xobs)^2);
    R2=sqrt(zobs(n)^2+(xsource+xobs)^2);
    gfree(n)=E0/(120*pi)*exp(-i*k0*R1)/R1;
    gref(n)=REF*E0/(120*pi)*exp(-i*k0*R2)/R2;
    gtot(n)=gfree(n)+gref(n); gpf(n)=gtot(n)/gfree(n);
end
% ------------------- MoM method -------------------
for n=1:Nz      % Impedance matrix for PEC Earth
    for m=1:Nz
        if m==n
          imp_matrix(m,n)=0.5;
        else
          imp_matrix(m,n)=0.0;
        end
  end
end
% Calculate initial voltages (a line souce at
(xsource,zsource))
for n=1:Nz
     zs(n)=(2*n-1)*delz/2; ds=sqrt(xsource^2+zs(n)^2);
     inc_mag(n)=(E0/
(120*pi))*exp(-i*k0*ds)/ds; V_inc(n)=-inc_mag(n);
end
ind_current=inv(imp_matrix)*transpose(V_inc);
for n=1:Nz   % RANGE PROFILE
  scat_mag(n)=0.0;
  for m=1:Nz
```

Table 13.2.
(Continued)

```
    dd=sqrt(xobs^2+(zobs(n)-zs(m))^2);
prinner=abs(xobs)/dd;
    scat_mag(n)=scat_mag(n)+ind_current(m)*besselh(1,2,k
0*dd)*prinner;
  end
  scat_mag(n)=0.25*delz*i*k0*scat_mag(n);
  dist=sqrt((xobs-xsource)^2+zobs(n)^2);
  inc_mag(n)=(E0/(120*pi))*exp(-i*k0*dist)/dist; %
isotropic radiator
  tot_mag(n)=inc_mag(n)+scat_mag(n); propf(n)=tot_
mag(n)/inc_mag(n);
end
plot(zobs,20*log10(abs(gpf)),zobs,20*log10(abs(propf)),
'r-');
xlabel('Distance [m]'); ylabel('Propagation Factor [dB]');
title('Propagation over flat-Earth: MoM vs. 2Ray');
% ---------------------- END ----------------------------
```

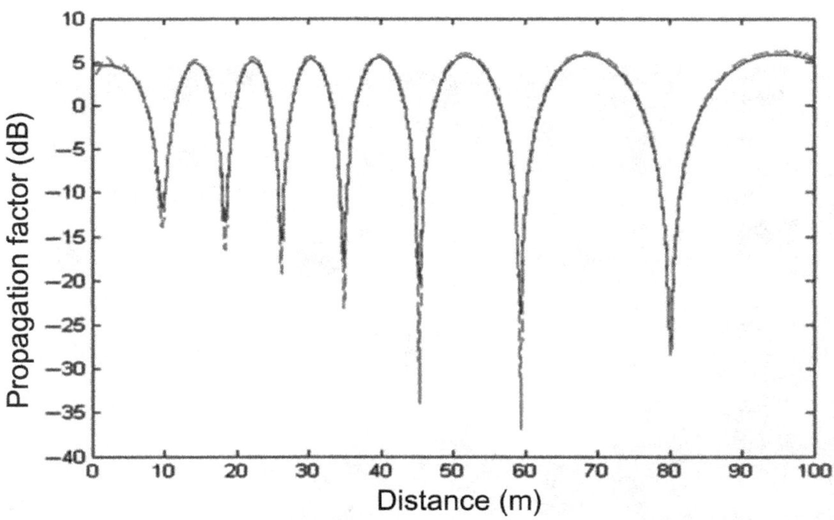

Figure 13.9. PF versus range: A comparison of MoM–2Ray models; solid: 2Ray, dashed: MoM.

length is $\lambda/10$. As observed, perfect agreement is obtained. Note that MoM suffers from abrupt truncation of the segments at the initial and last ranges (the line source illuminates segments at all ranges). This can be handled by taking into account a few extra segments before the initial range and after the last range.

13.5 MOM APPLIED TO RADIATION AND SCATTERING PROBLEMS

MoM method can be used to solve variety of complex radiation, propagation, and scattering problems. Three such problems are summarized here.

13.5.1 A Complex Antenna Structure

A typical long-range high frequency (HF) transmit antenna system is shown in Fig. 13.10. This antenna system transmits several kilowatt power to cover a range that extends up to 500 km in range and 100–120° in azimuth [12–16]. The system is specifically designed for maritime

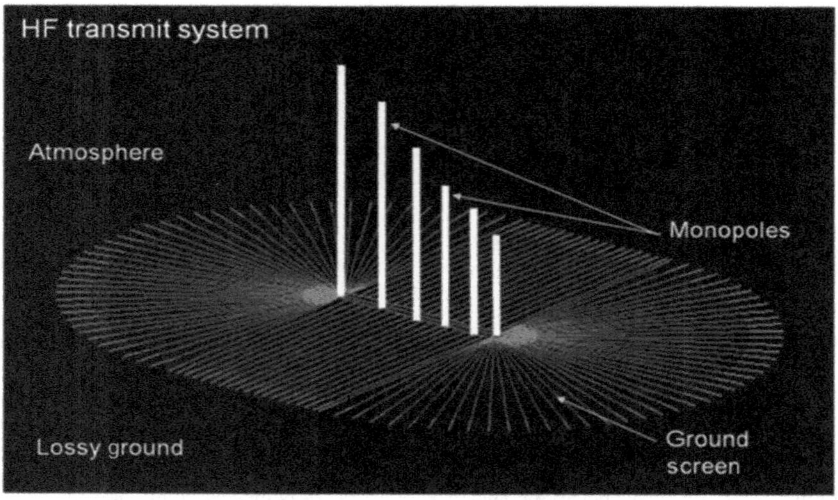

Figure 13.10. A typical transmit antenna system for HF radars and/or communication systems.

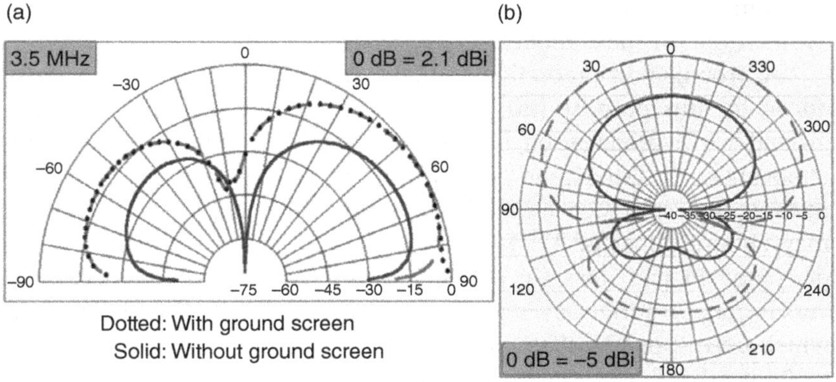

Figure 13.11. Vertical (a) and horizontal radiation patterns of the transmit antenna shown in Fig. 13.10 ($\varepsilon_g = 15$, $\sigma_g = 0.005$ S/m).

communication and/or surveillance [12–14]. As observed, there are six monopoles erected on top of a ground screen system. The ground screen with $\lambda/4$-long 120 wires compensates ground losses. Wire antenna systems such as this one are best handled with MoM.

One public MoM code for EM investigation is the numerical EM code (NEC) [17,18]. NEC is an outgrowth of a program developed in the 1970s, called the antenna modeling program (AMP). It has been developed at the Lawrence Livermore Laboratory, Livermore, California, under the sponsorship of the Naval Ocean Systems Center and the Air Force Weapons Laboratory. Public domain NEC2 is used in this study. NEC2 requires discrete wire-grid model of the antenna system under investigation. MoM (NEC2) simulation results of the complex antenna system shown in Fig. 13.10 at 3.5 MHz are plotted in Fig. 13.11. Patterns are normalized to 0 dB and exact values are indicated on the plots. Both vertical and horizontal radiation patterns show that using a ground screen increases the gain of the antenna system around 10 dB.

13.5.2 Ground Wave Propagation Modeling

Ground wave propagation over realistic irregular terrain profiles with lossy BC is of great importance in radar and/or communication system planning, coverage analysis, and link establishments [19–30]. These problems can only be solved via numerical propagation models. The MoM is one of these models.

A propagation prediction virtual tool GrMoMPE has been introduced for this purpose [20]. The package can be used to investigate ground wave propagation over nonflat terrain through homogeneous atmosphere via MoM and compare the results with another powerful method; the split-step parabolic equation method [22–24]. The user may choose either of TE/TM modes for the analysis. Results can be presented in path loss (in dBV/m) versus range or PF (in decibels) versus range. Variations may be "above ground" or "above terrain" which correspond to data recording on a helicopter (flying longitudinally at a fixed height above sea level) or on a truck (traveling longitudinally along the terrain profile), respectively.

An example is given in Fig. 13.12 where a typical 20-km-long nonflat propagation path with three quite smooth hills and a deep valley is drawn. The maximum height of the terrain is approximately 300 m.

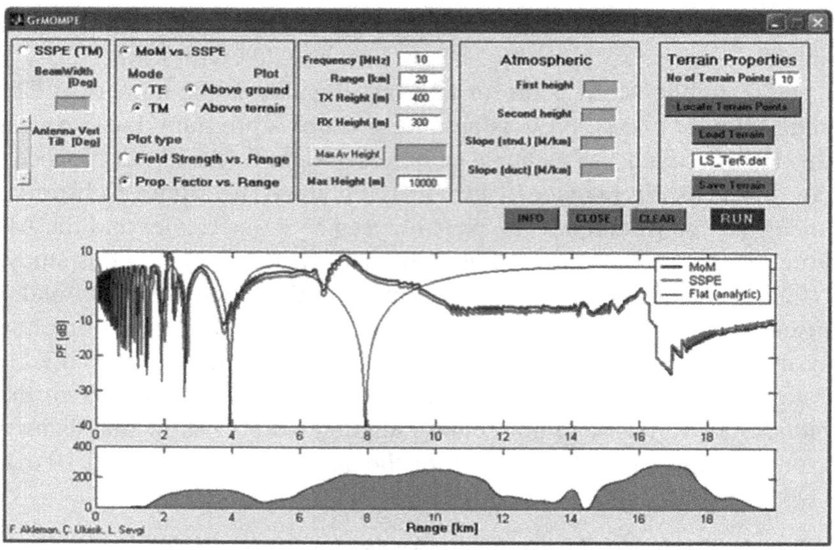

Figure 13.12. A typical terrain profile (with range in kilometers and height in meters) and a MoM versus SSPE comparison via signal strength (given as PF in decibels as a function of range in kilometers for a TM type of problem ($f = 10$ MHz, transmitter height of 400 m, receiver height above ground of 300 m).

The frequency is set to 10 MHz. The transmitter and receiver heights are 400 and 300 m, respectively.

The results belong to the TM case and PF versus range above the ground is plotted. The analytically exact result of the nonpenetrable flat earth is also plotted to show the effects of the terrain profile on signal strength. As expected, signal strength increases (decreases) in front of (behind) the hills. As observed, the agreement between MoM and SSPE results is quite impressive.

The second example presented in Fig. 13.13 belongs to a TE-type propagation scenario with two Gaussian hills for a 30-km-long propagation path. The first hill is approximately 4-km wide and 100-m high at 14-km distance. The second hill has the same width but is 200-m high at 22-km distance. The frequency is 10 MHz, and the transmitter height is 400 m. The two plots correspond to field strength versus range

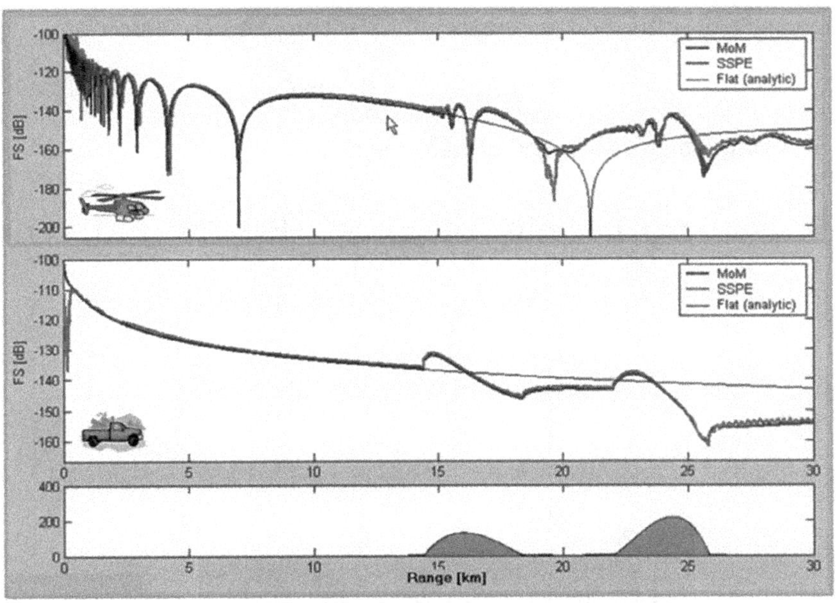

Figure 13.13. A propagation scenario with two smooth, 150-m high, 4-km wide and 200-m high, 4-km wide Gaussian-shaped hills, and MoM versus SSPE comparisons: signal strength in dBV/m as a function of range in kilometers for the TE type of problem ($f = 10$ MHz, maximum range of 30 km, transmitter height of 400 m). (Top) The results for a receiver height above ground of 400 m and (bottom) for a receiver height above terrain of 10 m.

at 400 m above the ground (top plot) and 10 m above the terrain (bottom plot). Note that direct application of MoM as described earlier is segment limited because of huge matrix inversion problems. As many as 6000–8000 segments can be used to obtain numerical results in a reasonable computation period (e.g., 5–15 min). Beyond that, acceleration approaches are used [25–30].

13.5.3 EM Scattering from Infinitely Long Cylinder

Consider a perfectly conducting cylinder excited by an impressed electric field E_z^i (see Fig. 13.14). The impressed field induces surface currents J_z on the conducting cylinder, which produce a scattered field E_z^s. The field due to J_z is given by

$$E_z^s(\vec{\rho}) \cong -\frac{k\eta}{4}\int_C J_z(\vec{\rho}\,')H_0^{(2)}(k|\vec{\rho}-\vec{\rho}\,'|)dl' \tag{13.42}$$

specialized to the cylinder surface C. The boundary condition is

$$E_z = E_z^i + E_z^s = 0 \quad \text{on } C. \tag{13.43}$$

Combining (13.42) and (13.43) yields

$$E_z^i(\vec{\rho}) - \frac{k\eta}{4}\int_C J_z(\vec{\rho}\,')H_0^{(2)}(k|\vec{\rho}-\vec{\rho}\,'|)dl' = 0 \quad \rho \text{ on } C, \tag{13.44a}$$

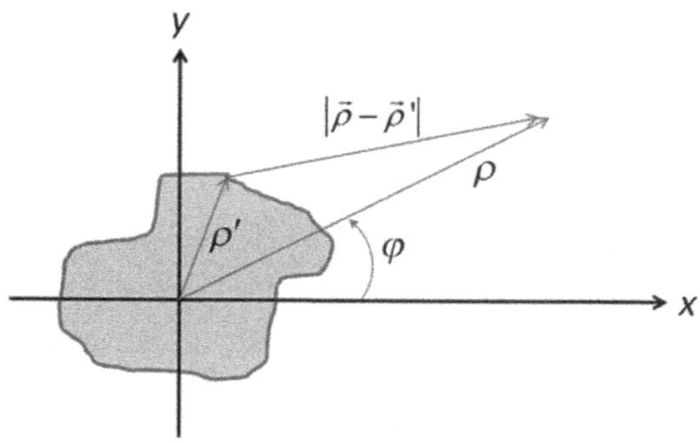

Figure 13.14. Geometry of the 2D scattering problem.

$$E_z^i(\vec{\rho}) = \frac{k\eta}{4} \int_C J_z(\vec{\rho}\,') H_0^{(2)}(k|\vec{\rho} - \vec{\rho}\,'|) dl' \quad \rho \text{ on } C \qquad (13.44b)$$

with $E_z^i(\vec{\rho})$ known and $J_z(\vec{\rho}\,')$ unknown.

According to MoM, the scattering contour C is divided into N segments. Segment lengths are small enough to assume constant incident-field-induced currents on them. In this case, $J_z(\vec{\rho}\,')$ in (13.44) may be moved outside the integral,

$$E_z^i(\vec{\rho}) = \sum_{n=1}^{N} \alpha_n \frac{k\eta}{4} \int_{\Delta C_n} H_0^{(2)}(k|\vec{\rho} - \vec{\rho}\,'|) dl' \quad \rho \text{ on } C. \qquad (13.45)$$

Here, ΔC_n are segment locations. Writing down (13.45) for all segments yields an $N \times N$ system,

$$[A_{nn}][\alpha_n] = [B_n], \qquad (13.46)$$

where α_n are unknown segment currents, $B_n = E_z^i(x_n, y_n)$ are incident fields at the center of each segment, and

$$A_{mn} = \frac{k\eta}{4} \int_{\Delta C_m} H_0^{(2)}\left(k\sqrt{(x_m - x')^2 + (y_m - y')^2}\right) dl'. \qquad (13.47)$$

An approximate solution of (13.47) is

$$A_{nn} \approx \frac{k\eta}{4} \Delta C_n \left\{1 - j\frac{2}{\pi} \ln\left(\frac{1.7810 k \Delta C_n}{4e}\right)\right\} \quad \text{for} \quad m = n, \qquad (13.48a)$$

$$A_{nn} \approx \frac{k\eta}{4} \Delta C_n H_0^{(2)}\left(k\sqrt{(x_m - x_n)^2 + (y_m - y_n)^2}\right) \quad \text{for} \quad m \neq n. \qquad (13.48b)$$

This completes MoM formulation for the TM_z-type 2D scattering form infinite-long PEC cylinder. Note that analytical solution of this problem (see Ref. 7, p. 610 (11–102)),

$$\sigma_{2D} = \frac{2\lambda}{\pi} \left| \sum_{n=-\infty}^{+\infty} \varepsilon_n \frac{J_n(ka)}{H_n^{(2)}(ka)} \cos(n\varphi) \right|^2, \qquad (13.49)$$

which is valid for $\varphi_{\text{inc}} = 0°$ and for small cylinders ($a << \lambda$). Here, $\sigma_{2\text{D}}$ is the radar cross section (RCS) in 2D, which is defined as the ratio of the scattered and incident powers in front of the object. Table 13.3 lists a short MATLAB script which uses (13.46), (13.47), and (13.48) and solves $N \times N$ system and compares MoM results with analytical solution. Typical outputs are presented in Fig. 13.15 and Fig. 13.16. As

Table 13.3.
A Short MoM Code for 2D Scattering from a PEC Cylinder

```
%% ---------------------------------------------------
%   Program: MoM_Scattering.m
%   Purpose: To calculate Bi-static RCS of an infinite
PEC
%                cylinder for TMz case on xy-plane
%% ---------------------------------------------------
clear all;clc; format long;
%freq=input('Frequency [Hz]: ');
Ro=.25;             %input('Radius of cylinder [m]: ');
N=250;              %input('Number of segments: ');
phi_inc=0;         %input('Incident Wave angle [Deg]: ');
k=10;               % k=2*pi*freq/3e8;   % Wave number
% CONSTANTS
no=377; gama=1.781; ds=(2*pi*Ro)/N;   % Length of
segment
phi_inc=phi_inc*pi/180; alpha=pi/N;   % alpha is mid-
segment angle
for m=1:N      % Segment (x,y) coordinates
   x(m)=Ro*cos(alpha);   y(m)=Ro*sin(alpha);
   alpha=(2*m-1)*alpha;
end
%% [A]mn matrix is formed
for m=1:N
 for n=1:N
    if m==n   % Diagonal elements
       A(m,n)=((no*k*ds)/4)*(1-1i*(2/pi)*((log((gama*k
*ds)/4))-1));
    else       % Other elements
       tt=sqrt(((x(n)-x(m))^2)+(y(n)-y(m))^2);
       A(m,n)=((no*k)/4)*ds*besselh(0,2,k*tt);
```

Table 13.3.
(Continued)

```
    end
  end
end
%% Calculate incident fields on every segment
for n=1:N
  E(n)=exp(1j*k*(x(n)*cos(phi_inc)+y(n)*sin(phi_inc)));
end
%% Solve for segment currents
J=inv(A)*E';    % Induced surface current density
%% Calculate MoM scattered fields
for m=1:360  % Angle Loop
  phi(m)=m*pi/180; tot=0;
  for n=1:N  % Segment Loop
    tot=tot+J(n)*ds*exp(1i*k*(x(n)*cos(phi(m)) . . .
        +y(n)*sin(phi(m))));
  end
  rcs(m)=(k*no^2/4)*abs(tot)^2/(2*pi*Ro); mx=max(rcs);
end
%% Calculate Analytic scattered fields
for m=1:360  % Angle Loop
  phi(m)=m*pi/180;   atot=0;
  for n=-200:200   % Analytic solution loop (Phi_
inc=180 degrees)
    if n==0; epr=1; else; epr=2; end
    atot=atot+epr*(besselj(n,k*Ro)/
besselh(n,2,k*Ro)). . .
          *cos(n*phi(m));
  end
  arcs(m)=4/k*abs(atot)^2/(2*pi*Ro); amx=max(arcs);
end
%% Plot the results
polar(phi,rcs/mx); hold on;
polar(phi,arcs/amx,'r-');
title('Normalized RCS of a circular cylinder, a=1,
N=200');
%% ---------------------- END ------------------------
```

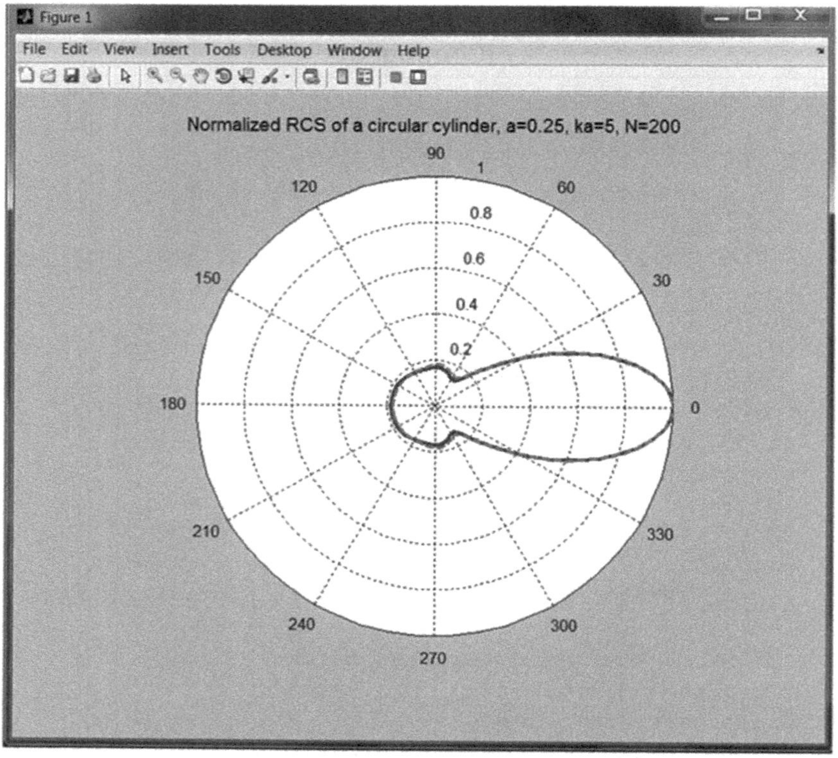

Figure 13.15. Polar plot showing MoM versus analytic results ($a = 0.25$ m).

observed, the agreement in Fig. 13.15 is very good; this is merely because the radius of the cylinder used in computations does not violate small cylinder restriction. On the other hand, a larger cylinder used in Fig.13.16 results in the displayed discrepancies between results.

Note that the MoM code in Table 13.3 is general. The user can change the angle of illumination, the size of the cylinder, and so on. Also, it can easily be modified to handle infinite-long PEC cylinders with arbitrary cross section. This can be achieved by only giving the segment coordinates of the selected cylinder with arbitrary cross section. For example, a piece of code, as given in Table 13.4, may be used for the segment generation of a square cylinder.

A MATLAB-based MoM code which calculates and plots bistatic RCS of both circular and square cylinders is given in Table 13.5. A typical simulation result obtained using this code is given in Fig. 13.17.

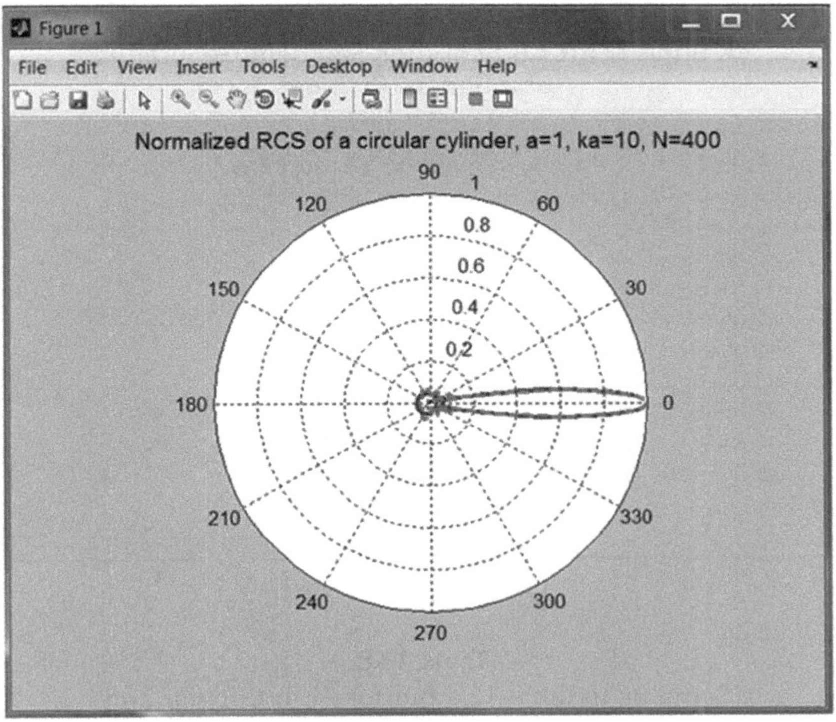

Figure 13.16. Polar plot showing MoM versus analytic results ($a = 1$ m).

Here RCS versus angle for the illumination from 60° is shown. Note that the illumination angle is defined in a way that angle of incidence equal to the angle of scattering for the backscattering RCS case (i.e., for a 60° incidence angle, plane wave hits the object located on the origin from 240°, as shown in Fig. 13.17).

13.5.4 3D RCS Modeling

A highly effective and free RCS prediction virtual tool, MGL-RCS, has been introduced [31–33]. The whole RCS prediction process including discrete MoM as well as finite-difference time-domain (FDTD) target model creation has been automated. FDTD simulations are performed directly with MGL-RCS tool. Any open-access NEC can be used for MoM simulations using the discrete models created with MGL-RCS. A few discrete MoM models created via MGL-RCS are shown in Fig. 13.18.

Table 13.4.
A Short MATLAB Script for Segment Generation of a Square Cylinder

```
%% -------------------------------------------------------
N=400;            %input('Number of segments: ');
aa=1; daa=4*aa/N;  % Square cylinder sizes
for m=1:N/4
    x2(m)=-aa/2+(m-1)*daa;
    y2(m)=-aa/2;
    x2(m+N/4)=aa/2;
    y2(m+N/4)=-aa/2+(m-1)*daa;
    x2(m+N/2)=aa/2-(m-1)*daa;
    y2(m+N/2)=aa/2;
    x2(m+3*N/4)=-aa/2;
    y2(m+3*N/4)=aa/2-(m-1)*daa;
end
%% -------------------- END -----------------------
```

Table 13.5.
Script as in Table 12.4 but for Both Circular and Square Cylinders

```
%% -------------------------------------------------------
%  Program: MoM_Scattering2.m
%  Purpose: To calculate 2D RCS of infinite-long
%              PEC circular and square cylinders (TMz)
%% -------------------------------------------------------
clear all;clc; format long;
Ro=0.5; N=400; phi_inc=60; k=10; no=377; gama=1.781;
ds=(2*pi*Ro)/N; phi_inc=phi_inc*pi/180; alpha=pi/N;
aa=1; daa=4*aa/N;  % Square cylinder sizes
for m=1:N     % Segment coordinates of circular cylinder
   phi=(2*m-1)*alpha; x1(m)=Ro*cos(phi);
y1(m)=Ro*sin(phi);
end
for m=1:N/4     % Segment coordinates of square cylinder
   x2(m)=-aa/2+(m-1)*daa; y2(m)=-aa/2;
   x2(m+N/4)=aa/2; y2(m+N/4)=-aa/2+(m-1)*daa;
   x2(m+N/2)=aa/2-(m-1)*daa; y2(m+N/2)=aa/2;
```

Table 13.5.
(Continued)

```
  x2(m+3*N/4)=-aa/2; y2(m+3*N/4)=aa/2-(m-1)*daa;
end
figure(1); plot(x1,y1,x2,y2, 'r-')  % Plot cylinders
xlim([-1 1]); ylim([-1 1]);
%% ============ Circular Cylinder
=========================
% [A]mn matrix is formed
for m=1:N
 for n=1:N
   if m==n  % Diagonal elements
     A(m,n)=((no*k*ds)/4)*(1-1i*(2/pi)*((log((gama*k
*ds)/4))-1));
   else     % Other elements
     tt=sqrt(((x1(n)-x1(m))^2)+(y1(n)-y1(m))^2);
     A(m,n)=((no*k)/4)*ds*besselh(0,2,k*tt);
   end
 end
end
for n=1:N % Calculate incident fields on segments
 E(n)=exp(1i*k*(x1(n)*cos(phi_inc)+y1(n)*sin(phi_inc)));
end
J=inv(A)*E'; % Solve for segment currents
for m=1:360  % Calculate MoM scattered fields
  phi(m)=m*pi/180; tot=0;
  for n=1:N  % Segment Loop
    tot=tot+J(n)*ds*exp(1i*k*(x1(n)*cos(phi(m)). . .
        +y1(n)*sin(phi(m))));
  end
  rcs1(m)=(k*no^2/4)*abs(tot)^2/(2*pi*Ro);
mx1=max(rcs1);
end
%% ============ Square Cylinder
=========================
for m=1:N
 for n=1:N
   if m==n  % Diagonal elements
     A(m,n)=((no*k*ds)/4)*(1-1i*(2/pi)*((log((gama*k
*ds)/4))-1));
   else     % Other elements
```

(Continued)

Table 13.5.
(Continued)

```
      tt=sqrt(((x2(n)-x2(m))^2)+(y2(n)-y2(m))^2);
      A(m,n)=((no*k)/4)*ds*besselh(0,2,k*tt);
    end
  end
end
for n=1:N % Calculate incident fields on segments

E(n)=exp(1i*k*(x2(n)*cos(phi_inc)+y2(n)*sin(phi_inc)));
end
J=inv(A)*E'; % Solve for segment currents
for m=1:360  %% Calculate MoM scattered fields
  phi(m)=m*pi/180; tot=0;
  for n=1:N  % Segment Loop
    tot=tot+J(n)*ds*exp(1i*k*(x2(n)*cos(phi(m)). . .
    +y2(n)*sin(phi(m))));
  end
  rcs2(m)=(k*no^2/4)*abs(tot)^2/(2*pi*Ro);
mx2=max(rcs2);
end
% Plot the results
figure(2)
polar(phi,rcs1/mx1); hold on; polar(phi,rcs2/mx2,'r-');
legend('Circular', 'Square')
title('Normalized RCS of cylinders, a=0.5, ka=5,
N=400');
%%----------------------- END ---------------------------
```

Note that FDTD is a time-domain method and yields both near and far fields in the time domain all around the target under investigation. It yields broadband transient EM effects if a Gaussian pulsed-plane wave is used as the incident wave. Once the user has this information, offline DFT may be applied to this time-domain data and a number of plots may be generated at different frequencies. On the other hand, NEC2 simulations must be repeated for each frequency, since MoM is a frequency-domain method. Finally, note that NEC2 package used in this paper gives the RCS as $\sigma_{\mathrm{dB}} = 10\log_{10}(\sigma/\lambda^2)$, therefore, one needs to add $10\log_{10}(\lambda^2)$ to the NEC2 results.

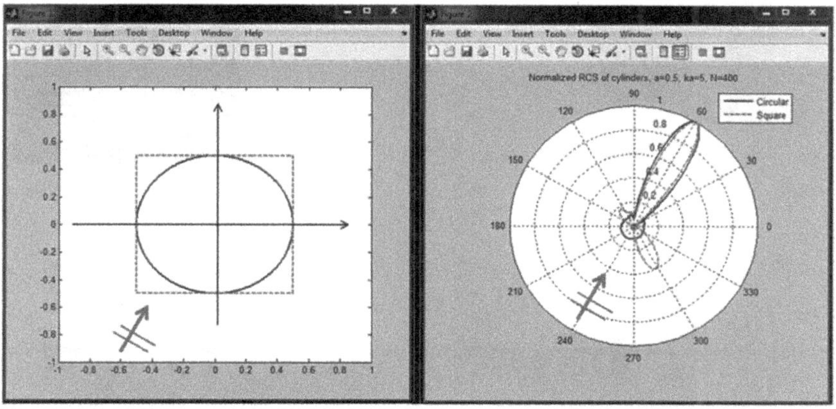

Figure 13.17. RCS versus angle of PEC cylinders ($a = 1$ m, $\varphi_{\text{inc}} = 0°$).

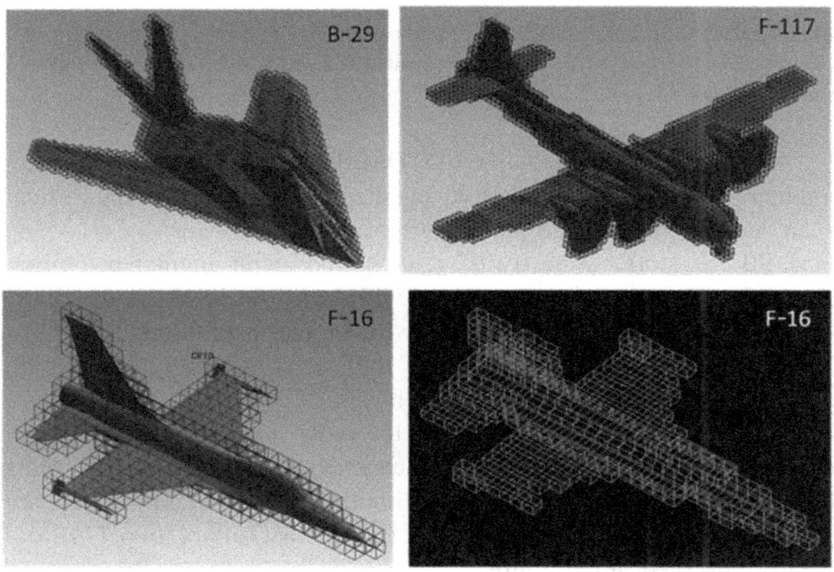

Figure 13.18. Typical discrete MoM models of air targets.

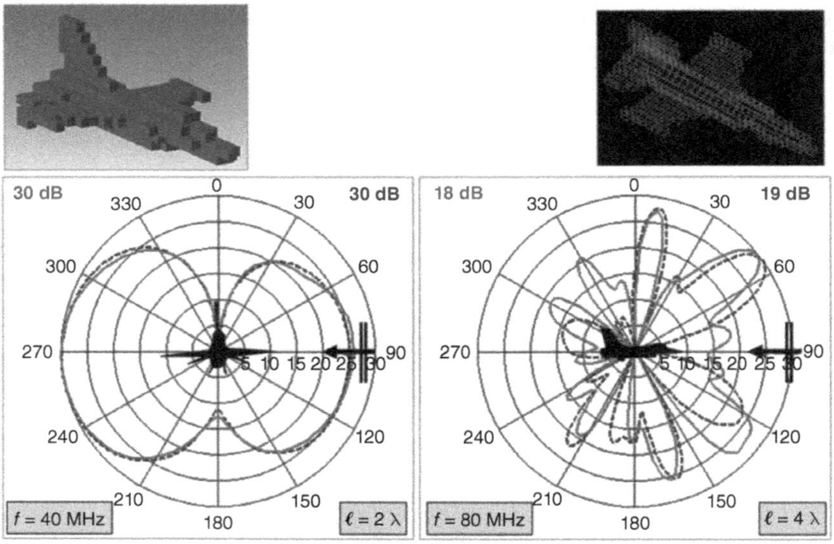

Figure 13.19. (Top) Discrete FDTD and MoM models of 15-m-long F-16 aircraft. (Bottom) Horizontal and vertical RCS variations computed with both MoM and FDTD methods (bistatic case).

Figure 13.19 belongs to the FDTD versus NEC comparisons, for an angular bistatic RCS predictions using the same target but for the vertical scan (i.e., $\sigma_{\varphi\varphi}$-case). The parameters of Fig. 13.19a are $f = 40$ MHz, $\theta_i = 90°$, $\varphi_i = 0°$, $\varphi_s = 0°$, $0° \leq \theta_s \leq 360°$ with angular resolution $\Delta\theta = 3°$. The parameters of Fig. 13.19b are $f = 80$ MHz, $\theta_i = 90°$, $\varphi_i = 90°$, $\varphi_s = 90°$, $0° \leq \theta_s \leq 360°$ with angular resolution $\Delta\theta = 3°$.

One nice feature of this virtual tool is its capability of creating multitargets from different 3DS graphics files. Figure 13.20 shows two F16 aircrafts: one, 12 m, on top of the other and side by side (15 m apart). Monostatic copolarized ($\sigma_{\varphi\varphi}$ case) RCS versus frequency plots of one F16 and two F16 predicted with the MoM (NEC2) package are given in Fig. 13.21. The effects of the interactions of two F-16 are clearly observed in the figure.

13.6 MOM APPLIED TO WEDGE DIFFRACTION PROBLEM

MoM has been successfully applied onto different EM problems [11,34,35]. A novel two-step MoM [35] approach has just been

Figure 13.20. Discrete MoM models of two F-16 flying together.

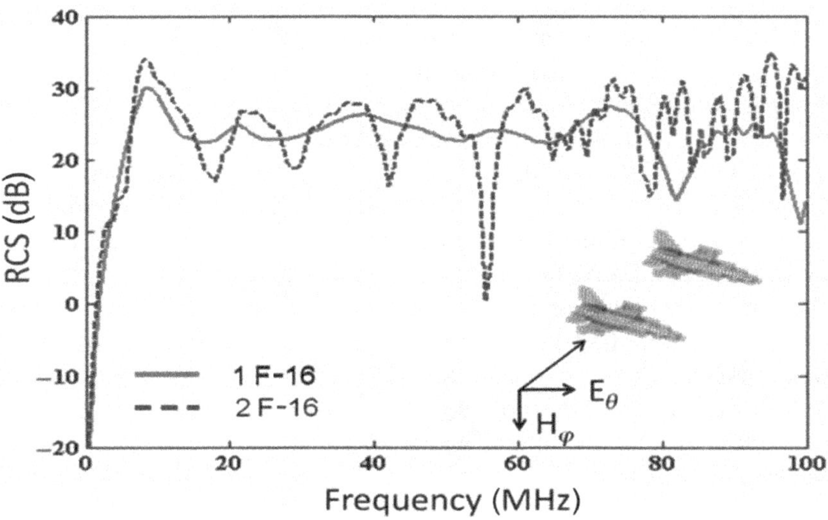

Figure 13.21. RCS versus frequency of single F-16 compared with side-by-side two F-16 15 m apart (monostatic case).

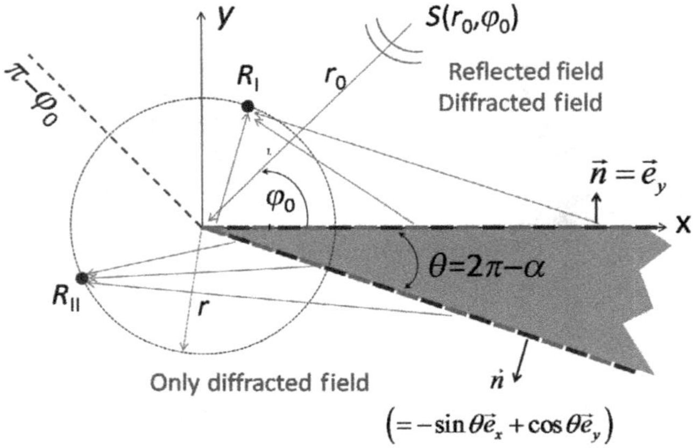

Figure 13.22. Wedge geometry and MoM modeling of scattered fields for the line source illumination. MoM-computed scattered fields contain reflected and diffracted fields ($\theta = 2\pi - \alpha$).

introduced for the extraction and visualization of diffracted fields on the canonical wedge scattering problem (see Chapter 10 for the definition of the diffraction problem and various high frequency asymptotic (HFA) solutions.

Figure 13.22 shows the 2D geometry of the semi-infinite wedge with PEC boundaries and exterior angle α. Wedge is located in a homogenous medium and illuminated by a cylindrical wave diverging from the line source $S(r_0, \varphi_0)$ and receiver point is given by $R(r, \varphi)$, where r, φ, z are the polar coordinates. The z-axis is aligned along the edge of the wedge. The angle φ is measured from the top face of the wedge. The MoM-based scattering model requires first analytical derivation of the 2D Green's function. Then, both surfaces of the wedge are discretized and replaced with a number of neighboring segments. The segment lengths are specified according to the wave frequency. As a rough criterion, the length of each segment should be equal to or less than one-tenth of the wavelength ($\lambda/10$). The number of segments on both surfaces are N; therefore, the total number of segments is $2N$. Note that although it is an infinite wedge, a rough truncation with ten to hundred wavelengths from the tip may be enough/required for the numerical simulations, depending on the polarization as well as other parameters. Note that each segment acts as a line source with omnidirectional radiation pattern on xz-plane for the TE polarization. Therefore, all segment currents are in parallel (and are perpendicular to the paper). Their mutual coupling (i.e., the impedance matrix) and scattered fields depend only on the distance. On the other hand, each segment acts as a short dipole for the TM polarization. Both their coupling and scattered fields are orientation (angle) dependent.

Incident fields in MoM modeling are injected analytically using the Green's function solution of the problem at hand; therefore, the scattered fields accumulated from individual source-induced segment currents contain only reflected and diffracted fields. Figure 13.22 shows the critical angle $\varphi = \pi - \varphi_0$ (reflection shadow boundary, RSB), which divides the whole scattering region into two: with (right) and without (left) reflected fields. As shown in the picture, diffracted fields exist everywhere and reflected plus diffracted fields exist only in the region $(0 \leq \varphi < \pi - \varphi_0)$.

The MoM procedure of the wedge diffraction is implemented as follows (see (13.31), (13.32), (13.33), (13.34), (13.35), (13.36), (13.37), (13.38), (13.39), (13.40), and (13.41) for both TE and TM sets of MoM equations).

- First, incident fields upon segments are calculated using the Green's function of the problem.
- The impedance matrix is formed.
- Then $2N \times 2N$ matrix system is solved and source-induced segment currents are obtained.
- Finally, scattered fields on the chosen observation points are calculated from the superposition of segment radiations using the Green's function. Total fields are obtained by adding the direct wave from the source to the receiver.

The diffracted-only fields can be obtained using the MoM procedure if reflected fields in region $(0 \leq \varphi < \pi - \varphi_0)$ are subtracted. The reflected fields in region $(0 \leq \varphi < \pi - \varphi_0)$ can be extracted as follows.

- First, infinite-plane geometry shown in Fig. 13.23 is taken into account and standard four-step MoM procedure explained earlier is applied. The MoM-computed scattered fields contain only reflected fields since there is no edge/tip discontinuity.

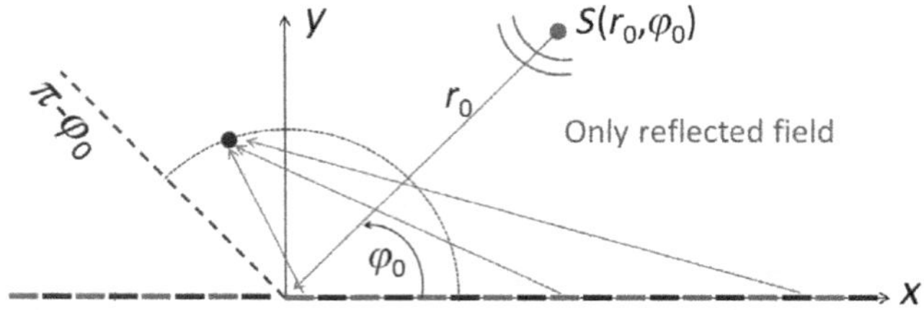

Figure 13.23. Infinite-plane geometry and MoM modeling of scattered fields for the line source illumination. MoM-computed scattered fields contain only reflected fields.

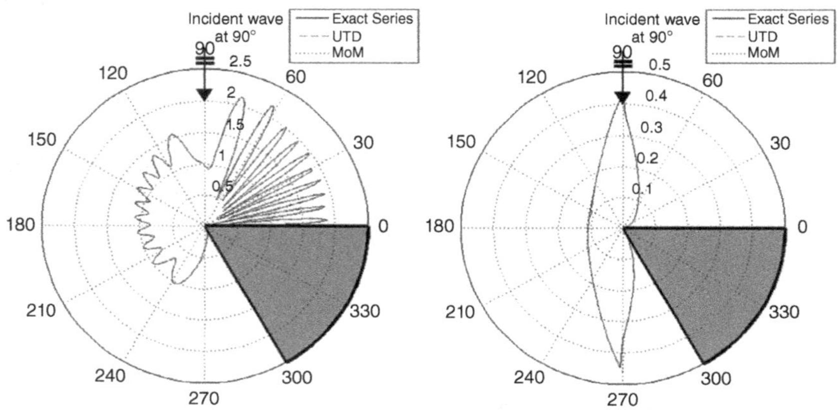

Figure 13.24. Wedge scattering for TM/SBC, exact, UTD, and MoM solutions: $\alpha = 300°$, $r = 50$ m, $r_0 = 100$ m, $\varphi_0 = 90°$, and $f = 30$ MHz. (Left) Total fields versus Angle; (right) diffracted fields versus angle.

- The scattered fields obtained with this geometry is subtracted from the scattered fields of the wedge geometry in region $(0 \leq \varphi < \pi - \varphi_0)$ and reflected fields are all eliminated. The result yields diffracted-only fields.

Examples using this novel MoM procedure are shown in Fig. 13.24, Fig. 13.25, Fig. 13.26, and Fig. 13.27, where MoM-extracted diffracted fields for both polarizations are compared against the analytical reference solution as well as the uniform theory of diffraction (UTD) model.

A nonpenetrable wedge with 60° interior angle is taken into account in Fig. 13.24. Soft boundary condition (SBC, i.e., TM_z case) is assumed.

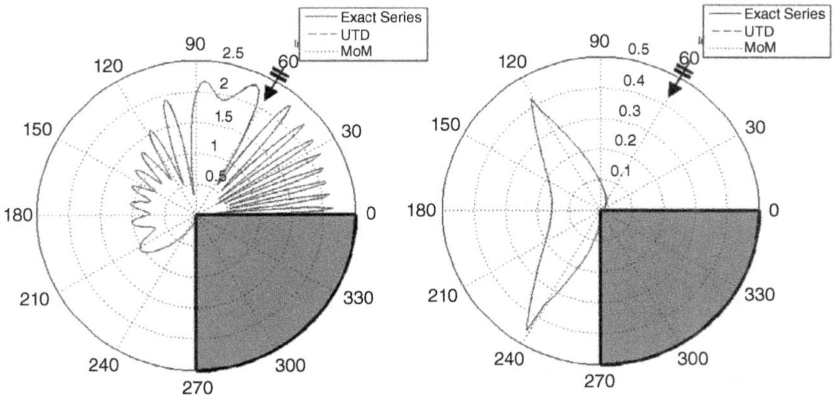

Figure 13.25. Wedge scattering for TM/SBC, exact, UTD, and MoM solutions: $\alpha = 270°$, $r = 50$ m, $r_0 = 100$ m, $\varphi_0 = 60°$, and $f = 30$ MHz. (Left) Total fields versus Angle; (right) diffracted fields versus angle.

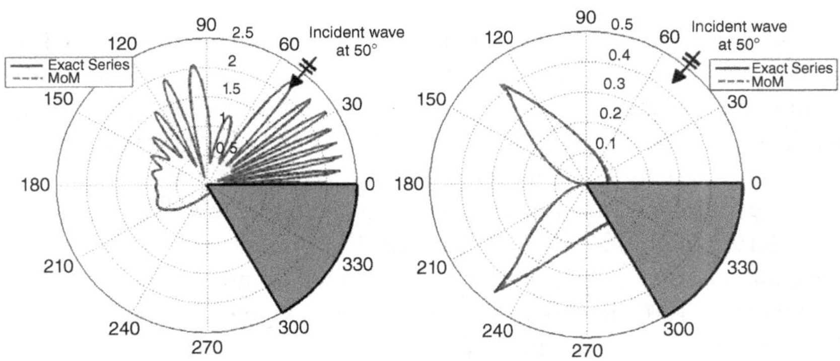

Figure 13.26. Wedge scattering for TE/HBC, exact solution versus MoM: $\alpha = 3000°$, $r = 50$ m, $r_0 = 100$ m, $\varphi_0 = 50°$, and $f = 30$ MHz. (Left) Total fields versus Angle; (right) diffracted fields versus angle.

Total and diffracted fields versus angle around the tip of the wedge on a circle with 5λ-radius are plotted. The wedge is illuminated by a line source located 10λ distance with $\varphi_0 = 90°$. Infinite wedge faces are truncated in 10λ and segment lengths are chosen $\lambda/10$. MoM results are compared with both exact series representation and UTD model. Total field versus angle plots on the left show the three regions clearly. As observed, very good agreement is obtained with MoM modeling even with these rough discretization parameters.

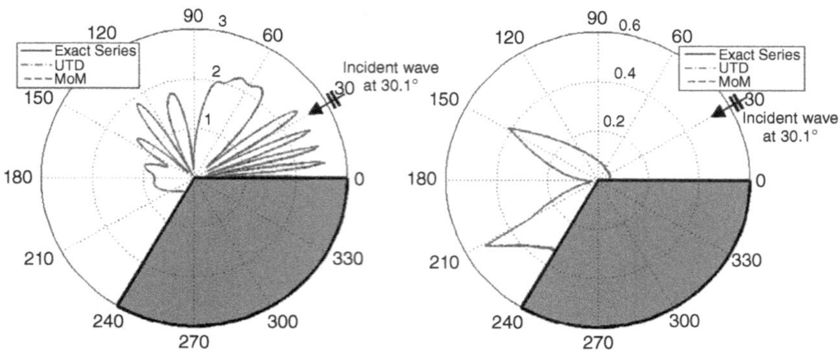

Figure 13.27. Wedge scattering for TE/HBC, exact, UTD, and MoM solutions: $\alpha = 240°$, $r = 50$ m, $r_0 = 100$ m, $\varphi_0 = 30°$, and $f = 30$ MHz. (Left) Total fields versus Angle; (right) diffracted fields versus angle.

A nonpenetrable wedge with 90° interior angle is used in Fig. 13.25. All the parameters kept the same as in Fig.13.24 except the illumination angle ($\varphi_0 = 60°$). Again, total and diffracted fields versus angle around the tip of the wedge are plotted and very good agreement is obtained among different models.

The other two examples belong to the other polarization (hard boundary condition, HBC/TE). In Fig. 13.26, a nonpenetrable wedge with 60° interior angle is taken into account. Infinite wedge faces are truncated in 100λ in this case and segment lengths are chosen, $\lambda/20$. Source and receiver distances from the tip of the wedge are kept the same. The illumination angle is 50°. Total and diffracted fields versus angle around the tip of the wedge on a circle are plotted. MoM results are compared with exact series representation. Again, very good agreement is obtained between the results.

The last example in Fig. 13.27 belongs to a nonpenetrable wedge with 120° interior angle. HBC is used. The illumination angle is 30°. Total and diffracted fields versus angle around the tip of the wedge are plotted using three different models. As observed, the agreement among the models is very good.

Table 13.6 lists a short MATLAB code for the generation of diffracted fields versus angle plots using the MoM procedure explained here.

Table 13.6.
A MoM Code for the Generation of Wedge-Diffracted Fields

```
function WedgeDiffMoM
clc, clear all, close all, format long
%% Input Parameters
fr = 30; r = 50; r0 = 100; alfa = 300; inc = 150;
sl=10; wl=80; % Segment and Wedge Length: lamda/sl,
wl*lamda
% START PROGRAM
I=momfindcurrentsflat(fr,r0,inc,sl,wl);
ang=linspace(0,alfa,302); ang=ang(2:end-1);
MOMSCTref=zeros(size(ang));
if inc<alfa-180&&inc>0, % SSI
   for i=1:length(ang),
       if ang(i)<180-inc
           MOMSCTref(i)=momanalysisflat(I,fr,r,ang(i),sl
,wl);
       end
   end
elseif inc>alfa-180&&inc<180, % DSI
   for i=1:length(ang),
       if ang(i)<180-inc,
         MOMSCTref(i)=momanalysisflat(I,fr,r,ang(i),sl,wl);
       end
       if ang(i)>2*alfa-inc-180
         MOMSCTref(i)=momanalysisflat(I,fr,r,a
lfa-ang(i),sl,wl);
     end; end; end
% MoM for wedge
utot=ones(size(ang));udif=ones(size(ang)); I=momfindcurre
nts(fr,alfa,r0,inc,sl,wl);
for i=1:length(ang),
   [diff_mag,tot_mag]=momanalysis(I,fr,alfa,r0,inc,r,ang(i
),sl,wl);
   utot(i)=tot_mag; udif(i)=diff_mag;
end
udif=udif-MOMSCTref; Ar=ang*pi/180;
%  Plot results
figure,polar(Ar,abs(utot));hold on
figure,polar(Ar,abs(udif));hold on
```

(Continued)

Table 13.6.
(Continued)

```
% ------------------------------
function I=momfindcurrentsflat(fr,r0,inc,sl,wl)
lam=300/fr; k0=2*pi/lam; zs=r0*cosd(inc);
xs=r0*sind(inc);
dr=lam/sl; rr=(-wl*lam:dr:wl*lam);
rr=0.5*(rr(1:end-1)+rr(2:end));
Nz=length(rr); zter=rr;   xter=zeros(size(zter));
imp_matrix=0.5*eye(Nz);   % Calculate the impedance
matrix
ds=sqrt((xter-xs).^2+(zter-zs).^2);
inc_magx=besselh(0,2,k0*ds);
I=imp_matrix\(-inc_magx.');   % Calculate currents on
each segment
% ------------------------------
function scat_mag=momanalysisflat(I,fr,r,ang,sl,wl)
lam=300/fr; k0=2*pi/lam; % wavenumber in free space
zobs=r*cosd(ang); xobs=r*sind(ang);
dr=lam/sl; rr=(-wl*lam:dr:wl*lam); rr=0.5*(rr(1:end-
1)+rr(2:end)); Nz=length(rr);
zter=rr;   xter=zeros(size(zter));
%% Calculate scattering fields
scat_mag=0;
for n=1:Nz
  dd=sqrt((xobs-xter(n)).^2+(zobs-zter(n)).^2);
  prinner=(xobs-xter(n))./dd;
  scat_mag=scat_mag+I(n).*besselh(1,2,k0*dd).*prinner;
end
scat_mag =  0.25*dr*1i*k0*scat_mag;   % scattering fields
at (zobs,xobs)
% ------------------------------
function I=momfindcurrents(fr,alfa,r0,inc,sl,wl)
lam=300/fr; k0=2*pi/lam; zs=r0*cosd(inc);
xs=r0*sind(inc);
dr=lam/sl; rr=(0:dr:wl*lam); rr=0.5*(rr(1:end-
1)+rr(2:end)); Nz=length(rr); zter1=rr;
xter1=zeros(size(zter1));
zter2=rr*cosd(alfa); xter2=rr*sind(alfa);
%% Calculate the impedance matrix
imp_matrix11=0.5*eye(Nz); imp_matrix22=0.5*eye(Nz);
```

Table 13.6.
(Continued)

```
imp_matrix12=zeros(Nz); imp_matrix21=zeros(Nz);
for m=1:Nz
  for n=1:Nz
    R =
sqrt((xter1(m)-xter2(n))^2+(zter1(m)-zter2(n))^2);
 prinner=-cosd(alfa)*(xter1(m)-xter2(n))/R+sind(alfa)*(zter1(m)-
              zter2(n))/R;
    imp_matrix12(m,n) = 0.25*dr*1i*k0*besselh(1,2,k0*R)*
prinner;
    R =
sqrt((xter2(m)-xter1(n))^2+(zter2(m)-zter1(n))^2);
    prinner=(xter2(m)-xter1(n))/R;
    imp_matrix21(m,n) = 0.25*dr*1i*k0*besselh(1,2,k0*R)*
prinner;
    end
end
%% Initial voltages on the surface from source at
(zs,xs)
% upper part at phi=0
ds=sqrt((xter1-xs).^2+(zter1-zs).^2);  inc_magx0=besselh
(0,2,k0*ds);
% lower part at phi=alfa
ds=sqrt((xter2-xs).^2+(zter2-zs).^2);  inc_magxa=besselh
(0,2,k0*ds);
V_inc1=-inc_magx0.';  V_inc2=-inc_magxa.';  V =
 V_inc1;V_inc2];
Z=[imp_matrix11,imp_matrix12;imp_matrix21,imp_matrix22];
I=Z\V; %% Calculate currents on each segment
% ------------------------------
function [diff_mag,tot_mag]=momanalysis(I,fr,alfa,r0,inc,
r,ang,sl,wl)
lam=300/fr; k0=2*pi/lam; zs=r0*cosd(inc);
xs=r0*sind(inc);
zobs=r*cosd(ang); xobs=r*sind(ang);
dr=lam/sl; rr=(0:dr:wl*lam);
rr=0.5*(rr(1:end-1)+rr(2:end));
Nz=length(rr); zter1=rr; xter1=zeros(size(zter1));
zter2=rr*cosd(alfa); xter2=rr*sind(alfa);
scat_mag=0; % Calculate scattering fields
```

(*Continued*)

Table 13.6.
(Continued)

```
for n=1:Nz
   dd=sqrt((xobs-xter1(n)).^2+(zobs-zter1(n)).^2);
   prinner=(xobs-xter1(n))/dd;
   scat_mag=scat_mag+I(n).*besselh(1,2,k0*dd).*prinner;
   dd=sqrt((xobs-xter2(n)).^2+(zobs-zter2(n)).^2);
   prinner=-cosd(alfa)*(xobs-xter2(n))/
dd+sind(alfa)*(zobs-zter2(n))/dd;
   scat_mag=scat_mag+I(Nz+n).*besselh(1,2,k0*dd).*prin
ner;
end
scat_mag =  0.25*dr*1i*k0*scat_mag; % scattering fields
at (zobs,xobs)
dist=sqrt((xobs-xs)^2+(zobs-zs).^2); inc_mag=besselh(0,2
,k0*dist);
tot_mag=inc_mag+scat_mag;
if inc<alfa-180&&inc>0, % SSI
      if ang>=180+inc, diff_mag=tot_mag; else diff_
mag=scat_mag; end
elseif inc>alfa-180&&inc<180, % DSI
      diff_mag=scat_mag;
end
% ------------------------ END --------------------------
```

Note that different discretizations are required for the TM (SBC) and TE (HBC) polarizations. Approximately, 10λ-long wedge sides are enough for the TM polarization, and this results in 200×200 matrix system. On the other hand, up to 100λ-long wedge sides (even more) may be required for the TE polarization where 2000×2000 matrix system is of interest. Nevertheless, both discrete systems can be solved within a minute with a regular student computer.

Note also that the *diffraction coefficient* has a *physical meaning* only for high frequency fields and far away from the wedge when $kr_0 >> 1$, $kr >> 1$, and $r >> \lambda$. This term is an HFA notion. Therefore, quantities presented in Fig. 13.24, Fig. 13.25, Fig. 13.26, and Fig. 13.27 are normalized diffracted fields; but diffraction coefficients can directly be obtained using the definition for the diffraction coefficients.

The critical question is "the real potential and applicability of the MoM method to fit HFA domain for RCS applications." How far can one go from the edge? One needs to reach at least $r = 10\lambda - 20\lambda$ (or maybe much more) to speak about possibility to obtain some reasonable values for diffraction coefficients. For an object of a size of several wavelengths (say, $l = 10 - 50\lambda$), the size of MoM space reaches to hundreds by hundreds even in 2D which necessitates several thousand segments with a rough segment discretization of $\lambda/10$. This is within the range of applicability with a regular student computer with 2–4 GB RAM and a few gigahertz speed. On the other hand, much larger memories and much higher speeds are essential in 3D MoM modeling. Fortunately, there are alternatives and hybridization approaches to overcome these problems.

13.7 MoM APPLIED TO WEDGE WAVEGUIDE PROBLEM

The same MoM procedure given in Section 13.6 can also be applied to propagation modeling inside a wedge waveguide with nonpenetrable boundaries. Analytical modeling in terms of different mode summation models are discussed in Chapter 9. The Green's function representations inside the wedge waveguide in terms of normal modes, adiabatic modes, and intrinsic modes are given with (9.27), (9.41), and (9.49), respectively. Also, Table 9.1 list a simple MATLAB code for the Green's function representation in terms of normal modes for the TE (Neumann—HBC) case.

A simple MATLAB code for the MoM computation of the line source response inside the wedge waveguide for HBC is listed in Table 13.7. Here, (13.37), (13.38), (13.39), (13.40), and (13.41) are used. This code can easily be modified for the SBC problem by replacing (13.37), (13.38), (13.39), (13.40), and (13.41) with (13.32), (13.33), (13.34), (13.35), and (13.36). The wedge boundaries are discretized and N-segments are used for the upper and lower boundaries. Then, the standard MoM procedure is applied with $2N \times 2N$ system, and fields inside any observation point is obtained. Figure 13.27, Figure 13.28, Figure 13.29, and Figure 13.30 belong to comparisons between mode summation and MoM models. As observed, the agreement is very good.

Table 13.7.
A Wedge Waveguide MoM Code for HBC

```
function WedgeGuideMoMHBC
clc, clear all, close all, format long
%% Input Parameters
% frequency [MHz], % wedge angle [Deg], % source
position [m,m], observer [m,m]: TE (HBC)
fr = 15;  alfa = 15; xs = 1000; ys = 15; xo = 150; yo
= 2;

if ys>xs*tand(alfa), error(['ys must be smaller than
',num2str(xs*tand(alfa))]),end
if yo>xo*tand(alfa), error(['yo must be smaller than
',num2str(xo*tand(alfa))]),end
xx=linspace(xo,xs,501); yy=linspace(0,xo*tand(alfa),501);
k=2*pi*fr/300; sl = 10;  wl = 80;  % (lamda/sl) &
(wl*lamda)
Uv=NaN(size(yy)); Uvmom=Uv; Uh=NaN(size(xx)); Uhmom=Uh;
[I,Z]=momfindcurrents(fr,alfa,xs,ys,sl,wl);      % MoM
Solution
Uvmom=NaN(size(yy));
for i=1:length(yy),
    Uvmom(i)=momanalysis(I,fr,alfa,xs,ys,xo,yy(i),sl
,wl);
end
Uhmom=NaN(size(xx));
for i=1:length(xx),
    Uhmom(i)=momanalysis(I,fr,alfa,xs,ys,xx(i),yo,sl
,wl);
end
plotfigure(xx,yy,Uvmom,Uhmom);
% --------------------------------------------------
function [I,Z]=momfindcurrents(fr,alfa,xs,ys,sl,wl)
lam=300/fr; k0=2*pi/lam; dr=lam/sl; rr=(0:dr:wl*lam);
rr=0.5*(rr(1:end-1)+rr(2:end)); N=length(rr); xter1=rr;
yter1=zeros(size(xter1)); xter2=rr*cosd(alfa);
yter2=rr*sind(alfa);
imp_matrix11=0.5*eye(N); imp_matrix22=0.5*eye(N);
imp_matrix12=zeros(N);   imp_matrix21=zeros(N);
for m=1:N
 for n=1:N
```

Table 13.7.
(Continued)

```
    R=sqrt((yter1(m)-yter2(n))^2+(xter1(m)-xter2(n))^2);
    prinner=-cosd(alfa)*(yter1(m)-yter2(n))/
R+sind(alfa)*(xter1(m)-xter2(n))/R;
    imp_matrix12(m,n)=0.25*dr*1i*k0*besselh(1,2,k0*R)*pri
nner;

    R=sqrt((yter2(m)-yter1(n))^2+(xter2(m)-xter1(n))^2);
    prinner=(yter2(m)-yter1(n))/R;
    imp_matrix21(m,n)=0.25*dr*1i*k0*besselh(1,2,k0*R)*pri
nner;
    end
end
%% Initial voltages on the surface coming from source at
(xs,ys)
ds=sqrt((yter1-ys).^2+(xter1-xs).^2); inc_
magx0=besselh(0,2,k0*ds);     % first part at phi=0
ds=sqrt((yter2-ys).^2+(xter2-xs).^2); inc_
magxa=besselh(0,2,k0*ds);     % second part at phi=alfa
V_inc1=-inc_magx0.';  V_inc2=-inc_magxa.'; V =
 V_inc1;V_inc2];
Z=[imp_matrix11,imp_matrix12;imp_matrix21,imp_matrix22];
I=Z\V;
% ----------------------------------------------------
function I2=momfindcurrent(Z,I,fr,alfa,sl,wl)
lam=300/fr; k0=2*pi/lam; dr=lam/sl; rr=(0:dr:wl*lam);
rr=0.5*(rr(1:end-1)+rr(2:end)); N=length(rr); xter1=rr;
yter1=zeros(size(xter1)); xter2=rr*cosd(alfa);
yter2=rr*sind(alfa);
%% Calculate scattering fields
sx0=zeros(size(xter1)); sxa=zeros(size(xter2));
yobs=yter2; xobs=xter2;
for n=1:N
     dd=sqrt((yobs-yter1(n)).^2+(xobs-xter1(n)).^2);
     prinner=(yobs-yter1(n))./dd;
     scat=I(n).*besselh(1,2,k0*dd).*prinner;
sxa=sxa+scat;
end
yobs=yter1; xobs=xter1;
for n=1:N
   dd=sqrt((yobs-yter2(n)).^2+(xobs-xter2(n)).^2);
```

(Continued)

Table 13.7.
(Continued)

```
  prinner=-cosd(alfa)*(yobs-yter2(n))./
dd+sind(alfa)*(xobs-xter2(n))./dd;
  scat=I(N+n).*besselh(1,2,k0*dd).*prinner;
  sx0=sx0+scat;
end
sx0=0.25*dr*1i*k0*sx0; sxa=0.25*dr*1i*k0*sxa;
V_inc1=-sx0.'; V_inc2=-sxa.'; V=[V_inc1;V_inc2]; I2=Z\V;
% ---------------------------------------------------
function tot_mag=momanalysis(I,fr,alfa,xs,ys,xobs,yobs,sl
,wl)
lam=300/fr; k0=2*pi/lam; dr=lam/sl; rr=(0:dr:wl*lam);
rr=0.5*(rr(1:end-1)+rr(2:end)); N=length(rr); xter1=rr;
yter1=zeros(size(xter1)); xter2=rr*cosd(alfa);
yter2=rr*sind(alfa);
scat_mag=0;      % Calculate scattering fields
for n=1:N
  dd=sqrt((yobs-yter1(n)).^2+(xobs-xter1(n)).^2);
  prinner=(yobs-yter1(n))./dd;
  scat=I(n).*besselh(1,2,k0*dd).*prinner;
scat_mag=scat_mag+scat;

  dd=sqrt((yobs-yter2(n)).^2+(xobs-xter2(n)).^2);
  prinner=-cosd(alfa)*(yobs-yter2(n))./
dd+sind(alfa)*(xobs-xter2(n))./dd;
  scat=I(N+n).*besselh(1,2,k0*dd).*prinner;
  scat_mag=scat_mag+scat;
end
scat_mag=0.25*dr*1i*k0*scat_mag; dist=sqrt((yobs-ys).^2+
(xobs-xs).^2);
inc_mag=besselh(0,2,k0*dist); tot_mag=(inc_mag+scat_
mag)/(4*1i);
% ---------------------------------------------------
function plotfigure(xx,yy,Uver,Uhor)
scrsz = get(0,'ScreenSize');
figure('Position',round([scrsz(3)/50 scrsz(4)/5 scrsz(3)/4
scrsz(4)/2.5]),'Color',[1 1 1]);
p1=plot(abs(Uver),yy,'-');hold on,ylabel('Height [m]');xlab
el('Field')
figure('Position',round([scrsz(3)/50 scrsz(4)/5 scrsz(3)/3
scrsz(4)/2.7]),'Color',[1 1 1]);
p1=plot(xx,abs(Uhor),'-');hold on,xlabel('Range [m]');
ylabel('Field')
%  -------------------- END --------------------
```

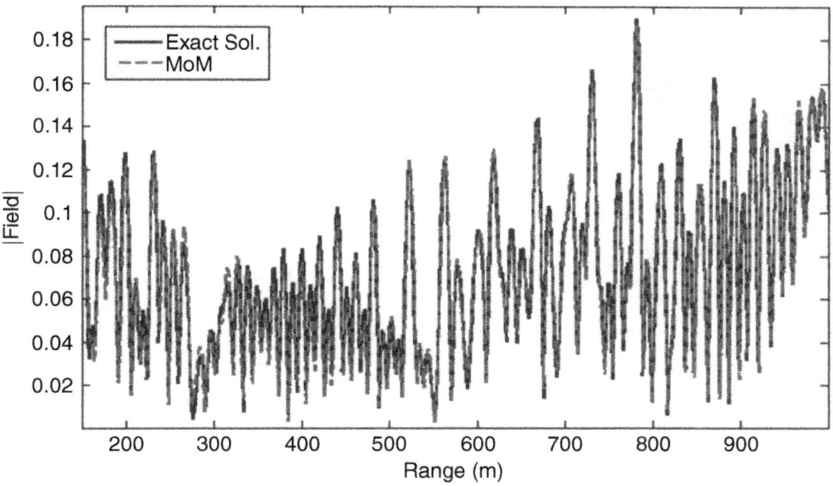

Figure 13.28. Wave field versus range computed with NM and MoM models. The parameters are $f = 15$ MHz, $\alpha = 15°$, $x_s = 1000$ m, $y_s = 15$ m, $x_o = 150$ m, $y_o = 2$ m (TM Pol/SBC).

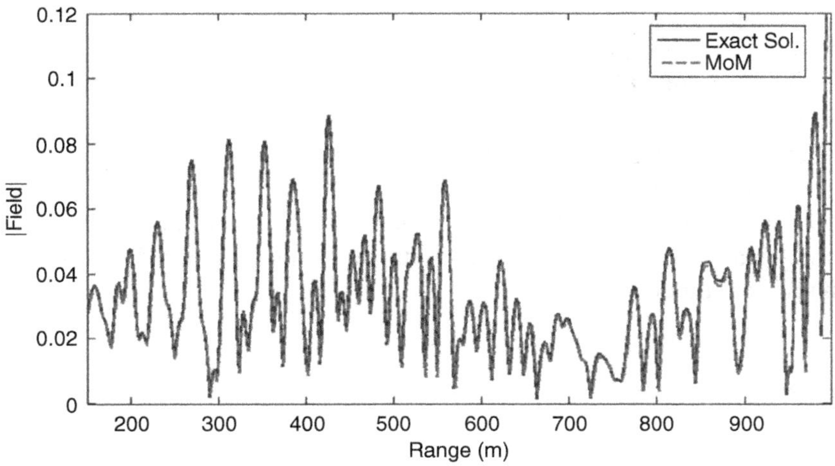

Figure 13.29. Wave field versus range computed with NM and MoM models. The parameters are $f = 15$ MHz, $\alpha = 15°$, $x_s = 1000$ m, $y_s = 15$ m, $x_o = 150$ m, $y_o = 2$ m (TE Pol/HBC).

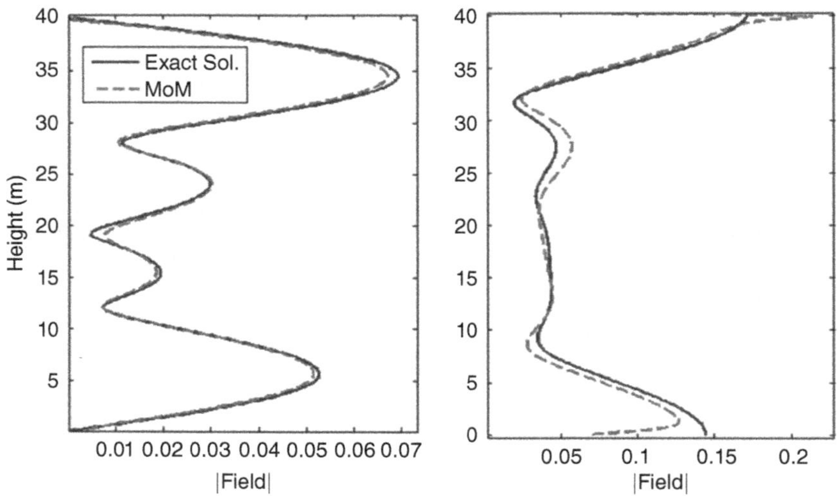

Figure 13.30. Wave field versus height computed with NM and MoM models. The parameters are $f = 15$ MHz, $\alpha = 15°$, $x_s = 1000$ m, $y_s = 15$ m, $x_o = 150$ m, $y_o = 2$ m. (Left) TM Pol/SBC; (right) TE Pol/HBC.

13.8 CONCLUSIONS

Method of moment (MoM) is reviewed in this chapter [35]. First, the method is briefly discussed. Steps to approximate the unknown function in linear operator equations in terms of a known finite series are outlined. The FS expansion is also revisited in this context. A few simple examples are included to better understand the MoM procedure. Finally, complex radiating and scattering problems that cannot be handled via analytical methods are simulated with MoM and compared with other numerical models.

REFERENCES

1. R. F. Harrington, *Field Computation by Moment Method*, 1st ed. 1968, IEEE Press, New York, 1993.
2. M. N. O. Sadiku, *Numerical Techniques in Electromagnetics*, 2nd ed., CRC Press, New York, 2001.

3. L. V. Kantorovich, V. I. Krylov, *Approximate Methods of Higher Analysis*, translated from Russian by C.D. Benster, John Wiley, New York, 1964.
4. Y. U. Vorobev, *Method of Moments in Applied Mathematics*, translated from Russian. Gordon & Breach, New York, 1965.
5. B. J. Strait, *Applications of the Method of Moments to Electromagnetics*, SCEEE Press, St. Cloud, FL, 1980.
6. E. Arvas, R. F. Harrington, "Computation of the Magnetic Polarizability of Conducting Disks and the Electric Polarizability of Apertures," *IEEE Transactions on Antennas and Propagation*, **31** (5), 1983, pp. 719–725.
7. C. A. Balanis, *Advanced Engineering Electromagnetics*, John Wiley, New York, 1989.
8. J. Moore, P. Pizer, eds., *Moment Methods in Electromagnetics: Techniques and Applications*, Research Studies Press, Letchworth, 1984.
9. R. F. Harrington, et al., eds., *Lectures on Computational Methods in Electromagnetics*, SCEEE Press, St. Cloud, FL, 1981.
10. L. Sevgi, "Numerical Fourier Transforms DFT and FFT," *IEEE Antennas and Propagation Magazine*, **49** (3), 2007, pp. 238–243.
11. G. Apaydin, L. Sevgi, "Method of Moment (MoM) Modeling for Resonating Structures: Propagation inside a Parallel Plate Waveguide," *ACES, International Journal on Applied Computational Electromagnetics*, **27** (10), 2012, pp. 842–849.
12. L. Sevgi, A. M. Ponsford, H. C. Chan, "An Integrated Maritime Surveillance System Based on Surface Wave HF Radars, Part I – Theoretical Background and Numerical Simulations," *IEEE Antennas and Propagation Magazine*, **43** (4), 2001, pp. 28–43.
13. M. Ponsford, L. Sevgi, H. C. Chan, "An Integrated Maritime Surveillance System Based on Surface Wave HF Radars, Part II – Operational Status and System Performance," *IEEE Antennas and Propagation Magazine*, **43** (5), 2001, pp. 52–63.
14. L. Sevgi, *Complex Electromagnetic Problems and Numerical Simulation Approaches*, IEEE & John Wiley Press, New York, 2003.
15. L. Sevgi, "Numerical Simulation Approaches for Phased Array Design," *ACES Journal of Applied Computational Electromagnetic Society*, special issue on Phased Array Design (invited tutorial), **21** (3), 2006, pp. 206–217.
16. L. Sevgi, "Modeling and Simulation Strategies in High Frequency Surface Wave Radars," (special issue on HF Radars) *TJEECS, Turkish Journal of Electrical Engineering and Computer Sciences*, **18** (3), 2010, pp. 399–407.
17. G. J. Burke, A. J. Poggio, *Numerical Electromagnetic Code (NEC)—Method of Moments*, Lawrence Livermore National Laboratory, January 1981.
18. J. W. Rockway, et al., *The MININEC System: Microcomputer Analysis of Wire Antennas*, Artech House, Norwood, MA, 1988.
19. L. Sevgi, F. Akleman, L. B. Felsen, "Groundwave Propagation Modeling: Problem-Matched Analytical Formulations and Direct Numerical Techniques," *IEEE Antennas and Propagation Magazine*, **44** (1), 2002, pp. 55–75.

20. F. Akleman, L. Sevgi, "A Novel MoM- and SSPE-Based Groundwave Propagation Field Strength Prediction Simulator," *IEEE Antennas and Propagation Magazine*, **49** (5), 2007, pp. 69–82.
21. L. Sevgi, "Groundwave Modeling and Simulation Strategies and Path Loss Prediction Virtual Tools," (special issue on Electromagnetic Wave Propagation in Complex Environments: A Tribute to L. B. Felsen), *IEEE Transactions on Antennas and Propagation*, **55** (6), 2007, pp. 1591–1598.
22. M. Levy, *Parabolic Equation Methods for Electromagnetic Wave Propagation*, IEE Institution of Electrical Engineers, London, 2000.
23. G. Apaydın, L. Sevgi, "A Novel Split-Step Parabolic Equation Package for Surface Wave Propagation Prediction along Multi-Mixed Irregular Terrain Paths," *IEEE Antennas and Propagation Magazine*, **52** (4), 2010, pp. 90–97.
24. G. Apaydin, L. Sevgi, "Numerical Investigations of and Path Loss Predictions for Surface Wave Propagation over Sea Paths Including Hilly Island Transitions," *IEEE Transactions on Antennas and Propagation*, **58** (4), 2010, pp. 1302–1314.
25. J. T. Johnson, R. T. Shin, J. C. Eidson, L. Tsang, J. A. Kong, "A Method of Moments Model for VHF Propagation," *IEEE Transactions on Antennas and Propagation*, **45** (1), 1997, pp. 115–125.
26. D. Holliday, L. L. DeRaad Jr., G. J. St.-Cyr, "Forward-Backward Method for Scattering from Imperfect Conductors," *IEEE Transactions on Antennas and Propagation*, **46** (1), 1998, pp. 101–107.
27. H.-T. Chou, J. T. Johnson, "Formulation of Forward-Backward Method using Novel Spectral Acceleration for the Modeling of Scattering from Impedance Rough Surface," *IEEE Transactions on Geoscience and Remote Sensing*, **38** (1), 2000, pp. 605–607.
28. M. R. Pino, R. J. Burkholder, F. Obelleiro, "Spectral Acceleration of the Generalized Forward-Backward Method," *IEEE Transactions on Antennas and Propagation*, **50**, 2002, pp. 785–797.
29. C. A. Tunc, A. Altintas, V. B. Erturk, "Examination of Existent Propagation Models over Large Inhomogeneous Terrain Profiles Using Fast Integral Equation Solution," *IEEE Transactions on Antennas and Propagation*, **53** (9), 2005, pp. 3080–3083.
30. C. Alp Tunç, V. Ertürk, F. Akleman, A. Altıntaş, L. Sevgi, "Fast Integral Equation Solutions: Application to Mixed-Path Terrain Profiles and Comparisons with Parabolic Equation Method," *Complex Computing Networks*, Springer in Physics Series, **104**, 2006, pp. 55–64.
31. Ç. Uluışık, G. Çakır, M. Çakır, L. Sevgi, "Radar Cross Section (RCS) Modeling and Simulation: Part I—Definitions, Strategies, and Canonical Examples," *IEEE Antennas and Propagation Magazine*, **50** (1), 2008, pp. 115–126.
32. G. Çakır, M. Çakır, L. Sevgi, "Radar Cross Section (RCS) Modeling and Simulation: Part II—A Novel FDTD-Based RCS Prediction Virtual Tool," *IEEE Antennas and Propagation Magazine*, **50** (2), 2008, pp. 81–94.

33. G. Çakır, L. Sevgi, "Radar Cross Section (RCS) Analysis of High Frequency Surface Wave Radars," (special issue on HF Radars) *TJEECS, Turkish Journal of Electrical Engineering and Computer Sciences*, **18** (3), 2010, pp. 457–467.
34. G. Apaydin, L. Sevgi, "A Canonical Test Problem for Computational Electromagnetics (CEM): Propagation in a Parallel Plate Waveguide," *IEEE Antennas and Propagation Magazine*, **54** (4), 2012, pp. 290–315.
35. E. Arvas, L. Sevgi, "A Tutorial on the Method of Moments," *IEEE Antennas and Propagation Magazine*, **54** (3), 2012, pp. 260–275.

CHAPTER 14

Finite-Difference Time-Domain Method

This chapter is devoted to the most widely used numerical model: the finite-difference time-domain (FDTD) method [1]. First, one-dimensional (1D) FDTD is discussed. Plane wave and the transmission line (TL) problems are reviewed. Then, two-dimensional (2D) FDTD models are presented.

14.1 FDTD REPRESENTATION OF EM PLANE WAVES

A point source radiates electromagnetic (EM) fields isotropically and generates spherical waves. In most EM problems, such as antenna and radar cross-section (RCS) analyses, far fields are of interest (distances are much larger than EM wavelength) and EM wave front may be assumed locally plane. This is true when the distance between EM source and the observer goes to infinity.

A plane wave is a nonphysical wave where electric field, magnetic field, and direction of propagation are perpendicular to each other. In free space, it has a *wave impedance* of

$$Z_0 = \frac{|\vec{E}|}{|\vec{H}|} = \sqrt{\frac{\mu_0}{\varepsilon_0}} = 120\pi = 377\,\Omega, \tag{14.1}$$

Electromagnetic Modeling and Simulation, First Edition. Levent Sevgi.

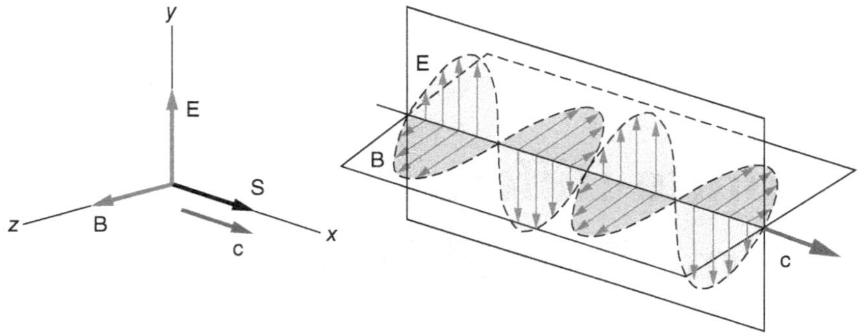

Figure 14.1. A plane wave propagating along x-direction with the speed of light.

which is also called the *characteristic impedance* of the free space. Electric and magnetic field components and the propagation direction are shown in Fig. 14.1 EM plane wave behaviors both in time and frequency domains will be investigated in this chapter.

14.1.1 Maxwell Equations and Plane Waves

The two Maxwell equations

$$\nabla \times \vec{E} = -\frac{\partial}{\partial t}\vec{B}, \tag{14.2a}$$

$$\nabla \times \vec{H} = \vec{J} + \frac{\partial}{\partial t}\vec{D} \tag{14.2b}$$

are the first-order vector forms which couples time variations of electric (magnetic) fields to the spatial variations of magnetic (electric) fields in free space in four-dimensional space. These two vector equations may be reduced to six scalar forms,

$$\begin{vmatrix} \vec{e}_x & \vec{e}_y & \vec{e}_z \\ \partial/\partial x & \partial/\partial y & \partial/\partial z \\ E_x & E_y & E_z \end{vmatrix} = -\mu\frac{\partial}{\partial t}(H_x\vec{e}_x + H_y\vec{e}_y + H_z\vec{e}_z), \tag{14.3a}$$

via

$$-\mu\frac{\partial}{\partial t}(H_x\vec{e}_x+H_y\vec{e}_y+H_z\vec{e}_z)=\vec{e}_x\left(\frac{\partial}{\partial y}E_z-\frac{\partial}{\partial z}E_y\right)-\vec{e}_y\left(\frac{\partial}{\partial x}E_z-\frac{\partial}{\partial z}E_x\right)$$
$$+\vec{e}_z\left(\frac{\partial}{\partial x}E_y-\frac{\partial}{\partial y}E_x\right). \tag{14.3b}$$

Here, $(\vec{e}_x, \vec{e}_y, \vec{e}_z)$ are unit vectors along x-, y-, and z-directions, respectively. Term by term matching of both sides results in

$$\mu\frac{\partial}{\partial t}H_x=\frac{\partial}{\partial z}E_y-\frac{\partial}{\partial y}E_z, \tag{14.3c}$$

$$\mu\frac{\partial}{\partial t}H_y=\frac{\partial}{\partial x}E_z-\frac{\partial}{\partial z}E_x, \tag{14.3d}$$

$$\mu\frac{\partial}{\partial t}H_z=\frac{\partial}{\partial y}E_x-\frac{\partial}{\partial x}E_y. \tag{14.3e}$$

Plane waves, for example, propagating along z-direction have equiphase planes that extend infinitely along x- and y-directions. They do not have longitudinal components E_z and H_z. Therefore,

$$\frac{\partial}{\partial x}\equiv 0, \frac{\partial}{\partial y}\equiv 0 \tag{14.4}$$

can be used in (14.3). Moreover, without loss of generality, one can assume that electric field is parallel to the x-axis (i.e., $E_y = 0$). Under these conditions, plane wave equations are obtained as

$$\frac{\partial}{\partial z}E_x=-\mu\frac{\partial}{\partial t}H_y, \tag{14.5a}$$

$$\frac{\partial}{\partial z}H_y=-\varepsilon\frac{\partial}{\partial t}E_x. \tag{14.5b}$$

As observed in (14.5), plane wave propagation along z-direction with electric field parallel to x-axis has the field components of $E_x(z, t)$ and $H_y(z, t)$.

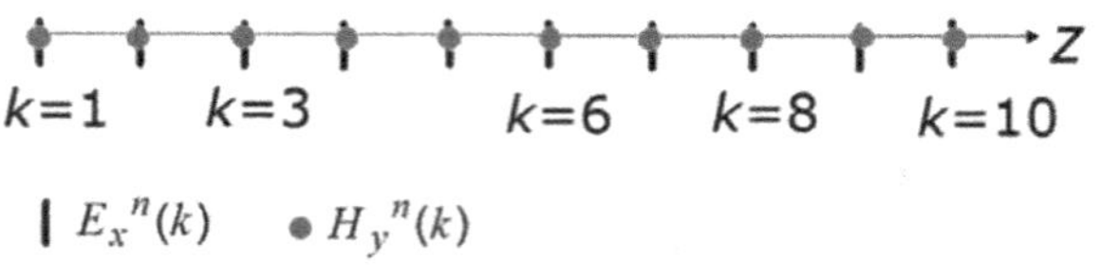

Figure 14.2. 1D FDTD and discretization in space (along z-axis).

14.1.2 FDTD and Discretization

One-dimensional (1D) FDTD representation of plane waves can be obtained by the discretization of (14.5) via finite-difference (FD) approach. Assume k and n, respectively, represent spatial and time indices and write $z = k \times \Delta z$, $t = n \times \Delta t$, where Δz and Δt are spatial and time steps, respectively. Electric and magnetic field components, in this case, are

$$E_x(z,t) = E_x(k,n) = E_x^n(k), \tag{14.6a}$$

$$H_y(z,t) = H_y(k,n) = H_y^n(k). \tag{14.6b}$$

The discretization and field components along z-direction are given in Fig. 14.2.

The discrete spatial points along z-direction between left and right boundaries are separated by Δz, while time instants are separated by Δt. The FDTD discretization of the first-order coupled equations are obtained via

$$\frac{\partial}{\partial z} E_x(z) \Rightarrow \frac{E_x^n(z+\Delta z) - E_x^n(z-\Delta z)}{2\Delta z} \tag{14.7a}$$

or, in terms of indices,

$$\frac{\partial}{\partial z} E_x(k) \Rightarrow \frac{E_x^n(k+1) - E_x^n(k-1)}{2\Delta z} \tag{14.7b}$$

in space, and

$$\frac{\partial}{\partial t} E_x \Rightarrow \frac{E_x^n(k) - E_x^{n-1}(k)}{\Delta t} \tag{14.7c}$$

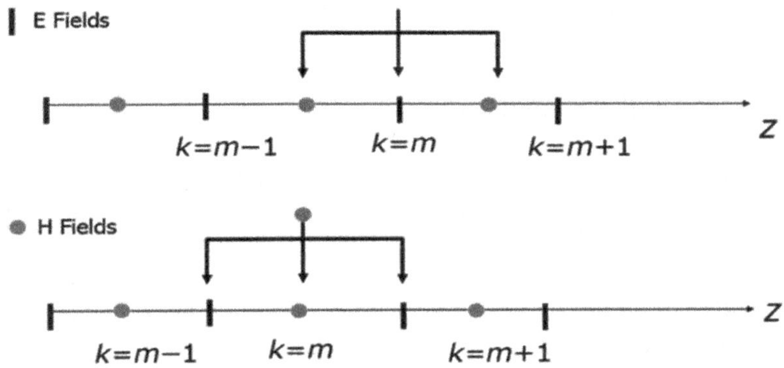

Figure 14.3. Leap-frog FDTD discretization.

in time as

$$E_x^n(k) = E_x^{n-1}(k) - \frac{\Delta t}{2\varepsilon_0 \Delta z}\left(H_y^n(k+1) - H_y^n(k-1)\right), \qquad (14.8a)$$

$$H_y^n(k) = H_y^{n-1}(k) - \frac{\Delta t}{2\mu_0 \Delta z}\left(E_x^n(k+1) - E_x^n(k-1)\right). \qquad (14.8b)$$

These iterative equations show that electric (magnetic) field at a node in the space at any time instant depends on its previous time value at the same node and two nearby (left and right nodes) magnetic (electric) fields at the same time instant. This means that electric (magnetic) fields at the same node are not used for magnetic (electric) field computations. With the help of this knowledge, magnetic field components are shifted by $\Delta z/2$ so that leap-frog scheme as shown in Fig. 14.3 can be obtained. In this case, (14.8) can be modified accordingly,

$$E_x^n(k) = E_x^{n-1}(k) - \frac{\Delta t}{\varepsilon_0 \Delta z}\left(H_y^n(k) - H_y^n(k-1)\right), \qquad (14.9a)$$

$$H_y^n(k) = H_y^{n-1}(k) - \frac{\Delta t}{\mu_0 \Delta z}\left(E_x^n(k) - E_x^n(k-1)\right) \qquad (14.9b)$$

and form the fundamental 1D iterative FDTD equations. Iterative equations belong to open-form solutions; therefore, they are always conditionally stable. One needs to parameterize the problem in a way to obey stability conditions. The physical restriction of the stability condition is that time and spatial discrete steps cannot be specified independently. Mathematically, one needs to derive a condition between Δt and Δz.

14.1.2.1 Stability Condition

The stability condition in 1D FDTD representations is nothing but the physical condition of wave scattering. For example, EM waves travel with the speed of light, so that in any Δt time step, the waves propagate a distance of $c \times \Delta t$. Since FDTD calculations are repeated at every Δt, the distance traveled during this time step should not be larger than the spatial step size Δz in order for FDTD be able to follow the wave phenomena between discrete spatial nodes. Because of this requirement, $\Delta z = c \times \Delta t$ should be satisfied. In other words, once the space is discretized, the maximum allowable time step must be $\Delta t = \Delta z/c$. In general, the stability condition of 1D FDTD is

$$\Delta t \leq \Delta z / c, \tag{14.10a}$$

which restricts arbitrary choice of time step and relates it to the spatial discretization.

Stability ensures that the total energy (energy in all grid points) in the FDTD during the time simulation is bounded (i.e., $\sum_{\substack{\text{FDTD}\\\text{cells}}} E^{n+1} \leq \sum_{\substack{\text{FDTD}\\\text{cells}}} E^{n}$). Under the condition $\Delta x = \Delta y = \Delta z = \Delta$, the stability condition in 1D in (14.10a) becomes $\Delta t \leq \Delta / c\sqrt{2}$ and $\Delta t \leq \Delta / c\sqrt{3}$, in 2D and 3D, respectively. Note that the stability condition in (14.10a) excludes the stability limit $\Delta t = \Delta/c$ because this is the weak stability limit. Due to numerical round off and/or truncation errors, it may go unstable. On the other hand, different FDTD simulation tests show that it should be as close to weak stability limit as possible (e.g., $\Delta t = 0.99\Delta/c$).

It should be noted that electric and magnetic field components are shifted in space by $\Delta z/2$; therefore, there should also be a time shift of $\Delta t/2$. Although same n is used to represent the time index, always remember that magnetic fields are calculated $\Delta t/2$ later if they both have the same time index n. The problem is then reduced to the choice of the spatial discretization rules. This condition is known as the numerical dispersion condition.

14.1.2.2 Numerical Dispersion

Equations (14.9) represent time-domain wave scattering in free space. There may be two cases for the determination of EM sources. The source may generate a continuous wave (CW) signal or a short pulse. They represent single frequency and broad band, respectively. Whichever it is, there is always a maximum frequency (minimum wavelength $\lambda_{\min}$) of interest. The rule is that spatial discretization step Δz should be much less than the minimum wavelength wave component. In general, a full cycle of this component should be sampled at least 6–10 points. In practice, the rough criteria is

$$\Delta z = \lambda_{\min} / 10. \tag{14.10b}$$

The number of samples per minimum wavelength may be increased depending on the size of the problem, memory and speed of the computer at hand, and the resolution and accuracy requirements. In practical problems, it may be any value between $\Delta z = \lambda_{\min}/10$ and $\Delta z = \lambda_{\min}/150$.

14.1.2.3 Excitation

Electromagnetic waves may be injected at any node $k = k_s$ via a time dependent source function $g(n)$ as

$$E_x^n(k_s) = E_x^n(k_s) + g(n), \tag{14.11a}$$

which is called soft injection (k_s may be any value between $k = 1$ and $k = ke$, i.e., the first and last nodes). For example, a Gaussian pulse function may be used for this purpose as

$$g(t) = \exp\left\{-\left(\frac{t - t_0}{T}\right)^2\right\}, \tag{14.11b}$$

Where t_0 and T, respectively, are time delay and pulse width. On contrary, a Gaussian shaped pulse may directly be injected at the beginning (before time simulations start) at the space,

$$g(z) = \exp\left\{-\left(\frac{z - z_s}{Z_g}\right)^2\right\}, \tag{14.12}$$

where z_s and Z_g are the location of the center of the pulse and spatial pulse width. Figure 14.4 shows Gaussian pulse injection via (14.11) in a 100 m space with $ke = 100$.

As observed in Fig. 14.4, the injection starts as a kind of impulse, then a Gaussian shaped pulse is formed and split into two waves propagating leftward (-z) and rightward (+z). As will be explained later, the space is terminated with a perfectly reflecting boundary at

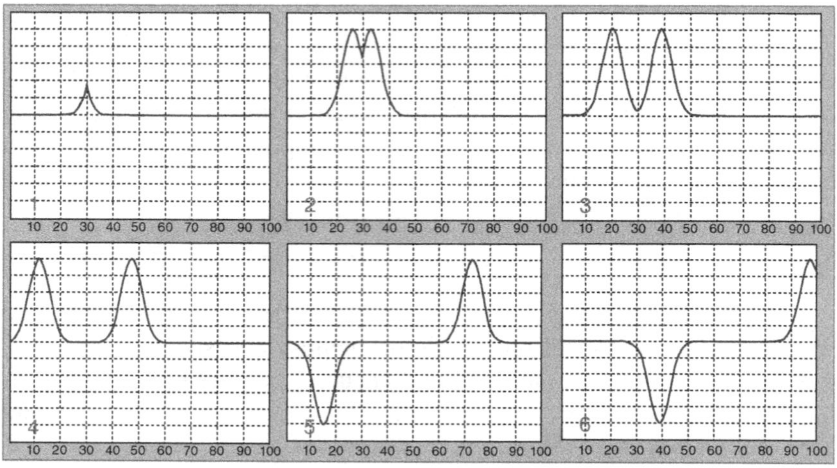

Figure 14.4. FDTD and Gaussian plane wave formation ($k_s = 35$, $T = 25\Delta t$).

left, but it opens to infinity at right. The opposite-phased reflection at left causes the left-propagating pulse turns upside down and become right-propagating pulse. The pulses disappear at the right end as time goes by.

As given in (14.12), the pulse may be directly located on the space: (with $z_s = k_s \times \Delta z$ and $Z_g = K_g \times \Delta z$),

$$E_x^0(k) = \exp\left\{-\left(\frac{k-k_s}{K_g}\right)^2\right\}, \tag{14.13a}$$

$$H_y^0(k) = E_x^0(k) / \eta_0; \eta_0 = Z_0 = \sqrt{\mu_0 / \varepsilon_0} = 377\ \Omega. \tag{14.13b}$$

Figure 14.5 belongs to an example for this kind of excitation. As observed, there is only a right propagating pulse which leaves the space as time goes by. The reason is that the pulse is located in space as both electric and magnetic field components (i.e., as a plane wave) via (14.13b).

14.1.2.4 Parameter Selection

The simulation of pulse propagation inside a finite space necessitates a good selection of pulse parameters in terms of discrete Δz and Δt. One way in parameter selection is to follow these steps:

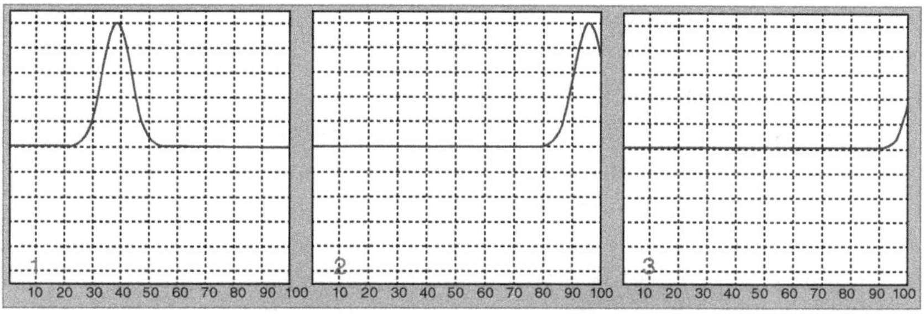

Figure 14.5. FDTD simulations and locating a Gaussian pulse in space.

- Specify Δz and ke and determine Δt by using the stability condition. Then, calculate pulse duration in time from the numerical dispersion relation. For example, if 10-sample per minimum wavelength ($\lambda_{min}/10$) is specified, then maximum frequency can be calculated from $B = f_{max} = c/(10 \times \Delta z)$. The time-domain Gaussian pulse may then be specified via two parameters $\alpha = 3.3B^2$ and $\phi = 4/\sqrt{\alpha}$ as $g(t) = \exp\{-\alpha(1 - \phi)^2\}$.
- A second method is to choose maximum frequency (or Gaussian bandwidth) first and then do the discretization accordingly. If, for this purpose, B or f_{max} is given the spatial step size may be chosen as $\Delta z = c/(10 \times B)$. The other two parameters may then be selected same.
- A third method is to choose $\alpha = 16(\phi \times \Delta t)^2$ and $\phi = 32$ (it may also be $\phi = 16$ or $\phi = 64$), and then specify the Gaussian pulse as $g(t) = \exp\{-\alpha(1 - \phi \times \Delta t)^2\}$.

14.1.2.5 Boundary Conditions

Any numerical computation must be finite in space and time. Therefore, a special treatment is required at the first ($k = 1$) and last ($k = ke$) nodes in space. As shown in Fig. 14.3, a computation at a node in space requires field values at both left and right nodes. For example, magnetic field computations at node $k = 2$ require electric field values at $k = 1$ and $k = 3$. This causes a problem at first and last nodes because left node of $k = 1$ and right node of $k = ke$ are missing. If no action is taken, then these field values are automatically assumed zero. No electric fields at left and/or right means the realization of Dirichlet-type boundary conditions (BCs). This corresponds to perfectly reflecting boundary terminations at the boundaries.

Usually, two types of boundary terminations are of interest: a finite region with Dirichlet BC and an open region that extends infinity (i.e., backreflected waves do not occur). The problem of giving a feeling of an open region is one of major numerical simulation research topics in computational EM and is difficult in 2D and 3D problems.

For example, in one-dimensional FDTD, if time and space steps are chosen to satisfy $\Delta z = c \times \Delta t$, obviously waves travel one node at one time step. In this case, it is easy to satisfy backreflection-free open region termination. For example, for the electric field at node $k = ke$ at time instant n, one only needs to use the electric field at node $k = ke - 1$ (at one node left) at time instant $n-1$, without doing any computations. Similarly, electric field node $k = 1$ at time instant n is equated to the electric field at node $k = 2$ at time instant $n-1$. In other words,

$$E_x^n(1) = E_x^{n-1}(2), \tag{14.14a}$$

$$E_x^n(ke) = E_x^{n-1}(ke-1) \tag{14.14b}$$

are used at the ends.

In general, if $\Delta t < \Delta z/c$, one can use different terms of the wave equation,

$$\left(\frac{\partial}{\partial z} - \frac{1}{c}\frac{\partial}{\partial t}\right)\left(\frac{\partial}{\partial z} + \frac{1}{c}\frac{\partial}{\partial t}\right)E_x = 0, \tag{14.15}$$

for the nodes $k = 1$ and $k = ke$. The first and second terms on the left correspond to right ($+z$) and left ($-z$) propagated waves, respectively. If right-propagated (i.e., backreflected) waves at the first node ($k = 1$) is going to be eliminated, then

$$\left(\frac{\partial}{\partial z} - \frac{1}{c}\frac{\partial}{\partial t}\right)E_x = 0 \tag{14.16a}$$

should be satisfied. Similarly, for the last node ($k = ke$),

$$\left(\frac{\partial}{\partial z} + \frac{1}{c}\frac{\partial}{\partial t}\right)E_x = 0 \tag{14.16b}$$

should be used to eliminate left-propagated waves so that free-space condition is going to be satisfied. The discrete iterative forms of these two equations may be given as (with $p = c \times \Delta t/\Delta x$)

$$E_x^n(1) = E_x^{n-1}(1)\left(\frac{1}{1-p}\right) + E_x^n(2)\left(\frac{p}{1-p}\right), \tag{14.17a}$$

$$E_x^n(ke) = E_x^{n-1}(ke)\left(\frac{1}{1+p}\right) + E_x^n(ke-1)\left(\frac{p}{1+p}\right). \tag{14.17b}$$

It should be noted that the iterative equations are repeatedly calculated in loops from the first node to the last one. A careful observation will show that while (14.17a) is suitable to this calculation, (14.17b) is not; therefore, the degree of backreflected wave absorption at the left node is not the same with the right node.

14.1.3 A One-Dimensional FDTD MATLAB Script

Following the procedures and steps explained earlier, a few line computer code can easily be prepared:

- supply input parameters and calculate others via stability and numerical dispersion relations,
- calculate electric fields at inner nodes,
- satisfy termination conditions,
- calculate magnetic fields, and
- repeat the same procedure for further time steps.

A sample code written in MATLAB is listed in Table 14.1. This code may directly be run to obtain pulsed plane wave propagation in Fig. 14.4 and Fig. 14.5.

14.1.4 MATLAB-Based FDTD1D Package

Based on the simple MATLAB code given in Table 14.1, a very effective user-friendly virtual tool is designed for undergraduate EM lectures [2]. The field components, 1D FDTD space, and termination options are presented in Fig. 14.6. The propagation medium may be free space, partially or fully dielectric filled, finite, semi-infinite, and infinite.

Table 14.1.
A Simple 1D FDTD MATLAB Code

```
% -----------------------------------------------------
% Program: LS_FDTD1D.m
% -----------------------------------------------------
clc; clear all;
Exzoldr=0;  % 1-time step previous value at RB(k=ke-1)
Exzoldl=0;  % 1-time step previous value at LB(k=ke-1)
% -----------------------------------------------------
nt=500; zmax=100; ke=100; kc=30; ep0=1/(36*pi*1e9);
mu0=4*pi/1e7; c=3e8;
lbc=input('Left Boundary: (1) PEC, (0) Free-space = ');
rbc=input('Right Boundary: (1) PEC, (0) Free-space = ');
tt=ke/2; t0=10; delz=floor(zmax/ke); dt=delz/c;
band=c/(10*delz); alfa=3.3*band*band; shift=4./
sqrt(alfa);
for k=1:ke    %  put zeros into arrays
  Ex(k)=0.0; Hy(k)=0.0; z(k)=k;
end
t=0;
for n=1:nt    % FDTD time loop starts
    t=n*dt;
    for k=2:ke-1      % Ex is calculated alon z
      Ex(k)=Ex(k)+dt/(ep0*delz)*(Hy(k-1)-Hy(k));
    end
    Ex(kc)=Ex(kc)+exp(-alfa*(t-shift)^2);
    if lbc==0; Ex(1)=Exoldl;  end
    if rbc==0; Ex(ke)=Exoldr; end
    for k=1:ke-1    % Hy is calculated along z
      Hy(k)=Hy(k)+dt/(mu0*delz)*(Ex(k)-Ex(k+1));
    end
    Exzoldr=Ex(ke-1); Exoldrt=Ex(ke); Exoldl=Ex(2);
    plot(z, Ex); xlim([1,ke]); ylim([-1.1,1.1]); grid;
    xlabel('X-axis [m]'); ylabel('Ex [V]'); pause(0.0001)
end
% ---------------------- END ----------------------
```

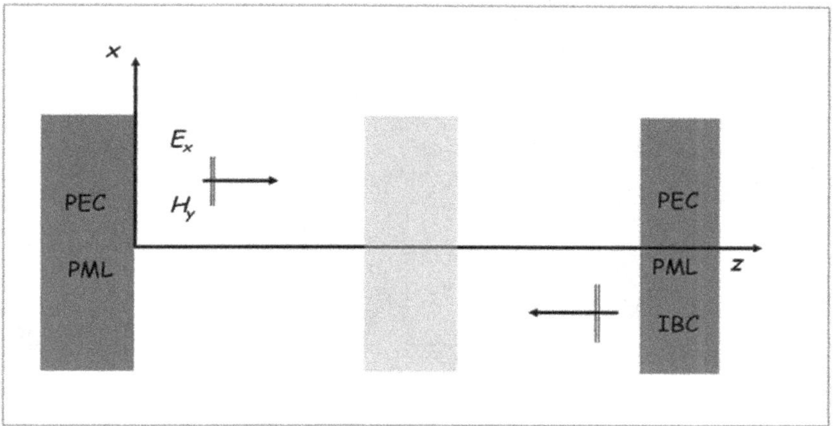

Figure 14.6. FDTD space, $\pm z$ propagating waves, and termination types.

Perfectly reflecting (perfectly electrical conductor [PEC]) and open boundary terminations (such as perfectly matched layer [PML]) may be chosen at left. Impedance-type BC is also implemented in addition to these two at right termination.

The front panel of FDTD1D is shown in Fig. 14.7. It is horizontally divided into two. Parameters grouped and located into four blocks are supplied at top; electric field versus range is displayed at bottom. The medium parameters (relative permittivity, permeability, and conductivity) are supplied in the first blocks. The user also specifies the termination conditions at left and right within this block.

The second-left parameter block is reserved for the slab that can be located vertically in the region. The user either supply data from related data boxes there or locate, shrink/extend the slab manually by using the mouse.

The third parameter block is used for the source specification. The user may specify the location of the source either by supplying its distance from the left end in terms of number of cells or may do it directly by the mouse. The source may be a Gaussian pulse injection at anywhere in the region, a plane wave entering from left or right, a CW, or rectangular pulse injections.

The last parameter block is for the operational parameters. The number of time steps (i.e., the simulation duration) is supplied as an integer from a data box which becomes a timer and counts back to zero

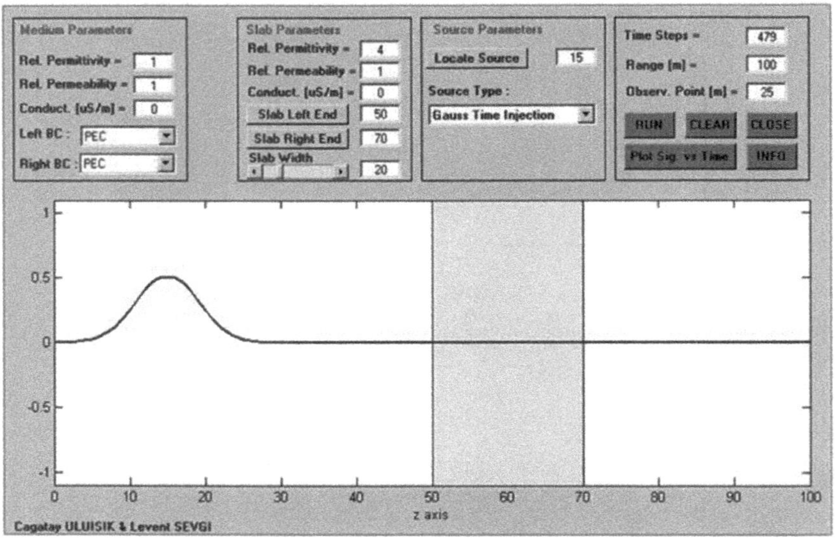

Figure 14.7. Front panel of FDTD 1D.

once the simulation is started. The command buttons are used to initiate or terminate the simulation. The "Plot Sg.vs.Time" button is used for offline separate graphs of signal versus time and its spectrum (i.e., signal vs. frequency) at the chosen observation point.

14.1.4.1 Sample Scenarios

The first scenario belongs to a finite lossless region where plane waves travel with the speed of light. Figure 14.8, Figure 14.9, Figure 14.10, and Figure 14.11 belong to this scenario at different time instants. Here, a Gaussian source is injected at node 30 inside 100 m region for a total of $500\Delta t$ simulation duration. The plots in Fig. 14.8, Fig. 14.9, Fig. 14.10, and Fig. 14.11 correspond to $t = 27\Delta t$, $t = 65\Delta t$, $t = 117\Delta t$, and $t = 167\Delta t$, respectively. The injected pulse, the two left- and right-propagated detached pulses, and reflections from left and right ends are clearly observed in these figures.

Since the number of nodes is 100, it requires $100\Delta t$ for a pulse to travel from one end to the other. This means there will be more than five travels between the ends. During these five travels, left and right propagating waves also interfere five times. The arrows show the direction of the traveling individual pulses.

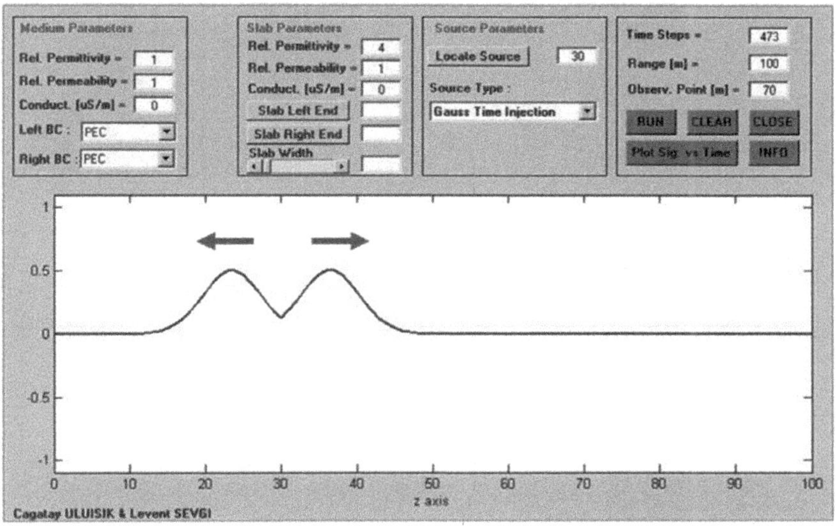

Figure 14.8. Finite region and plane wave propagation (at 27th time step).

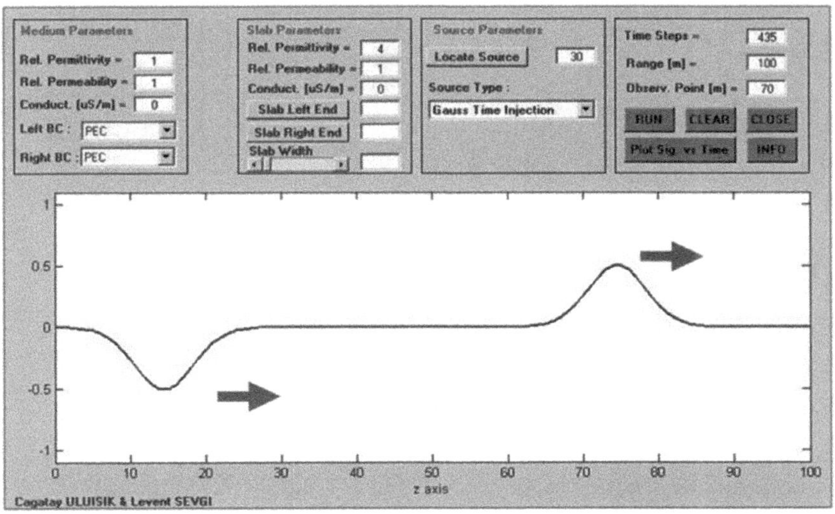

Figure 14.9. Finite region and plane wave propagation (at 65th time step).

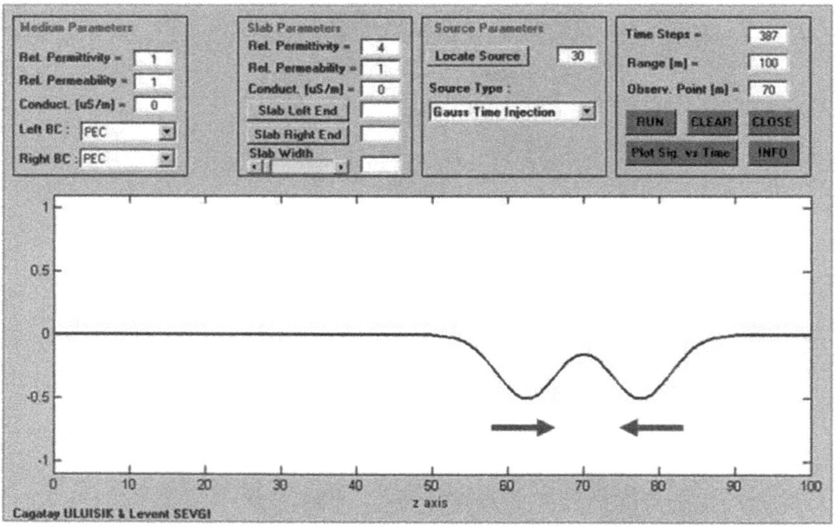

Figure 14.10. Finite region and plane wave propagation (at 117th time step).

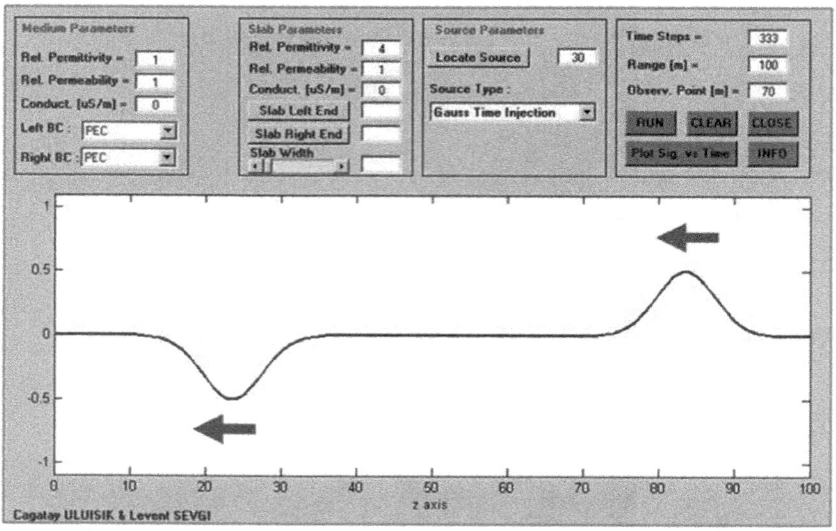

Figure 14.11. Finite region and plane wave propagation (at 167th time step).

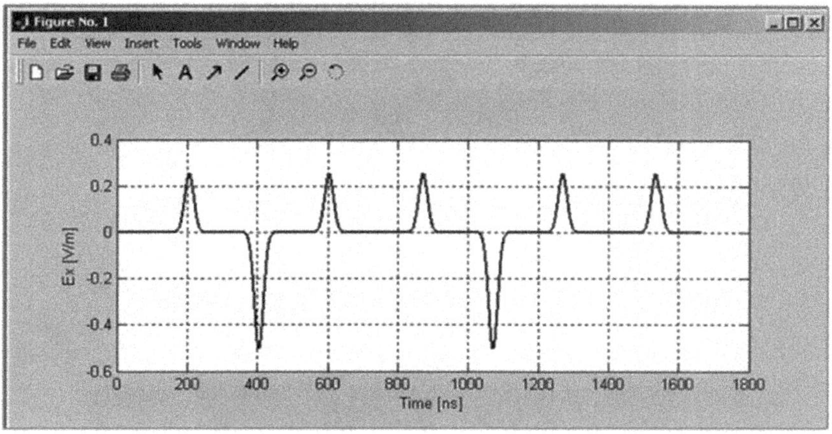

Figure 14.12. Signal versus time at node 70 for the source injection at node 30.

The signal versus time of this scenario at the chosen observation point after the simulation ends is shown in Fig. 14.12. The region is 100-m long so the time step is nearly 3.3 ns. The first pulse belongs to the pulse that directly travels between the source and observer. The second one is the accumulation of oppositely propagating left-reflected and right-reflected pulses. Since they are both once reflected, they both have equal but negative signs and add up around 400 ns. The reader may easily identify each pulse in this figure.

The virtual tool records time history of the plane waves at the observation point in a file as a two-column data. Taking frequency transform of this time data yields resonance characteristics of the region. The closed region in this scenario forms a kind of 1D resonator and its resonance frequencies can be calculated analytically as $f_{rm} = c/2D$ (Hz), where D is the length of the finite region. $D = 100$ m in this example, therefore, the first (dominant) resonance frequency is 1.5 MHz, and the rest are integer multiples of this frequency (i.e., 3.0, 4.5, 6.0 MHz, etc.). Figure 14.13 shows the discrete Fourier transform (DFT) of the data presented in Fig. 14.12.

As observed in Fig. 14.13, some of the resonances can be distinguished but some cannot. For example, although not as peak as the second resonance at 3.0 MHz, the first one at 1.5 MHz is also observable. The third resonance at 4.5 MHz is obviously missing. One should be well equipped with both wave representations in closed regions in

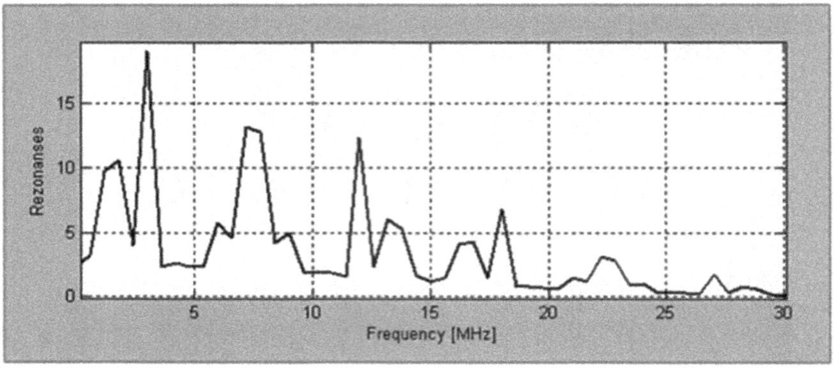

Figure 14.13. Resonance frequencies of the 1D 100-m-long resonator.

terms of eigenfunctions (also called modes of the structure) and eigenvalues and Fourier transform theory in a discrete environment in terms of fast Fourier transform (FFT) and DFT. Without understanding source-free wave representations in terms of wave equations (eigenvalue problem) and its close relations with the source-driven (TL) problem, and also the Green's function representation modeled via the Sturm–Liouville equations, it is extremely difficult to interpret the simulation results.

The resonances occur in Fig. 14.13 belong to the modes of the structure

$$E_{xm}(z) = \sqrt{\frac{2}{D}} \sin\left(\frac{m\pi}{D} z\right), \tag{14.18}$$

which are orthonormal wave objects, and can be used to represent any type of source function (including delta-Dirac-type excitations). But, in order to observe the resonance of a mode in the frequency spectrum, (a) it should be excited and (b) it should be observed. This means both the source and the observation points must not be located at null points (i.e., zero crossing) of that mode. For example, (14.18) tells us that the null points of the dominant mode ($m = 1$) are at both left and right terminations, and its amplitude is maximum at midrange (i.e., $z = D/2$). On the other hand, the second mode ($m = 2$) has null points at both ends as well as at min range. This is same for all other higher even-order modes (i.e., $m = 4$, $m = 6$, etc.) This means if either of the source or observation points is located at midpoint (i.e., at $z = 50$ m), then these modes either are not excited or not observed. After this information, the explanation of the reason why 4.5 MHz resonance is not observed is left to the reader.

Selecting the locations of the source and observer so as to excite and observe the modes of interest is important as explained earlier. Another important issue in observing the resonances is the discrete time–frequency parameter requirements. Since these are well explained in Chapter 5, they are not included here. But, it should be mentioned that maximum frequency of interest and frequency resolution requirements in terms of time steps and signal observation period should be well understood by the reader, in addition to periodicity problems which appear as spectrum aliasing and leakage because of discretization in both time and frequency domains. Another version of Fig. 14.13 is given in Fig. 14.14 after doubling the number of time steps from $500\Delta t$ to $1000\Delta t$. Now, try to explain why all resonances appear as clean peaks in this case.

The second scenario is pictured in Fig. 14.15. Here a 20-m long lossless dielectric layer with $\varepsilon_r = 4$ and $\mu_r = 1$ is located between 50th and 70th nodes. Again, left and right ends use perfectly reflecting-type terminations. The source is at 15th node. The picture belongs to pulse traveling along the region after $119\Delta t$. The time variation of the signal 10 m away from the source is plotted in Fig. 14.16. The user should be

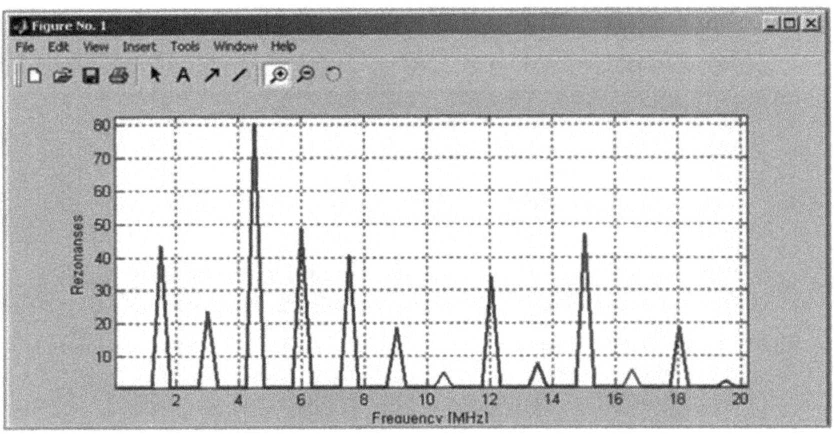

Figure 14.14. Resonance frequencies of the 1D 100-m-long resonator after $1000\Delta t$.

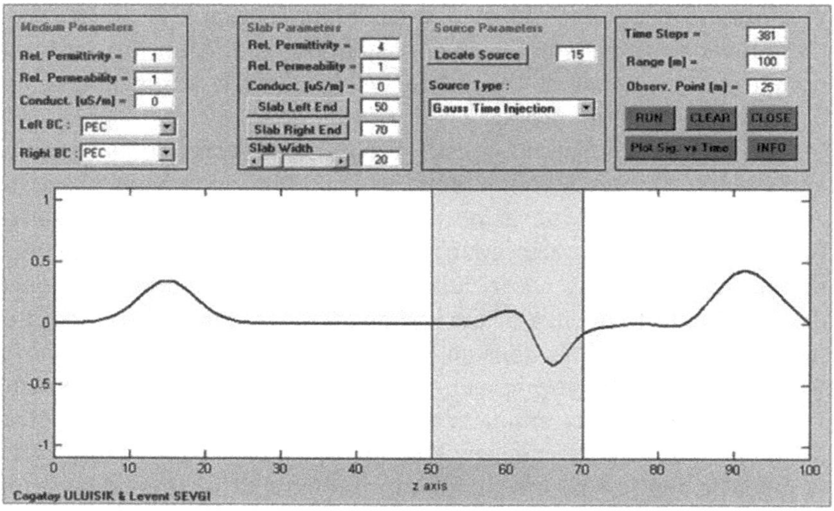

Figure 14.15. Second scenario and traveling pulses at 119th time step.

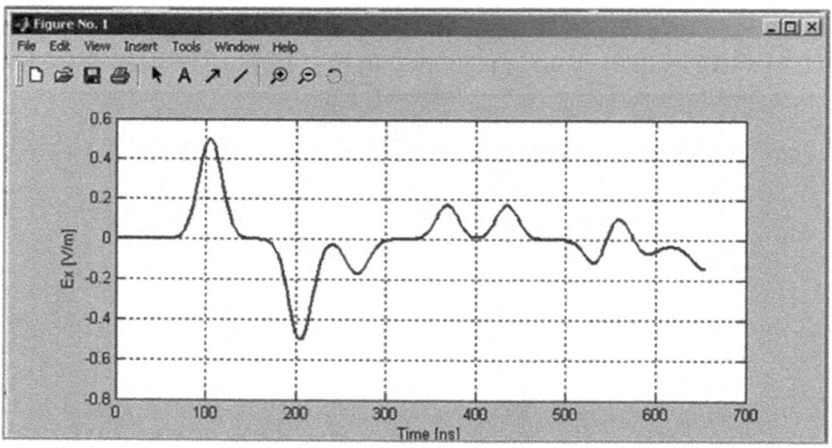

Figure 14.16. Signal versus time at the 25th node plotted after the simulation ends.

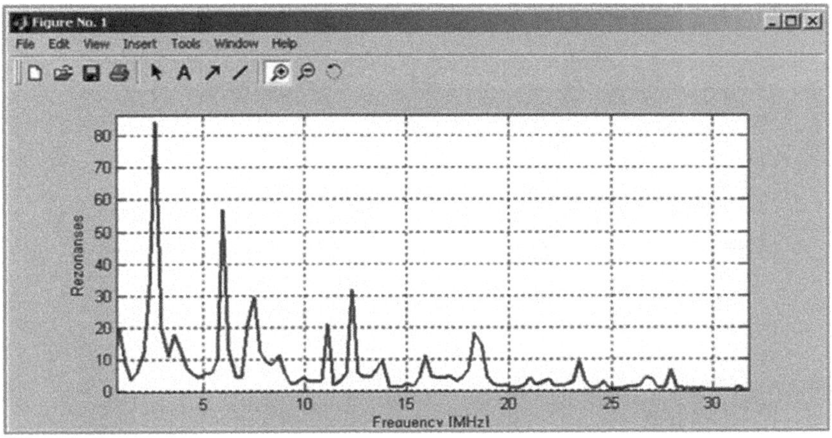

Figure 14.17. Frequency spectrum of the second scenario after $1000\Delta t$.

able to identify each pulse by locating the source and observer and taking the distances between them, the width of the slab, and distances to the boundaries. For example, in this scenario, reflections occur at four points: total reflections at left and right boundaries and partial reflections (and transmissions) at both edges of the dielectric slab. Finally, the frequency spectrum of this time record is given in Fig. 14.17. The reader should be able to explain continuous spectrum around the resonances caused by the second layer located in the region.

The third scenario belongs to a rectangular pulse excitation. Figure 14.18 shows time variation at a chosen observation point and its resonances.

The last two examples, given in Fig. 14.19 and Fig. 14.20, belong to CW time injection. In this example, the program excites first and second modes of a 100-m long finite region to illustrate time-domain behavior of a resonance (any other mode can be located by changing lines 940–948 of the FTD1D.m file). The reader may observe that the resonating wave does not propagate; it only resonates up and down.

FDTD1D may be used to make movie files of the simulations. The reader may activate lines 999–1003 to automatically capture frames during the simulation and then use Windows MovieMaker utility to make the movie file.

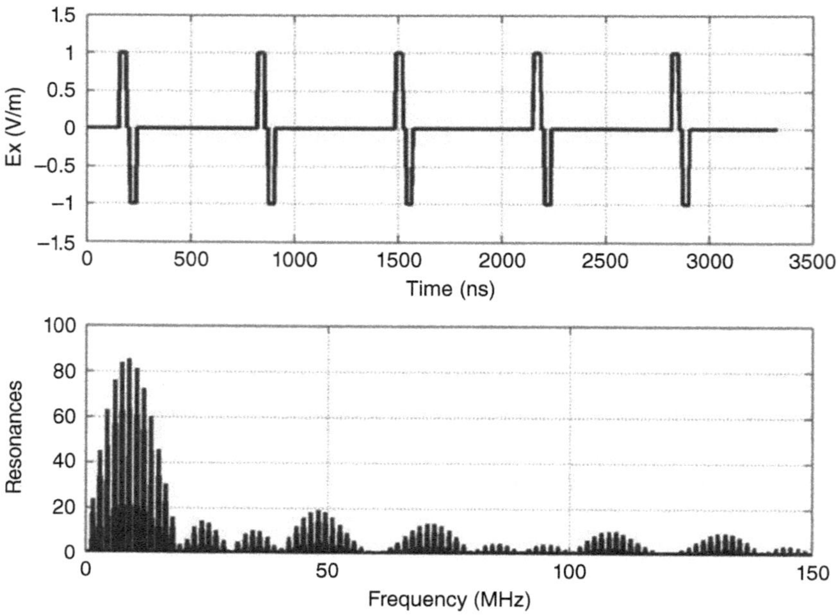

Figure 14.18. (Top) Rectangular pulse observed at 95 m of a 100-m-long region, source being at 30 m, (bottom) the frequency spectrum and resonances.

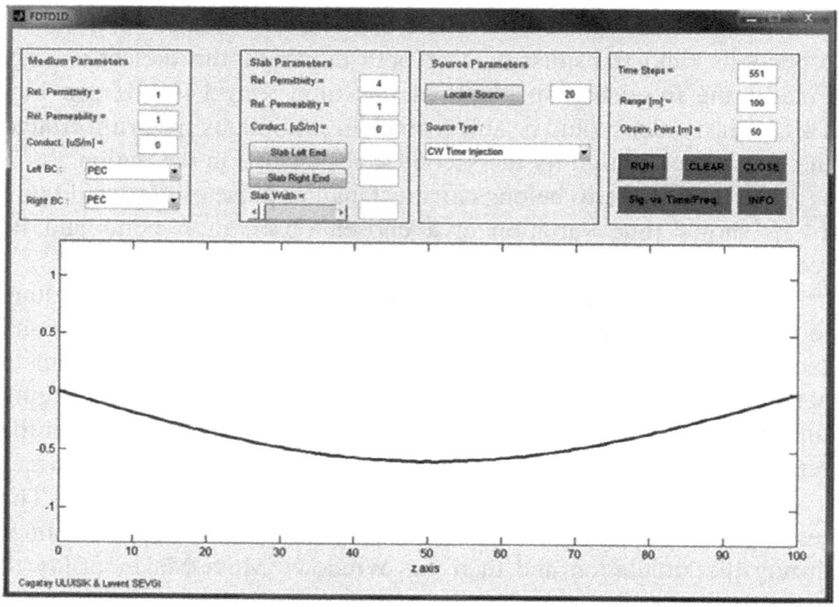

Figure 14.19. First mode (resonance) of the structure.

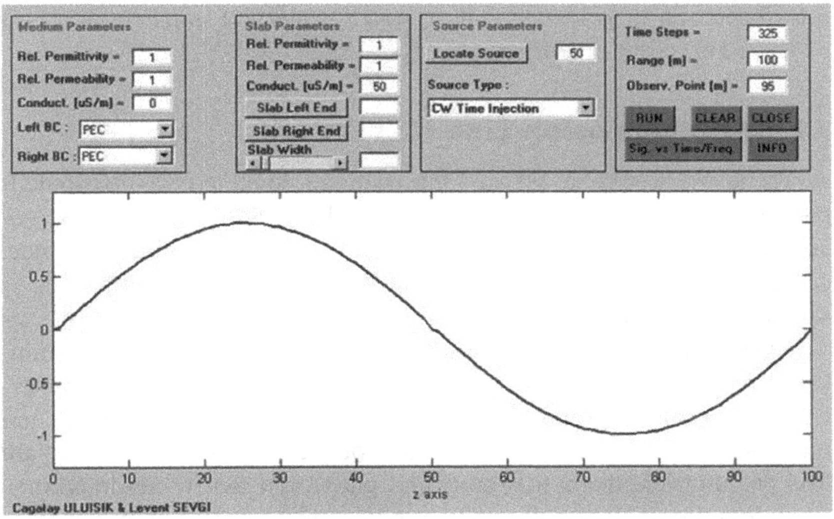

Figure 14.20. Second mode (resonance) of the structure.

14.2 TRANSMISSION LINES AND TIME-DOMAIN REFLECTOMETER

A TL is a special conductor of some length used to connect an electric source to a load [3]. It may be a pair of wires, coaxial cables, waveguides of arbitrary cross sections, microstrip lines, planar or coplanar circuits, and so on. Although there are differences among them and some may require different approaches, a TL made from two conductors (a pair of parallel wires) is accepted as the fundamental structure and the theory is established according to this assumption. The theory, basic concepts, and characteristic equations explained in the next section may be obtained either from EM wave theory or independently from the circuit theory (Chapter 6, Section 6.12 briefly reviews the analogy between wave and circuit models of the basic TL). The other three sections discuss the time-domain reflectometer concept, which have long and widely been used in fault location and identification in TLs. A very effective MATLAB package TDRMeter [4] has been prepared and discussed in details in these sections. Although summarized in these sections when necessary, in order to follow these sections easily, the reader is advised to review basic Fourier and Laplace transformation theories as well as steady-state and transient analyses of wave and circuit theories.

14.2.1 Transmission Line (TL) Theory

A TL is more than a set of long, parallel line; it is a distributed-parameter physical system. They are used to transmit electric energy and communication signals from one point to the other, that is, connect sources to loads. A typical two-wire TL located along z-direction is sketched in Fig. 14.21.

A TL has two sets of constitutive parameters. The primary parameters are listed in Table 14.2. The secondary parameters, the characteristic impedance-Z_0 and the complex propagation constant-γ, are mentioned in Fig. 14.21. The real parts of the propagation constant are loss per unit length (α [dB/km]) and phase constant (β [radian/km]). For a loss-free TL, $\alpha = 0$ and only phase shift occurs along the line.

There is a direct analogy between a uniform TL and the uniform plane wave, which will be summarized in the next section. The field distributions for the uniform plane wave and for the uniform TL are

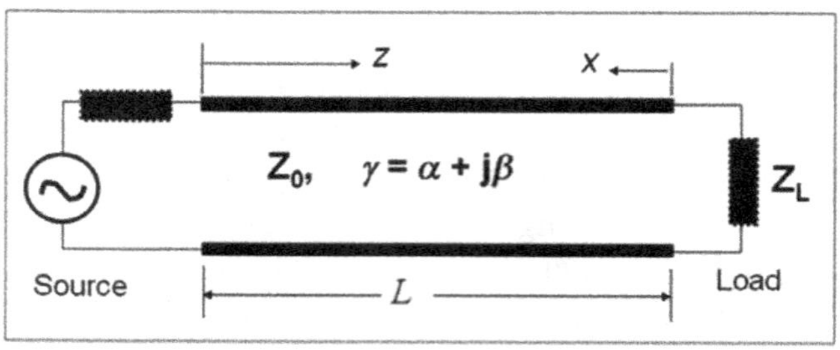

Figure 14.21. A sketch of a two-wire TL.

Table 14.2.
Primary Parameters of a TL

Parameter	Symbol	Unit
Unit-length resistance	R	Ω/m
Unit-length conductivity	G	S/m
Unit-length inductance	L	H/m
Unit-length capacitance	C	F/m

both known as transverse EM (TEM) waves, where both electric and magnetic fields are perpendicular to the direction of propagation. That is, they both lie in the transverse plane. Because of this analogy, TEM wave approach may be used in analyzing TLs, which is the field approach. Another approach is the circuit (or network) approach where voltages and currents in terms of two independent variables, time and distance, can be used to establish sets of differential equations.

It is customary to assume the TL along z-axis so that time-domain TL equations can be given as

$$-\frac{\partial}{\partial z}v(z,t)=Ri(z,t)+L\frac{\partial}{\partial t}i(z,t), \tag{14.19a}$$

$$-\frac{\partial}{\partial z}i(z,t)=Gv(z,t)+C\frac{\partial}{\partial t}v(z,t). \tag{14.19b}$$

The minus (−) sign on the left represents voltage and current drop toward the load. These are the coupled TL equations. The steady-state sinusoidal forms of these equations are

$$\frac{d}{dz}V(z)=-(R+j\omega L)I(z), \tag{14.20a}$$

$$\frac{d}{dz}I(z)=-(G+j\omega C)V(z), \tag{14.20b}$$

where $\omega = 2\pi f$ is the angular frequency in radians/s. By using these first-order coupled equations, second-order decoupled TL equations can be obtained as

$$\frac{d^2}{dz^2}V(z)=\gamma^2V(z), \tag{14.21a}$$

$$\frac{d^2}{dz^2}I(z)=\gamma^2I(z), \tag{14.21b}$$

with

$$\gamma=\sqrt{(R+j\omega L)(G+j\omega C)}=\alpha+j\beta \tag{14.21c}$$

being the complex propagation constant. For a lossless TL (i.e., $R = G = 0$), complex propagation constant reduces to $\gamma = j\beta$. In this case, the solution can be written as

$$V(z) = V_1 e^{-j\beta z} + V_2 e^{j\beta z}, \tag{14.22a}$$

$$I(z) = \frac{1}{Z_0}\left(V_1 e^{-j\beta z} - V_2 e^{j\beta z}\right), \tag{14.22b}$$

where Z_0 is the characteristic impedance given as

$$Z_0 = \sqrt{\frac{R + j\omega L}{G + j\omega C}} \rightarrow \sqrt{\frac{L}{C}} \quad (\text{when } R = G = 0), \tag{14.23}$$

and V_1 and V_2 are complex constants that are going to be determined from the BCs at the source end ($z = 0$) and load end ($z = L$). It is practical to measure distances from the load; therefore, Equations (14.22a) and (14.22b) reduce to

$$V(x) = V^+ e^{j\beta x} + V^- e^{-j\beta x}, \tag{14.24a}$$

$$I(x) = \frac{1}{Z_0}\left(V^+ e^{j\beta x} - V^- e^{-j\beta x}\right) \tag{14.24b}$$

by replacing $z = L - x$ and using $V^+ = V_1\exp(-j\beta L)$ and $V^- = V_2\exp(-j\beta L)$. Equations (14.24a) and (14.24b) express voltage and current at any junction in terms of traveling voltage and current waves toward the load (i.e., $V^+\exp(j\beta x)$) and toward the source (i.e., $V^- e^{-j\beta x}$). These two terms are also called as incident and reflected voltages, respectively.

Two important parameters in TL theory are the voltage reflection coefficient (Γ) and the standing wave ratio (SWR or s). The voltage reflection coefficient at the load is defined as

$$\Gamma(x = 0) = \Gamma_0 = \frac{V^-}{V^+} = \frac{Z_L - Z_0}{Z_L + Z_0} \tag{14.25}$$

and can be found at any point along the TL from

$$\Gamma(x) = \Gamma_0 e^{-j2\beta x}. \tag{14.26}$$

SWR is defined as a ratio of maximum voltage to the minimum voltage along the line as

$$\text{SWR} = s = \frac{V_{\max}}{V_{\min}} = \frac{1 + |\Gamma_0|}{1 - |\Gamma_0|} \quad \Rightarrow \quad |\Gamma_0| = \frac{s - 1}{s + 1}. \tag{14.27}$$

Finally, impedance of a lossless TL at any distance x, with characteristic impedance-Z_0 and propagation constant-β, terminated by a load impedance of Z_L is given as

$$Z_{in} = Z_0 \frac{Z_L + jZ_0 \tan(\beta x)}{Z_0 + jZ_L \tan(\beta x)}. \tag{14.28}$$

Understanding the physical meanings of these parameters is very important in dealing with numerical simulation techniques; therefore, they are to be shown later.

14.2.1.1 Characteristic Impedance

Characteristic impedance is an important parameter in TL analysis. As shown in (14.23), it is a parameter related to the primary line constants and is real for an ideal (loss-free) TL. It is defined as *the ratio of voltage divided by current at any point along an infinite-length homogeneous TL* or *the ratio of voltage divided by current at any point along a reflection-free homogeneous TL.*

As understood from (14.28), the impedance seen at a distance x from the end of a TL terminated with Z_L depends on the characteristic impedance, load impedance, and the distance. It may take any complex value even when characteristic and load impedances are real. It is important to note that the impedance of a line at any point reduces to the characteristic impedance when the load impedance is real and equal to the characteristic impedance.

14.2.1.2 Voltage Reflection Coefficient

A discontinuity occurs when an impedance Z (not equal to Z_0) is inserted at any point along a TL. A discontinuity means an occurrence of an impedance value different from characteristic impedance of the line. Voltage reflection coefficient, given in (14.25) and (14.26), states that voltage reflection occurs at these discontinuities and is directly determined by the load and the TL characteristic impedances. A voltage at any point is the sum of incident and reflected voltages. When incident and reflected voltages are in (opposite) phase, a voltage maximum (minimum) occurs. There is no reflected voltage when the TL is terminated by a load impedance equal to the TL's characteristic impedance.

Reflection coefficient is a complex number with a modulus in the range of [0,1]. The extreme cases are as follows.

- When $Z_L = Z_0$, reflection coefficient is zero ($\Gamma = 0$). The power along the line totally transferred to the load. This is called *matched transmission line*.
- When $Z_L = 0$ (short circuited [SC]), reflection coefficient is minus one ($\Gamma = -1$). In this case, the power along the line totally reflects back (with a phase shift of 180°).
- When $Z_L \to \infty$ (open circuited [OC]), reflection coefficient is one ($\Gamma = 1$). In this case, the power along the line totally reflects back (without phase shift).

14.2.1.3 Standing Wave Ratio

Having a discontinuity along a line means voltage waves traveling in both directions. These voltage waves interfere and voltage maxima and minima occur. These waves are called standing waves having a SWR calculated from (14.27). SWR is a real number and is between $[1, \infty)$:

- SWR = 1 means there is no impedance discontinuity along the TL (matched line)
- SWR $\to \infty$ shows the TL is OC or SC. Voltage fluctuation along the TL is highest and full reflection occurs.

The input impedance of a SC or OC TL is pure imaginary, either inductive or capacitive, and depends on the length of the line. This is important since any reactance and/or capacitance can be realized from an OC or SC TL by just adjusting the TL length. The voltage, current, and impedance along a TL are periodic with a period of half wavelength. Any reactance and/or capacitance obtained via a SC line with length l can also be obtained via OC line with length $l + \lambda/4$. These properties of SC and OC terminated TL are used in impedance matching, filter, coupler, and power divider designs.

14.2.2 Plane Wave–Transmission Line Analogy

There is a close analogy between the propagation of plane waves in a homogeneous medium and the propagation of voltage and current along a uniform TL. This analogy can be used not only as an aid to understand these concepts but also to obtain the solutions to realistic problems. A background in EM wave theory, for example, certainly helps to write down the solutions to EM wave problems or vice versa.

For a uniform TL having primary parameters R, L, C, and G per unit length, the coupled TL equations of the voltage and current in differential forms given in (14.19) can be rearranged as

$$\frac{\partial v(z,t)}{\partial z} + L\frac{\partial i(z,t)}{\partial t} + Ri(z,t) = 0, \quad (14.29a)$$

$$\frac{\partial i(z,t)}{\partial z} + C\frac{\partial v(z,t)}{\partial t} + Gv(z,t) = 0, \quad (14.29b)$$

while Maxwell equations for a uniform plane wave propagating along the $+z$ direction inside a homogeneous medium and having only E_x and H_y components can be written as

$$\frac{\partial E_x(z,t)}{\partial z} + \mu\frac{\partial H_y(z,t)}{\partial t} = 0, \quad (14.30a)$$

$$\frac{\partial H_y(z,t)}{\partial z} + \varepsilon\frac{\partial E_x(z,t)}{\partial t} + \sigma E_x(z,t) = 0. \quad (14.30b)$$

Inspecting (14.29) and (14.30) shows that the quantities in Table 14.3 are analogous.

If the voltages and currents, and electric and magnetic fields are sinusoidal in time, (14.29) and (14.30) become

$$\frac{dV(z)}{dz} + (R + j\omega L)I(z) = 0, \quad \frac{dE_x(z)}{dz} + (0 + j\omega\mu)H_y(z) = 0, \quad (14.31a)$$

Table 14.3.
Analogous Parameters

Circuit Model	Plane Wave Model
V (V)	E (V/m)
I (A)	H (A/m)
C (F/m)	ε (F)
L (H/m)	μ (H)
G (S/m)	σ (S/m)
R (Ω/m)	$1/\sigma$

$$\frac{dI(z)}{dz}+(G+j\omega C)V(z)=0, \quad \frac{dH_y(z)}{dz}+(\sigma+j\omega\varepsilon)E_x(z)=0 \tag{14.31b}$$

in frequency domain, and the second-order uncoupled equations become

$$\frac{d^2V(z)}{dz^2}-\gamma^2V(z)=0, \quad \frac{d^2E_x(z)}{dz^2}-\gamma^2E_x(z)=0, \tag{14.32a}$$

$$\frac{d^2I(z)}{dz^2}-\gamma^2I(z)=0, \quad \frac{d^2H_y(z)}{dz^2}-\gamma^2H_y(z)=0, \tag{14.32b}$$

where

$$\gamma^2=(R+j\omega L)(G+j\omega C) \quad \gamma^2=(j\omega\mu)(\sigma+j\omega\varepsilon) \tag{14.33}$$

in the TL and plane wave representations, respectively. The solution of (14.32) is

$$V(z)=A_1\cosh(\gamma z)+B_1\sinh(\gamma z), \quad E_x(z)=A_1\cosh(\gamma z)+B_1\sinh(\gamma z), \tag{14.34a}$$

$$I(z)=A_2\cosh(\gamma z)+B_2\sinh(\gamma z), \quad H_y(z)=A_2\cosh(\gamma z)+B_2\sinh(\gamma z). \tag{14.34b}$$

The characteristic impedance of the TL (Z_0) and characteristic or intrinsic impedance of the propagation medium (η_0) are

$$Z_0=\sqrt{\frac{R+j\omega L}{G+j\omega C}}, \quad \eta_0=\sqrt{\frac{j\omega\mu}{\sigma+j\omega\varepsilon}}. \tag{14.35}$$

For the special case of a *lossless* line ($R = G = 0$) or lossless (*nonconducting*) ($\sigma = 0$) medium,

$$Z_0=\sqrt{\frac{L}{C}}\text{ for TL}, \quad \eta_0=\sqrt{\frac{\mu}{\varepsilon}}\text{ for plane waves}, \tag{14.36a}$$

$$\gamma=\alpha+j\beta\Rightarrow(\alpha=0)\Rightarrow\beta=\omega\sqrt{LC} \quad \text{or} \quad \beta=\omega\sqrt{\varepsilon\mu}. \tag{14.36b}$$

Finally, wave propagation along a lossless and in a lossless medium will be

$$v=\frac{1}{\sqrt{LC}}=3\times10^8\text{ m/s} \quad \text{or} \quad v=\frac{1}{\sqrt{\varepsilon_0\mu_0}}=3\times10^8\text{ m/s}. \tag{14.37}$$

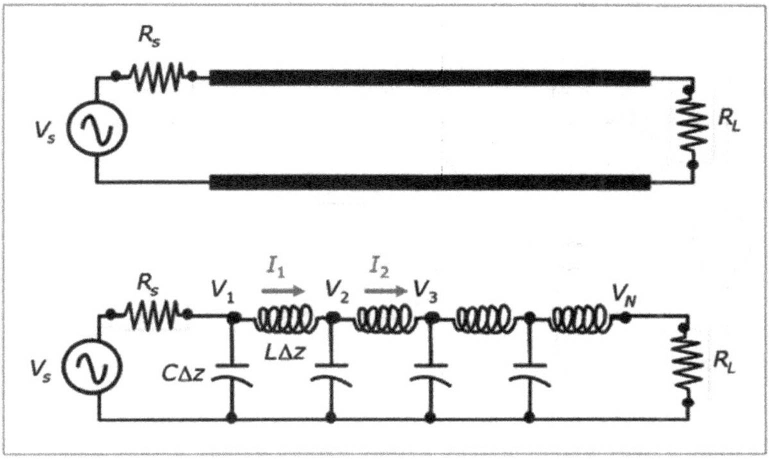

Figure 14.22. The two-wire TL and its discretized models.

14.2.3 FDTD Representation of TL Equations

The initial TL equations given in (14.19) can be discretized directly according to the classical finite-difference approach step by step starting from the lossless TLs.

14.2.3.1 Lossless TL

The lossless ($R = 0$, $G = 0$) TL and its discrete model are shown in Fig. 14.22. The finite-length TL is divided into N equal pieces which are cascaded consecutively.

Although the discrete model in Fig. 14.22 is suitable for the FDTD discretization, it is slightly modified as given in Fig. 14.23 in order to use full symmetry in each unit cells.

One to one analogy between EM plane waves in free space and lossless ($R = 0$, $G = 0$) TL may easily be observed. This means electric and magnetic fields, as well as permittivity and permeability of the medium can directly be replaced with their dual parameters listed in Table 14.3 in order to obtain the FDTD iterative equations for the lossless TL,

$$v^n(k) = V^{n-1}(k) - \frac{\Delta t}{C\Delta z}\left(i^n(k) - i^n(k-1)\right), \qquad (14.38a)$$

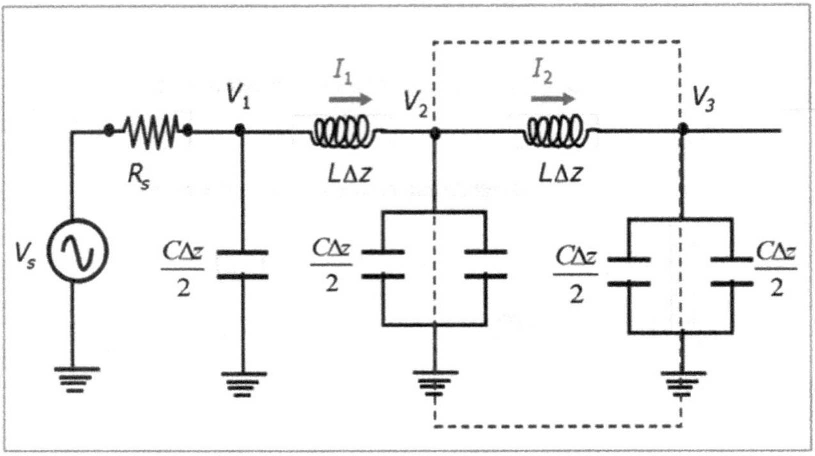

Figure 14.23. A full-symmetric-discrete FDTD model of the two-wire TL.

$$i^n(k) = i^{n-1}(k) - \frac{\Delta t}{L\Delta z}\left(v^n(k) - v^n(k-1)\right). \tag{14.38b}$$

Here, the integers k and n represent the node number and the time step, respectively, so that physical space and time values are specified via $v(z, t + \Delta t/2) = v^n(k)$ and $i(z, t) = v^{n+1/2}(k)$ by using $x = k \times \Delta z$ and $t = n \times \Delta t$, where Δz and Δt are the discretization lengths in space and time, respectively. It should be kept in mind that the stability and numerical dispersion conditions

$$\Delta t \le \Delta z / c, c = \frac{1}{\sqrt{LC}}, \Delta z \le \lambda_{\min} / 10 \tag{14.39}$$

are also valid here. Since $Z_0 = \sqrt{L/C}$ for the lossless TL, (14.38) can be modified as

$$v^n(k) = V^{n-1}(k) - Z_0\left(i^n(k) - i^n(k-1)\right), \tag{14.40a}$$

$$i^n(k) = i^{n-1}(k) - \frac{1}{Z_0}\left(v^n(k) - v^n(k-1)\right). \tag{14.40b}$$

The same leap-frog scheme is used for this case, too (see Fig. 14.24); therefore, voltage and current nodes are shifted $\Delta z/2$ in space, which means there is a $\Delta t/2$ time delay between voltages and currents although the node number n is same.

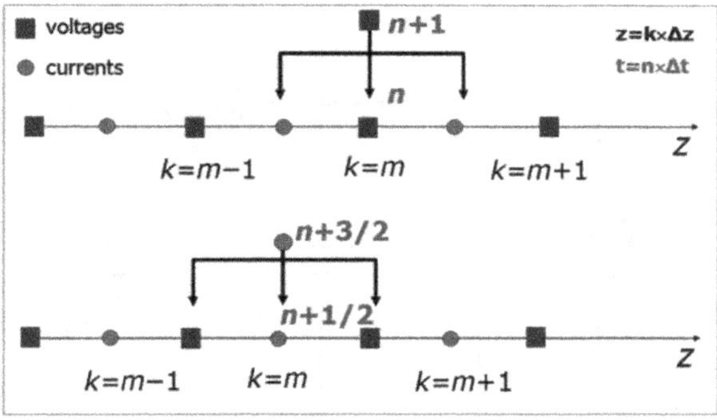

Figure 14.24. Leap-frog FDTD scheme of the two-wire TL model.

14.2.3.2 FDTD Representations for the Lossy TL

The finite-difference discretization in this case can directly be applied on to (14.19) or (14.29) and

$$v^n(k)=\left[\frac{\frac{C}{\Delta t}-\frac{G}{2}}{\frac{C}{\Delta t}+\frac{G}{2}}\right]v^{n-1}(k)-\left[\frac{1}{\frac{C}{\Delta t}+\frac{G}{2}}\right]\frac{\left(i^n(k)-i^n(k-1)\right)}{\Delta x} \qquad (14.41a)$$

and

$$i^n(k)=\left[\frac{\frac{L}{\Delta t}-\frac{R}{2}}{\frac{L}{\Delta t}+\frac{R}{2}}\right]i^{n-1}(k)-\left[\frac{1}{\frac{L}{\Delta t}+\frac{R}{2}}\right]\frac{\left(v^n(k)-v^n(k-1)\right)}{\Delta x} \qquad (14.41b)$$

can be obtained. Note that these equations directly reduce to (14.38) or (14.40) if lossless conditions ($R=0$, $G=0$) are applied. Inhomogeneity of a TL is introduced by replacing the primary parameters R, L, C, and G in (8) with R', L', C', and G', where $R' = R + R_f$, $G' = G + G_f$, $L' = L + L_f$, and $C' = C + C_f$ (R_f, G_f, L_f, and C_f being the unit-length parameters of the fault).

The nodes of the TL extend from $k = 1$ to $k = N$. Since iterative equations at each node require values of both left and right neighboring nodes, special treatments are required for the end nodes. As in FDTD representations of plane waves described in Section 14.1, if no action is taken, it corresponds to SC terminations at both source and load ends. This works if the source voltage is applied or injected accordingly.

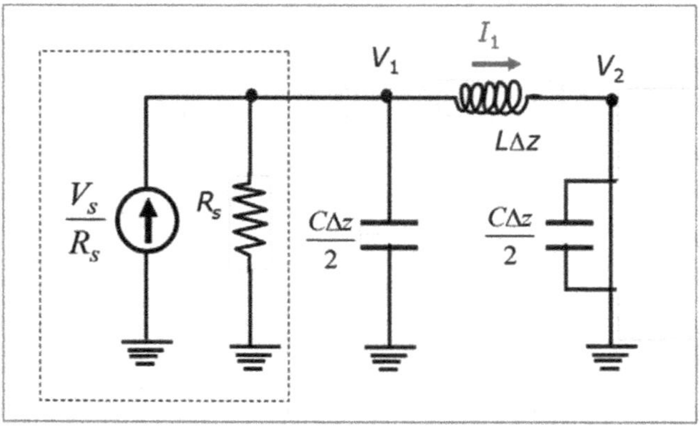

Figure 14.25. Discretization at the source end via Norton equivalent.

14.2.3.3 Source Injection

A voltage source $v(z, t)$ with the internal resistance R_s (see Fig. 14.25) can be included to the iterative TL equations by replacing the Thevenin source with its Norton equivalent.

The Kirchhoff's current law (KCL) at node1 becomes

$$\frac{v_s^n}{R_s} = i^n(1) + C\frac{\Delta z}{2}\frac{\partial v^n(1)}{\partial t} + \frac{v^n(1)}{R_s}. \tag{14.42}$$

By using

$$\frac{\partial v^n(1)}{\partial t} = \frac{v^{n+1/2}(1) - v^{n-1/2}(1)}{2}, \; v^n(1) = \frac{v^{n+1/2}(1) + v^{n-1/2}(1)}{2}, \tag{14.43}$$

one obtains

$$v^{n+1/2}(1) = \left[\frac{R_s\frac{C\Delta z}{2\Delta t} - \frac{1}{2}}{R_s\frac{C\Delta z}{2\Delta t} + \frac{1}{2}}\right] v^{n-1/2}(1) - \left[\frac{1}{R_s\frac{C\Delta z}{2\Delta t} + \frac{1}{2}}\right]\{R_s\} i^n(1) - \frac{\left(v_s^{n+1/2} + v_s^{n-1/2}\right)}{2} \tag{14.44}$$

for the source node ($k = 1$).

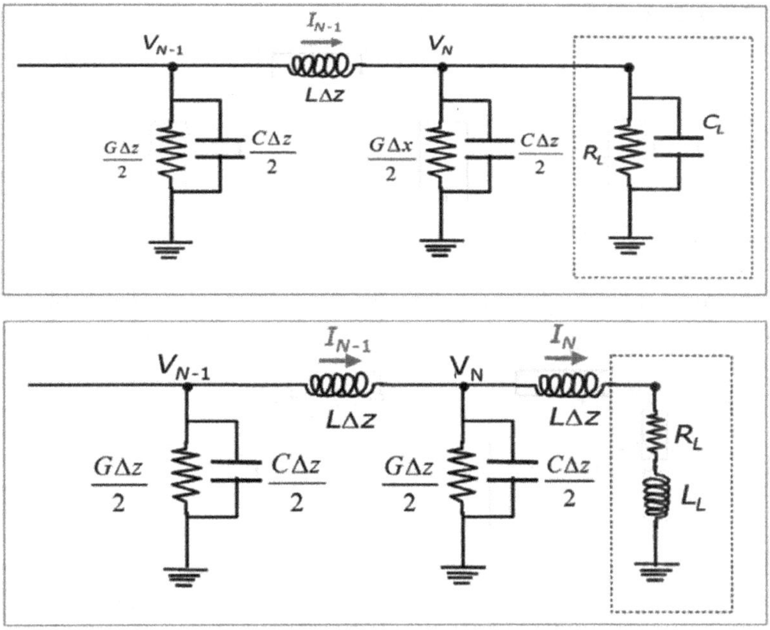

Figure 14.26. (Top) Parallel *RC* load and (bottom) serial *RL* load.

14.2.3.4 Complex Termination

Similar procedures may be followed for the last node ($k = N$) for various complex terminations. For example, parallel RC and serial RL terminations, respectively, are shown in Fig. 14.26.

14.2.3.5 Simple R_L Termination

If a resistive load R_L is connected to the TL, the iterative equation at $k = N$ becomes

$$v^{n+1/2}(N) = \left[\frac{R_L \dfrac{C\Delta z}{2\Delta t} - \dfrac{1}{2}}{R_L \dfrac{C\Delta z}{2\Delta t} + \dfrac{1}{2}}\right] v^{n-1/2}(N) + \left[\frac{1}{R_L \dfrac{C\Delta z}{2\Delta t} + \dfrac{1}{2}}\right] R_L i^n(N-1). \tag{14.45}$$

It should be noted that (14.45) reduces to $v^n(N) = -v^{n-1}(N)$ and $v^n(N) = v^{n-1}(N)$ for the SC ($R_L = 0\ \Omega$) and OC ($R_L \rightarrow \infty$) terminations, respectively.

14.2.3.6 Parallel R_LC_L Termination

The KCL for the last node for this termination (see Fig. 14.23) is

$$i^n(N-1) = v^n(N)\left(\frac{G\Delta z}{2} + \frac{1}{R_L}\right) + \frac{C\Delta z}{2}\frac{\partial v^n(N)}{\partial t} + C_L\frac{\partial v^n(N)}{\partial t}, \quad (14.46)$$

and, via finite-difference discretization, its iterative form follows (14.47). It should be noted that choosing $R_L = 0\ \Omega$ here must reduce to SC termination. Since (14.29) does not directly yield this result, it must be replaced with $v^n(N) = -v^{n-1}(N)$ if $R_L = 0\ \Omega$ is selected.

$$v^{n+1/2}(N) = \left[\frac{\dfrac{C\Delta z + 2C_L}{2\Delta t} - \dfrac{1}{2}\left(\dfrac{G\Delta z}{2} + \dfrac{1}{R_L}\right)}{\dfrac{C\Delta z + 2C_L}{2\Delta t} + \dfrac{1}{2}\left(\dfrac{G\Delta z}{2} + \dfrac{1}{R_L}\right)}\right] v^{n-1/2}(N) + \left[\frac{1}{\dfrac{C\Delta z + 2C_L}{2} + \dfrac{1}{2}\left(\dfrac{G\Delta z}{2\Delta t} + \dfrac{1}{R_L}\right)}\right] i^n(N-1). \quad (14.47)$$

14.2.3.7 Serial R_LL_L Termination

Writing Kirchhoff's voltage law (KVL) for the serial RL termination and following the same procedure results in

$$i^{n+1}(N) = \left[\frac{\dfrac{L\Delta z + 2L_L}{2\Delta t} - \dfrac{1}{2}\left(\dfrac{R\Delta z}{2} + R_L\right)}{\dfrac{L\Delta z + 2L_L}{2\Delta t} + \dfrac{1}{2}\left(\dfrac{R\Delta z}{2} + R_L\right)}\right] i^n(N) + \left[\frac{1}{\dfrac{L\Delta z + 2L_L}{2\Delta t} + \dfrac{1}{2}\left(\dfrac{R\Delta z}{2} + R_L\right)}\right] v^{n+1/2}(N) \quad (14.48)$$

at the last node, as if an extra current path is added (see Fig. 14.26). Therefore, if (14.48) is used for termination, calculation of the voltage at the last node via

$$v^n(N) = \left[\frac{\frac{C}{\Delta t} - \frac{G}{2}}{\frac{C}{\Delta t} + \frac{G}{2}}\right] v^{n-1}(N) - \left[\frac{1}{\frac{C}{\Delta t} + \frac{G}{2}}\right] \frac{\left(i^n(N) - i^n(N-1)\right)}{\Delta z} \tag{14.49}$$

is also required.

14.2.3.8 Parallel $R_L L_L C_L$ Termination

Connecting an inductance and a capacitor together at the load results in a kind of a resonance circuit. The situation in this case is a little bit different since it requires a solution of a time-domain second-order differential equation. In this case, writing the KCL at the last node results in

$$i^n(N-1) = v^n(N)\left(\frac{G\Delta z}{2} + \frac{1}{R_L}\right) + \frac{C\Delta z}{2}\frac{\partial v^n(N)}{\partial t} + C_L \frac{\partial v^n(N)}{\partial t} + i_L^n(t), \tag{14.50}$$

which requires addition of the inductance current in the form of

$$i_L(t) = \frac{1}{L_L}\int_0^t v_L(t)dt + i_0. \tag{14.51a}$$

Therefore, time-domain accumulation of the voltage at the last node during the time-domain simulations is required in order to perform this integration numerically. Assuming zero initial inductor current, using

$$\frac{1}{L}\int_0^t v^n(t)dt \rightarrow \frac{\Delta t}{L_L}\sum_{k=0}^{n} v^{n-1/2}(N) + \frac{\Delta t}{L_L}\left\{\frac{v^{n+1/2}(N) + v^{n-1/2}(N)}{2}\right\}, \tag{14.51b}$$

and rearranging (14.50) yield

$$v^{n+1/2}(N)=\left[\frac{\dfrac{C\Delta z+2C_L}{2\Delta t}-\dfrac{1}{2}\left(\dfrac{G\Delta z}{2}+\dfrac{1}{R_L}+\dfrac{\Delta t}{L_L}\right)}{\dfrac{C\Delta z+2C_L}{2\Delta t}+\dfrac{1}{2}\left(\dfrac{G\Delta z}{2}+\dfrac{1}{R_L}+\dfrac{\Delta t}{L_L}\right)}\right]v^{n-1/2}(N)$$

$$+\left[\frac{1}{\dfrac{C\Delta z+2C_L}{2\Delta t}+\dfrac{1}{2}\left(\dfrac{G\Delta z}{2}+\dfrac{1}{R_L}+\dfrac{\Delta t}{L_L}\right)}\right]\{i^n(N-1)-sk\} \tag{14.52a}$$

for the parallel *RLC* termination with

$$sk=\frac{\Delta t}{L_L}\sum_{k=1/2}^{n-1/2}v^k(N), \tag{14.52b}$$

sk being the inductor current calculated from the time integral of the voltage.

14.2.3.9 Serial $R_L L_L C_L$ Termination

Writing the KVL at the last node for the serial resonance termination yields

$$\begin{aligned} v^{n+1/2}(N)=i^{n+1/2}(N)\left(R_L+\frac{R\Delta z}{2}\right)+\frac{L\Delta z}{2}\frac{\partial i^{n+1/2}(N)}{\partial t} \\ +L_L\frac{\partial i^{n+1/2}(N)}{\partial t}v_C(t), \end{aligned} \tag{14.53a}$$

with

$$v_C(t)=\frac{1}{C_L}\int_0^t i_C(t)dt+v_0. \tag{14.53b}$$

Again, assuming no stored capacitive energy at the load (i.e., $v_0=0$), using

$$\frac{1}{C_L}\int_0^t i^n(t)dt\rightarrow\frac{\Delta t}{C_L}\sum_{k=0}^{n}i^{n-1}(N)+\frac{\Delta t}{C_L}\left\{\frac{i^{n+1}(N)+i^{n-1}(N)}{2}\right\}, \tag{14.54}$$

and rearranging (14.53) results in

$$i^{n+1}(N) = \left[\frac{\dfrac{L\Delta z + 2L_L}{2\Delta t} - \dfrac{1}{2}\left(\dfrac{R\Delta z}{2} + R_L + \dfrac{\Delta t}{2C_L}\right)}{\dfrac{L\Delta z + 2L_L}{2\Delta t} + \dfrac{1}{2}\left(\dfrac{R\Delta z}{2} + R_L + \dfrac{\Delta t}{2C_L}\right)}\right] i^n(N)$$

$$+ \left[\frac{1}{\dfrac{L\Delta z + 2L_L}{2\Delta t} + \dfrac{1}{2}\left(\dfrac{R\Delta z}{2} + R_L + \dfrac{\Delta t}{2C_L}\right)}\right] \left\{v^{n+1/2}(N) + sk\right\} \tag{14.55a}$$

for the serial *RLC* termination with

$$sk = \frac{\Delta t}{C_L} \sum_{k=1}^{n-1} i^k(N). \tag{14.55b}$$

It should be noted that (14.53) requires calculation of the voltage at the last node via

$$v^{n+1/2}(N) = \left[\frac{\dfrac{C}{\Delta t} - \dfrac{G}{2}}{\dfrac{C}{\Delta t} + \dfrac{G}{2}}\right] v^{n-1/2}(N) - \left[\frac{1}{\dfrac{C}{\Delta t} + \dfrac{G}{2}}\right] \frac{\left(i^n(N) - i^n(N-1)\right)}{\Delta z}. \tag{14.56}$$

A short MATLAB script *TRLine.m* is written to simulate source-driven time-domain TL equations under various terminations explained earlier. Table 14.4 lists the script written only for the resistive load (compare this code with the plane wave code given in Table 14.1). The user supplies the unit-length TL parameters, select one of three different source types (Gaussian, rectangular, or trapezoidal pulse), and specify the termination type; then time-domain simulations can be done. Movie visualization of the pulse propagation and reflections along the TL is possible during the simulations. Also, time history of the signal recorded at the specified observation point can be plotted offline. Figure 14.27 shows examples of both plots; voltage distribution

Table 14.4.
The TRLine.m MATLAB Script

```
% ----------------------------------------------------------
%  Program: TRLine.m
% ----------------------------------------------------------
clc; clear all;
vzoldr=0; % 1-time step previous value at RB (k=ke-1)
vzoldl=0; % 1-time step previous value at LB (k=ke-1)
lbc=input('Left  Boundary: (1) PEC, (0) Free-space = ');
rbc=input('Right Boundary: (1) PEC, (0) Free-space = ');
% ------------ Simulation parameters ------------
nt=500; zmax=100; ke=100; kc=30; eta0=50;
c0=100e-12;l0=eta0^2*c0;
c=1/sqrt(c0*l0); tt=ke/2; t0=10; delz=floor(zmax/ke);
dt=delz/c;
band=c/(10*delz); alfa=3.3*band*band; shift=4./
sqrt(alfa);
% ------ Array initialization ------------
for k=1:ke; vz(k)=0.0; iz(k)=0.0; z(k)=k; end
t=0;
for n=1:nt    % FDTD time loop starts
    t=n*dt;
    for k=2:ke-1    % Vz is calculated along z
      vz(k)=vz(k)+eta0*(iz(k-1)-iz(k));
    end
    source=exp(-alfa*(t-shift)^2); vz(kc)=vz(kc)+source;
    if lbc==0; vz(1)=vzoldl; end
    if rbc==0; vz(ke)=vzoldr; end
    for k=1:ke-1    % Iz is calculated along z
      iz(k)=iz(k)+(vz(k)-vz(k+1))/eta0;
    end
    vzoldr=vz(ke-1); vzoldrt=vz(ke); vzoldl=vz(2);
    plot(z, vz); xlim([1,ke]); ylim([-1.1,1.1]);
    xlabel('TL [m]'); ylabel('Voltage [V]'); grid;
pause(0.01);
end
% ------------------------ END ------------------------
```

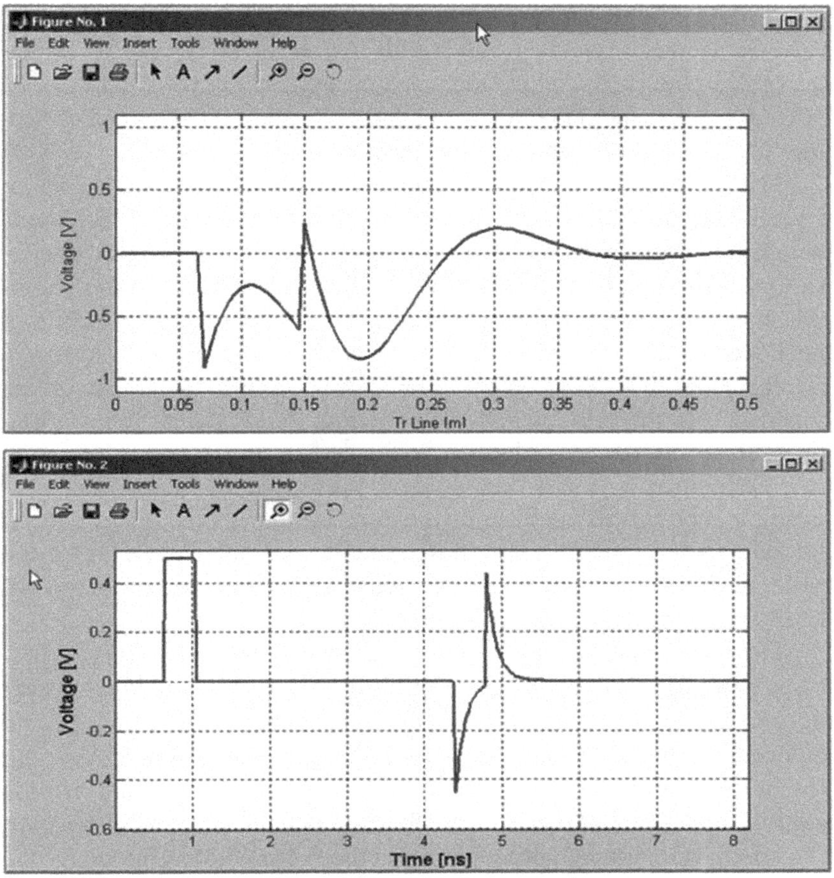

Figure 14.27. (Top) Voltage along the TL and (bottom) signal versus time at a chosen observation point.

along the TL at a given instant (top) and signal versus time observed at the specified point (bottom).

14.2.4 MATLAB-Based TDRMeter Package

A virtual tool—time-domain reflectometer (TDRMeter)—based on the discretization of time-domain TL equations in terms of finite-difference approximation is introduced here [4]. TDRMeter has been employed in a variety of fields for determining the electrical parameters of materials over a wide frequency bandwidth. An early form of TDRMeter, dating from the 1930s, which most people are familiar with, is RADAR. Other forms of the TDRMeter are LIDAR, coax TDRMeter, optical fiber OTDR, and broadband impulse radars.

The TDRMeter is used to locate and identify faults in all types of metallic paired cable. It can locate major or minor cabling problems including sheath faults, broken conductors, water damage, loose connectors, crimps, cuts, smashed cables, shorted conductors, system components, and a variety of other fault conditions. The basic principle of the TDRMeter is fairly simple. A pulse is generated and sent down the TL. If the line is terminated by a load that is equal to the characteristic impedance of the line, no reflected wave or echo will be returning to the generator or source, otherwise reflection occurs. By measuring the time difference between the incident pulse and the echo (reflected pulse), it is possible to determine the distance to the fault that produced the echo pulse. Calculations must include the velocity factor of the cable. Also, detailed analysis of the echo signal can reveal additional details of the faults or reflecting objects.

The speed and accuracy of the TDRMeter makes it today's preferred method of cable fault location. Although today's instruments are more user friendly, a good understanding of the basic principles and applications of a TDRMeter is essential to successful troubleshooting. The pulse generated by the TDRMeter takes a certain amount of time and thus distance to launch. This distance is known as the *blind spot*. The length of the blind spot varies with the pulse width. The longer the pulse width, the larger the blind spot. It is more difficult to locate a fault contained within the blind spot. If a fault is suspected within the first section of the cable, it is advisable to add a length of cable (jumper) between the TDRMeter and the cable being tested. Any faults that may have been hidden in the blind spot can now easily be located (only remember to take the length of the jumper cable into account). It is best if the jumper cable has the same characteristic impedance with the cable under test.

The TDRMeter package is a multipurpose program with the front panel given in Fig. 14.28. There are four input data blocks on top of the panel. The user supplies the unit-length TL parameters on the left (from which the characteristic impedance is automatically displayed). The midleft block is reserved for the generator parameters. The user

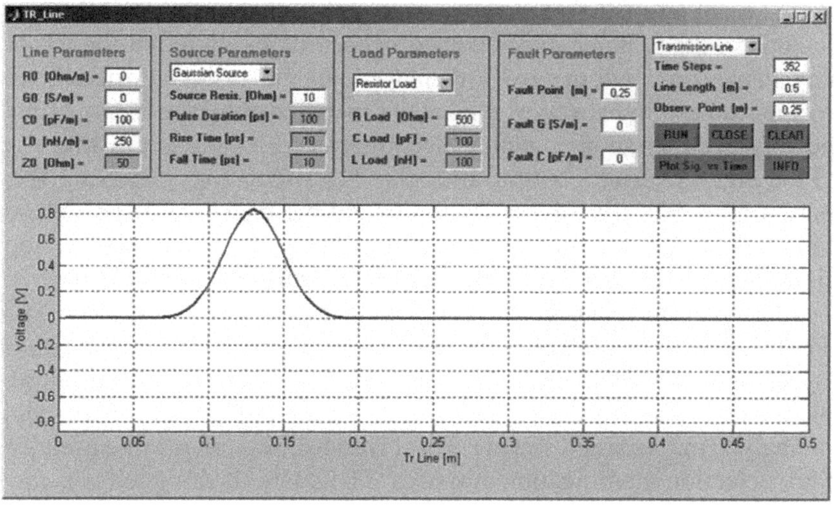

Figure 14.28. Front panel of the MATLAB-based TDRMeter package.

may select one of three different source types—*Gaussian pulse*, *rectangular pulse*, and a *trapezoidal pulse*—from the pop-up menu and supply pulse duration and rise/fall times (if the source is trapezoidal). Internal source resistor is also supplied inside this block. The pulse length of the Gaussian voltage source is automatically selected according to user-specified line length and other dicsretization parameters. The user may choose long pulse durations (i.e., at least longer than the total TL travel time) and simulate a *step voltage source*. The third block on midright is used for the specification of the load; the selection in the pop-up menu of this block includes a resistive load, a parallel combination of resistor/capacitor, a serial combination of a resistor/inductor, and serial and parallel resonance terminations. Based on the selection, the user is asked to supply *RLC* elements of the load in the activated data boxes. The last block at the right is used for the fault specification. The user is allowed to change unit-length admittance and/or conductance (the user can define and add other types of faults by modifying the M-file). The TL length, observation point, and number of simulation steps are supplied at the top right of the front panel together with the runtime buttons.

The output of the TDRMeter is given via two different plots. The incident voltage pulse and reflected echoes (if exist) along the TL at any instant are automatically displayed on the reserved window as movie frames. Signal versus time at the specified observation point may be displayed in another plot which becomes visible by pressing the "Plot Sig. vs. Time" button after the TD simulations. Pressing this button automatically saves signal versus time data into a file named "SigvsTime.dat" for further offline signal analysis. The counter at top right displays the number of time steps left during the simulation.

The pop-up menu at the top right contains a key selection. The TDRMeter can be used either as a *TL simulator* or a *TD Reflectometer*. By default, the selection is TL simulator. The user specifies all input parameters and runs the tool to visualize time-domain TL effects and or analyze the recorded signals. The TDRMeter option may be used for fault detection/identification purposes. TDRMeter is designed in a way that the user specifies only the TL and generator parameters. Once the TDRMeter option is selected, the termination and fault blocks disappear. The load is selected automatically and randomly, and a fault is introduced at an arbitrary point along the TL, so the user can find out its type and/or numerical values (if possible) by only observing/analyzing the output plots, for example, signal versus time on a selected TL point. The user may do blind tests and observe problems along the TL after specifying the TL and source parameters and starting the simulations by pressing the "Run" button. The results of these blind tests (i.e., randomly generated parameters) reappear when the user reselects the "TR Line" option at the end of the simulations. In this way, the user can check his results against the data displayed on the front panel of the TDRMeter.

The plot in Fig. 14.28 belongs to a 50 Ω, 0.5 m loss-free homogeneous (uniform) TL excited with a Gaussian pulse source having 10-Ω internal resistor, and terminated by a resistive load of 100 Ω (the TL is fault-free, i.e., uniform). The simulation time step is 400. Since unit-length inductance and capacitor values are 250 nH/m and 100 pF/m, respectively, the speed of voltage and current waves along the line is calculated from the limiting case to be $v = 2$ m/s. The TL is divided into 100 nodes; therefore, $\Delta x = 0.5/100 = 5$ mm. The time step Δt is calculated to be $\Delta t = \Delta x/v = 25$ ps. With this choice, the voltage (or current) pulse propagates one node at a time; therefore, it will take $100\Delta t$ for the injected pulse to reach the load. The total of $400\Delta t$ will

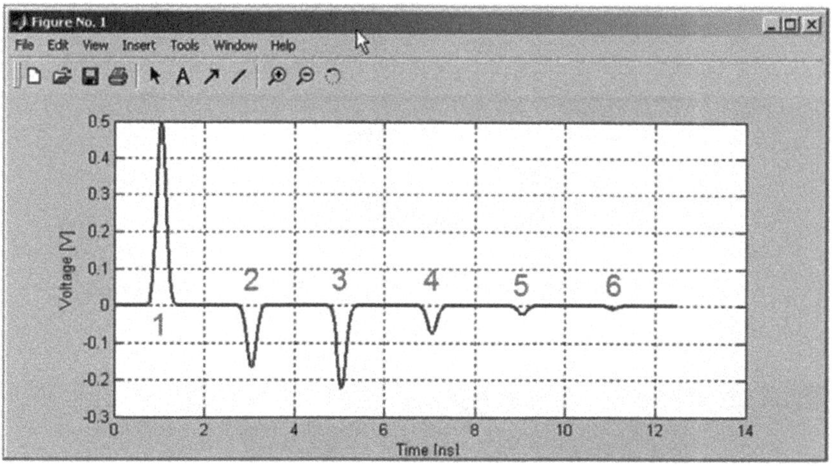

Figure 14.29. Signal versus time plot of a 50 Ω, 0.5 m lossless SC TL excited by a Gaussian pulse generator with 50 Ω-internal resistor with $G_f = 4$ S/m at 0.3 m.

result in two reflections from the load and two reflections from the source.

Figure 14.29 shows signal versus time plot (at 0.1 m) of a typical scenario where a 50 Ω, 0.5 m loss-free SC TL excited by a Gaussian pulse generator with 50 Ω internal resistor (i.e., matched source) with a G-type fault at 0.3 m ($G_f = 4$ S/m). In this case, incident voltage pulse reflects not only at the load but also from the fault point in either direction.

The sketch of this scenario is drawn in Fig. 14.30 with the identification of the first three echoes (identification of the other three echoes is left to the reader).

The TDRMeter tool can be used to teach, understand, and visualize the TD pulse characteristics and echoes from various discontinuities/terminations, and fault types and locations can be predicted. The simplest method is to predict the length of the TL (for the case of resistive discontinuities/termination) by measuring the transit time between the incident pulse and the echo (if separated in time). Distinguishing the end- and fault-reflected pulses in time, measuring delays among them, and marking the maximum amplitudes are enough for this purpose. An example is given in Fig. 14.31.

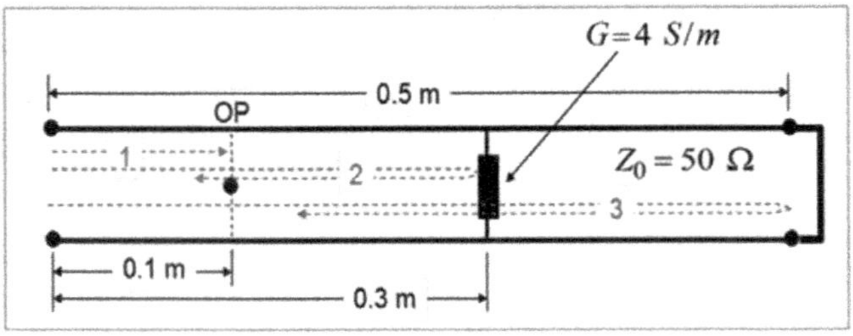

Figure 14.30. The scenario of the echoes observed in Fig. 14.29.

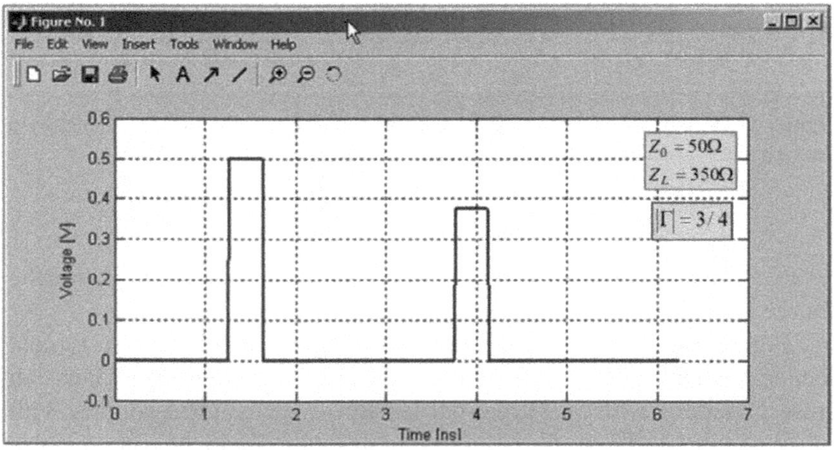

Figure 14.31. Signal versus time for the resistive termination showing the incident rectangular pulse and the echo (reflected pulse) along a 50 Ω, 0.5 m TL (pulse length = 400 ps, $R_s = 50\ \Omega$, $R_L = 350\ \Omega$).

A 50 Ω, 0.5 m loss- and fault-free uniform TL, matched generator at the left end, and $R_L = 350\ \Omega$ resistive load at the right end are excited with a rectangular pulse having 400 ps pulse duration. The figure shows signal versus time recorded at midpoint of the TL. The time delay between the two pulses gives the distance from the observation point to the end of the TL. The transit time to the first pulse gives the distance from the generator to the observation point. The ratio of the pulse amplitudes is 0.75 and corresponds to the modulus of the voltage reflection coefficient, which can easily be verified from

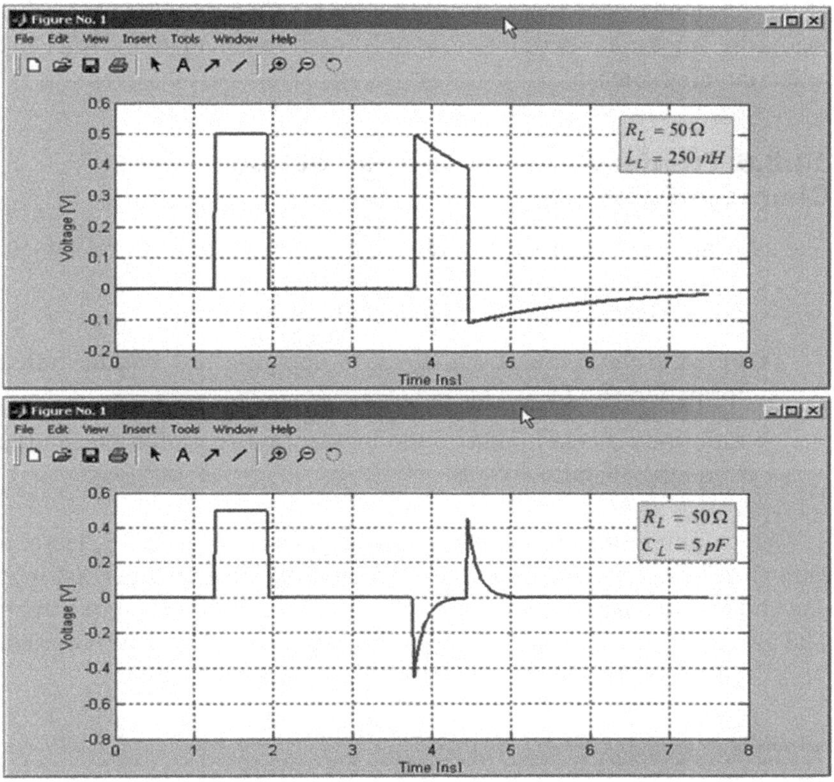

Figure 14.32. Signal versus time at midpoint of a 50 Ω, 0.5 m TL under (a) serial *RL* termination (R_L = 50 Ω, L_L = 250 nH) and (b) parallel *RC* termination (R_L = 50 Ω, C_L = 5 pF). Matched termination is used at the source (R_s = 50 Ω). Pulse length = 400 ps.

$$\Gamma_L = \frac{Z_L - Z_0}{Z_L + Z_0} = \frac{350 - 50}{350 + 50} = 0.75.$$

Unfortunately, this method does not work for complex faults and/or terminations as given in Fig. 14.32a,b, which show signal versus time variations for serial *RL* (R_L = 50 Ω, L_L = 250 nH) and parallel *RC* (R_L = 50 Ω, C_L = 5 pF) complex loads, respectively (other parameters are $R_s = Z_0$ = 50 Ω, TL length = 0.5 m, rectangular pulse duration = 700 ps). Although the shape of the echoes gives clues on the type of termination (e.g., inductive load basically affects the DC portion of the echo pulse, but rise and fall times are affected by the capacitive load), the amplitudes certainly cannot be used to calculate/measure the reflection coefficient.

14.2.5 Fourier Analysis and Reflection Characteristics

The standard Fourier procedure for the complex loads is as follows:

- discriminate incident and reflected pulses in time,
- move to the frequency domain by applying fast Fourier transformation (FFT),
- ratio the reflected pulse to the incident pulse within the source frequency band.

An example of the *FFT procedure* is given in Fig. 14.33. Here, a fault-free 0.5-m-long 50 Ω TL is connected to a matched generator at one end and to a parallel *RC* load ($R_L = 50\ \Omega$, $C_L = 10$ pF) at the other. The plot in Fig. 14.32b (bottom one) shows signal versus time recorded

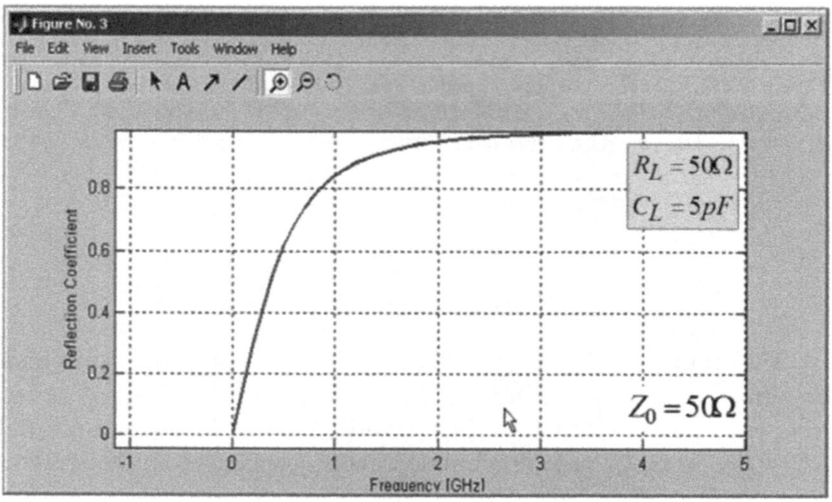

Figure 14.33. Voltage reflection coefficient versus frequency of Fig 14.32b obtained with the FFT procedure. Solid: TD simulation result; dashed: analytical exact solution.

at midpoint of the TL where one can easily discriminate incident and load-reflected pulses. Applying the FFT procedure yields the voltage reflection coefficient versus frequency as given in Fig. 14.33. The solid and dashed lines in the figure represent the results of the FFT procedure and the analytical exact solution obtained from analytical calculations, respectively. Excellent agreement shows the power of the procedure. As expected, the capacitor is OC at DC and low frequencies and reflection coefficient is zero (since $R_L = Z_0 = 50\ \Omega$). On the other hand, the capacitor acts as SC and the modulus of the reflection coefficient approaches to 1 (actually, taking the real part after the FFT yields $\Gamma = -1$).

Another example is given in Fig. 14.34. The same fault-free, 0.5-m-long, 50 Ω TL is connected to a matched generator at one end and

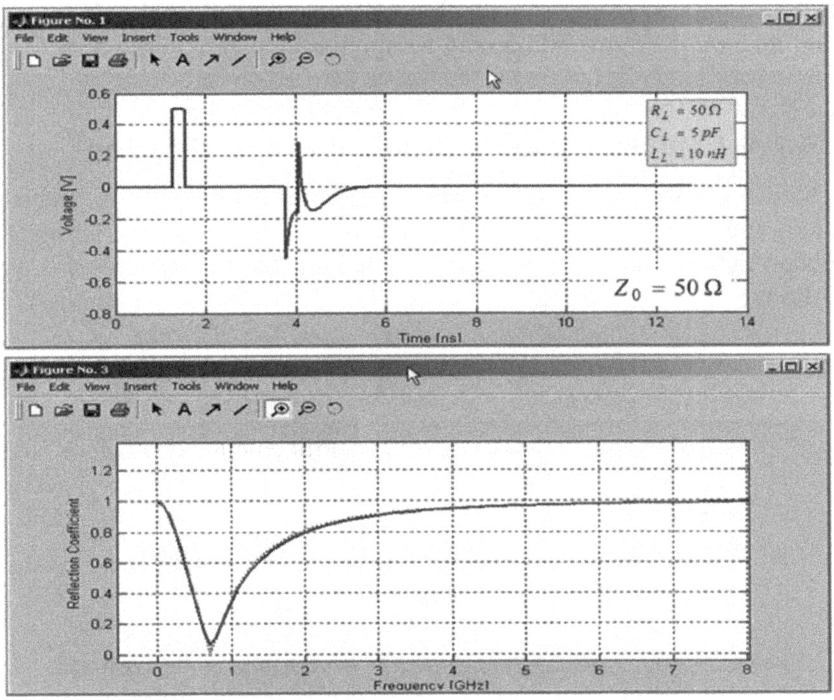

Figure 14.34. (a) Signal versus time at midpoint of a 50 Ω, 0.5 m TL. Load: parallel *RLC* ($R_L = 50\ \Omega$, $C_L = 5$ pF, $L_L = 10$ nH). Source: $R_s = 50\ \Omega$. Pulse length = 400 ps. (b) Voltage reflection coefficient versus frequency obtained with the FFT procedure. Solid: TD simulation result; dashed: analytical exact solution.

to a parallel resonance circuit at the other ($R_L = 50\ \Omega$, $C_L = 5$ pF, $L_L = 50$ nH). Figure 14.34a shows signal versus time recorded at mid-point. The voltage reflection coefficient versus frequency obtained after the procedure explained earlier is shown in Fig. 14.34b. The solid and dashed lines in the figure represent offline FFT results and analytical exact solution, respectively. As expected, the reflection is minimal at the resonance frequency of about 711 MHz calculated from $f_r = \left(2\pi\sqrt{L_L C_L}\right)^{-1}$.

It should be noted that one can still apply the FFT procedure even when the pulses cannot be discriminated in time. In this case, the TD simulations should be repeated twice; first along the TL with the unknown termination, then along the TL with matched termination. The incident plus reflected pulses are recorded in the first simulation, while only the incident pulse exists and is recorded in the second run. Subtracting the second from the first will yield the reflected-only pulse, and then the FFT procedure may be applied.

14.2.6 Laplace Analysis and Fault Identification

The shape of the echo may be used to predict the nature of the mismatch along the TL [5]. One method is to use the *Laplace transformation* to obtain time variation of the echo $s_r(t)$ analytically. The procedure is as follows:

- derive the Laplace transform of the generator (e.g., V_i/s for the step voltage V_i),
- write down the voltage reflection coefficient in the *s-domain* (for example, $\Gamma(s) = (R_L + s_L L - Z_0)/(R_L + s_L L + Z_0)$) for a serial *RL* termination),
- multiply these two, take the inverse Laplace transform of the product, and derive $s_r(t)$ analytically.

For example, $s_r(t)$ for the *serial RL termination* is derived as

$$s_r(t) = V_i\left[\left(1+\frac{R_L - Z_0}{R_L + Z_0}\right) + \left(1-\frac{R_L - Z_0}{R_L + Z_0}\right)e^{-t/\tau}\right], \tau = \frac{L_L}{R_L + Z_0}. \tag{14.57}$$

Here τ is the time constant. Similarly, the $s_r(t)$ for *the parallel RC termination* is

$$s_r(t) = V_i\left[\left(1+\frac{R_L - Z_0}{R_L + Z_0}\right)\left(1-e^{-t/\tau}\right)\right], \tau = \frac{R_L Z_0}{R_L + Z_0}\times C_L. \quad (14.58)$$

Another method—without resorting the Laplace transform—is to observe the echo signal in time at the time limits, that is, at $t = 0$ and when $t \rightarrow \infty$ (here, $t = 0$ corresponds to the instant that the incident pulse hits the load). For example, the inductance of a *RL* load combination initially acts as an infinite impedance (OC) and full reflection occurs at $t = 0$. Its current builds up exponentially with time and acts SC as $t \rightarrow \infty$. Figure 14.35 sketches this scenario. A step voltage with amplitude V_i becomes $2 \times V_i$ when hits the serial RL pair connected parallel to the TL marked as $t = 0$ in the sketch.

The voltage then exponentially decays and approaches to

$$V_i\left(1+\frac{R_L - Z_0}{R_L + Z_0}\right) \quad (14.59)$$

as $t \rightarrow \infty$. An example obtained with the TDRMeter is given in Fig. 14.36. Here, signal versus time of a fault-free, uniform, 0.5-m-long

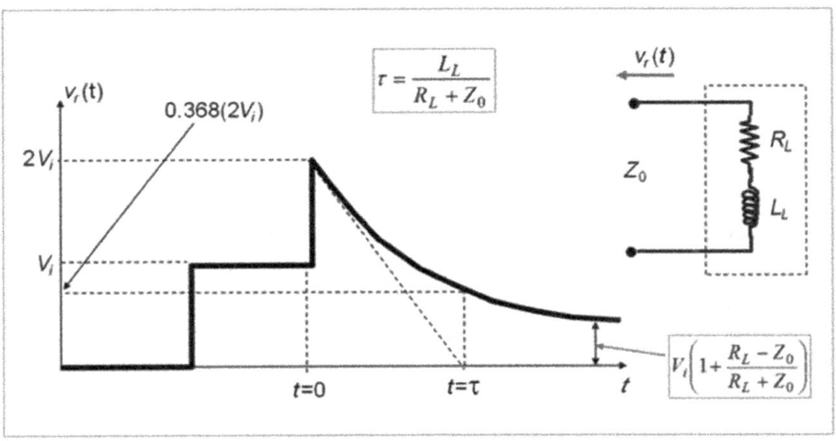

Figure 14.35. The sketch of the Laplace procedure; prediction of the inductive load from the time signature of the echo by measuring the time constant τ.

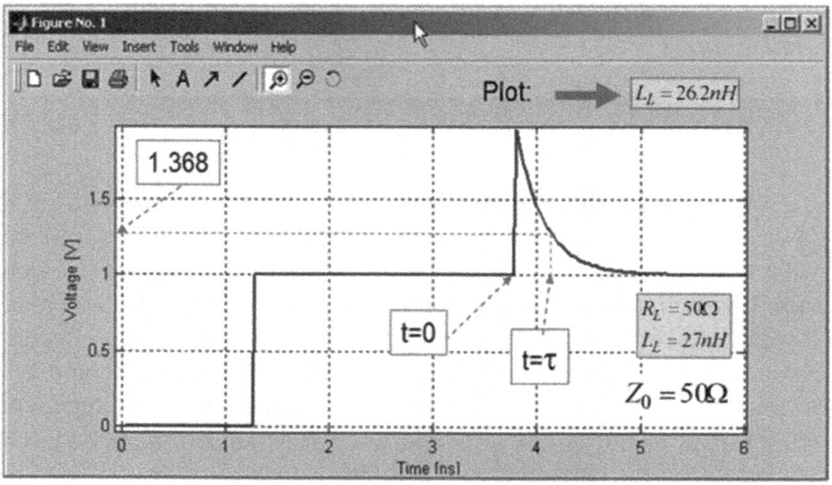

Figure 14.36. Signal versus time (step response) at midpoint of a 50 Ω, 0.5 m TL under serial *RL* termination ($R_L = 50\ \Omega$, $L_L = 27$ nH). Matched termination is used at the source ($R_s = 50\ \Omega$). The value of the inductor calculated from the plot is 26.2 nH.

50 Ω TL, connected to an ideal generator ($R_s = 0\ \Omega$) at the left end, and a serial RL load at the other with $R_L = 50\ \Omega$ and $L_L = 27$ nH, is shown. The response of the *unit-amplitude step voltage* at midpoint of the TL is recorded and is plotted in the figure.

Since $R_s = Z_0 = 50\ \Omega$, the final voltage value is also 1 V. The time delay between 2 V (at 3.8 ns) and 1.368 V (at 4.062 ns) is read from the "Sigvs.Time.dat" file and found to be $\tau = 0.22$ ns. Using the time constant τ equation given inset of Fig. 14.35, the inductance value is calculated to be $L_L = 26.2$ nH.

The capacitor value for the parallel *RC* combination may also be predicted from the time signature of the echo. A parallel capacitor load initially acts as an SC termination so full reflection with 180° phase difference occurs at $t = 0$. The capacitor voltage builds up exponentially with time and acts OC as $t \to \infty$. Figure 14.37 sketches this scenario. A step voltage with amplitude V_i drops to 0 V when hits the parallel *RC* pair (marked as $t = 0$ in the sketch) (note that $V_i = R_s/(R_s + Z_0)$). The voltage increases exponentially and approaches to

$$V_i\left(1+\frac{R_L - Z_0}{R_L + Z_0}\right) \tag{14.60}$$

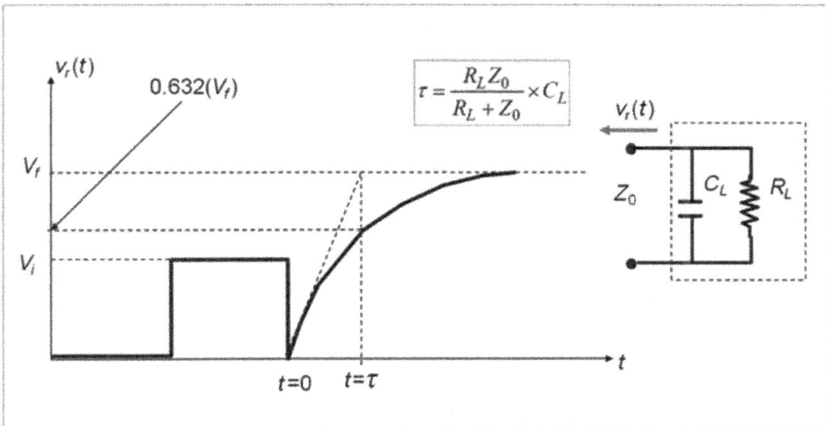

Figure 14.37. The sketch of the Laplace procedure: Prediction of the capacitive load from the time signature of the echo by measuring the time constant τ.

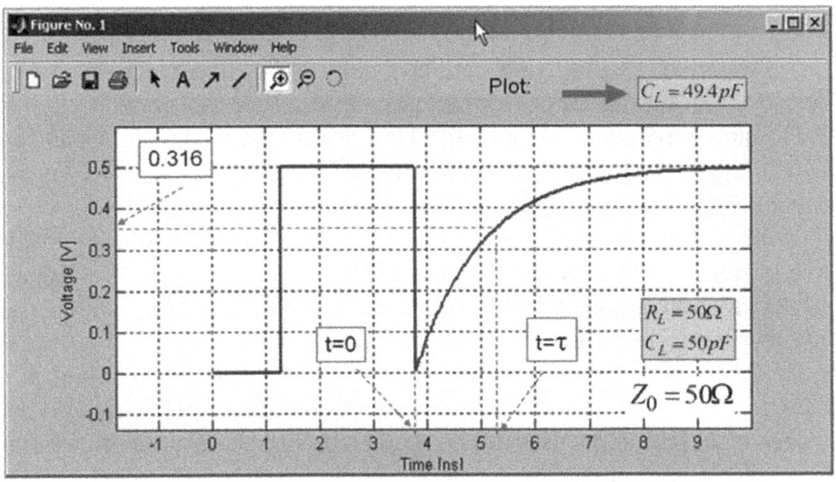

Figure 14.38. Signal versus time (step response) at midpoint of a 50 Ω, 0.5 m TL under parallel *RC* termination ($R_L = 50$ Ω, $C_L = 50$ pF). Matched termination is used at the source ($R_s = 50$ Ω). The value of the capacitor calculated from the plot is 49.4 pF.

as $t \rightarrow \infty$. The TDRMeter result for this scenario is given in Fig. 14.38. Here, signal versus time of a fault-free, uniform, 0.5-m-long 50 Ω TL, connected to a $R_s = 50$ Ω generator, and a parallel *RC* load ($R_L = 50$ Ω, $C_L = 50$ pF), is shown at midpoint. Since $R_s = Z_0 = 50$ Ω, the initial and final voltages are both 0.5 V. The time instants marked with $t = 0$

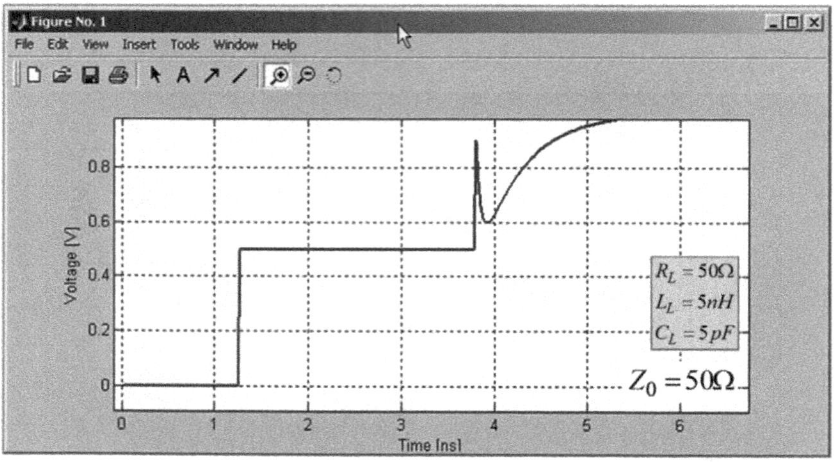

Figure 14.39. Signal versus time at midpoint of a 50 Ω, 0.5 m TL for a step voltage source under serial resonance termination ($Z_L = 50$ Ω, $L_L = 5$ nH, $C_L = 5$ pF, $R_s = 50$ Ω).

and $t = \tau$ in the figure are extracted from the "Sigvs.Time.dat" file as 3.775 and 5.01 ns, respectively. This results in a time constant of $\tau = 0.245$ ns. Using the τ equation in Fig. 14.36 inset, the value of the capacitor is calculated to be $C_L = 49.4$ pF.

Very often, the TLs suffer more than individual inductive or capacitive termination effects, and resonances occur. Figure 14.39 and Figure 14.40 illustrate signal versus time variations of terminations with serial and parallel resonance circuits, respectively. For these kinds of terminations the Laplace transformation procedure may also be applied and analytical expressions of the echo signal may be derived. Although the values of *LC* elements may not be predicted from the exponential variations directly, as done for the serial *RL* and parallel *RC* terminations earlier, the values may still be predicted by the application of different curve fitting methods and/or novel algorithm such as artificial intelligence, genetic algorithms, and so on.

The TDRMeter virtual tool can be used to teach, understand, test, and visualize the TD TL characteristics; echoes from various discontinuities and terminations. Inversely, the types and locations of fault can be predicted from the recorded data. Any electronic device and/or circuit includes TL lines (i.e., coaxial, two-wire cables,

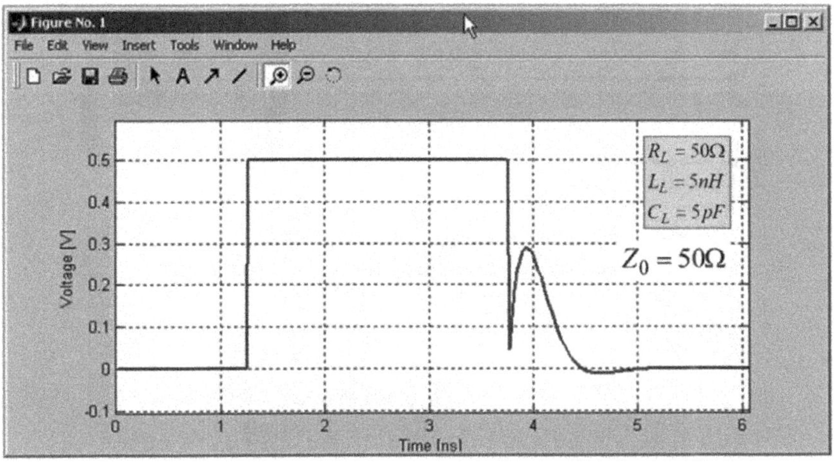

Figure 14.40. Signal versus time at midpoint of a 50 Ω, 0.5 m TL for a step voltage source under parallel resonance termination (Z_L =50 Ω, L_L = 5 nH, C_L = 5 pF, R_s = 50 Ω).

parallel plate lines, microstrip lines, etc.); therefore, have transient EM characteristics.

A wide range of EM problems may be simulated with the TDRMeter virtual tool, such as conducted emissions, common and differential mode effects, and so on, both in time and frequency domains. Also, reflections due to various types of transmission structures, impedance effects of vias, signal and power line couplings, and so on, on printed circuits with smaller and smaller dimensions make signal integrity one of the most important and complex EM problems. Therefore, the virtual tool may be particularly helpful for a student or circuit designer in the evaluation of signal integrity.

Although the virtual tool presented here is a very effective tool for training young EM engineers in fault detection, estimation, and identification, practical problems require a good amount of experience. For example, see the last scenario shown in Fig. 14.41. Can you identify the location and type of the fault by analyzing time variation of the echoes only?

As explained earlier, TDRMeter can also be used in fault location and identification training. Two examples are given for two different fault scenarios in Fig. 14.42, Fig. 14.43, Fig. 14.44, and Fig. 14.45. In

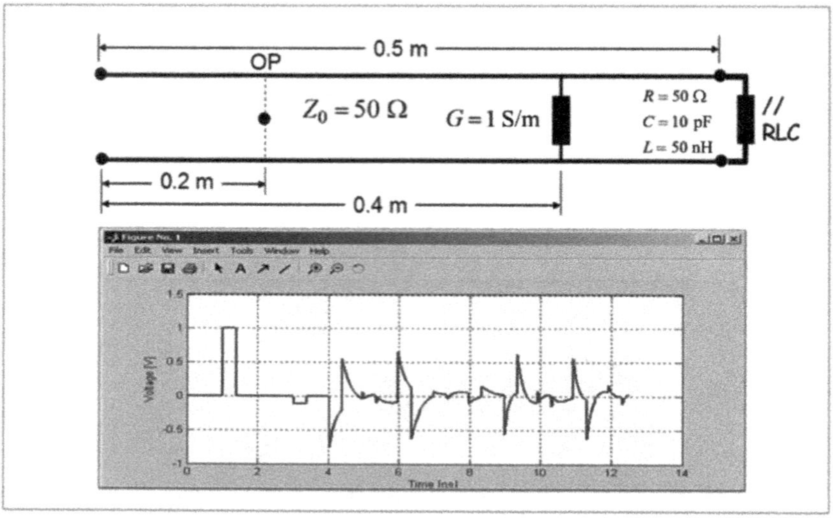

Figure 14.41. The last scenario and time variation of the echoes.

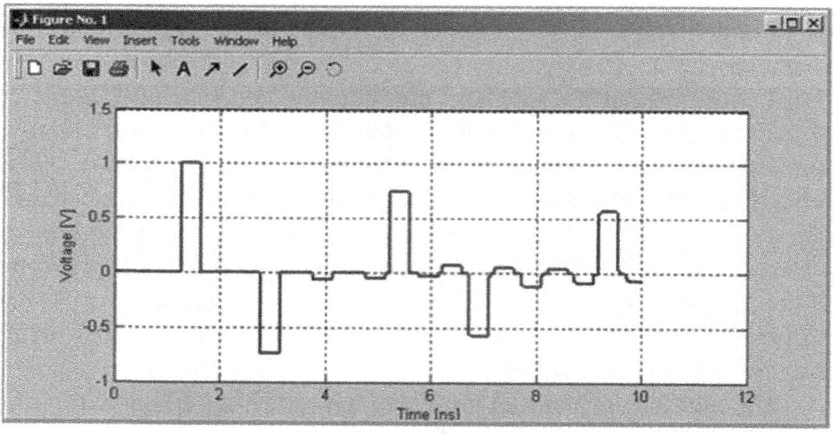

Figure 14.42. Signal versus time at midpoint of the TL for an unknown fault.

Fig. 14.42, time history of a fault is observed at midpoint of a TL with an unknown termination and fault.

First of all, the pulse shapes are all rectangular, which means most probably the termination and fault are both resistive. Second, there is a strong reflection from the load with 180° phase shift, and this means

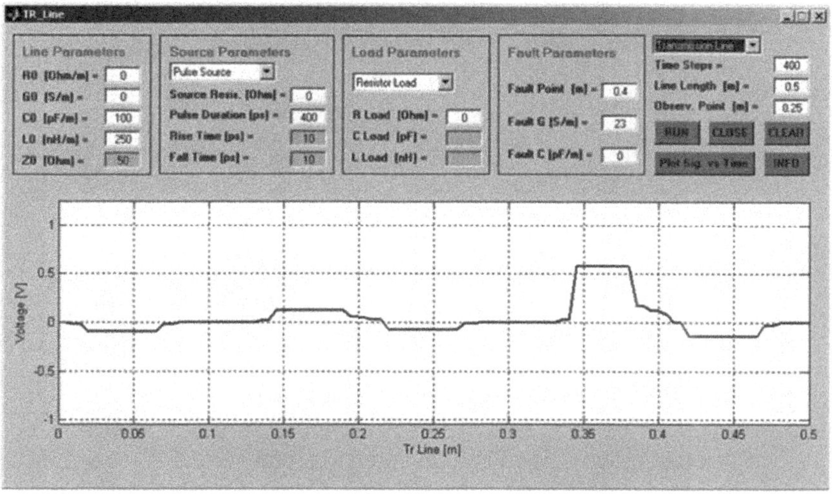

Figure 14.43. The scenario of time the signal observed in Fig. 14.42.

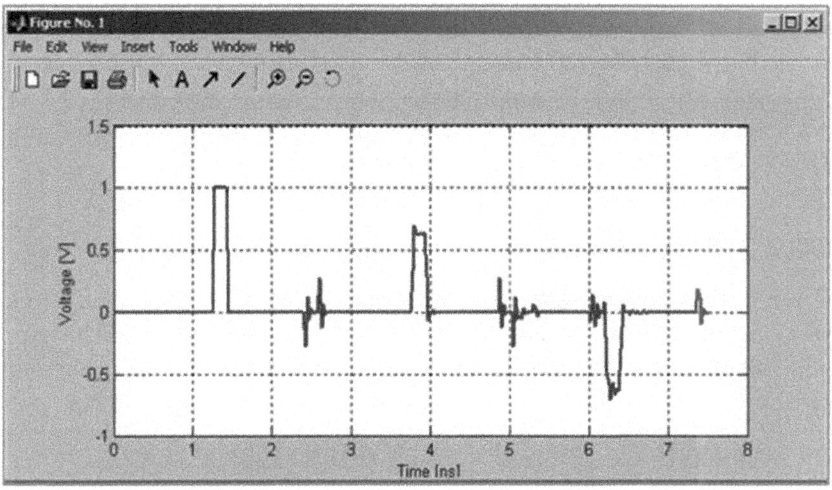

Figure 14.44. Signal versus time of the second fault scenario.

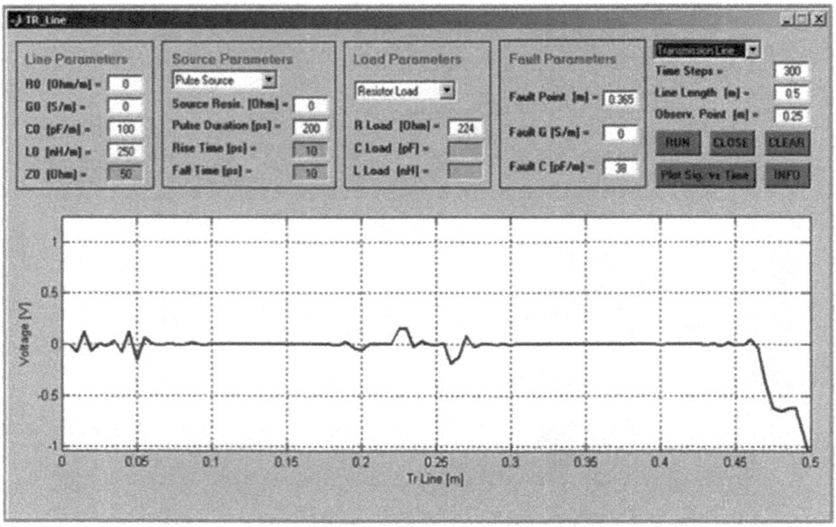

Figure 14.45. The scenario of time the signal observed in Fig. 14.44.

the load is far less than the characteristic impedance of the TL. Time delay between reflected pulses can be used to identify fault location. The scenario of this termination and fault is presented in Fig. 14.43. The plots for the second fault scenario are given in Fig. 14.44 and Fig. 14.45. The comments of these plots are left to the reader.

14.2.7 Step Response

Wave reflection process may be best understood analytically via showing the step response of a lossless TL connected to a Thevenin equivalent source and terminated by a resistive load. Voltage reflection coefficients at the load and source, respectively, for this case are

$$\Gamma_L = \frac{R_L - Z_0}{R_L + Z_0}, \Gamma_S = \frac{R_S - Z_0}{R_S + Z_0}. \tag{14.61}$$

The unit step source voltage

$$v_S(t) = V_0 U(t), U(t) = \begin{cases} 1 & \text{for} \quad t > 0 \\ 0 & \text{for} \quad t < 0 \end{cases}. \tag{14.62}$$

Initially, the TL acts as a resistor having a value of the characteristic impedance connected to the source via the source resistor connected serially to the TL. This means the first voltage pulse launched on to the TL at the input is going to be

$$v_1^+(z,t) = V_0 \frac{Z_0}{R_s + Z_0} U\left(t - \frac{z}{v}\right) = V_1^+ U\left(t - \frac{z}{v}\right), \tag{14.63}$$

where v is the velocity of the voltage wave along the line. If D is the length of the TL and $t_D = D/v$ is the end-to-end total travel time along the TL, then total voltage after infinite number of reflections from both ends will be

$$\begin{aligned} v(z,t) = V_0 \frac{Z_0}{R_s + Z_0} \Bigg[& U\left(t - \frac{z}{v}\right) + \Gamma_L U\left(t - 2t_D + \frac{z}{v}\right) + \Gamma_L \Gamma_S U\left(t - 2t_D - \frac{z}{v}\right) \\ & + \Gamma_L^2 \Gamma_S U\left(t - 4t_D + \frac{z}{v}\right) + \Gamma_L^2 \Gamma_S^2 U\left(t - 4t_D - \frac{z}{v}\right) \\ & + \Gamma_L^3 \Gamma_S^2 U\left(t - 6t_D + \frac{z}{v}\right) + \Gamma_L^3 \Gamma_S^3 U\left(t - 6t_D - \frac{z}{v}\right) \ldots \Bigg] \end{aligned} \tag{14.64}$$

The steady-state voltage value may be obtained by replacing the infinite term summation with its limit form when $t \to \infty$. In this case, the unit functions drop and (14.64) becomes

$$v(z, t \to \infty) = V_0 \frac{Z_0}{R_s + Z_0} \left[1 + \Gamma_L + \Gamma_L \Gamma_S + \Gamma_L^2 \Gamma_S + \Gamma_L^2 \Gamma_S^2 + \Gamma_L^3 \Gamma_S^2 + \ldots\right], \tag{14.65}$$

$$v(z, t \to \infty) = V_0 \frac{Z_0}{R_s + Z_0} \frac{1 + \Gamma_L}{1 - \Gamma_L \Gamma_S}, \tag{14.66}$$

$$v(z, t \to \infty) = V_0 \frac{R_L}{R_s + R_L}. \tag{14.67}$$

The steady-state current will then be derived as

$$i(z, t \to \infty) = \frac{V_0}{R_s + R_L}. \tag{14.68}$$

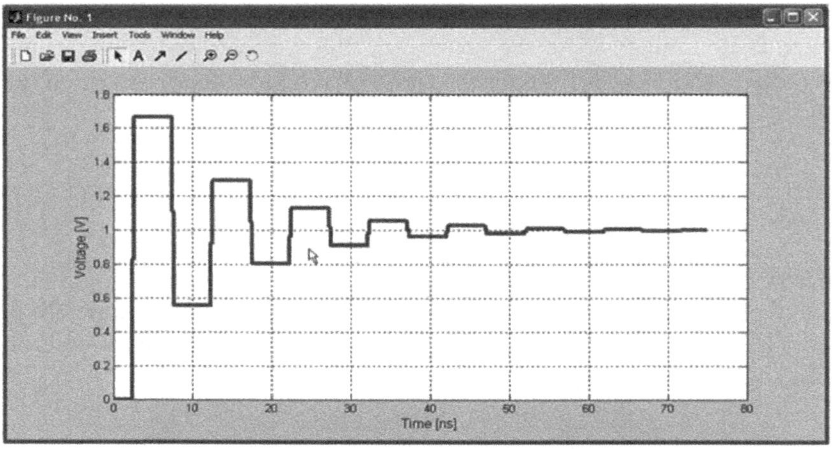

Figure 14.46. Step response closes the end of a 1 m long lossless OC TL ($Z_0 = 50\ \Omega$, $R_s = 10\ \Omega$). The total simulation step is 3000.

The step response of a TL can also be simulated via the TDRMeter package. In order to do that, the reader only needs to select rectangular pulse excitation with infinitely long (e.g., 1×10^6 ns) pulse duration. An example for the step response of a 50 Ω, OC TL is shown in Fig. 14.46. It is obvious from (14.67) that when TL is OC, the load voltage is equal to the source voltage. Since $V_0 = 1$ V in this example, the termination voltage approaches to this value as $t \rightarrow \infty$. Moreover, source resistor is less than the characteristic impedance of the TL; therefore, overdamping occurs at the load as observed in the figure.

Another example is given in Fig. 14.47 for the same parameters except source resistor is much higher than the TL's characteristic impedance. In this case, the voltage again approaches source voltage by gradual increase as $t \rightarrow \infty$. The final example (see Fig. 14.48) belongs to SC termination with source resistor much less than the TL's characteristic impedance.

Transmission lines (TLs) are investigated briefly in this section. A simple FDTD-based MATLAB script is introduced for the visualization of pulsed voltages and/or currents directly in time domain. Also, a comprehensive EM virtual tool TDRMeter is designed for the investigation of various characteristics of the TL. The TDRMeter virtual tool can be used to teach, understand, test, and visualize the TD TL

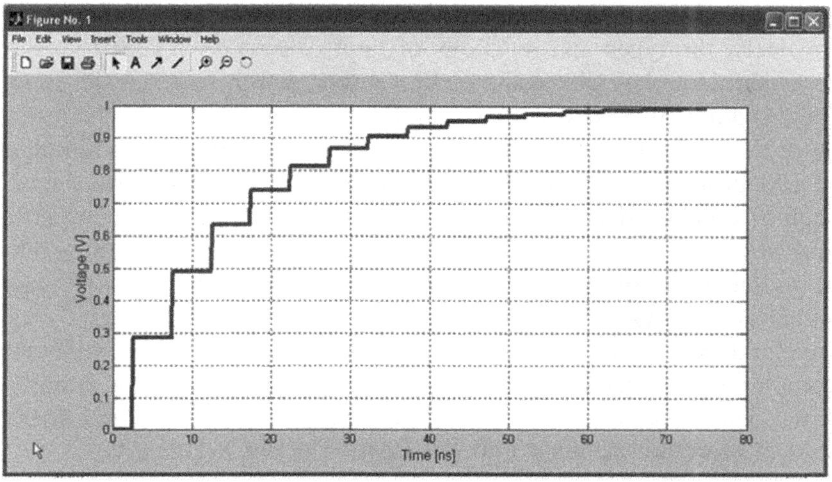

Figure 14.47. Step response close the end of a 1-m-long lossless OC TL ($Z_0 = 50\ \Omega$, $R_s = 300\ \Omega$). The total simulation step is 3000.

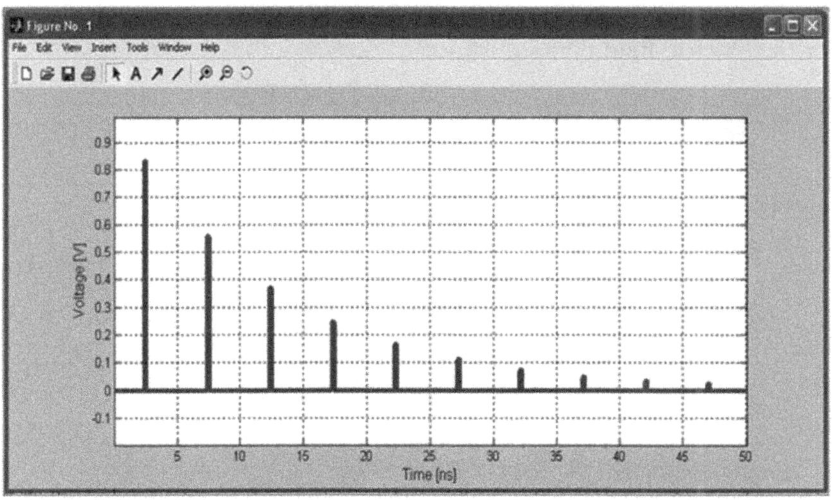

Figure 14.48. Step response close the end of a 1-m-long lossless SC TL ($Z_0 = 50\ \Omega$, $R_s = 10\ \Omega$). The total simulation step is 3000.

characteristics, echoes from various discontinuities, and terminations. Inversely, the types and locations of fault can be predicted from the recorded data. Any electronic device and/or circuit includes TL lines (i.e., coaxial, two-wire cables, parallel plate lines, microstrip lines, etc.); therefore, they have transient EM characteristics. A wide range of EM problems may be simulated with the TDRMeter virtual tool, such as conducted emissions, common and differential mode effects, and so on, both in time and frequency domains. Also, reflections due to various types of transmission structures, impedance effects of vias, signal and power line couplings, and so on, on printed circuits with smaller dimensions make signal integrity one of the most important and complex EM (especially electromagnetic compatibility [EMC]) problems. Therefore, the virtual tool may be particularly helpful for a student or circuit designer in the evaluation of signal integrity.

14.3 1D FDTD WITH SECOND-ORDER DIFFERENTIAL EQUATIONS

The second-order form for this plane wave representation, for example, for the electric field component, is

$$\left(\frac{\partial^2}{\partial z^2} - \frac{1}{c^2}\frac{\partial^2}{\partial t^2}\right)E_x(z,t) = 0, \tag{14.69}$$

with the boundary and initial conditions,

$$E_x(0,t) = g_1(t), \tag{14.70a}$$

$$E_x(Z_m,t) = g_2(t), \tag{14.70b}$$

$$E_x(z,0) = f_1(z), \tag{14.70c}$$

$$\frac{\partial E_x(z,0)}{\partial t} = f_2(z). \tag{14.70d}$$

Using numerical equivalents of the second-order differential operators in (14.69),

$$\frac{\partial^2}{\partial z^2}E_x^n(k) \Rightarrow \frac{E_x^n(k-1) - 2E_x^n(k) + E_x^n(k+1)}{\Delta z^2}, \tag{14.71a}$$

$$\frac{\partial^2}{\partial t^2}E_x^n(k) \Rightarrow \frac{E_x^{n-1}(k)-2E_x^n(k)+E_x^{n+1}(k)}{\Delta t^2}, \tag{14.71b}$$

with $p = c\Delta t/\Delta x$, yields

$$E_x^1(k)=(1-p^2)E_x^0(k)+\Delta t f_2(z)+\frac{p^2}{2}\left(E_x^0(k-1)+E_x^0(k+1)\right) \quad (n=1), \tag{14.72}$$

$$E_x^{n+1}(k)=2(1-p^2)E_x^n(k)-E_x^{n-1}(k)+p^2\left(E_x^n(k)+E_x^n(k+1)\right) \quad (n\geq 2). \tag{14.73}$$

As observed, the FDTD time loop starts after the second time step. This is merely because of the use of second-order differential operators. The user has to supply the first two steps manually. Since

$$E_x^0(k)=f_1(z), \tag{14.74a}$$

$$\frac{\partial E_x^0(k)}{\partial t}=f_2(z)=\frac{E_x^1(k)-E_x^{-1}(k)}{2\Delta t}, \tag{14.74b}$$

functions f_1 and f_2 have to be injected manually for the first two time steps.

The generic FDTD form in (14.73) uses values of three different time instants. This means that the values of previous two tie instants have to be stored. The procedure will be as follows:

- define these three time instant values as

$$E_x^{n+1}(k)=E_{\text{new}}(k),\ E_x^{n-1}(k)=E_{\text{old}}(k),\ E_x^n(k)=E(k), \tag{14.75}$$

- use these definitions in (14.72)

$$E(k)=(1-p^2)E_{\text{old}}(k)+2f_2(z)+\frac{p^2}{2}\left(E_{\text{old}}(k-1)+E_{\text{old}}(k+1)\right), \tag{14.76}$$

- assign $E_{\text{old}}(k) = E(k)$,
- start FDTD time loop with $n = 2,3,4, \ldots, N$ for the following three operations:

$$E_{\text{new}}(k) = 2(1-p^2)E(k) - E_{\text{old}}(k) + p^2(E(k)+E(k+1)),$$
$$E_{\text{old}}(k) = E(k), E(k) = E_{\text{new}}(k)$$

Time injection of the pulsed plane wave is quite different in the second-order wave equation. The pulsed plane wave may be injected from left or right (i.e., as $E_x(0, t) = g(t)$) or $E_x(Zm, t) = g(t)$). Alternatively, the pulsed plane wave may be located in space at the beginning. In this case, if Gauss function

$$g(z) = \exp\left\{-\left(\frac{z-z_s}{Z_g}\right)^2\right\} \rightarrow E_x^0(k) = \exp\left\{-\left(\frac{k-k_s}{K_g}\right)^2\right\} \tag{14.77a}$$

is used at the first time step, then once-differentiated Gaussian function

$$\begin{aligned} g'(z) &= -2\left(\frac{z-z_s}{Z_g}\right)\exp\left\{-\left(\frac{z-z_s}{Z_g}\right)^2\right\} \\ &\rightarrow E_x^1(k) = -2\left(\frac{k-k_s}{K_g}\right)\exp\left\{-\left(\frac{k-k_s}{K_g}\right)^2\right\} \end{aligned} \tag{14.77b}$$

must be used at the next (second) time step. Here, $z_s = k_s\Delta z$ is the location of pulse maximum and $Z_g = K_g\Delta z$ is spatial pulse extend. The rest is handled via iterations listed earlier. Table 14.5, Table 14.6, and Table 14.7 belong to the same MATLAB script which simulates 1D FDTD using the wave equation (14.73) but for different excitations. In Table 14.5, a spatial Gaussian pulse is injected before the FDTD time loop starts. Therefore, the pulse will directly be separated into two pulses propagating left and right once this code is run. The code in Table 14.6 injects two pulses from left and right boundaries. They may or may not have the same phase (this is achieved by just adding a negative sign to one of them). Finally, the code in Table 14.7 injects the pulse in the time domain at the specified source node (as done) in Section 14.1.2.

Note that (14.73) may also be used to model the oscillations of a taut string stretched between two fixed points ($0 \le z \le Z_m$) when $E_x(z, t)$ is replaced with the vertical displacement function $U(z, t)$ and

Table 14.5.
A MatLab Script for 1D FDTD with Wave Equation (Spatial Source Injection)

```
%% ====================================================
%     PROGRAM: LS_WAVE_1
%     AUTHOR:  LEVENT SEVGI
%     PURPOSE: FDTD SIMULATION OF 1D WAVE EQUATION IN THE
TIME DOMAIN
%     EQUATION       : C^2 * U_xx - u_tt = 0,      0 < X <
a,   t > 0
%     INITIAL COND.: U(x,0) = 0,    U_t(x,0) = 0
%     BOUND. COND. : U(0,t) = f_1(t), U(a,t) = f_2(t)
(ONCE-DIFF. GAUSS)
%     Source Injection: Spatial
%% ====================================================
clear all; clc;
eps0=8.82e-12;  mu0=pi*4e-7; eta=sqrt(mu0/eps0); c=1./
sqrt(eps0*mu0);
% Get input parameters
a = input('Length of X space,  a [m] : ');
n = input('Number of X points, n [ ] : ');
kmax = input('# of time iterations, kxmax [ ] : ');
lbc=input('Left Boundary Termination: (1) PEC, (0) Free-
space = ');
rbc=input('Right Boundary Termination: (1) PEC, (0)
Free-space = ');
% Calculate other parameters (using stability condition)
delx=a/(n-1); delt=delx/c; p=c*delt/delx;   %0.5;
ns=n/4; m=n/5; nt0=m/2.; x=0:delx:a;
% Prepare 1D arrays for the iterations
u=zeros(1,n);           % Current time values
ue=zeros(1,n);          % Previous time values
uy=zeros(1,n);          % Next time values
uoldl=0.0; uoldr=0.0;
% Source parameters
band=c/(10*delx); alfa=3.3*band*band; shift=4./
sqrt(alfa);
%% ---------------------------------------------------
ns=n/4; nT=n/5;   % Locate Gaussian pulse in space (at
t=0)
```

(*Continued*)

Table 14.5.
(Continued)

```
for i=2:n-1
  f1=exp(-(i-ns)^2/nT); u(i)=f1;
end
for i=2:n-1 % First Time step
  f2= -2*(i-ns)*exp(-(i-ns)^2/nT)/nT;
  uy(i)=(1-p^2)*u(i)+delt*f2+0.5*p^2*(u(i-1)+u(i+1));
end
ue=u; u=uy;
%% ----------------------------------------------------
for k=1:kmax    % Time Iteration starts
  t=k*delt;
  if lbc==0; u(1)=uoldl; end
  if rbc==0; u(n)=uoldr; end
  for i=2:(n-1)
    uy(i)=2*(1-p^2)*u(i)-ue(i)+p^2*(u(i-1)+u(i+1));
  end
  uoldr=u(n-1); uoldl=u(2);
  ue=u; u=uy;
% Plot wave vs. x-axis
  plot(x,u); xlabel('x axis'); xlabel('Wave Amplitude');
  xlim([0,a]); ylim([-1.1,1.1]); legend('Wave
Propagation', 4);
  pause(0.001)
end
% ----------------------- END -----------------------
```

$c = Tg/w$ (T, w, and g are the force, weight for unit string length, and the gravitational acceleration). The examples presented earlier show that the term numerical propagation in electrostatics, in the iterative forms of discrete Poisson and Laplace equations, refers to the convergence of the solution not EM wave propagation. On the other hand, in EM, numerical propagation refers to real wave propagation since Maxwell equations in their general forms model EM waves.

14.4 TWO-DIMENSIONAL (2D) FDTD MODELING

Most of EM problems including radiation and beam forming/beam steering, scattering, and propagation modeling are handled in 2D.

Table 14.6.

A MatLab Script for FDTD with 1D Wave Equation (Source Injection in Time from Left and Right Nodes)

```
%% ===========================================
%  PROGRAM: LS_WAVE_2
%  Source Injection: From boundary nodes
%% ===========================================
clear all; clc;
eps0=8.82e-12;  mu0=pi*4e-7; eta=sqrt(mu0/eps0); c=1./
sqrt(eps0*mu0);
% Get input parameters
a=1; n=100; ns=n/4; m=n/5; nt0=m/2; kmax=400; lbc=1;
rbc=1;
% Calculate other parameters (using stability condition)
delx=a/(n-1); delt=delx/c; p=c*delt/delx; x=0:delx:a;
% Prepare 1D arrays for the iterations
u=zeros(1,n);           % Current time values
ue=zeros(1,n);          % Previous time values
uy=zeros(1,n);          % Next time values
uoldl=0.0; uoldr=0.0;
% Source parameters
band=c/(10*delx); alfa=3.3*band*band; shift=4./
sqrt(alfa);
%% -----------------------------------------
for k=1:kmax    % Time Iteration starts
  t=k*delt;
  u(1)=-(0.5*k/nt0)*exp(-16.*(k*delt-nt0*delt)^2/
(m*delt)^2); %f_1(t)
  u(n)=(0.5*k/nt0)*exp(-16.*(k*delt-nt0*delt)^2/
(m*delt)^2);  %f_2(t)
  if lbc==0; u(1)=uoldl; end
  if rbc==0; u(n)=uoldr; end
  for i=2:(n-1)
    uy(i)=2*(1-p^2)*u(i)-ue(i)+p^2*(u(i-1)+u(i+1));
  end
  uoldr=u(n-1); uoldl=u(2);
% Replace the arrays
  ue=u; u=uy;
% Plot wave vs. x-axis
  plot(x,u); xlabel('x axis'); xlabel('Wave Amplitude');
  xlim([0,a]); ylim([-1.1,1.1]); legend('Wave Propagation', 4);
  pause(0.001)
end
% ----------------------- END -----------------------
```

Table 14.7.
A MatLab Script for FDTD with 1D Wave Equation (Source Injection in Time from a Given Node)

```
%% ==================================================
%   PROGRAM: LS_WAVE_3
%   Source Injection: From boundary nodes
%% ==================================================
clear all; clc;
eps0=8.82e-12;  mu0=pi*4e-7; eta=sqrt(mu0/eps0); c=1./
sqrt(eps0*mu0);
% Get input parameters
a=1; n=100; ns=n/4; m=n/5; nt0=m/2; kmax=400; lbc=1;
rbc=1;
% Calculate other parameters (using stability condition)
delx=a/(n-1); delt=delx/c; p=c*delt/delx; x=0:delx:a;
% Prepare 1D arrays for the iterations
u=zeros(1,n);          % Current time values
ue=zeros(1,n);         % Previous time values
uy=zeros(1,n);         % Next time values
uoldl=0.0; uoldr=0.0;
% Source parameters
band=c/(10*delx); alfa=3.3*band*band; shift=4./
sqrt(alfa);
%% -----------------------------------------------
for k=2:kmax    % Time Iteration starts
  t=k*delt;
  u(1)=0.0; u(n)=0.0;
  if lbc==0; u(1)=uoldl; end
  if rbc==0; u(n)=uoldr; end
  for i=2:(n-1)
    uy(i)=2*(1-p^2)*u(i)-ue(i)+p^2*(u(i-1)+u(i+1));
  end
  uoldr=u(n-1); uoldl=u(2);
% Excitation
  uy(ns)=uy(ns)+0.075*(t-shift)*exp(-alfa*(t-shift)^2)/
delt;
  ue=u; u=uy;
% Plot wave vs. x-axis
  plot(x,u); xlabel('x axis'); xlabel('Wave Amplitude');
  xlim([0,a]); ylim([-1.1,1.1]); legend('Wave
Propagation', 4);
  pause(0.001)
end
% ----------------------- END-----------------------
```

Therefore, 2D FDTD models and a generic package (virtual tool) are introduced in this section.

Maxwell equations

$$\nabla\times\vec{E} = -\mu\frac{\partial\vec{H}}{\partial t}, \tag{14.78a}$$

$$\nabla\times\vec{H} = \varepsilon\frac{\partial\vec{E}}{\partial t} + \sigma\vec{E} + \vec{J} \tag{14.78b}$$

reduce to two sets of scalar equations (i.e., TM_z and TE_z) in two-dimension if $\partial/\partial_y \equiv 0$.

SET #1: TM_z

$$\frac{\partial H_y}{\partial z} = -J_x - \varepsilon\frac{\partial E_x}{\partial t} + \sigma E_x, \tag{14.79a}$$

$$\frac{\partial E_x}{\partial z} - \frac{\partial E_z}{\partial x} = -\mu\frac{\partial H_y}{\partial t}, \tag{14.79b}$$

$$\frac{\partial H_y}{\partial x} = J_z + \varepsilon\frac{\partial E_z}{\partial t} + \sigma E_z. \tag{14.79c}$$

SET #2: TE_z

$$\frac{\partial E_y}{\partial z} = \mu\frac{\partial H_x}{\partial t}, \tag{14.80a}$$

$$\frac{\partial H_x}{\partial z} - \frac{\partial H_z}{\partial x} = J_y + \varepsilon\frac{\partial E_y}{\partial t} + \sigma E_y, \tag{14.80b}$$

$$\frac{\partial E_y}{\partial x} = -\mu\frac{\partial H_z}{\partial t}. \tag{14.80c}$$

As can be seen from these two sets of equations, only a *y*-directed electrical line source (LS) is enough to excite TE_z-type solutions. On the other hand, either *x*- or *z*-directed electric dipoles (or both) are required to excite TM_z-type solutions. Therefore, these two can be investigated separately. Figure 14.49 illustrates the two-dimensional computation volume and location of field components for the FDTD model and for both sets of equations.

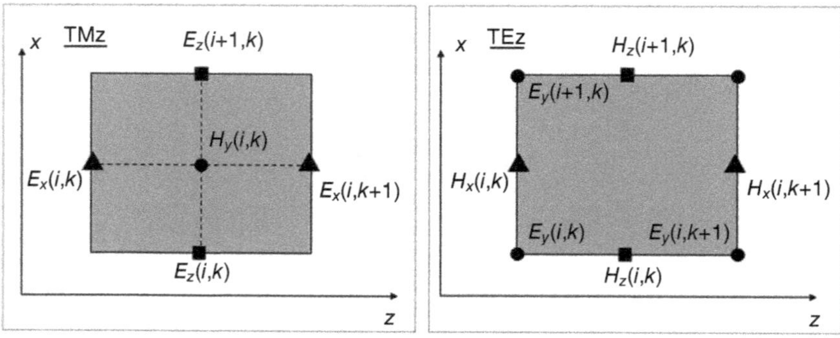

Figure 14.49. Reduced 2D FDTD components on *xz*-plane.

14.4.1 Field Components and FDTD Equations

14.4.1.1 TM$_z$ Case

The FDTD iterative equations for this type of problem are

$$E_x^n(i,k) = \frac{2\varepsilon - \sigma\Delta t}{2\varepsilon + \sigma\Delta t} E_x^{n-1}(i,k) - \frac{2\Delta t}{(2\varepsilon + \sigma\Delta t)\Delta z}\left[H_y^{n-1/2}(i,k) - H_y^{n-1/2}(i,k-1)\right], \tag{14.81a}$$

$$H_y^{n+1/2}(i,k) = H_y^{n-1/2}(i,k) + \frac{\Delta t}{\mu_0 \Delta x}\left[E_z^n(i,k) - E_z^n(i-1,k)\right] - \frac{\Delta t}{\mu_0 \Delta z}\left[E_x^n(i,k) - E_z^n(i,k-1)\right], \tag{14.81b}$$

$$E_z^n(i,k) = \frac{2\varepsilon - \sigma\Delta t}{2\varepsilon + \sigma\Delta t} E_z^{n-1}(i,k) + \frac{2\Delta t}{(2\varepsilon + \sigma\Delta t)\Delta x}\left[H_y^{n-1/2}(i,k) - H_y^{n-1/2}(i-1,k)\right], \tag{14.81c}$$

which, for example, can be used basically for the vertically polarized ground wave propagation problem. Majority of the low frequency (LF), medium frequency (MF), high frequency (HF), and very high frequency (VHF) communication systems use vertical monopoles erected above the ground and/or a building.

14.4.1.2 TE$_z$ Case

The FDTD equations for the TE$_z$-type problem are

$$E_y^n(i,k) = \frac{\varepsilon}{\varepsilon + \sigma\Delta t} E_y^{n-1}(i,k) - \frac{\Delta t}{(\varepsilon + \sigma\Delta t)\Delta x}\left[H_z^{n-1/2}(i+1,k) - H_z^{n-1/2}(i,k)\right] + \frac{\Delta t}{(\varepsilon + \sigma\Delta t)\Delta z}\left[H_x^{n-1/2}(i,k+1) - H_x^{n-1/2}(i,k)\right], \quad (14.82a)$$

$$H_x^{n+1/2}(i,k) = H_x^{n-1/2}(i,k) - \frac{\Delta t}{\mu\Delta z}\left[E_y^n(i,k) - E_y^n(i,k-1)\right], \quad (14.82b)$$

$$H_z^{n+1/2}(i,k) = H_z^{n-1/2}(i,k) + \frac{\Delta t}{\mu_0\Delta x}\left[E_y^n(i,k) - E_y^n(i-1,k)\right], \quad (14.82c)$$

which, for example, can be used in horizontally polarized wave propagation over the Earth's surface. Communication systems for ultrahigh frequency (UHF) frequencies and above use horizontally polarized waves. Today's electronic systems mostly use both polarizations.

14.4.2 FDTD-Based Virtual Tool: MGL2D Package

The virtual tool MGL2D is prepared in C++ in order to simulate both TE$_z$- and TM$_z$-type 2D EM problems [6]. The front panel of MGL2d is shown in Fig. 14.50. The front panel of MGL2D is horizontally divided into two; the top blocks are reserved for the user-supplied parameters. The large simulation window shows the online wave scattering on a chosen scenario. The user first specifies the range and height of the simulation window. Since a 500 × 250 cell space is reserved in the virtual tool to optimize the speed, it is advised to choose these values with the range/height ratio of 2.

The second-order MUR-type BC is used for the termination of the simulation window at four edges. The "+" and "–" signs represent free-space and nonpenetrable termination at each edge, respectively. The third block OBSTACLE is used for material location. PEC and/or lossy materials with different shapes can be located at anywhere on the simulation window by just using the mouse. Buttons are used for rectangular, triangular, and elliptical shaped obstacles. Selecting one of

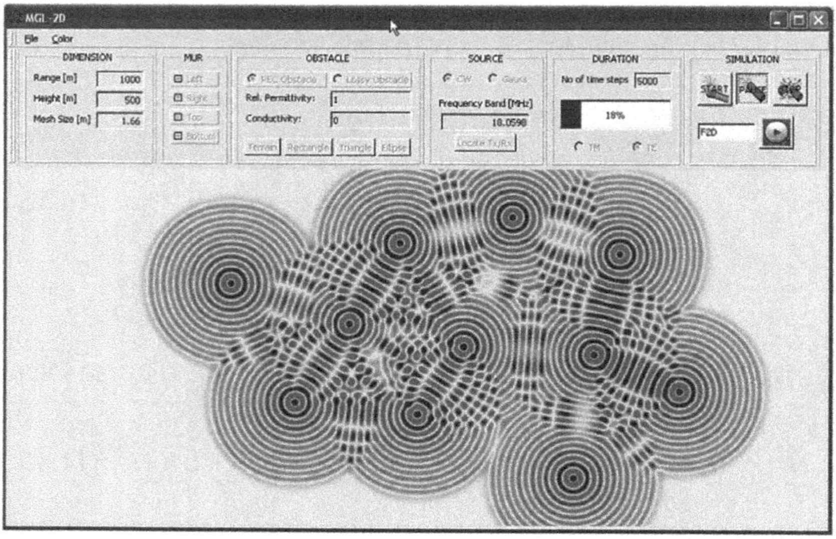

Figure 14.50. The front panel of the MGL2D virtual tool.

them and clicking and dragging the mouse at anywhere is enough to locate an obstacle with these shapes. The size of the obstacle can be adjusted by the mouse. Different colors are used for different materials. For example, PEC objects are drawn in black. Any other lossy object is in orange from dark to light, depending on the supplied loss of the material. The left and right buttons are used for locating and removing an object fully or partially. Two excitation types are reserved for the source: a CW or a pulsed Gaussian source. The frequency of the CW signal and the band of the Gaussian pulse are determined automatically from the cell sizes. The user also specifies the type of the problem and the number of time steps for the simulation. The control buttons are used to start, pause, and restart the simulation. The most beautiful task of this virtual tool is that the user can record video clips of the simulation. Finally, any user-prepared scenario can be saved and reloaded for further analyses.

Any 2D EM problem can be simulated in the MGL2D. The vertical and horizontal coordinates of the simulation window correspond to x- and z-axes, respectively. The user only needs to design his/her scenario by using a mouse. Multisource and multiobservation points can arbitrarily be located. For example, the scenario in Fig. 14.50 belongs to 11 CW LSs located at arbitrary points in free space. The 2D cylindrical wave spread and the interference among sources make the source points appear nicely.

14.4.3 Characteristic Examples

In order to show the capabilities of MGL2D, different examples are included in this section. The user may repeat these tests and also video capture wave scattering for each case.

14.4.3.1 Wireless Systems and Rural Propagation

Propagation engineers need to know ground wave propagation losses along nonflat terrain profiles in rural regions in order to predict link budget of a radio communication system, to design a wireless network, to optimize wireless cellular plans, and so on; therefore, they require powerful radiowave field strength prediction tools. MGL2D can effectively be used for this purpose. An example is given in Fig. 14.51. The

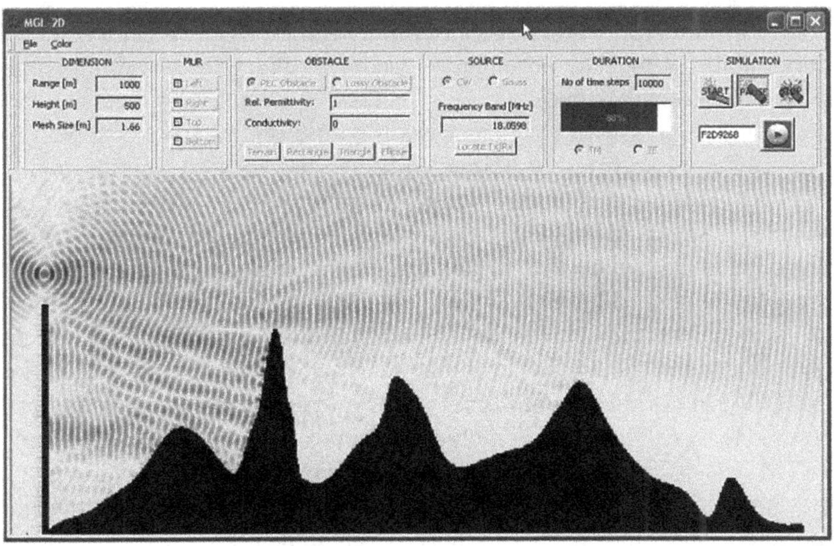

Figure 14.51. Rural propagation modeling: User-specified 1-km-long terrain profile and 18 MHz CW signal scattering.

user only needs to select *Terrain* from the obstacle block and draw the terrain by clicking on a number of points on the screen. MGL2D uses a cubic-spline interpolation algorithm and matches the best terrain profile which crosses all the marked points. Here, a user-drawn 1-km-long terrain profile is shown in a 1 km × 0.5 km rectangular window. The highest point of the terrain is about 250 m and an 18 MHz CW source is 325 m above the ground. Late time response is shown in the figure. Wave scattering, shadowing, and interference caused by multi-reflected wave components are clearly observed in the figure.

Any type of penetrable or nonpenetrable terrain profiles may be drawn and propagation along these profiles can visually be observed. The shadowing effects along deep valleys can also be investigated in the frequency domain. To do that, the user needs to use a pulse source and locate observation points on suitable regions, so that time-domain behaviors recorded on these observation points can be transformed into the frequency domain using a short discrete or fast Fourier scripts.

14.4.3.2 Propagation in Urban Areas: Vertical Projection

MGL2D can also be used in propagation modeling in urban regions. Roughly speaking, urban propagation modeling may be grouped into two, propagation along vertically projected regions, in order to incorporate ground reflections and building heights, or propagation along horizontally projected propagation paths so that effects among high rises can be investigated.

A vertically projected propagation path along a street is shown in Fig. 14.52. Five different-height buildings are located longitudinally. All the buildings are nonpenetrable, except the left wall of the second building. An 18 MHz CW signal has been generated and early time response is shown in the figure. The scenario belongs to a TE_z-type problem. It should be noted that the excitation for TE_z- and TM_z-type problems are totally different. The excitation is performed via the E_y component for the TE_z-type problem, which means a LS is used. For the TM_z-type problem, the signal may be injected via E_x or E_z components. Here, E_x is chosen which corresponds to a vertical short dipole in TM_z case.

Any scenario may be saved for further uses. It may then be reloaded, modified, and other scenarios may be created. Figure 14.53 shows a slightly modified version of the scenario shown in Fig. 14.52 with a late time response. The fourth building is replaced with a kind of pen-

Figure 14.52. Outdoor propagation modeling along a vertically projected user-specified street after 1000 time steps.

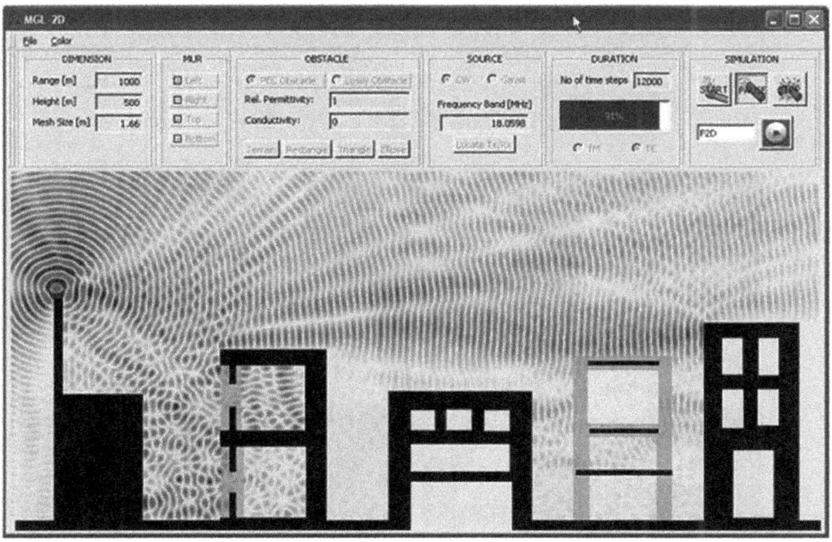

Figure 14.53. Outdoor propagation modeling along a vertically projected user-specified street after 11,000 time steps.

etrable three-floor building with nonpenetrable floors. The penetration of the signal through the walls is clearly observed in this scenario.

14.4.3.3 Propagation in Urban Areas: Horizontal Projection

Propagation along horizontally projected regions is important in downtowns with high buildings. For example, having a transmitter and receiver pair somewhere between fifth and tenth floor of a 20-floor building lets us neglect ground reflections. Therefore, the propagation medium may be thought as a kind of tunnel among the buildings. A typical example is shown in Fig. 14.54, where a number of different shaped nonpenetrable buildings are located on each side of a 100-m-long street. A transmitter is located on the corner of the first building on top left. Figure 14.55 shows late time signal scattering for the same scenario. Although most of the region is shaded by these buildings, the signal reaches almost everywhere via multireflections as well as diffractions.

Indoor propagation has increasingly gained attention especially in parallel to the development of the novel home and office wireless

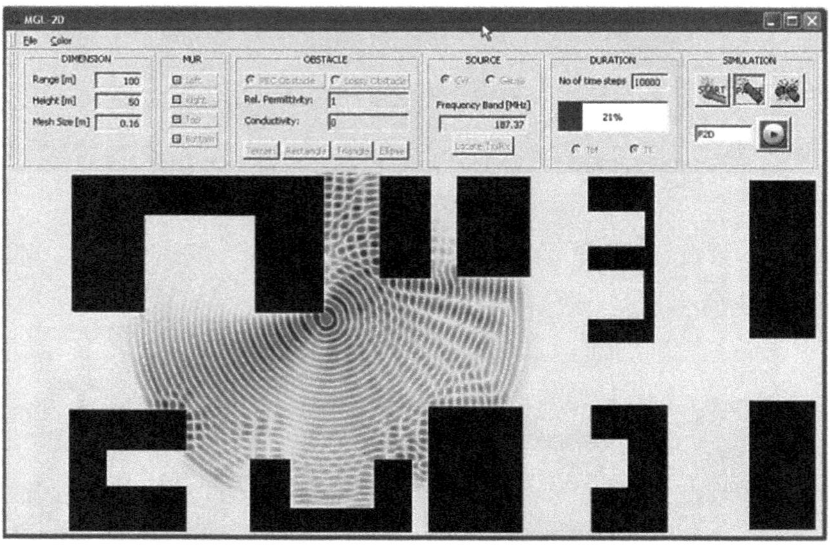

Figure 14.54. Outdoor propagation modeling along a horizontally projected user-specified street after 2100 time steps.

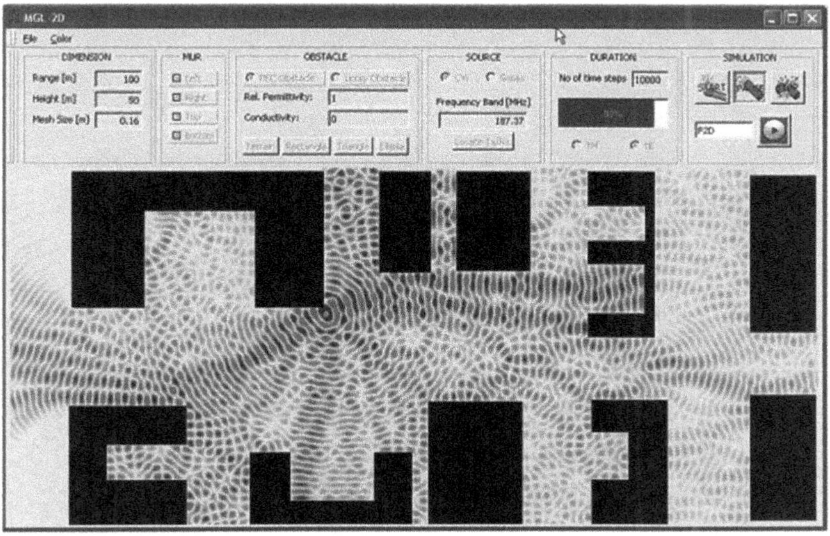

Figure 14.55. Outdoor propagation modeling along a horizontally projected user-specified street after 8600 time steps.

systems, such as WiFi, Bluetooth, and so on. MGL2D may also be used to simulate indoor propagation in two-dimension. An example scenario is given in Fig. 14.56. Here, a top view of a two-room building is pictured together with various loss-free and lossy objects. The CW signal source is located at top right. The TE_z type is chosen. It should be noted that the CW signal frequency is specified by MGL2D automatically from "the cell size equal to one over ten of the wavelength" criterion. The 2D simulation window is again 1 km × 0.5 km and the corresponding CW signal frequency is 18 MHz.

14.4.3.4 Indoor Propagation

Reducing the sizes of the simulation window to 10 m × 5 m will increase the CW signal frequency to 1.8 GHz. Therefore, the size of a room, a street, or a terrain is not that important in the visualization of EM effects. Using the frequency scaling, any size can be chosen to visualize EM effects. It should also be noted that it is not possible to increase the signal frequency without increasing the number of cells of the simulation window. A 600 × 300 window is fixed in MGL2D;

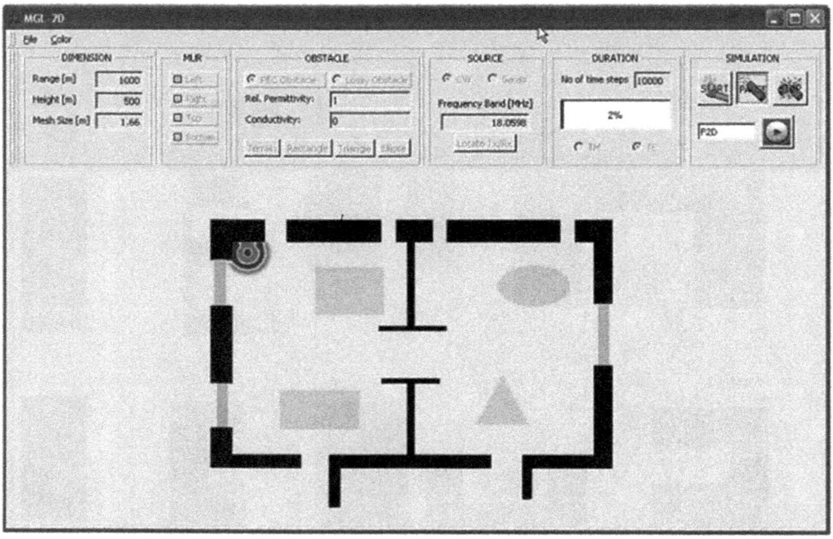

Figure 14.56. Indoor propagation modeling: Top view of a typical two-room apartment with various penetrable and nonpenetrable objects. The source is located inside left room.

therefore, in order to double the signal frequency keeping the object sizes same, one need to change the simulation window to 1200 × 600 and this is not possible at this stage.

Figure 14.57 shows the late time response in the same scenario given in Fig. 14.56. As observed, although the source is masked by a variety of objects, it is still possible to cover almost all the region via multireflected and edge- and tip-diffracted waves. The simulations may be repeated with different color scales. Each color represents different signal strengths. In some cases, different tones of different colors are used to represent different signal strengths. In others, different colors are used for different signal strengths.

Using the tones for different signal strengths does not always yield a nice visualization. For example, edge- and tip-diffracted waves are always much weaker than direct and reflected waves (usually one or two orders, sometimes may be up to three orders). In order to better visualize these much weaker diffracted signals, different colors may be used. An example is shown in Fig. 14.58 which exactly belongs to the same scenario given in Fig. 14.57. Obviously, changing the colors may result in strong visualization of weak signals.

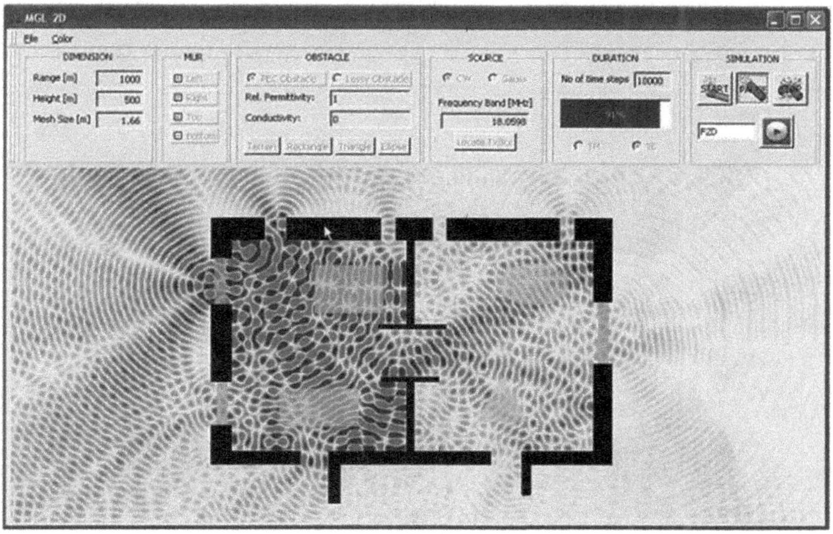

Figure 14.57. Indoor propagation modeling: Late time response of a CW signal and the coverage.

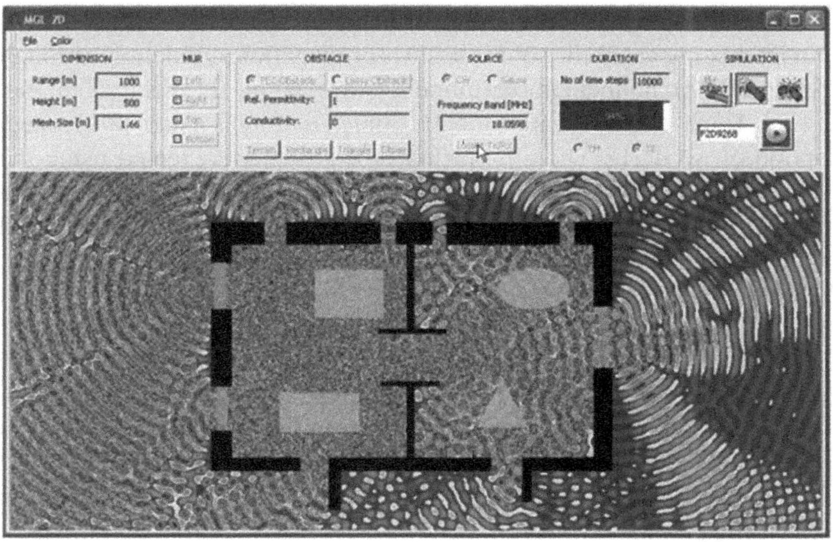

Figure 14.58. Indoor propagation modeling: Same as in Fig. 14.56 but with a different color scale.

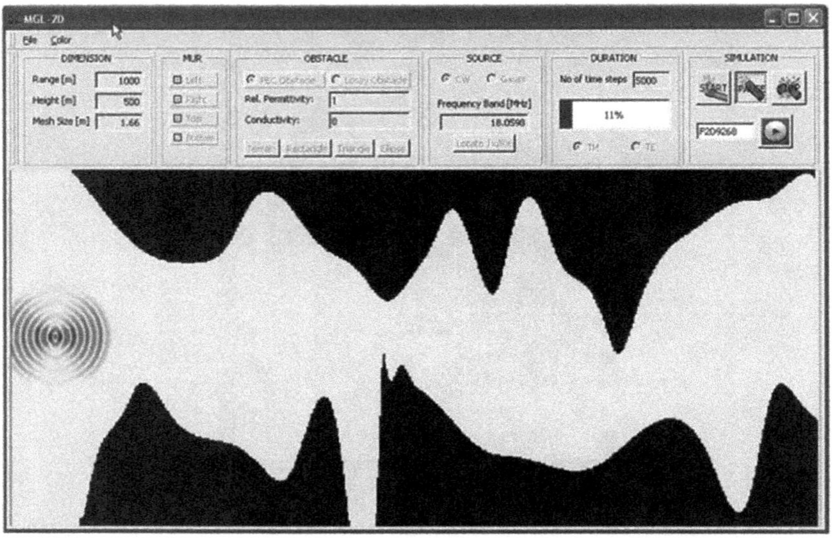

Figure 14.59. Propagation through a user-specified tunnel and a CW source location.

14.4.3.5 Propagation through Tunnels

An interesting capability of MGL2D is propagation simulations through penetrable or nonpenetrable tunnels. An example is pictured in Fig. 14.59. Here, a kind of nonsmooth tunnel is drawn using the *Terrain* option of MGL2D. The top portion of the tunnel is created using the right button of the mouse, while the bottom profile is drawn using the left button. The CW source and early time response are also observed in the figure. The late time response of this scenario is shown in Fig. 14.60.

MGL2D is also capable of simulating closed structures. For example, two rectangular resonators are shown in Fig. 14.61. Here the simulation window is 100 m × 50 m; therefore, the sizes of the resonators are approximately 25 m × 10 m. They both are excited by Gaussian pulses and time-domain responses are recorded inside each. The short MATLAB script listed in Table 14.8 is used to plot both time variations inside the resonators and frequency responses. The time variations of both recorded signals are plotted in Fig. 14.62. Their frequency responses are given in Fig. 14.63.

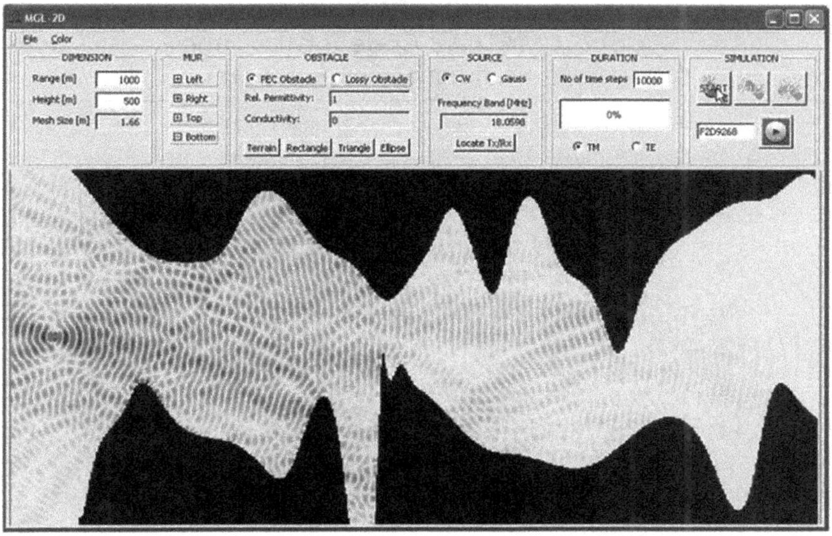

Figure 14.60. Propagation through a user-specified tunnel: Late time response.

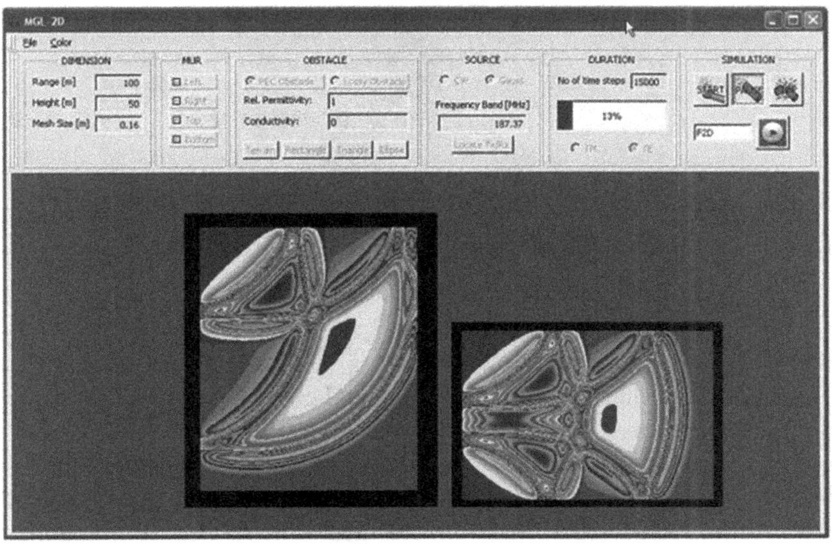

Figure 14.61. Two rectangular resonators and pulse scattering inside.

Table 14.8.
A Short MATLAB Code for the Time and Frequency Responses

```
% ------------------------------------------------------------
% Program: MGLFFT.m
% ------------------------------------------------------------
clc; clear all;
obs1=load('obs1.dat'); obs2=load('obs2.dat');
figure(1)
plot(obs1(:,1)*1e9,obs1(:,2),'Linewidth',2)
hold on; plot(obs2(:,1)*1e9,obs2(:,2),'r-','Linewidth',2)
xlabel('Time [Nanosec]');
ylabel('Ey(t)');
title('Field components at two observation points');
% --------
dt=obs1(2,1)-obs1(1,1);
[N M]=size(obs1); fmax=1/(2*dt);
for n=1:N
   tt(n)=(n-1)*dt;
   f(n)=-fmax+2*(n-1)*fmax/(N-1);
end
X=fft(obs1(:,2));  X1=fftshift(X);
X=fft(obs2(:,2));  X2=fftshift(X);
% ---
figure(2); plot(f/1e9,abs(X1),'b','Linewidth',2)
hold on; plot(f/1e9,abs(X2),'r-','Linewidth',2)
xlabel('Frequency [GHz]');
ylabel('Amplitude [dB]');
title('Frequency Responses of the resonators');
% ----------------------- END -----------------------
```

It is known that locations of source and observer inside a resonator directly affect the excitation and/or observation of the resonators' modes. The user may change source/observer locations inside the resonator and observe resonance shifts in the frequency domain. Even the sizes of the resonators may be changed and resonance frequency changes may be investigated. For example, keeping the width same, one can gradually change the length of the resonator and observe the resonances.

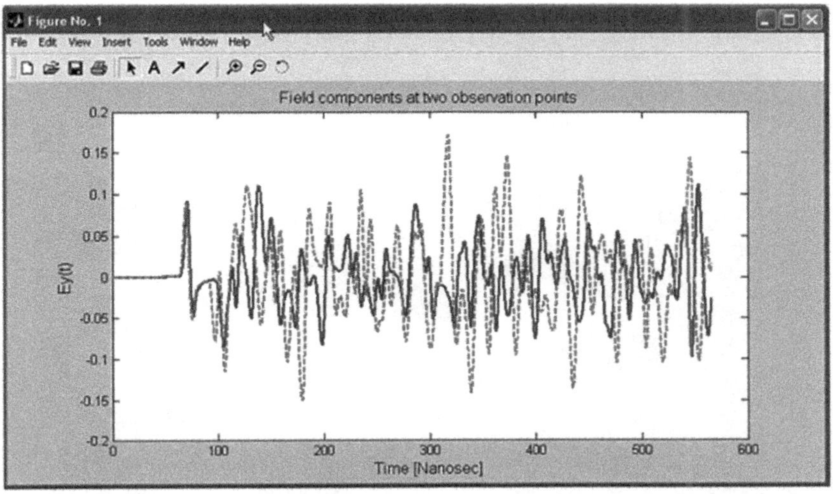

Figure 14.62. Pulse scattering inside the resonator recorded at two different observation points.

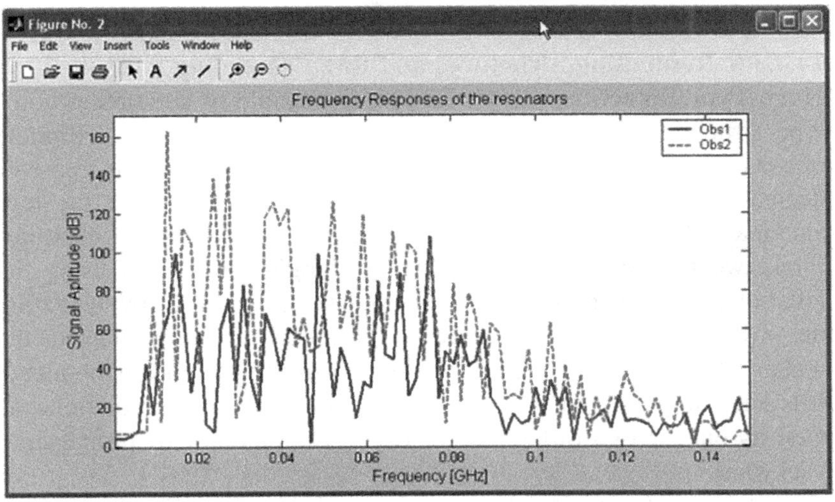

Figure 14.63. Fourier transform of the recorded signals shown in Fig. 14.61. The observed resonance frequencies determined by the sizes of the resonator.

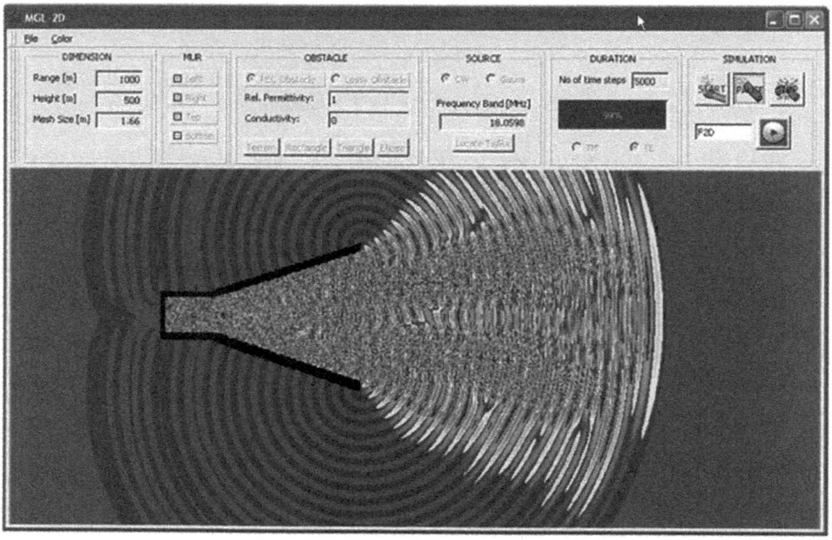

Figure 14.64. A horn-type element excited with a CW source inside.

14.4.3.6 Resonating Structures and EMC Modeling

Horn-Type Structures. MGL2D is also capable of simulating horn-type structures. Although only rectangular, triangular, and elliptical objects can be located on the simulation window, a gradual increase or decrease in the cross section may also be drawn. For example, the horn antenna in Fig. 14.64 is drawn using *Rectangle* and *Triangle* buttons. First, two upside-down nonpenetrable triangles are drawn using the left button of the mouse. Then inside is made empty by repeating the procedure with the right button. The procedure is repeated for the rectangular sections and the structure shown in the figure is obtained. It is amazing to see the strong forward and weakly side radiations, as well as one or two order weaker edge-diffracted waves propagating backward.

Another interesting radiating structure is shown in Fig. 14.65. This is a horn-type structure connected to a parallel plate open-ended waveguide. There is a kind of parabolic reflector in front of the horn. Both forward and backward radiations as well as diffractions are observed in the figure.

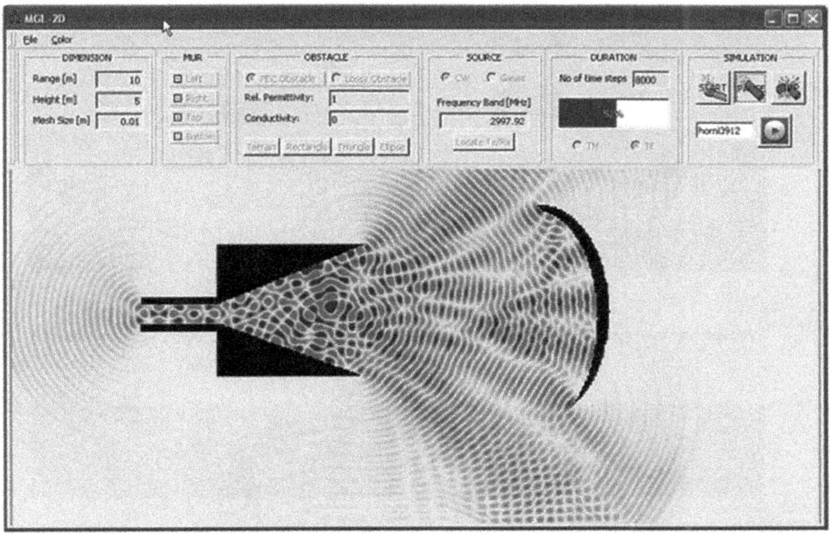

Figure 14.65. A horn-type element and a kind of parabolic reflector in the front.

Periodic Structures. It is also possible to model periodic structures in MGL2D. Two examples are given in this subsection. First a two-section parallel plate waveguide with nonpenetrable boundaries is created. The sections are coupled via small aperture-type discontinuities. This is a 100 m $\times$ 50 m structure with 0.16 m cell size, and this corresponds to a CW signal with approximately 187 MHz. The front panel of the MGL2D for this scenario is given in Fig. 14.66. The strong propagation along the top waveguide and weakly coupling to the bottom waveguide while propagating is observed in the figure. As done with the resonators, time frequency analysis along the waveguide can be performed using the same MATLAB script given in Table 14.8.

Another periodic structure simulated in MGL2D is presented in Fig. 14.67. This is a 1-m-long parallel plate waveguide with periodic triangular barriers asymmetrically located on top and bottom waveguide walls. In order show the effects of time-domain CW signal propagation inside this periodic parallel plate waveguide, an identical waveguide without periodic triangular barriers is also created. Both waveguides are fed with a TE_z-type source. The frequency is nearly 3 GHz. The plots show wave propagation inside the waveguides after

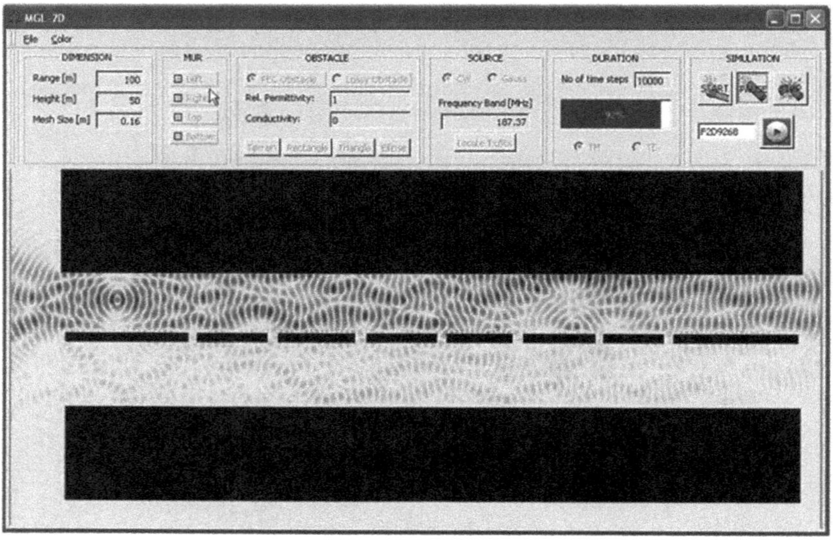

Figure 14.66. A double-layered parallel plate waveguide with coupling apertures in the middle.

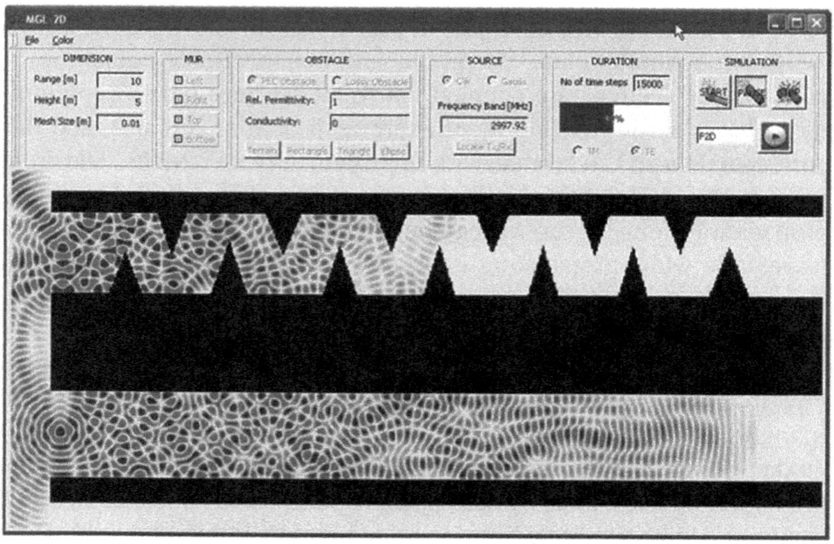

Figure 14.67. A parallel plate waveguide with asymmetrically located periodic triangular barriers.

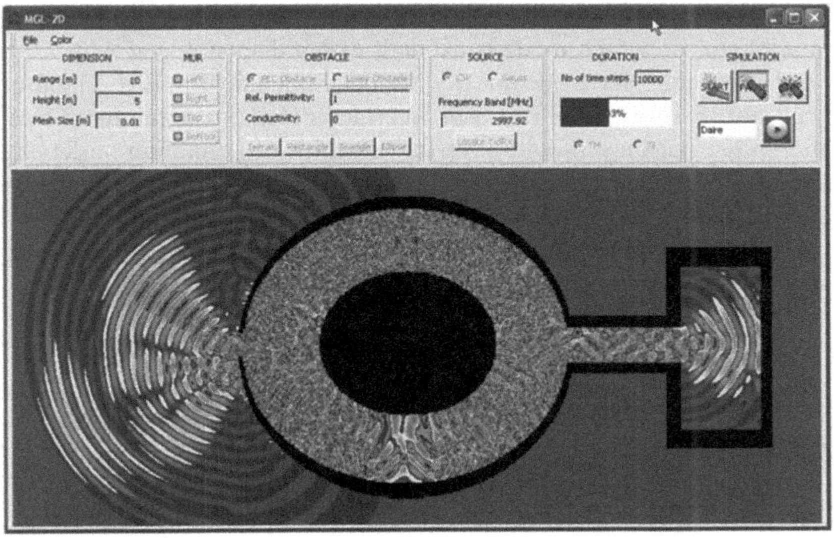

Figure 14.68. User-drawn three-part closed resonator.

about 7500 time steps. As observed, the waveguide with periodic barriers acts as a delay line. These types of waveguides act as delay lines as long as source and observer inside the waveguide do not have a clear line-of-sight (LOS) connection.

Other Complex Structures. The final example in Fig. 14.68 belongs to a three-part complex resonator. The first part is a cylindrical resonator and circular resonances are determined by the length of the peripheral. It is connected to a rectangular resonator via a short parallel plate waveguide. The width and length of the rectangular resonator determine the resonances. The short parallel plate waveguide also transmits energy from one end to the other via the modes (i.e., resonances) excited inside. Therefore, if one locates a Gaussian pulse source somewhere inside and observes the signal some other point inside, he/she could get the resonance characteristics. The only thing to do is to use the MATLAB script listed in Table 14.8 in order to plot resonances versus frequency.

It should be very educating if one tries to predict resonances of this structure and compares the predicted values with the resonances obtained via the simulation. Changing the location of both source and the observer, and/or using multisource and multiobservers inside it is possible to see which mode is excited, which does not, and why.

Virtual tools have become very effective in teaching EM. The user need not to know graphical user interface details, programming tips, and so on, but they should have the capability of handling variety of different problems and should give enough mathematical explanations as well as physical insights. They have been used more and more not only in classical lectures but also in novel and alternative teaching approaches such as *problem-based learning*, *inquiry-based teaching*, and so on.

14.5 CANONICAL 2D WEDGE SCATTERING PROBLEM

The 2D FDTD model may also be used in scattering investigations [7]. Chapter 10 presents the canonical wedge scattering problem and reviews basic high frequency asymptotic (HFA) models. Here, the same problem is taken into account and solved numerically using a 2D FDTD model. A MATLAB-based diffraction tool for the investigation of scattered fields around a 2D PEC wedge is also introduced. The results of the FDTD simulations are compared with analytical exact as well as some HFA techniques. A major numerical simulation problem, FDTD staircase modeling problem, is removed by the application of Dey–Mittra conformal approach.

14.5.1 Problem Postulation

The 2D wedge diffraction is a canonical problem and plays a fundamental role in the construction of HFA techniques [1–8]. It has been revisited several times for the last couple of decades. One of these revisits is presented in Ref. 9, where several HFA techniques, such as geometric optics (GO), physical optics (PO), physical theory of diffraction (PTD), uniform theory of diffraction (UTD), and parabolic equation (PE), are compared with each other and advantages/disadvantages are listed. A MATLAB-based HFA numerical tool has also been introduced [8]. Then, a novel, finite-difference time-domain (FDTD)-based multistep numerical diffraction coefficient calculation method has been introduced in Ref. 9.

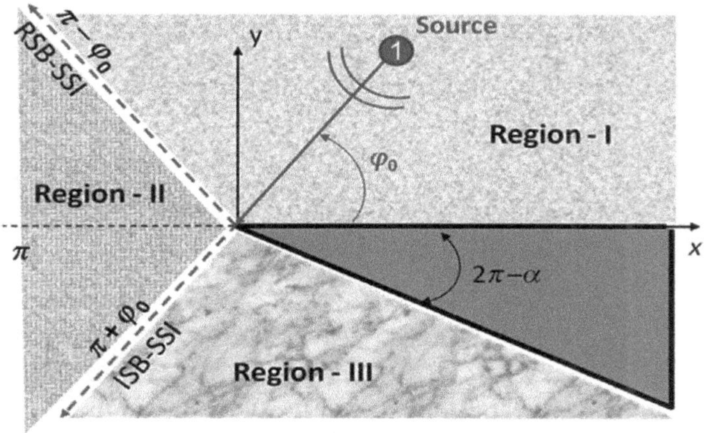

Figure 14.69. The 2D PEC wedge problem, line source (LS) illumination, and three characteristic regions separated by RSB and ISB.

A novel MATLAB-based WedgeFDTD package is introduced in this section. The package uses the multistep FDTD approach presented in Ref. 9 and results are compared with available exact and HFA models. Deficiencies of staircase discretization are removed by the application of Dey–Mittra conformal technique [10].

The 2D wedge scattering scenario is pictured in Fig. 14.69. The polar coordinates r, φ, z are used throughout the paper. The z-axis is aligned along the edge of the wedge. The angle φ is measured from the top face of the wedge. The exterior angle of the wedge equals α. The wedge is illuminated by a LS. Source and observer points are (r_0, φ_0) and (r, φ), respectively.

The scenario in Fig. 14.69 ($0 < \varphi_0 < \pi$) belongs to the single side illumination (SSI), where *the top face is always illuminated.* The three regions are separated by two critical boundaries: *incident shadow boundary* (ISB) ($\varphi = \pi + \varphi_0$) and *reflection shadow boundary* (RSB) ($\varphi = \pi - \varphi_0$). All field components—incident, reflected, and diffracted fields—exist in region I. In region II, only incident and diffracted fields exist. Only diffracted fields exist in region III.

If the source is in ($\alpha - \pi < \varphi_0 < \pi$), *both faces are always illuminated* and this is called double side illumination (DSI). In this case, all three field components exist in regions ($0 \leq \varphi < \pi - \varphi_0$) and

($2\alpha - \pi - \varphi_0 \leq \varphi < \alpha$). There is no reflections in ($\pi - \varphi_0 \leq \varphi < 2\alpha - \pi - \varphi_0$).

The field (for $\exp(-i\omega t)$) outside the wedge satisfies

$$\left(\frac{\partial^2}{\partial r^2}+\frac{1}{r}\frac{\partial}{\partial r}+\frac{1}{r^2}\frac{\partial^2}{\partial\varphi^2}+k^2\right)u=\frac{I_0}{r}\delta(r-r_0)\delta(\varphi-\varphi_0), \tag{14.83}$$

the BC

$$u_s=0 \quad \text{or} \quad \partial u_k\,/\,\partial n=0 \text{ on } \varphi=0,\alpha, \tag{14.84a}$$

and the Sommerfeld's radiation condition (SRC) at infinity,

$$\underset{r\to\infty}{\mathrm{Lim}}\sqrt{kr}\left(\frac{du}{dr}-iku\right)=0. \tag{14.84b}$$

Function u_s represents the z-component of electric field intensity E_z (TM), while function u_h is the z-component of magnetic field intensity H_z (TE), which, in acoustics, refer to acoustically soft (soft boundary condition, SBC) and hard (hard boundary condition, HBC) wedges, respectively.

14.5.2 Review of Analytical Models

The problem has analytical exact as well as HFA solutions. Only analytical exact solution by integral and UTD model are given; the rest can be found in Chapter 10 and in Refs. 7–9 and 11.

14.5.2.1 Exact Solution by Integral

The diffracted field solutions with SBC and HBC for both SSI and DSI are presented by Bowman and Senior in handbook [12] as

$$\begin{aligned}u_s^{d,BS}&=\{V_d(-\pi-\varphi+\varphi_0)-V_d(\pi-\varphi+\varphi_0)\}\\&\quad-\{V_d(-\pi-\varphi-\varphi_0)-V_d(\pi-\varphi-\varphi_0)\},\end{aligned} \tag{14.85a}$$

$$\begin{aligned}u_s^{d,BS}&=\{V_d(-\pi-\varphi+\varphi_0)-V_d(\pi-\varphi+\varphi_0)\}\\&\quad+\{V_d(-\pi-\varphi-\varphi_0)-V_d(\pi-\varphi-\varphi_0)\},\end{aligned} \tag{14.85b}$$

where

$$V_d(\beta)=\frac{1}{2\pi n}\int_0^{\infty} H_0^{(1)}[kR(it)]\frac{\sin(\beta/n)}{\cosh(t/n)-\cos(\beta/n)}dt. \tag{14.85c}$$

$$n=\alpha/\pi,\ R(\eta)=\sqrt{r^2+r_0^2+2rr_0\cos(\eta)}.$$

Total fields $u_{s,h}^{t,BS}$ can be obtained by adding GO fields

$$u_{s,h}^{t,BM}/u_0=\left(u_{s,h}^{d,BM}+u_{s,h}^{GO}\right)/u_0, \tag{14.86}$$

where GO field that under LS excitation is given as follows:

$$u_{s,h}^{GO}=\begin{cases} H_0^{(1)}(kR_1)\pm H_0^{(1)}(kR_2), & 0\le\varphi<\pi-\varphi_0 \\ H_0^{(1)}(kR_1), & \pi-\varphi_0\le\varphi<\pi+\varphi_0. \\ 0, & \pi+\varphi_0\le\varphi\le\alpha \end{cases} \tag{14.87a}$$

for SSI and

$$u_{s,h}^{GO}=\begin{cases} H_0^{(1)}(kR_1)\pm H_0^{(1)}(kR_2), & 0\le\varphi<\pi-\varphi_0 \\ H_0^{(1)}(kR_1), & \pi-\varphi_0\le\varphi<2\alpha-\pi-\varphi_0. \\ H_0^{(1)}(kR_1)\pm H_0^{(1)}(kR_3), & 2\alpha-\pi-\varphi_0\le\varphi\le\alpha \end{cases} \tag{14.87b}$$

for DSI.

14.5.2.2 The UTD Model

The UTD diffracted fields are in the form of

$$u_{s,h}^{d,\mathrm{UTD}}=u_0 d_{s,h}^{\mathrm{UTD}}\frac{e^{-jkr}}{\sqrt{r}}, \tag{14.88}$$

where $u_0=H_0^{(2)}(kr_0)$ and the time dependence is $\exp(j\omega t)$ [13]. According to the UTD, the diffraction coefficients for SBC and HBCs with LS as follows:

$$d_s^{UTD}(r,\varphi,\varphi_0,k) = \frac{-e^{-j\pi/4}}{2n\sqrt{2\pi k}}\left[\begin{array}{l}\cot\left(\dfrac{\pi-\xi^-}{2n}\right)F\left(kLg^+(\xi^-)\right)+\cot\left(\dfrac{\pi+\xi^-}{2n}\right)F(kLg^-(\xi^-)) \\ -\cot\left(\dfrac{\pi-\xi^+}{2n}\right)F\left(kLg^+(\xi^+)\right)-\cot\left(\dfrac{\pi+\xi^+}{2n}\right)F\left(kLg^-(\xi^+)\right)\end{array}\right], \tag{14.89a}$$

$$d_h^{\mathrm{UTD}}(r,\varphi,\varphi_0,k) = \frac{-e^{-j\pi/4}}{2n\sqrt{2\pi k}}\left[\begin{array}{l}\cot\left(\dfrac{\pi-\xi^-}{2n}\right)F\left(kLg^+(\xi^-)\right)+\cot\left(\dfrac{\pi+\xi^-}{2n}\right)F\left(kLg^-(\xi^-)\right) \\ +\cot\left(\dfrac{\pi-\xi^+}{2n}\right)F\left(kLg^+(\xi^+)\right)+\cot\left(\dfrac{\pi+\xi^+}{2n}\right)F\left(kLg^-(\xi^+)\right)\end{array}\right], \tag{14.89b}$$

Where $\xi^+ = \varphi + \varphi_0$, $\xi^- = \varphi - \varphi_0$, and the Fresnel function is

$$F(X)=2j\sqrt{X}e^{jX}\int_{\sqrt{X}}^{\infty} e^{-j\tau^2}d\tau. \tag{14.90}$$

L, $g^{\pm}$ are determined as given in Ref. 13 by

$$L=\frac{rr_0}{(r+r_0)},\quad g^{\pm}(\xi)=2\cos^2\left(\frac{2n\pi N^{\pm}-\xi}{2}\right). \tag{14.91}$$

Here, $N^{\pm} = (\pm\pi + \xi)/(2n\pi)$ are the integers which most nearly satisfy the last equation given in (14.91). Note that the cotangent functions in (14.89) become singular at ISB/RSB and are replaced with

$$\cot\left(\frac{\pi\pm\beta}{2n}\right)F\left(kLg^{\pm}(\xi)\right)\simeq n\left[\sqrt{2\pi kL}\,\mathrm{sgn}(\varepsilon)-2kL\varepsilon e^{-j\frac{\pi}{4}}\right]e^{-j\frac{\pi}{4}} \tag{14.92}$$

for small ε. The UTD-based total fields are then obtained by adding the GO fields appropriately,

$$u_{s,h}^{t,\mathrm{UTD}}/u_0 = u_{s,h}^{d,\mathrm{UTD}}/u_0 + u_{s,h}^{\mathrm{GO}}, \tag{14.93}$$

with the GO solutions given as for SSI ($0 < \varphi_0 < \alpha - \pi$) ,

$$u_{s,h}^{\mathrm{GO}} = \begin{cases} e^{jkr\cos(\varphi-\varphi_0)} \pm e^{jkr\cos(\varphi+\varphi_0)}, & 0 \le \varphi < \pi - \varphi_0 \\ e^{jkr\cos(\varphi-\varphi_0)}, & \pi - \varphi_0 \le \varphi < \pi + \varphi_0 \\ 0, & \pi + \varphi_0 \le \varphi \le \alpha \end{cases} \quad (14.94a)$$

and for DSI ($\alpha - \pi < \varphi_0 < \pi$),

$$u_{s,h}^{\mathrm{GO}} = \begin{cases} e^{jkr\cos(\varphi-\varphi_0)} \pm e^{jkr\cos(\varphi+\varphi_0)}, & 0 \le \varphi < \pi - \varphi_0 \\ e^{-jkr\cos(\varphi-\varphi_0)}, & \pi - \varphi_0 \le \varphi < 2\alpha - \pi - \varphi_0. \\ e^{jkr\cos(\varphi-\varphi_0)} \pm e^{jkr\cos(2\alpha-\varphi-\varphi_0)}, & 2\alpha - \pi - \varphi_0 \le \varphi \le \alpha \end{cases} \quad (14.94b)$$

Two examples are given in Fig. 14.70 and Fig. 14.71, where analytical exact solution is compared with the UTD model. Both total and diffracted field variations around the wedge are plotted. These figures belong to DSI/HBC and SSI/SBC, respectively. Angles of incidences in these plots are 110° and 90°, respectively. As observed, results agree very well [8].

14.5.3 The FDTD Model

The FDTD method is a pure numerical technique where Maxwell equations are directly discretized in the time domain. Since its first

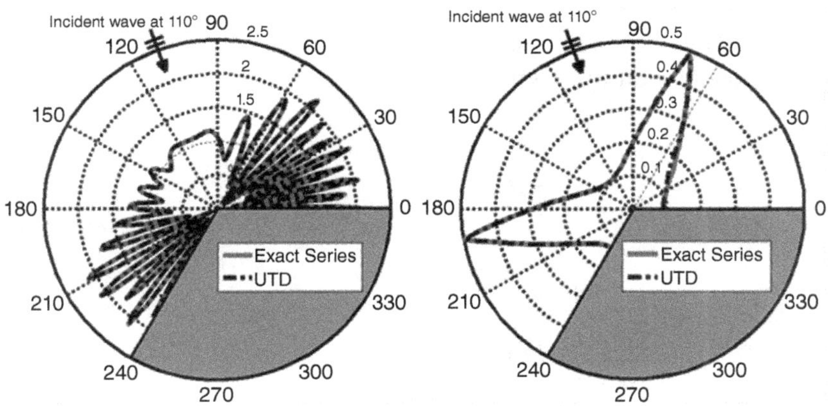

Figure 14.70. (Left) Total and (right) diffracted fields around the PEC wedge. Exact by series versus UTD solution (HBC, $\alpha = 240°, f = 30$ MHz, $r = 50$ m, $kr = 31.4$).

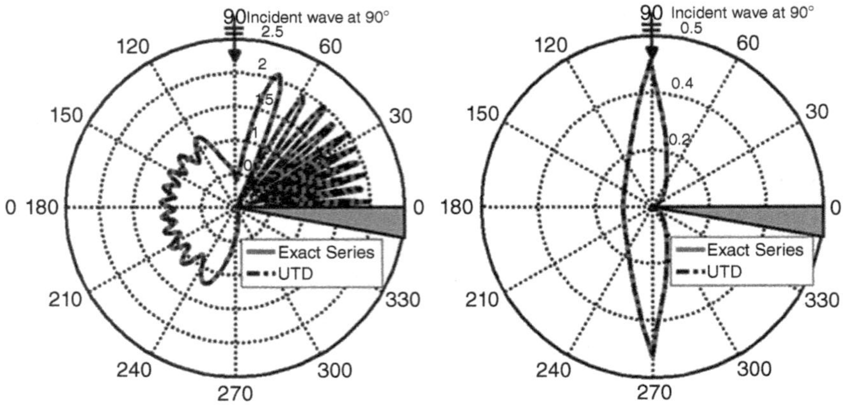

Figure 14.71. (Left) Total and (right) diffracted fields around the PEC wedge. Exact by series versus UTD solution (SBC, $\alpha = 350°$, $f = 30$ MHz, $r = 50$ m, $kr = 31.4$).

introduction [1], it has been widely used in variety of EM problems. A general, multistep FDTD-based diffraction model under the LS illumination is introduced in Ref. 9, where diffracted fields under any *source/observer locations* can be extracted. Results are calibrated via comparisons against analytical exact solutions as well as the UTD and PE models.

The FDTD-based wedge scattering model uses one of the two cases presented in Fig. 14.72. Here, a LS is injected to the $E_z(H_z)$ component in the TM_z (TE_z) model.

The 2D FDTD (TM_z type, i.e., soft BC) equations (with H_x, H_y, and E_z components) on the discrete xy-domain are

$$H_x^{\tilde{n}}(i,j) = \frac{2\mu - \sigma_m \Delta t}{2\mu + \sigma_m \Delta t} H_x^{\tilde{n}-1}(i,j) + \frac{1}{2\mu + \sigma_m \Delta t} \frac{2\Delta t}{\Delta y} \left[E_z^n(i,j+1) - E_z^n(i,j) \right], \tag{14.95a}$$

$$H_y^{\tilde{n}}(i,j) = \frac{2\mu - \sigma_m \Delta t}{2\mu + \sigma_m \Delta t} H_y^{\tilde{n}-1}(i,j) + \frac{1}{2\mu + \sigma_m \Delta t} \frac{2\Delta t}{\Delta x} \left[E_z^n(i+1,j) - E_z^n(i,j) \right]. \tag{14.95b}$$

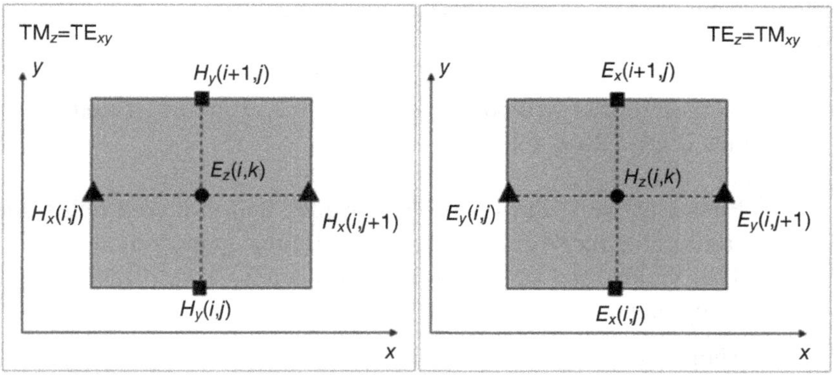

Figure 14.72. 2D FDTD cells on the *xy*-plane and locations of the field components.

$$E_z^n(i,j) = \frac{2\varepsilon - \sigma_e \Delta t}{2\varepsilon + \sigma_e \Delta t} E_z^{n-1}(i,j) + \frac{1}{2\varepsilon + \sigma_e \Delta t} \frac{2\Delta t}{\Delta x} \left[H_x^{\tilde{n}}(i,j) - H_x^{\tilde{n}}(i-1,j) \right] + \frac{1}{2\varepsilon + \sigma_e \Delta t} \frac{2\Delta t}{\Delta y} \left[H_y^{\tilde{n}}(i,j) - H_y^{\tilde{n}}(i,j-1) \right]. \tag{14.95c}$$

The 2D FDTD (TE$_z$ type, i.e., hard BC) equations (with E_x, E_y, and H_z components) on the discrete *xy*-domain are

$$E_x^n(i,j) = \frac{2\varepsilon - \sigma_e \Delta t}{2\varepsilon + \sigma_e \Delta t} E_x^{n-1}(i,j) + \frac{1}{2\varepsilon + \sigma_e \Delta t} \frac{2\Delta t}{\Delta y} \left[H_z^{\tilde{n}}(i,j+1) - H_z^{\tilde{n}}(i,j) \right], \tag{14.96a}$$

$$E_y^n(i,j) = \frac{2\varepsilon - \sigma_e \Delta t}{2\varepsilon + \sigma_e \Delta t} E_y^{n-1}(i,j) + \frac{1}{2\varepsilon + \sigma_e \Delta t} \frac{2\Delta t}{\Delta x} \left[H_z^{\tilde{n}}(i+1,j) - H_z^{\tilde{n}}(i,j) \right], \tag{14.96b}$$

$$H_z^{\tilde{n}}(i,j) = \frac{2\mu - \sigma_m \Delta t}{2\mu + \sigma_m \Delta t} H_z^{\tilde{n}-1}(i,j) + \frac{1}{(2\mu + \sigma_m \Delta t)} \frac{2\Delta t}{\Delta x} \left[E_x^n(i,j) - E_x^n(i-1,j) \right] + \frac{1}{(2\mu + \sigma_m \Delta t)} \frac{2\Delta t}{\Delta y} \left[E_y^n(i,j) - E_y^n(i,j-1) \right]. \tag{14.96c}$$

Here, *i* and *j* are labels of the discrete mesh points, Δx and Δy are the mesh sizes along *x* and *y* axes, respectively, and Δt is the time step.

Medium parameters are permittivity (ε), permeability (μ), and conductivity (σ). Note that $\tilde{n} = n + 1/2$ refer to $\Delta t/2$ time delay between electric and magnetic field computation time instants [1].

According to the method presented in Ref. 9 the diffracted fields are extracted for SSI as follows.

- First, FDTD is run for the PEC wedge and scattered fields are stored at n-receivers located on a circular path around the tip. This yields total fields; incident, reflected, and diffracted field components in region II; only the diffracted fields in region III.
- Then, the wedge is removed, FDTD is run for free space, and scattered fields are stored with the same receivers. This yields incident fields all around.
- Finally, bottom face of the wedge is removed and top face is stretched to take up the entire transversal section of simulation space. Then, FDTD is run for the third time. The stored fields yield reflected fields on the upper half-plane.

The three pulses—the incident $u_h^{\text{inc,FDTD}}$, reflected $u_h^{\text{Refl,FDTD}}$, and diffracted $u_h^{\text{d,FDTD}}$—are obtained using the three-step stored data. The related diffracted fields in the frequency domain are then extracted by the application of fast Fourier transform (FFT) as $u^d(f) = \text{FFT}\{u^{d,\,FDTD}(t)\}$. For DSI, a replica of third step is required for the bottom face of the wedge. The numerical diffraction coefficient is obtained via

$$d_h^{\text{FDTD}} = \frac{\text{FFT}\left\{u_h^{d,\text{FDTD}}\right\}}{\text{FFT}\left\{u_h^{\text{inc,FDTD}}\right\}} \frac{\sqrt{r}}{e^{ikr}}. \tag{14.97}$$

14.5.4 Discretization and Dey–Mittra Approach

The FDTD procedure summarized earlier is calibrated against analytical exact as well as HFA models [10]. It is shown that the method works very well under arbitrary source/observer locations with the standard Yee model using the staircase discretization, except for angles near the bottom face of the wedge. In staircase approximation, when the center of cubic cell is embedded inside the PEC object, all surrounding electric fields of this cell are set to zero. The TM mode fits perfectly with the staircase approximation because tangential electric fields coin-

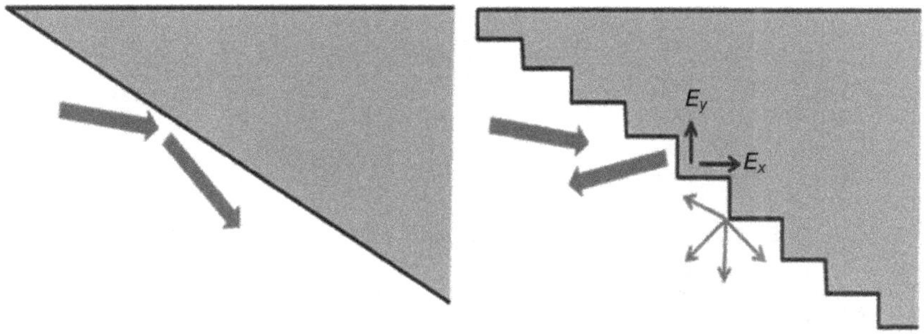

Figure 14.73. Physical and nonphysical case for the TE_z problem (thick arrows show incident and reflected waves; thin arrows show tip-diffracted waves).

cide with the third dimension. On the other hand, the TE mode is problematic as illustrated in Fig. 14.73. As observed, the staircase approximation leads to spurious diffracted fields which can erroneously alter results significantly.

In general, one can neglect staircase approximation error by selecting sampling resolution greater than $\lambda/60$. However, this increases computational burden proportionally and it is not suitable method for dealing with large problems and problems where phase information is crucial (as many as $\lambda/100$ may not be adequate).

Several methods are proposed to overcome staircase problems. Nonuniform gridding based on using smaller cells around the object is one approach, but the payoff is the increase in total number of FDTD cells and memory. Another approach is to use conformal FDTD models based on integration contour deformation around curved object being modeled and applying Faraday's law. Then, cell shape is changed to fit boundaries of modeled object. Among existing conformal models, Dey–Mittra technique described in Ref. 10 is found to be the most effective one here for the wedge problem. Fundamental steps of the Dey–Mittra technique are to be summarized later through cases shown in Fig. 14.74. In Fig. 14.74a, slanted object crosses slightly into top right cell. If the parameter $s/\Delta x\Delta y$ (s is the small area in the figure) is less than R_1 (which is the key parameter specifying numerical stability), one can neglect this penetration and set all four surrounding electric fields to zero for the top right cell. On the other hand, if Faraday's law

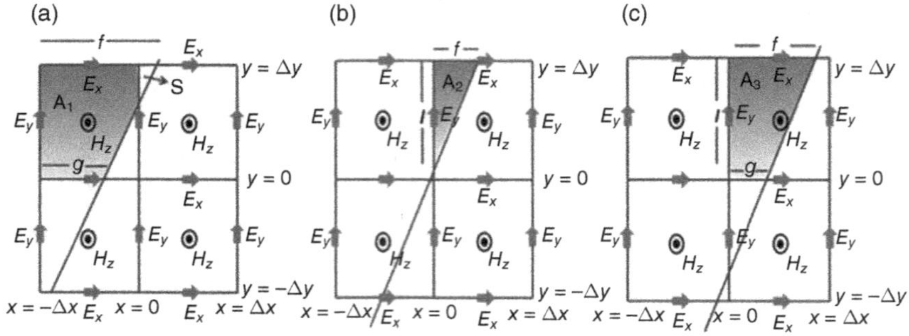

Figure 14.74. Dey–Mittra conformal FDTD scenarios.

is applied and integrated along the contour of area A_1 for top left cell, one can obtain

$$H_z^{n+1/2}(-\Delta/2, \Delta y/2) = H_z^{n-1/2}(-\Delta/2, \Delta y/2) + \frac{\Delta t}{\mu_0 A_1}\begin{pmatrix} E_y^n(-\Delta x, \Delta y/2)\Delta y + \\ E_x^n(-\Delta x/2, \Delta y)f - E_x^n(-\Delta x/2, 0)g \end{pmatrix}. \tag{14.98}$$

In Fig. 14.74b, the ratio of $A_2/\Delta x\Delta y$ is larger than R_1. In this case, applying Faraday's law to the top right cell and integrating over the contour of area A_2 gives the update equation for H_z on the top right cell despite the fact that it is in PEC. Noting that two electric field components reside in PEC and set to zero, we have

$$H_z^{n+1/2}(\Delta/2, \Delta y/2) = H_z^{n-1/2}(\Delta/2, \Delta y/2) + \frac{\Delta t}{\mu_0 A2}\left(E_y^n(0, \Delta y/2)l + E_x^n(\Delta x/2, \Delta y)f\right). \tag{14.99}$$

In Fig. 14.74c, applying Faraday's law and integrating along over the contour of area A_3 gives

$$H_z^{n+1/2}(\Delta/2, \Delta y/2) = H_z^{n-1/2}(\Delta/2, \Delta y/2) + \frac{\Delta t}{\mu_0 A_3}\begin{pmatrix} E_y^n(0, \Delta y/2)\Delta y \\ +E_x^n(\Delta x/2, \Delta y)f - E_x^n(\Delta x/2, 0)g \end{pmatrix}. \tag{14.100}$$

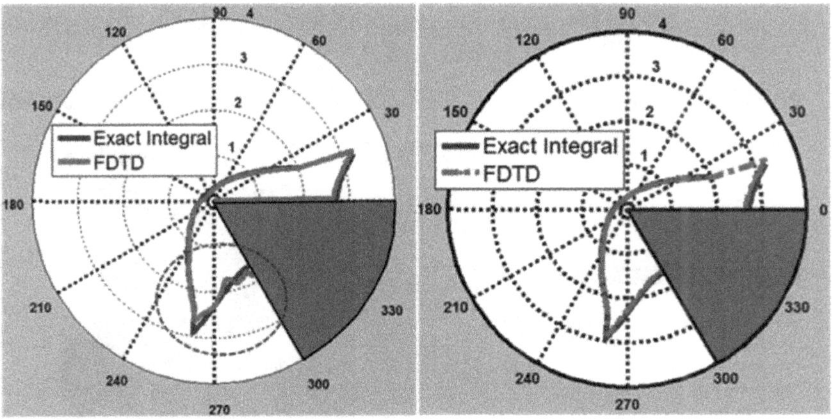

Figure 14.75. Exact solution versus FDTD for (a) staircase approximation (b) Dey–Mittra model.

Despite its accuracy of modeling curved or slanted objects, Dey–Mittra conformal technique can cause instabilities. For this reason, time step should be reduced depending on the required accuracy. Stability analyses of several conformal methods are investigated in Ref. 14. Here, selection of the key parameter $R_1 = 0.0025$ requires 30% reduction of time step below the normal limit of stability. Another handicap of using Dey–Mittra conformal technique is that it requires complex mesh generation for calculating intersection points of unit cells and boundaries of the object modeled.

Figure 14.75 compares Dey–Mittra and staircase discretization models. Here, a 60° wedge is illuminated with a LS at 160°. As observed in Fig. 14.75b, simulation results with Dey–Mittra conformal FDTD model agrees very well with analytical exact solution everywhere. On the other hand, staircase approximation significantly fails to yield correct behavior near the bottom face of the wedge (see Fig. 14.75a). Things will get worse when both faces of wedge illuminated (DSI). Dey–Mittra conformal FDTD method fits exact results well.

14.5.5 The WedgeFDTD Package and Examples

A novel MATLAB-based WedgeFDTD package with the front panel given in Fig. 14.76 (showing a time instant just before cylindrical waves

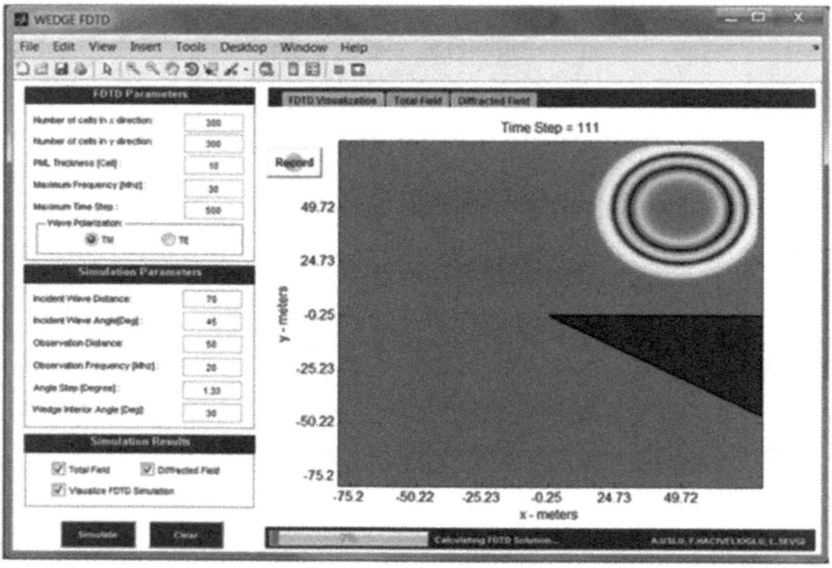

Figure 14.76. The front panel of WedgeFDTD package.

emanating from a LS hit the top of the wedge) is developed to simulate scattered fields for a 2D PEC wedge [11]. Results are compared with analytical exact integral method as well as the UTD model. The wedge is discretized with the Dey–Mittra approach. The program is designed with using both MATLAB and Java swing library components. The interface of the program is divided into two panels as left and right. The user can enter the FDTD and simulation parameters from the left panel. At the bottom of this panel, user may select simulation results and visualization of different wave pieces. Figure 14.77 shows a late time instant, showing all—incident, reflected, and diffracted—field components.

The right panel is reserved for the FDTD visualization and simulation results. Three tabs are placed at top for this purpose. The tabs are activated after the corresponding simulation type is clicked from the left panel. Also, user can select the color map used for FDTD visualization from the pop-up menu which is placed at top right of this panel. Video recording property is embedded in program via external Mencoder application to reduce video size. For this purpose, one can use the record

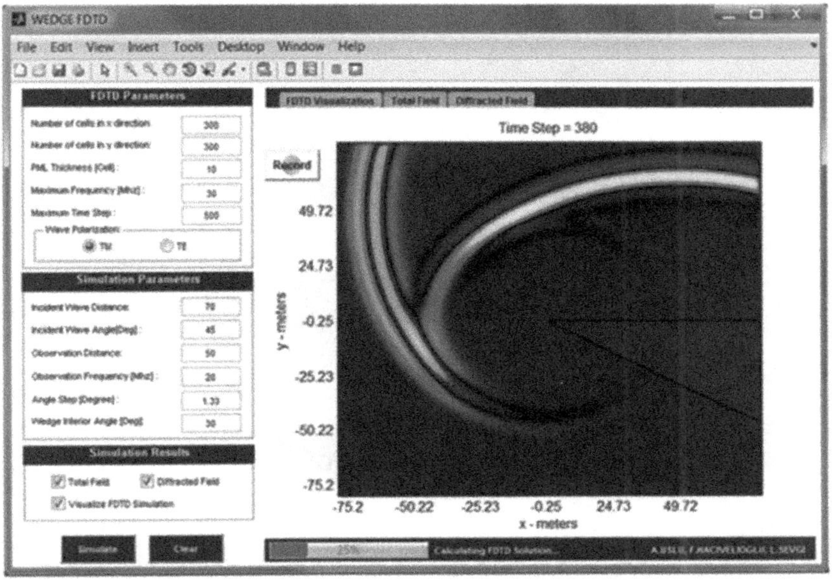

Figure 14.77. The front panel of WedgeFDTD package showing reflected and tip-diffracted waves.

button which placed at top left of this panel. Bottom panel is dedicated for progress information of the simulation and dynamically appears while the simulation is running. The slider which separates the left and right panels enables enlarging/compressing panel sizes via horizontal scrolling. User can also export simulation results by using the standard MATLAB figure toolbar.

A few examples are included here. The first example is a TM_z EM or soft BC acoustic problem and belongs to a scenario with a 60° wedge apex angle, illuminated by a 20 MHz LS located 70 m from the tip with the incidence angle of 120°. Total and diffracted fields are shown in Fig. 14.78 and Fig. 14.79, respectively.

The second example is a TE_z EM or hard BC acoustic problem which belongs to a wedge with 30° interior angle, illuminated with a 30 MHz LS whose distance from the wedge is 80 m and the incidence angle is 45°W. The total and diffracted fields are observed at a distance of 50 m. Results are given in Fig. 14.80 and Fig. 14.81, respectively.

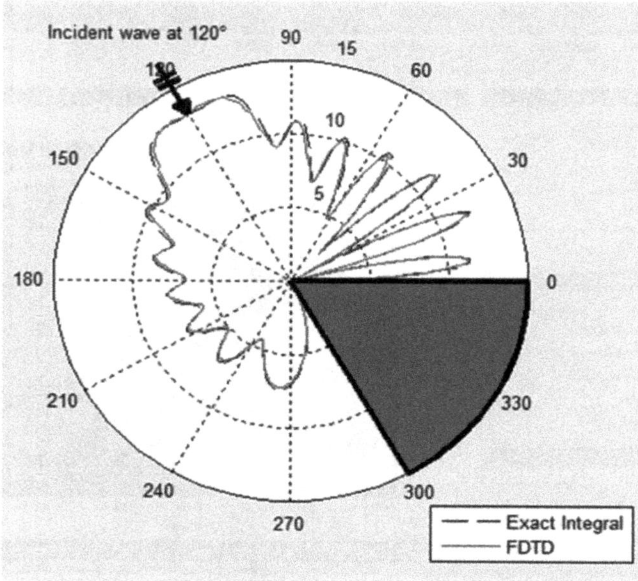

Figure 14.78. Total fields versus angle computed with both FDTD and exact integral models ($\alpha = 300°, f = 20$ MHz, $r = 70$ m, $\varphi_0 = 120°$).

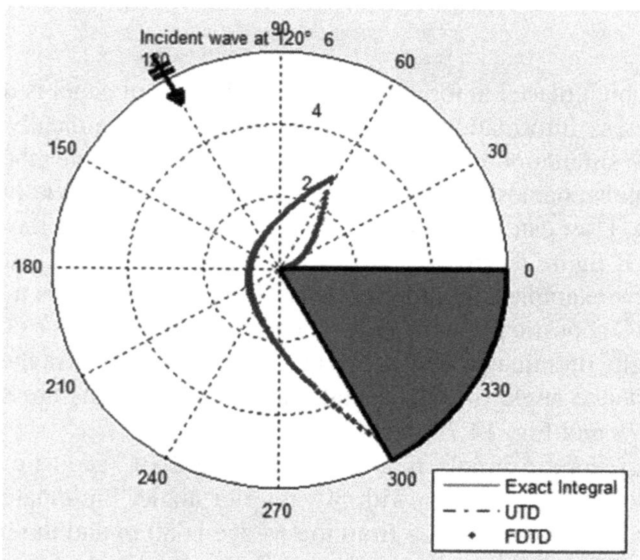

Figure 14.79. Diffracted fields versus angle solution computed with the FDTD, UTD, and exact integral models ($\alpha = 300°, f = 20$ MHz, $r = 70$ m, $\varphi_0 = 120°$).

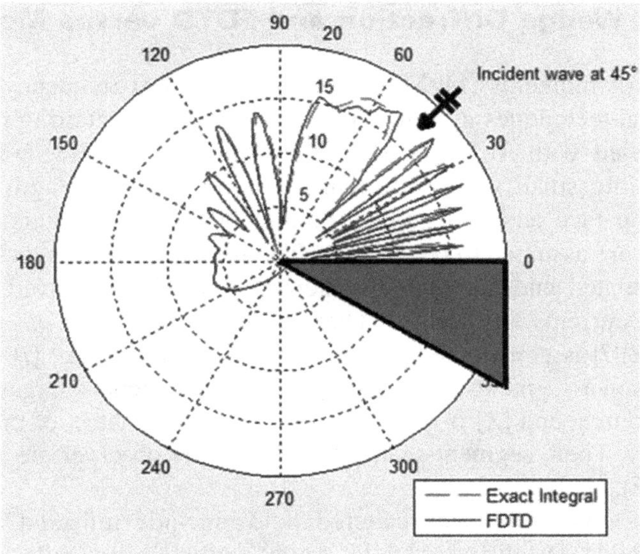

Figure 14.80. Total fields versus angle solution computed with FDTD, UTD, and exact integral models ($\alpha = 330°, f = 30$ MHz, $r = 80$ m, $\varphi_0 = 45°$).

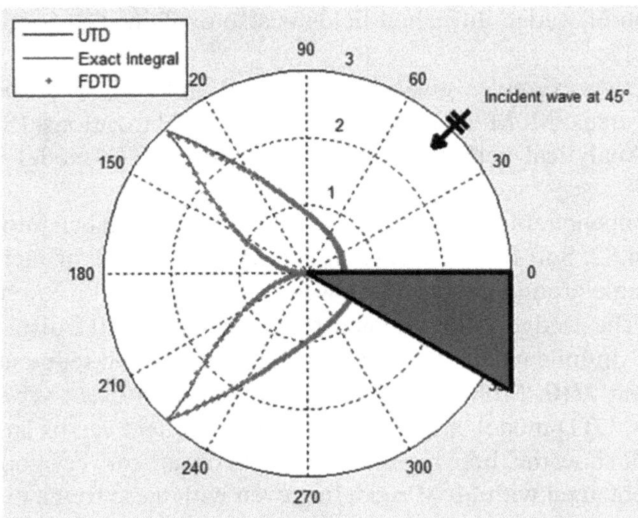

Figure 14.81. Diffracted fields versus angle solution computed with, FDTD, UTD, and exact integral models ($\alpha = 330°, f = 30$ MHz, $r = 80$ m, $\varphi_0 = 45°$).

14.5.6 Wedge Diffraction and FDTD versus MoM

Method of moments (MoM) is one of the earliest frequency domain numerical techniques used in EM [15–17]. Wedge scattering may also be modeled with MoM. In this model, the faces of the wedge are divided into small segments compared with the wavelength (N segments top face and N segments bottom face). The currents on each segment are assumed to be constant. The source-excited segment fields are calculated and the matrix system is numerically solved and the segment currents are derived. Then, a $2N \times 2N$ system of equations $[V] = [Z][I]$ is constructed and solved numerically. Here, $[I]$ contains the unknown segment currents, $[V]$ contains segment voltages excited by the source, and $[Z]$ is the $2N \times 2N$ impedance matrix of the wedge boundary. Then, segment-scattered fields at the observer are accumulated [17].

The MoM-computed scattered fields include reflected and diffracted fields. The diffracted fields can be extracted once reflected fields are available. The reflected-only fields can be obtained from an infinite-plane scenario (i.e., by taking $\alpha = \pi$ and repeating the MoM procedure) [17]. The necessary MoM equations for both polarizations are given in Chapter 13 (see (13.32), (13.33), (13.34), (13.35), (13.36), (13.37), (13.38), (13.39), (13.40), and (13.41)). The MoM procedure for the extraction of wedge diffracted fields is also explained in Section 13.6, Chapter 13.

The two examples given in Fig. 14.82 and Fig. 14.83 belong to FDTD versus MoM comparisons for both polarizations (SBC and HBC). Analytical reference solution and the UTD model are also included.

A nonpenetrable wedge with 60° interior angle is taken into account in Fig. 14.82. Soft BC (TM_z case) is assumed. Total and diffracted fields versus angle around the tip of the wedge on a circle with 5λ-radius are plotted. The wedge is illuminated by a LS located 10λ-distance with $\varphi_0 = 45°$. Infinite wedge faces are truncated in 10λ and segment lengths are chosen $\lambda/10$. MoM results are compared with exact series representation, UTD model, and FDTD model. Total field versus angle plots on the left show the three regions clearly. As observed, very good agreement is obtained with MoM modeling even with these rough discretization parameters.

The second example in Fig. 14.83 belongs to a nonpenetrable wedge with the same interior angle and illumination angle as earlier,

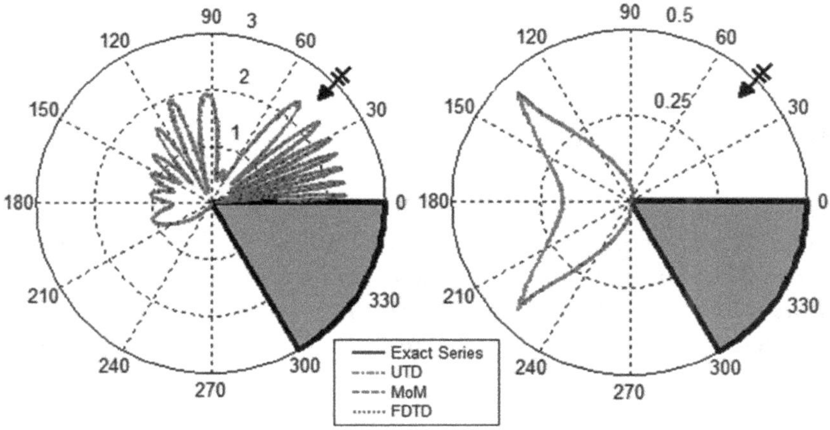

Figure 14.82. (Left) Total fields and (right) diffracted fields around PEC wedge as a function of angles at 30 MHz using exact series, UTD, MoM, and FDTD solutions: $\alpha = 300°$, $r = 50$ m, $r_0 = 100$ m, $\varphi_0 = 45°$ (TM/SBC case).

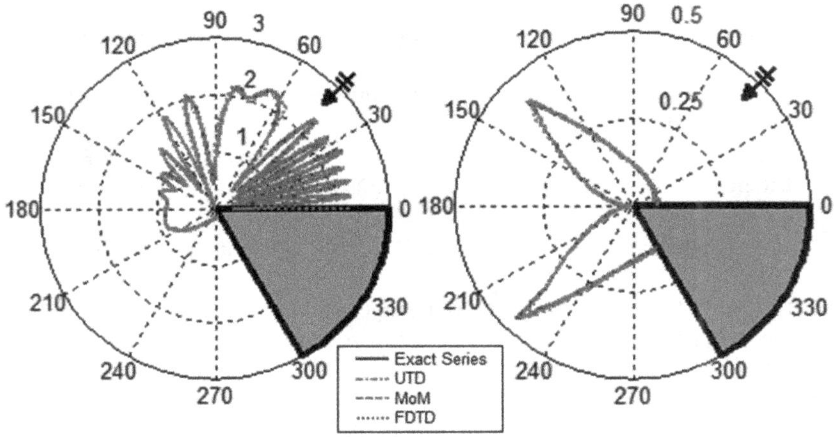

Figure 14.83. (Left) Total fields and (right) diffracted fields around PEC wedge as a function of angles at 30 MHz using exact series, UTD, MoM, and FDTD solutions: $\alpha = 300°$, $r = 50$ m, $r_0 = 100$ m, $\varphi_0 = 45°$ (TE/HBC case).

but HBC is used. As observed, the agreement among the models is also very good.

14.6 CONCLUSIONS

The well-known and widely used FDTD model is reviewed. Packages for plane wave propagation and TLs are introduced. A general-purpose 2D FDTD package is also introduced. A novel MATLAB-based tool is developed to investigate wedge scattering with the FDTD method. Diffracted fields are extracted and are compared with the results of HFA models.

REFERENCES

1. K. S. Yee, "Numerical Solution of Initial Boundary Value Problems Involving Maxwell's Equations in Isotropic Media," *IEEE Transactions on Antennas and Propagation*, **14**, 1966, pp. 302–307.
2. L. Sevgi, "Review of Discrete Solutions of Poisson, Laplace, and Wave Equations," *IEEE Antennas and Propagation Magazine*, **50** (1), 2008, pp. 246–254.
3. C. A. Balanis, *Advanced Engineering Electromagnetics*, 2nd ed., John Wiley & Sons, Inc., Hoboken, NJ, 2012.
4. L. Sevgi, Ç. Uluışık, "A Matlab-Based Transmission Line Virtual Tool: Finite-Difference Time-Domain Reflectometer," *IEEE Antennas and Propagation Magazine*, **48** (1), 2006, pp. 141–145.
5. D. V. Widder, *The Laplace Transform*, Dover Books on Mathematics, New York, 2010.
6. G. Çakır, M. Çakır, L. Sevgi, "A Multipurpose FDTD-Based Two Dimensional Electromagnetic Virtual Tool," *IEEE Antennas and Propagation Magazine*, **48** (4), 2006, pp. 142–151.
7. F. Hacıvelioğlu, L. Sevgi, P. Ya. Ufimtsev, "Electromagnetic Wave Scattering from a Wedge with Perfectly Reflecting Boundaries: Analysis of Asymptotic Techniques," *IEEE Antennas and Propagation Magazine*, **53** (3), 2011, pp. 232–253.
8. F. Hacıvelioğlu, M. A. Uslu, L. Sevgi, "A MATLAB-Based Simulator for the Electromagnetic Wave Scattering from a Wedge with Perfectly Reflecting Boundaries," *IEEE Antennas and Propagation Magazine*, **53** (4), 2011, pp. 234–243.
9. G. Çakır, L. Sevgi, P. Ya. Ufimtsev, "FDTD Modeling of Electromagnetic Wave Scattering from a Wedge with Perfectly Reflecting Boundaries: Comparisons

against Analytical Models and Calibration," *IEEE Transactions on Antennas and Propagation*, **60** (7), 2012, pp. 3336–3342.

10. S. Dey, R. Mittra, "A Locally Conformal FDTD Algorithm for Modeling Three-Dimensional Perfectly Conducting Objects," *IEEE Microwave and Guided Wave Letters*, **7** (9), 1997, pp. 273–275.
11. M. A. Uslu, L. Sevgi, "Matlab-Based Virtual Wedge Scattering Tool for the Comparison of High Frequency Asymptotics and FDTD Method," *ACES, International Journal on Applied Computational Electromagnetics*, **27** (9), 2012, pp. 697–705.
12. J. J. Bowman, T. B. A. Senior, P. L. E. Uslenghi, eds., *Electromagnetic and Acoustic Scattering by Simple Shapes*, Hemisphere Publishing Corporation, New York, 1987.
13. R. G. Kouyoumjian, P. H. Pathak, "A Uniform Geometrical Theory of Diffraction for an Edge in a Perfectly Conducting Surface," *Proceedings of the IEEE*, **62**, 1974, pp. 1448–1461.
14. C. J. Railton, J. B. Scheider, "An Analytical and Numerical Analysis of Several Locally Conformal FDTD Schemes," *IEEE Transactions on MTT*, **47** (1), 1999, pp. 56–66.
15. R. F. Harrington, *Field Computation by Moment Method*, 1st ed. 1968, IEEE Press, New York, 1993.
16. E. Arvas, L. Sevgi, "A Tutorial on the Method of Moments," *IEEE Antennas Propagation Magazine*, **54** (3), 2012, pp. 260–275.
17. G. Apaydin, L. Sevgi, "A Novel Wedge Diffraction Modeling Using Method of Moments (MoM)," *IEEE Transactions on Antennas and Propagation*, 2013 (unpublished).

CHAPTER 15

Parabolic Equation Method

This chapter reviews the parabolic equation method (PEM). The PEM represents one-way propagation and is widely used in two-dimensional (2D) radio wave/ground wave propagation modeling. It takes the earth's curvature, atmospheric refractivity variations, nonflat terrain scattering, and boundary losses into account. Both split-step PE (SSPE)- and finite element method (FEM)-based numerical solution approaches are introduced. Algorithms of both SSPE and FEMPE propagators are first tested on two simple 2D propagation problems and are calibrated against analytical reference solutions. Several simple MATLAB scripts are given for this purpose. Then, a radio wave propagation package GrSSPE is presented for the visualization of EM propagation over nonflat terrain through nonhomogeneous atmosphere for waves radiated by a horizontally oriented antenna over the ground. The package can be used as an engineering tool to test the effects of various user-specified longitudinal terrain profiles and refractivity variations, as well as guiding and antiguiding conditions on radio wave propagation, especially for cellular rural coverage planning, and as an educational aid (virtual tool) in EM lectures, for example, in antennas and propagation, wireless communication, and so on. Then, a problems 2D propagation problem is revisited. The 2D propagation problem over flat earth with a single-knife-edge-type obstacle is discussed and the four ray (4Ray) model is coded. Accurate source modeling is also discussed in this

Electromagnetic Modeling and Simulation, First Edition. Levent Sevgi.

chapter. The final canonical problem is the propagation modeling inside a 2D dielectric slab waveguide. This problem is modeled in terms of mode summation and SSPE models. A MATLAB-based package is also included.

15.1 INTRODUCTION

Radio engineers or site surveyors dream to have numerical propagation tools that calculate ground wave path loss between any two points marked on their digital maps. This necessitates the solution of EM wave equation in three dimensions (3D), which takes into account various EM effects, such as the irregular terrain profile, the vegetation, the earth's curvature, the atmospheric refractivity, the presence of buildings, cars, and other obstacles, and so on. Also, it must include all the relevant scattering components (e.g., multiple reflections and refractions, edge/tip diffractions, surface and/or leaky waves, etc.) that account for the path loss [1–3]. Unfortunately, this is not yet in sight.

Radio wave transmission over the earth's surface, which includes both ground and sky waves, is one of the most difficult and important propagation problems and has been a subject of interest from the beginning of the twentieth century. Sky wave is mostly affected by the upper atmosphere (ionosphere), while ground wave propagation characteristics depend on the lower and middle atmosphere (troposphere) parameters as well as those of the ground. Ground waves have three components which are categorized as direct waves, ground-reflected waves, and surface waves. A spherical earth, which may have various ground characteristics and above which exists a radially inhomogeneous atmosphere, is used as a canonical model to determine ground wave propagation characteristics between two points, which may be selected anywhere on or above the ground. The physical characteristics of the propagation depend on many parameters such as the operating frequency, medium parameters, transmitter and receiver locations, and the geometry (boundary conditions[BCs]) between them. Although it is required to model surface impedance as well as nonflat terrain for a realistic ground wave propagation analysis at high frequency (HF) and lower very high frequency (VHF), perfect electrical conducting (PEC) terrain is applicable for the frequencies above 100 MHz, which is important for a wide range of EM applications including mobile communication. In the case that the frequency is lower than 100 MHz, this approach also gives enough information for the propagation and scattering mechanisms for ground wave.

The problem is still challenging; a full-wave, observable-based, numerically computable analytical solution nor a 3D generally applicable, numerical solution has not appeared yet. The current numerical strategy is to take the 2D projections between the transmitter and the receiver, and reduce the 3D problem at hand into 2D range–height scenarios. Highly attractive 2D propagation prediction virtual tools have been developed for the last decade [4–11]. These virtual tools are based on various techniques, such as the parabolic equation (PE) [4], the method of moments (MoM) [5], PE + MoM [6], the finite-difference time domain (FDTD) [7], the transmission line matrix (TLM) [8], ray-mode summation [9,10], and ray tracing, ray shooting [11]. Among these, the PE-based propagation tools are speculated to be the most attractive and effective ones.

Two of these virtual tools (i.e., Refs. 4 and 6) are of special interest in terms of the topic discussed in this paper. The GrSSPE presented in Ref. 4 is a simple MATLAB-based ground wave propagation package for the visualization of EM propagation over irregular terrain through nonhomogeneous atmosphere for waves radiated by a horizontally oriented antenna over the ground. This package is revised and presented in Section 15.8. A MATLAB package, which modifies the MoM by the application of the forward/backward spectral acceleration (FBSA) [5] technique and integrates it with the SSPE method, is introduced in Ref. 6, with which precise SSPE versus MoM comparisons are possible. The reader who is interested in ground wave propagation prediction problems is strongly advised to read those papers and try the virtual tools presented there.

The PE was first applied to the underwater acoustic propagation modeling by Leontovich and Fock in the 1940s [12]. One major problem at that time was to detect the nuclear submarines as early as possible, and this certainly necessitates a good understanding of underwater acoustic propagation in deep as well as shallow waters with variable densities, bottom profiles, streams, and others. Later, it has become popular in EM wave propagation modeling through the earth-troposphere waveguides. Although Leontovich and Fock [12,13] were the pioneers, the PE approach has become famous after the introduction of the Fourier split-step SSPE algorithm [14]. Since then, the PE technique has been improved, accommodated with many auxiliary tools, and applied to variety of complex propagation problems [15–22]. The book by Levy [23] is a good source which discusses the PE modeling in detail and gathers a huge list of PE related studies. The reader is referred to a recent survey on the PE modeling [24].

The FEM has also been used in developing PE-based numerical propagation tools. Initially, the FEM-based PE models were also applied in underwater acoustic propagation prediction problems (e.g., see Ref. 25). A few FEM-based PE models have also appeared in EM wave propagation modeling for the last decade [26–31], but a systematic investigation with characteristic applications and reliable comparisons is still missing.

The aim of this chapter is to (1) review both the SSPE and FEMPE propagation approaches and introduce MATLAB-based simple routines, (2) present canonical test cases and do systematic comparisons between the SSPE and FEMPE methods, (3) calibrate both the SSPE and FEMPE tools against analytical exact (reference) solutions, and (4) introduce several simple MATLAB scripts in 2D propagation modeling.

Note that, apart from these studies, there exist several computer software programs, most of which have been developed for military purposes, for predicting radar coverage under the effect of environmental factors that influence refractivity. These are [32,33] Integrated Refraction Effects Prediction System (IREPS), Engineer's Refractive Effects Prediction System (EREPS), Tactical Electronic Support System (TESS), Advanced Refractive Effects Prediction System (AREPS), Tropospheric Electromagnetic Parabolic Equation Routine (TEMPER), and Terrain Parabolic Equation Model (TPEM). These programs implement the ray optics or *one-way* PE techniques or a hybrid model combining these methods.

15.2 THE PARABOLIC EQUATION (PE) MODEL

The standard parabolic wave equation may be obtained from the 2D Helmholtz equation,

$$\frac{\partial^2 U}{\partial x^2} + \frac{\partial^2 U}{\partial z^2} + k_0^2 n^2 U = 0, \tag{15.1a}$$

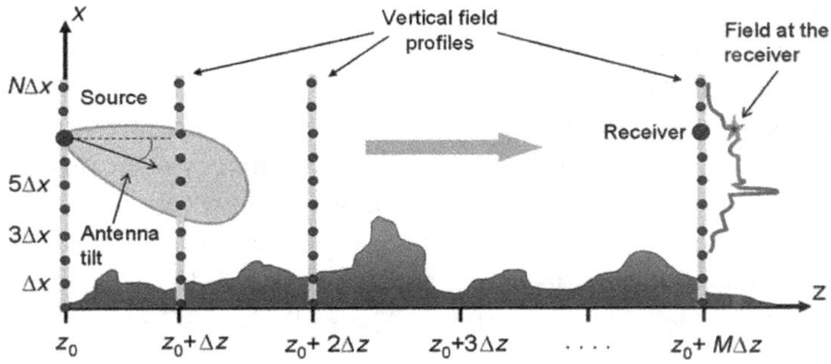

Figure 15.1. The 2D projected propagation region, x and z are the height and range coordinates, respectively. The region is N_x-segments vertically and N_z-segments longitudinally. An antenna pattern is located vertically at the initial range and height. SSPE and FEMPE are longitudinally marching methods and calculate the vertical field profiles at the next range step from the previous one. Both sweep the region longitudinally up to the desired range step by step.

by separating the rapidly varying phase term ($U(x, z) = u(z, x) \exp(-ik_0 z)$). The amplitude factor $u(z, x)$ varies slowly in range when the direction of propagation is predominantly along the z-axis (i.e., paraxial direction). Under $\exp(-i\omega t)$ time dependence, the PE is then given as

$$\frac{\partial^2 u}{\partial x^2} + 2ik_0 \frac{\partial u}{\partial z} + k_0^2 (n^2 - 1)u = 0. \tag{15.1b}$$

Here, $u = u(z, x)$ denotes the wave amplitude (either of the electric or magnetic field components depending on the type of the problem), x and z stand for the transverse (i.e., the height above the ground) and the longitudinal (i.e., the direction of propagation) coordinates, respectively (see Fig. 15.1), n is the refractive index, and $k_0 = 2\pi/\lambda$ is the free space wave number. The 2D propagation scenario is completed by choosing the appropriate transverse and longitudinal BC. The transverse BC of the ground (along x) is

$$\left(\alpha_1 \frac{\partial}{\partial x} + \alpha_2 \right) u(z, x) \bigg|_{x=0} = 0. \tag{15.2}$$

Here, α_1, α_2 are constants and $\alpha_i = 0$ (i = 1, 2) results in Dirichlet (horizontal polarization) and Neumann (vertical polarization) BCs for

PEC surface, respectively. The Cauchy-type BC is introduced with $\alpha_1 = 1,\ \alpha_2 = ik_0\sqrt{\gamma - 1}$ and $\alpha_1 = 1,\ \alpha_2 = ik_0\sqrt{\gamma - 1}/\gamma$ for horizontal and vertical polarizations, respectively, $\gamma = \varepsilon_r + i60\sigma\lambda$ is the relative complex dielectric constant in terms of relative permittivity (ε_r) and conductivity (σ) of the ground at range z. Since waves propagating upward either go to infinity or bent down because of the refractivity variations, a window function (i.e., hanning or hamming) can be applied to the vertical field profile above the selected height at each range in order to eliminate reflection effects. The radiation BC along z is

$$\left\{\frac{\partial}{\partial z} - ik_0\right\}u(z, x)\bigg|_{z\to\pm\infty} \to 0. \qquad (15.3)$$

Although the PE in 2D describes one-way propagation and cannot take the backscattered waves into account, this is not a serious restriction in order to investigate waves emanating from a transmit antenna and reach a receiver. It should be noted that the solution of (15.1b) in 2D yields waves that are attenuated by the square root of the distance as they propagate away from the transmitter (i.e., cylindrical wave spread for an infinite-length line source which allows the reduction of the 3D wave equation into 2D). Therefore, the user should divide the field strength results by the square root of distance in order to obtain path loss versus range variations in a realistic (3D) environment [17].

15.3 THE SPLIT-STEP PARABOLIC EQUATION (SSPE) PROPAGATION TOOL

The PE in (15.1b) turns out to be a first order ordinary differential equation that can be solved directly if a Fourier transform from the x-domain to the spectral k_x-domain is applied. Although, in real problems, n can be a function of height and/or range (i.e., $n = n(z, x)$), this approach is appropriate because the equation is solved at each small range step size Δ_z, which is chosen small enough so that within any Δ_z interval, refractive index can be assumed to be constant with respect to z. The numerical SSPE solution for $m = 1, 2, \ldots, M$ is given as

$$\begin{aligned} u(z_0 + m\Delta z, x) = \exp\left[i\frac{k_0}{2}(n^2 - 1)\Delta z\right]F^{-1} \\ \left\{\exp\left[-i\frac{k_x^{\,2}\Delta z}{2k_0}\right]F\left\{u(z_0 + (m-1)\Delta z, x)\right\}\right\}. \end{aligned} \qquad (15.4)$$

This equation can be used to calculate $u(z, x)$ along z with the steps of Δ_z, once the initial field distribution $u(z_0, x)$ is given. A one-dimensional (1D) array can be used to store the transverse field profiles at N_x vertical height points at N_z discrete ranges with replacement. The standard SSPE procedure is discussed in many of the studies listed in the references (e.g., Refs. 4, 6, and 17) but are summarized for the sake of completeness:

- **The SSPE Procedure:** An antenna pattern representing the initial height profile of the field $u(z_0, x)$ is injected first [4,17]. The initial profile is propagated longitudinally from z_0 to $z_0 + \Delta z$ for $m = 1$ via (15.4) and $u(z_0 + \Delta z, x)$ is obtained. This new height profile is then used as the initial profile for the next step and $u(z_0 + 2\Delta z, x)$ is obtained for $m = 2$ via (15.4). The procedure is applied repeatedly and the vertical field profiles are stored at each range step until the desired range.
- **Fast Fourier Transform (FFT) Requirements:** The SSPE sequentially operates between the x and k_x domains, which are the Fourier transform pairs. Since the transforms are implemented numerically, the domains are truncated at $\pm X_{\max}$ and $\pm k_{x\,\max}$. The transverse and longitudinal step sizes Δx and Δz, respectively, and the maximum height $X_{\max}$ are determined from the source and/or observation requirements as well as sampling necessities and aliasing effects. Once $X_{\max}$ is determined, $k_{x\,\max}$ is found from the Nyquist sampling criteria, $X_{\max} \times k_{x\,\max} = \pi L$, L being the transform size.
- **Boundary Implementation:** The SSPE cannot automatically handle the BCs at the surface. It is satisfied at the surface by extending the initial vertical profile from $[0 - X_{\max}]$ to $[-X_{\max}, X_{\max}]$ (odd and even symmetric for Dirichlet and Neumann BC, respectively) or by applying a sine or cosine FFT, respectively.
- The problem at hand has a vertically semiopen propagation region ($x \rightarrow \infty$); therefore, an abrupt truncation is required at certain height, which means the strong artificial reflections will occur if not taken care of. These nonphysical reflections can be removed by using artificial gradually absorbing layers above the height of interest. This is achieved by adding a small imaginary

part to the refractive index (by making n complex) or by windowing in order to eliminate reflection effects.

- **Irregular Terrain:** Irregular terrain is implemented within the SSPE algorithm via a terrain file and staircase approximation [4]. For a PEC terrain, the SSPE algorithm is implemented as follows. The 1D vertical field array starts from the bottom (i.e., just above the smooth earth's surface), and the longitudinally changing PEC terrain profile is injected into the computation subspace via either a terrain file or a terrain function; various techniques (e.g., staircase approximation or conformal mapping) are used in such applications. To include terrain effects at each range, array elements below the local terrain (i.e., field values inside the PEC domain) are forced to be zero. Only forward scattering effects are taken into account with this implementation.
- **Staircase terrain** modeling is simple, since neither analytical terrain function nor slope values are required; only the terrain height at each range step is needed. On each segment of constant height, the field is calculated in the usual way for staircase terrain, applying necessary BC at the ground. When terrain height changes, corner diffraction is ignored and the field is simply set to zero on vertical terrain facets, which is consistent with the assumption that the ground does not support propagation at the wavelength of interest. Since the computation height is not changed due to the terrain, there is no need to modify refractive index; therefore, it is also easy to implement the staircase terrain modeling into the SSPE algorithm. One just needs to take into account that ground does not support propagation under the constant height for each segment. Although staircase terrain cannot model the terrain as smooth as it must be, comparisons show that the results are quite satisfactory.
- **PEC ground** modeling is adequate for many applications (especially above 100–200 MHz), but a more accurate model is required below these frequencies. The surface impedance implementation (i.e., Cauchy-type BC) involves additional modifications, but is not too difficult. The discrete mixed Fourier transform (DMFT) presented in Ref. 17 extends the SSPE algorithm to account for the ground losses.

- **Source Injection:** The antenna pattern at the initial range is specified by giving the height x_s, the beamwidth θ_{bw}, and the tilt angle θ_{tilt}. It is specified in transverse wave number domain via

$$U(0,k_x)=f(k_x)\exp(-ik_x x_s)-f^*(-k_x)\exp(ik_x x_s)\ \text{(Dirichlet BC)}, \tag{15.5a}$$

$$U(0,k_x)=f(k_x)\exp(-ik_x x_s)+f^*(-k_x)\exp(ik_x x_s)\ \text{(Neumann BC)}, \tag{15.5b}$$

  where $U(0, k_x)$ is the Fourier transformed $u(z, x)$ at range z_0. A Gaussian antenna pattern is defined as

$$f(k_x)=\exp\left(\frac{-k_x^2 2\ln 2}{4k_0\sin(\theta_{bw}/2)}\right). \tag{15.5c}$$

- The tilt is introduced by replacing $f(k_x)$ with $f(k_x - k_0\sin\theta_{tilt})$. The initial field profile is then obtained via inverse FFT.
- For the discrete sine/cosine (DST/DCT)-based SSPE, the initial vertical source profile may directly be located on the spatial domain via

$$f(x)=\exp\left(\frac{-(x-x_s)^2}{w^2}\right)\exp(ik_0 x\sin\theta_{tilt}),\ w=\frac{2\ln 2}{k_0\sin\left(\frac{\theta_{bw}}{2}\right)}. \tag{15.5d}$$

- **Maximum Height:** Maximum available height is calculated depending on the choice of maximum slope of the terrain and elevation angle. Therefore, antenna tilt angle, which is permitted to be $-15°$ and $+15°$ here should be decided first. It is known that the relation between maximum height (x_{max}) and the maximum transverse wave number ($k_{x\,max}$)—depending on the FFT size—is used for computation. If the number of discretized height points over ground is L, this relation is $L\pi = x_{max}k_{x\,max}$, $k_{x\,max} = k\sin(\alpha_{max})$, where α_{max} is defined by the sum of elevation angle, maximum slope, and the desired maximum propagation angle that will be confined inside the validity region of the method. Here, an extra angle of 5° is added for convenience in the maximum propagation angle calculations. Inside the code, $L = N/2 = 512$ because instead of using a sine or cosine transform, FFT is used and computation space in x-axis is taken from $-x_{max}$ to x_{max}. Dirichlet

BC is used on the ground inside the code; therefore, the initial field distribution is taken as odd symmetric with respect to ground in order to force that $u(0, x) = 0$ at earth's surface. The height step size is calculated inside the code after maximum height is chosen, which should be lower than "maximum available height" for the validity of the method. The range step size is also based on α_{max} and defined as $\Delta z = \Delta x/\sin\alpha_{max}$. However, "observation step size," which shows the interval between observation ranges, should be chosen by the user.

Note that the SSPE in (15.4) is a narrow angle model and is valid near axial propagation direction. That is, vertical variations above 12°–15° cannot be handled. A wide angle model which is valid up to 40°–45° is as follows:

$$u(z+\Delta z, x) = \exp[ik_0(n-1)\Delta z]F^{-1}\left\{\exp\left[\frac{-ik_x^2\Delta z}{k_0\left(1+\sqrt{1-k_x^2/k_0^2}\right)}\right]F\{u(z,x)\}\right\}. \quad (15.6)$$

Table 15.1 and Table 15.2 lists simple MATLAB scripts for the implementation of the SSPE with both FFT and DST functions (i.e., under Dirichlet BC). It requires the initial field array U, initial range

Table 15.1.
A Simple FFT-Based SSPE Procedure

```
function FFTSSPE(U,Z1,Z2,DELZ,K0)
% ---------------------------------------------------
% Program: FftSSPE.m
% ---------------------------------------------------
clear all; clc; format long
NN=2*size(U); dx=X(2)-X(1); HMAX=X(NN); A0=1.2e-6;
alfa=1.e-6;
for k=1:NN/2   % Profile from -HMAX to Ground
     Field2(k)=complex(-U(k));
end
```

Table 15.1.
(Continued)

```
for k=1:NN/2  %Swap elements
     Field(k)=Field2(NN/2+1-k);
end
for k=NN/2+1:NN    % Profile from Ground to HMAX
     Field(k)=complex(U(-NN/2+k));
end
for k=1:NN   % HEIGHT, TRANSVERSE WAVENUMBER &
REFRACTIVITY
     X(k)=-2*HMAX+(k-1)*dx;
     KX(k)=-1/(2*dx)+(k-1)/(NN*dx);
     if  (X(k)>=0)
       REFR(k)=complex(1-A0*X(k));
     else
       REFR(k)=complex(1+A0*X(k));
  end
  CC1(k)=exp(-1i*DELZ*(KX(k)*2*pi)∧2/(2*K0));
  CC2(k)=exp(1i*DELZ*K0*(REFR(k)-1)/2);
end      if  (abs(X(k))>=HMAX)   % MAKE REFRACTIVITY
COMPLEX ABOVE HMAX/2
    REFR(k)=REFR(k)+1i*alfa*abs(X(k)-HMAX);
end
KK=(Z2-Z1)/DELZ; DMAX=1.+KK/20.; n=0; Z=Z1+DELZ;
while  (Z<=Z2)   % RANGE LOOP BEGINS
  n=n+1;
  Fld2=fftshift(Field); Field=fft(Fld2);
Fld2=fftshift(Field);
  Fld2=CC1.*Fld2; % multiply element by element
  Field=fftshift(Fld2); Fld2=ifft(Field);
Field=fftshift(Fld2);
  Field=CC2.*Field; Z=Z+DELZ;
  for k=1:NN/2
    XX(k)=(k-1)*dx; FSSPE(k)=0.05*n+abs(Field(NN/2+k));
  end
  figure(1);  % Plot Height vs. Field
  plot(FSSPE,XX); xlabel('Field'); ylabel('Height [m]');
end  % ------------------- END ----------------------
```

Table 15.2.
A Simple DST-Based SSPE Procedure

```
function DSTSSPE(U,Z1,Z2,DELZ,K0)
% -----------------------------------------------------
% Program: DstSSPE.m
% -----------------------------------------------------
clear all; clc; format long
NN=size(U); dx=X(2)-X(1); HMAX=X(NN); A0=1.2e-6;
alfa=1.e-6;
XX=0:dx:NN*dx; KX=linspace(0,pi/dx,NN).';
for k=1:NN % Height and Refractivity
   Field(k)=complex(U(k)); REFR(k)=complex(1-A0*X(k));
   CC1(k)=exp(-1i*DELZ*KX(k)^2/(2*K0));
   CC2(k)=exp(1i*DELZ*K0*(REFR(k)-1)/2);
   if (abs(X(k))>=HMAX)  % Make Refractivity complex
above HMAX
      REFR(k)=REFR(k)+1i*alfa*abs(X(k)-HMAX);
   end
end
n=0; Z=Z1;
while (Z<=Z2)  % Range Loop
   n=n+1; ZZ(n)=Z; Z=Z+DELZ;
   Field(2:end-1)=CC2(2:end-1).*idst(CC1(2:end-1).*dst(Fie
ld(2:end-1)));
end
figure;  % Plot Height vs. Field
plot(abs(Field),XX); xlabel('Field'); ylabel('Height [m]');
% ---------------------- END ------------------------
```

*Z*1, final range *Z*2, and the range increment DELZ, and the wave number k_0.

A simple example presented in Chapter 13 (see Section 13.4, Example 3) is revisited here to initially test the SSPE procedure. The flat earth with PEC ground is modeled with the DST-based SSPE method and compared against the two-ray analytical model. The MATLAB script does this comparison, as is listed in Table 15.3. An example output is presented in Fig. 15.2.

Table 15.3.
DST-Based SSPE versus 2Ray Model for Flat Earth with PEC Ground

```
% -------------------------------------------------------
%  Program: SSPEvs2Ray.m
% -------------------------------------------------------
clear all; clc;
freq=1e8; k0=2*pi*freq/3e8; lambda=3e8/freq;  REF=-1;
xs=150; TT=2; hmax=2000; Nx=1024; dx=2*hmax/Nx;
zlast=1e4; dz=lambda/2; Nz=zlast/dz; alfa=1.e-8;
X=0:dx:(Nx-1)*dx; Field0=exp(-(X-xs).∧2/TT∧2);
kX=linspace(0,pi/dx,Nx).';
%% ------------------ 2RAY PACKAGE -------------------
for n=1:Nx      % Vertical 2Ray profile at z=zlast
   xobs=X(n);
   R1=sqrt(zlast∧2+(xs-xobs)∧2);
R2=sqrt(zlast∧2+(xs+xobs)∧2);
   F2Ray(n)=exp(-1i*k0*R1)/sqrt(R1)+REF*exp(-1i*k0*R2)/
sqrt(R2);
end
%% ------------------ SSPE PACKAGE -------------------
for k=1:Nx % HEIGHT, TRANSVERSE WAVENUMBER &
REFRACTIVITY
  Field(k)=complex(Field0(k));    REFR(k)=complex(1);
  CC1(k)=exp(-1i*dz*kX(k)∧2/(2*k0)); CC2(k)=exp(1i*dz*k0
*(REFR(k)-1)/2);
  if (abs(X(k))>=hmax)   % Make Refractivity complex
above hmax
    REFR(k)=REFR(k)+1i*alfa*abs(X(k)-hmax);
  end
end
for n=1:Nz      % SSPE Range Loop

Field(2:end-1)=CC2(2:end-1).*idst(CC1(2:end-
1).*dst(Field(2:end-1)));
end
Rmax=max(abs(F2Ray)); Fmax=max(abs(Field));
plot(abs(F2Ray)/Rmax,X,abs(Field)/Fmax,X,'r-');
legend('2Ray','PEM'); xlabel('Height [m]');ylabel('Field
 ]'); ylim([0,hmax/3]);
title('Propagation over flat-Earth: PEM vs. 2Ray');
% ---------------------- END -----------------------
```

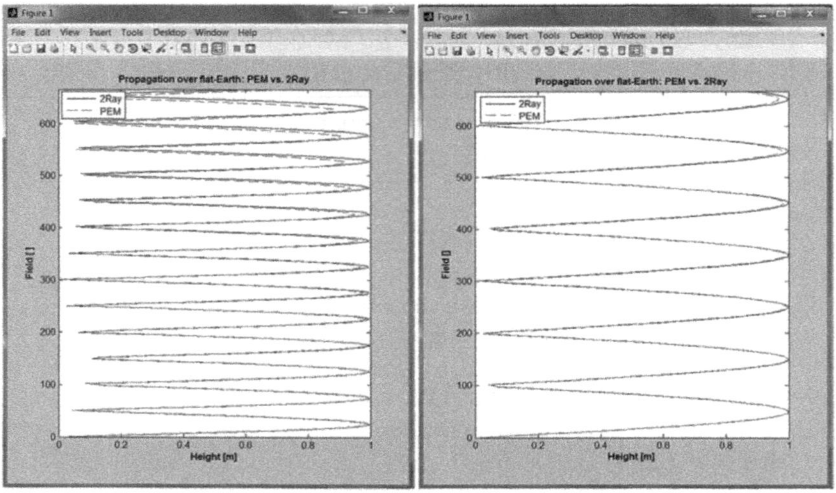

Figure 15.2. Vertical field above PEC flat earth at (left) 5 km and (right) 10 km.

15.4 THE FINITE ELEMENT METHOD-BASED PE PROPAGATION TOOL

The FEM-based longitudinally marching PE solution is constructed first by multiplying (15.1) with a smooth test function $v(x)$, while considering the BC in (15.2); integrating from $x = 0$ to $x = X_{\max}$ [29,30],

$$\int_0^{X_{\max}} \left\{ \frac{\partial^2 u(z,x)}{\partial x^2} + k_0^2\left(n^2(x) - 1\right)u(z,x) + 2ik_0 \frac{\partial u(z,x)}{\partial z} \right\} v(x)dx = 0, \tag{15.7}$$

and then using the integrating by parts rule,

$$\int_0^{X_{\max}} \left[-\frac{\partial u(z,x)}{\partial x}\frac{\partial v(x)}{\partial x} + k_0^2\left(n^2(x)-1\right)u(z,x)v(x) + 2ik_0\frac{\partial u(z,x)}{\partial z}v(x) \right] dx$$
$$+ \frac{\partial u(X_{\max})}{\partial x}v(X_{\max}) - \frac{\partial u(0)}{\partial x}v(0) = 0. \tag{15.8}$$

The last two terms of (15.8) must be taken into consideration according to the Cauchy BCs. They have no contributions for the homogeneous Dirichlet and Neumann BCs.

The idea of the FEM is to divide the domain, here the vertical profile $[0 - X_{\max}]$, into subdomains (called elements), in which basis functions are generally formed with the help of linear piecewise Lagrange polynomials as [29]

$$B_1^e(x) = \frac{x_2^e - x}{x_2^e - x_1^e},\; B_2^e(x) = \frac{x - x_1^e}{x_2^e - x_1^e}, \tag{15.9}$$

where e stands for the elements between nodes x_1^e and x_2^e. The field function $u(z, x)$ can be approximated by $u_{\mathrm{ap}}(z, x)$ at the discrete height points as

$$u_{\mathrm{ap}}(z, x) = \sum_{e=1}^{n_e} \sum_{j=1}^{2} c_j^e(z) B_j^e(x), \tag{15.10}$$

where n_e is the number of elements and $c_j^e(z)$ denotes the coefficients of unknown functions. If the test function is replaced with $B_m(x)$ for $m = 1, 2$, (15.8) becomes

$$\sum_{e=1}^{n_e} \left\{ \begin{array}{l} \displaystyle\sum_{j=1}^{2} c_j^e \int_{x_1^e}^{x_2^e} \frac{\partial B_j^e}{\partial x} \frac{\partial B_m^e}{\partial x} dx - k_0^2 c_j^e \int_{x_1^e}^{x_2^e} \left(n^2(x) - 1\right) B_j^e B_m^e dx \\ \displaystyle -2ik_0 \sum_{j=1}^{2} \frac{\partial c_j^e}{\partial z} \int_{x_1^e}^{x_2^e} B_j^e B_m^e dx \end{array} \right\} = 0. \tag{15.11}$$

The matrix form of (15.11) can then be represented as

$$(\mathrm{K} + \mathrm{L})c + \mathrm{M}\frac{\partial c}{\partial z} = 0, \tag{15.12}$$

or $\left[K_{mj}^e + L_{mj}^e\right]\left\{c_j^e\right\} + \left[M_{mj}^e\right]\left\{\dfrac{\partial c_j^e}{\partial z}\right\} = \{0\}$ for $e = 1, \ldots, n_e$, $m = 1, 2$, and $j = 1, 2$ with

$$K_{mj}^e = \int_{x_1^e}^{x_2^e} \frac{\partial B_m^e}{\partial x} \frac{\partial B_j^e}{\partial x} dx, \tag{15.13}$$

$$L_{mj}^{e} = -k_0^2 \int_{x_1^e}^{x_2^e} \left(n^2(x) - 1\right) B_m^e B_j^e dx, \tag{15.14}$$

$$M_{mj}^{e} = -2ik_0 \int_{x_1^e}^{x_2^e} B_m^e B_j^e dx. \tag{15.15}$$

The matrix **L** is determined once the refractivity is given. For example, if a linearly decreasing vertical refractivity is considered as

$$n^2(x) = 1 - a_0 x, \tag{15.16}$$

where a_0 is a positive constant, these elemental matrices between the nodes x_1^e and x_2^e are obtained as

$$\left[K^e\right] = \frac{1}{h^e}\begin{bmatrix} 1 & -1 \\ -1 & 1 \end{bmatrix}, \tag{15.17}$$

$$\left[L^e\right] = \frac{k_0^2 a_0 h^e}{12}\begin{bmatrix} 3x_1^e + x_2^e & x_1^e + x_2^e \\ x_1^e + x_2^e & x_1^e + 3x_2^e \end{bmatrix}, \tag{15.18}$$

$$\left[M^e\right] = -\frac{ik_0 h^e}{3}\begin{bmatrix} 2 & 1 \\ 1 & 2 \end{bmatrix}, \tag{15.19}$$

where $h^e = x_2^e - x_1^e$ shows the grid width.

Using the Crank–Nicholson approximation based on the improved Euler method for $k = 2, \ldots N_z$ [31],

$$c^k = c^{k-1} + \frac{\Delta z}{2}\left(\frac{\partial c^k}{\partial z} + \frac{\partial c^{k-1}}{\partial z}\right), \tag{15.20}$$

multiplying by M and eliminating the derivative terms by using the differential equation (15.12), one obtains

$$\mathrm{M}c^k = \mathrm{M}c^{k-1} + \frac{\Delta z}{2}\mathrm{M}\left(\frac{\partial c^k}{\partial z} + \frac{\partial c^{k-1}}{\partial z}\right), \tag{15.21}$$

$$\mathrm{M}c^k = \mathrm{M}c^{k-1} - \frac{\Delta z}{2}(\mathrm{K} + \mathrm{L})\left(c^k + c^{k-1}\right), \tag{15.22}$$

$$\left(\mathrm{M} + (\mathrm{K} + \mathrm{L})\frac{\Delta z}{2}\right)c^k = \left(\mathrm{M} - (\mathrm{K} + \mathrm{L})\frac{\Delta z}{2}\right)c^{k-1}, \tag{15.23}$$

which yields an unconditionally stable and accurate system with a discretization error of $O(\Delta z^2)$. The coefficients of the initial field c^1 at $z = 0$ are generated from the Gaussian antenna pattern specified by its height, beamwidth, and tilt. Although Crank–Nicholson gives fast solution, it should be noted that this method has disadvantages since oscillation occurs for large Δz.

15.5 ATMOSPHERIC REFRACTIVITY EFFECTS

The PEM has long been used in radio wave propagation modeling. Radio waves are directly affected by the atmospheric variations; therefore, propagation engineers/modelers shall take these variations into account with great care.

The earth's atmosphere is a collection of many gases together with suspended particles of liquids and solids, and its range and height-dependent variations are modeled in terms of refractivity in EM wave propagation. Because of the closeness of n to unity, it is customary to use the refractivity N defined as

$$N = (n-1)\times 10^6. \tag{15.24}$$

N is dimensionless, but is measured in "N units" for convenience. N depends on the pressure P (mbar), the absolute temperature T (K), and the partial pressure of water vapor e (mbar) as

$$N = 77.6\frac{P}{T} + 3.73\times 10^5\frac{e}{T^2}, \tag{15.25}$$

which is valid in Earth-troposphere waveguides and used in ground wave propagation modeling. If the refractive index were constant, radio waves would propagate in straight lines. Since n decreases with height, radio waves are bent downward toward the earth, so that the radio horizon lies further away than the optical horizon. It should be noted that the radio horizon effect is taken into account either by using N with a_e or by introducing a fictitious medium where N is replaced by the modified refractivity M,

$$M = N + \frac{x}{a}\times 10^6 = N + 157x, \tag{15.26}$$

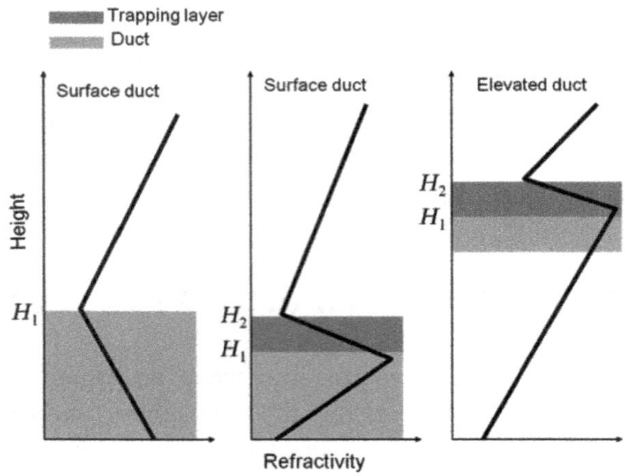

Figure 15.3. Surface and elevated duct formation because of vertical refractivity variations; (left and middle) surface duct and (right) elevated duct.

with the height x given in kilometers. In (15.26),

$$a = 6378\ \text{km}, \quad \frac{10^6}{a} = 157, \quad \text{and} \quad \frac{\partial M}{\partial x} = \frac{\partial N}{\partial x} + 157. \quad (15.27)$$

For the standard atmosphere (i.e., for a vertical linearly decreasing refractive index), N decreases by about 40 N unit/km, while M increases by about 117 N unit/km. Subrefraction (super-refraction) occurs when the rate of change in N with respect to height (i.e., $\partial N/\partial x$) is less (more) than 40 N unit/km.

A vertically decreasing (increasing) refractivity causes a wave bend toward (away from) the earth's surface. Piecewise linear, bilinear, and three-linear refractivity variations as given in Fig. 15.3 cause formation of surface and/or elevated ducts. Surface-based ducts occur when the air aloft is exceptionally warm and dry compared with the air at earth's surface, and may occur during day and night over the ocean and near land. The trapping layer may well rise above the ocean as one moves from the coastal region into the open ocean, extend hundreds of kilometers, and last for days. One special case of the surface ducts is the ocean evaporation duct which may exist almost all of the time, extend from a few meters above the ocean surface up to 10–30 m in heights (on a world rough average, it is 10–15 m), and be effective for EM propagation above 250–300 MHz.

A kind of inversion may occur between 30° north and south, when a layer of warm, dry air overlies a cool, moist layer of air. This is often called marine boundary layer, and may create a strong ducting condition above the marine boundary layer. Elevated ducts may extend from a few hundreds of meters above the surface to a few kilometers.

Mathematical representation of all these cases can be achieved via

$$n(x) = n_0 \begin{cases} 1 + a_0 x & x \leq H_1 \\ 1 + a_0 H_1 + b_0 x & H_1 < x \leq H_2, \\ 1 + a_0 H_1 + b_0 H_2 - a_0 x & x > H_2 \end{cases} \tag{15.28}$$

where n_0 is the refractivity at the surface, and the user may form any kind of refractivity by specifying first and second heights (in meters), H_1 and H_2, respectively, and the slopes a_0 and b_0 of the modified refractivity (therefore, supplied values are multiplied by 10^{-9}, i.e., $a_0 \rightarrow a_0 \times 10^{-9}$). For example, a vertically linear decreasing refractivity up to 1 km altitude (i.e., a surface duct) can be specified by various ways: by giving (1) $H_1 = 0$ and $H_2 = 1000$, and $a_0 = 0$, $b_0 < 0$, or (2) $H_1 = 1000$ and $H_2 = 10{,}000$, and $a_0 < 0$, $b_0 =$ arbitrary. Similarly, a bilinear refractivity up to 1 km (elevated duct) with the duct height at 600 m can be given by (1) $H_1 = 600$ and $H_2 = 1000$, and $a_0 > 0$, $b_0 < 0$, or (2) $H_1 = 0$ and $H_2 = 600$, and $a_0 < 0$, $b_0 > 0$, and so on.

15.6 A 2D SURFACE DUCT SCENARIO AND REFERENCE SOLUTIONS

Propagation over the PEC flat earth with a linearly decreasing vertical refractivity profile given in (15.16) is a canonical structure to generate the reference solutions [34,35]. In such a nonhomogeneous environment, the scalar wave function $u(z, x)$ satisfies the 2D wave equation,

$$\frac{\partial^2 u}{\partial x^2} + \frac{\partial^2 u}{\partial z^2} + k_0^2 n^2(x) u = 0, \tag{15.29}$$

subject to BC in (15.2) and the radiation BC in (15.3). This problem has analytical exact solutions in terms of Airy functions for the range-independent vertical refractive index. The modal series solution (under $\exp(-i\omega t)$ time dependence) can be expressed as

$$u(x,z)=\sum_{q=0}^{\infty}C_{0q}\psi_q(x)e^{i\beta_q z}, \tag{15.30}$$

where β_q is the longitudinal propagation constant for the related mode represented by index q. The new function $\psi_q(x)$ satisfies the 1D wave equation in the transverse domain,

$$\left[\frac{d^2}{dx^2}+k_0^2n^2(x)-\beta_q^2\right]\psi_q(x)=0. \tag{15.31}$$

Using $k_0^2n^2(x)-\beta_q^2=K_1x+K_2$ and $\rho=-K_1^{-2/3}(K_1x+K_2)$, the new variables are obtained for linearly decreasing vertical refractivity in (15.16) as $K_1=-a_0k_0^2, K_2=k_0^2-\beta_q^2, \rho=(a_0k_0^2)^{1/3}x-(a_0k_0^2)^{-2/3}(k_0^2-\beta_q^2)$; and the wave equation becomes the Airy equation [2,34],

$$\left[\frac{d^2}{d\rho^2}-\rho\right]\psi_q(\rho)=0. \tag{15.32}$$

The solution of this equation with BCs given in (15.2) and (15.3) is constructed in terms of modes that are represented by the first kind of Airy function $\mathrm{Ai}(\rho)$, with ρ being real,

$$\psi_q(\rho)=C_{0q}\mathrm{Ai}(\rho), \tag{15.33}$$

where C_{0q} is the normalization constant defined as

$$C_{0q}=\left[\int_0^{\infty}\mathrm{Ai}^2(\rho)d\rho\right]^{-1/2}. \tag{15.34}$$

The condition at the surface is then reduced to

$$\mathrm{Ai}'(-\sigma_q)+\alpha\mathrm{Ai}(-\sigma_q)=0, \tag{15.35}$$

$$\mathrm{Ai}(-\sigma_q)=0, \tag{15.36}$$

$$\mathrm{Ai}'(-\sigma_q)=0, \tag{15.37}$$

for the Cauchy-, Dirichlet-, and Neumann-type BCs, respectively. Here, the prime denotes the derivative with respect to the vertical coordinate. The Dirichlet BC is treated throughout this chapter. The longitudinal propagation constant for each mode is found as

$$\beta_q = \pm\left[k_0^2 - \sigma_q (a_0 k_0^2)^{2/3}\right]^{1/2}, \tag{15.38}$$

and the modes are trapped between the surface and their modal caustics determined from

$$X_{cq} = -\sigma_q\left[a_0 k_0^2\right]^{-1/3}. \tag{15.39}$$

Finally, the exact modal solution of Airy-type wave equation is then found to be

$$u(x,z) = \sum_{q=1}^{N} C_{0q}\mathrm{Ai}((a_0 k_0^2)^{1/3} x - \sigma_q) e^{i\beta_q z}. \tag{15.40}$$

The problem is then reduced to finding the modal excitation coefficients from a given antenna pattern using orthonormality property. A simple MATLAB script developed for the calculation of fields through an atmosphere characterized by a vertically decreasing linear refractivity profile is listed in Table 15.4. Figure 15.4, Figure 15.5, and

Table 15.4.
A Short MATLAB Mode Sum Code for the Surface Duct Problem

```
% -------------------------------------------------
%  Program: LINEAR.m
%  This program computes transverse Field Profile
(surface duct)
% -------------------------------------------------
clear all; clc; format long
% ----------- Program constants -----------
NX=1000; A0=1.2e-6; K0=2*pi; KOS=K0*K0;
P1=(A0*KOS)^.333334;
HH=500; TT=100; HMAX=750; delx=HMAX/(NX-1); z1=55e3;
z2=110e3;
% ----------- Roots of Ai(-TIG)=0 ----------
TIG=[2.33810741; 4.08794944; 5.52055983; 6.78670809;
7.94413359; 9.02265085;10.04017434; 11.00852430;
11.93601556; 12.82877675];
```

(Continued)

Table 15.4.
(Continued)

```
for k=1:NX     % Building transverse source profile
    XX(k)=(k-1)*delx; SGAUSS(k)=exp(-((XX(k)-HH)/TT)∧2);
    NMField(k)=0.0; NMFldz1(k)=0.0;NMFldz2(k)=0.0;
end
OK=1; mn=0;
while OK>0
   mn=mn+1;
   if (mn<=10) RT(mn)=TIG(mn); else RT(mn)=(1.5*(mn*pi
-pi/4))∧(2./3.); end
   XCQ=RT(mn)/P1; BETA(mn)=sqrt(K0S-RT(mn)*P1∧2);
   for k=1:NX    % Modal Normalisation constant
      CRT=-RT(mn)*(1-XX(k)/XCQ); RMODE(mn,k)=Airy(CRT);
RM(k)=Airy(CRT)∧2;
   end
   V0=1.0/(sqrt(trapz(XX,RM)));
   for k=1:NX   % Modal excitation constant
      SG(k)=SGAUSS(k)*V0*RMODE(mn,k);
   end
   EXC(mn)=V0*trapz(XX,SG); % EXCITATION (plus NORM)
FACTOR
   for k=1:NX    % Calculate fields at z=0, z=z1 and z=z2
      CRT=-RT(mn)*(1-XX(k)/XCQ); Mode(k)=EXC(mn)*Airy(
CRT);
      NMField(k)=NMField(k)+Mode(k);
      NMFldz1(k)=NMFldz1(k)+Mode(k)*exp(1i*BETA(mn)*z1);
      NMFldz2(k)=NMFldz2(k)+Mode(k)*exp(1i*BETA(mn)*z2);
   end
   subplot(1,4,1); plot(real(Mode),XX,'LineWidth',2);
   ylim([0,HMAX]); xlabel('Mode Profile'); ylabel('Height
 m]');
   legend('Mode No:', num2str(mn));
   subplot(1,4,2); xlim([0,1]); ylim([0,HMAX]);
   plot(SGAUSS,XX,abs(NMField),XX,'r-','LineWidth',2);
   xlabel('Field at z=0'); legend('Source', 'NM Sum');
   subplot(1,4,3); plot(abs(NMFldz1),XX,'k','LineWidth',2);
   ylim([0,HMAX]); xlabel('Field Profile'); legend('z=z1');
   subplot(1,4,4); plot(abs(NMFldz2),XX,'k','LineWidth',2);
   ylim([0,HMAX]); xlabel('Field Profile'); legend('z=z2');
pause
end    % ------------------- END ----------------------
```

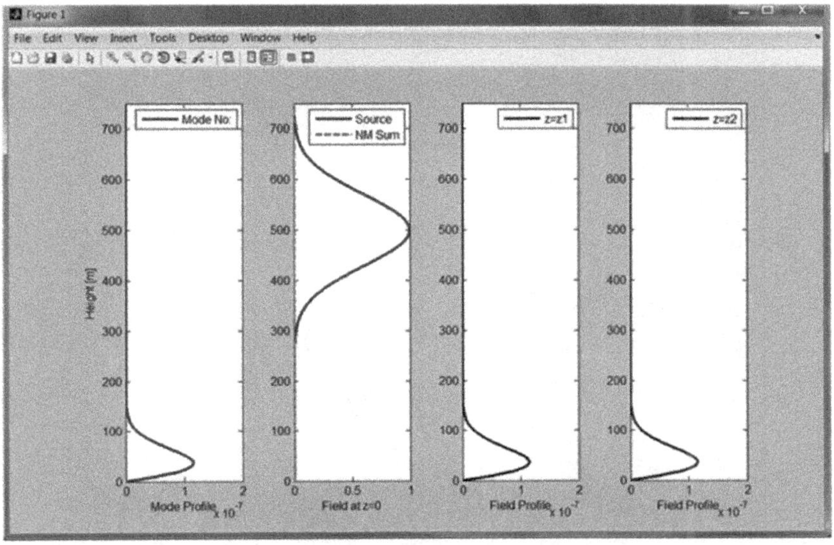

Figure 15.4. (L–R) First mode, initial vertical source, and fields at two ranges.

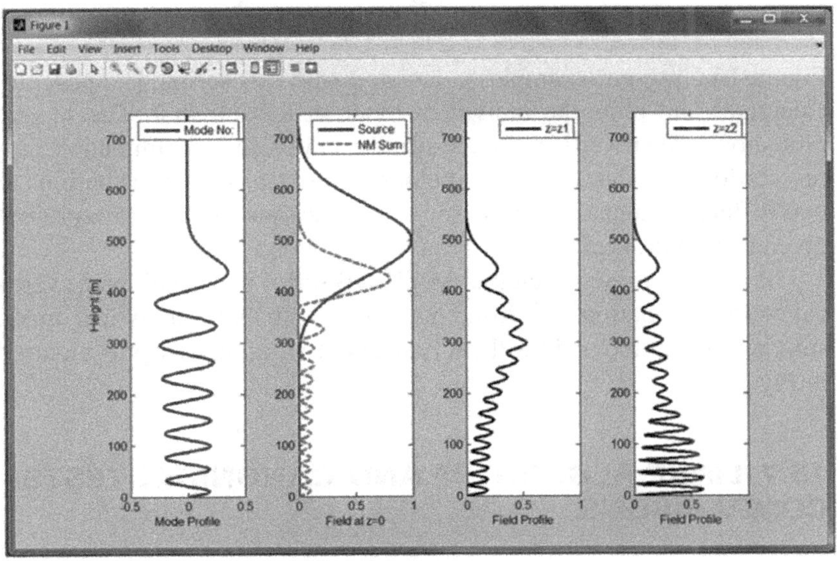

Figure 15.5. (L–R) Thirteenth mode, initial vertical source, and fields at two ranges.

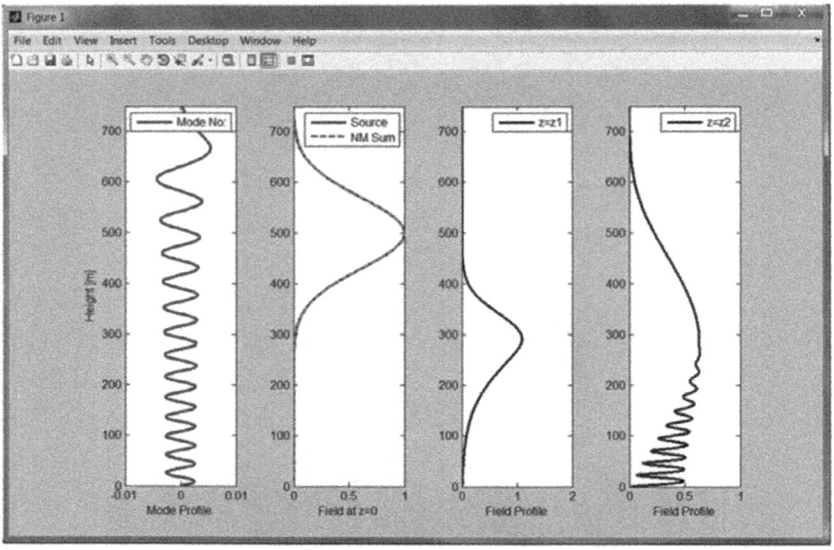

Figure 15.6. (L–R) Twentieth mode, initial vertical source, and fields at two ranges.

Figure 15.6 present examples calculated with this script. The program calculates and plots the individual mode at the left first. The second plot shows given vertical source profile and its mode summation representation. The last two plots belong to vertical field distributions at two different ranges. As observed, 20 modes are enough to represent given source profile.

Table 15.5 presents another MATLAB script which computes vertical field distributions at different ranges with both analytical mode summation method and SSPE. A typical output of this script is pictured in Fig. 15.7.

15.7 LINPE ALGORITHM AND CANONICAL TESTS/COMPARISONS

A novel MATLAB-based propagation tool LINPE is developed for the tests and comparisons of SSPE and FEMPE and for their calibration [33,36–39]. The flowchart of the LINPE is given in Fig. 15.8. The

Table 15.5.
A Short MATLAB Code for the Mode Sum and SSPE Methods

```
% -----------------------------------------------------------
% Program: LS_LINEAR.m
% Analytic vs. SSPE for vertically decreasing linear
profile
% -----------------------------------------------------------
clear all; clc; format long
% ------ Ai(-TIG)=0 -------
TIG=[2.33810741; 4.08794944; 5.52055983; 6.78670809;
7.94413359; . . .
    9.02265085; 10.04017434; 11.00852430; 11.93601556;
12.82877675];
NN=1024; JJ=NN/4; A0=1.2e-6; alfa=1.e-6; K0=2*pi;
K0S=K0*K0;
P1=(A0*K0S)∧.333334; Z1=0; Z2=2e5; DELZ=500; HH=200;
TT=100;
HMAX=750; MM=15;delx=HMAX/(NN-1);
% ----------------- Analytical Model ------------------
for k=1:NN % SET INITIALS TO ZERO FOR ALL ARRAYS
    XX(k)=0; SGAUSS(k)=0; SG(k)=0;
end
for l=1:MM
    RT(l)=0; EXC(l)=0; beta(l)=0;
end
for k=1:NN   % Profile from Ground to HMAX
    XX(k)=(k-1)*delx; SGAUSS(k)=exp(-((XX(k)-HH)/TT)∧2);
end
for k=1:MM    % Assing EIGENVALUES
   if (k<=10)
      RT(k)=TIG(k);
   else
      RT(k)=(1.5*(k*pi-pi/4))∧(2./3.);
   end
end
for MODE=1:MM   % Mode sum
   XCQ=RT(MODE)/P1; BETA(MODE)=sqrt(K0S-RT(MODE)*P1∧2);
   for k=1:NN  % calculate normalisation constant
      CRT=-RT(MODE)*(1-XX(k)/XCQ);
      RMODE(MODE,k)=Airy(CRT); RM(k)=Airy(CRT)∧2;
   end
```

(*Continued*)

Table 15.5.
(Continued)

```
  V0=1.0/(sqrt(trapz(XX,RM)));   for k=1:NN   %
Excitation constants
    SG(k)=SGAUSS(k)*V0*RMODE(MODE,k);
  end
  EXC(MODE)=V0*trapz(XX,SG); % EXCITATION (plus NORM)
FACTOR
end
for MODE=1:MM    % Mode sum
  XCQ=RT(MODE)/P1;
  for k=1:JJ
    Y(k)=(k-1)*(HMAX/(JJ-1));
    CRT=-RT(MODE)*(1-Y(k)/XCQ); FMODE(MODE,k)=EXC(MODE)*
Airy(CRT);
  end
end
% ------------------- SSPE PACKAGE --------------------
dx=4*HMAX/(NN-1);
for k=1:NN/2
  x=k*dx; A(k)=exp(-((x-HH)/TT)∧2);
end
for k=1:NN/2   % Profile from -HMAX to Ground
    Field2(k)=complex(-A(k));
end
for k=1:NN/2   %Swap elements
    Field(k)=Field2(NN/2+1-k);
end
for k=NN/2+1:NN    % Profile from Ground to HMAX
    Field(k)=complex(A(-NN/2+k));
end
for k=1:NN   % HEIGHT, TRANSVERSE WAVENUMBER &
REFRACTIVITY
    X(k)=-2*HMAX+(k-1)*dx;   KX(k)=-1/(2*dx)+(k-1)/
(NN*dx);
    if (X(k)>=0)
      REFR(k)=complex(1-A0*X(k));
    else
      REFR(k)=complex(1+A0*X(k));
  end
```

Table 15.5.
(Continued)

```
   CC1(k)=exp(-1i*DELZ*(KX(k)*2*pi)∧2/(2*K0));
   CC2(k)=exp(1i*DELZ*K0*(REFR(k)-1)/2);
end
if (abs(X(k))>=HMAX)   % MAKE REFRACTIVITY COMPLEX ABOVE
HMAX/2
    REFR(k)=REFR(k)+1i*alfa*abs(X(k)-HMAX);
end
KK=(Z2-Z1)/DELZ; DMAX=1.+KK/20.; n=0; Z=Z1+DELZ;
while (Z<=Z2)   % RANGE LOOP BEGINS
   n=n+1;
   for k=1:JJ
      CFUNC(k)=0;
      for MODE=1:MM
         CFUNC(k)=CFUNC(k)+FMODE(MODE,k)*exp(i*BETA(MODE
)*Z);
      end
   end
   Fld2=fftshift(Field); Field=fft(Fld2);
Fld2=fftshift(Field);
   Fld2=CC1.*Fld2; % multiply element by element
   Field=fftshift(Fld2); Fld2=ifft(Field);
Field=fftshift(Fld2);
   Field=CC2.*Field;
   Z=Z+DELZ;
   for k=1:JJ
      B(k)=0.05*n+abs(CFUNC(k));
S(k)=0.05*n+abs(Field(NN/2+k));
   end
   figure(1);  % Plot Height vs. Field
   set(1,'Position', [132 208 960 280]); grid;
   plot(B,Y,S,Y,'r-'); xlim([0,DMAX]); ylim([0,HMAX]);
hold on;
   xlabel('Propagation direction'); ylabel('Height [m]');
   title(['LINEAR PROFILE   Solid: NM   Dashed: SSPE']);
pause;
end % Range Loop ends
% ------------------------ END ------------------------
```

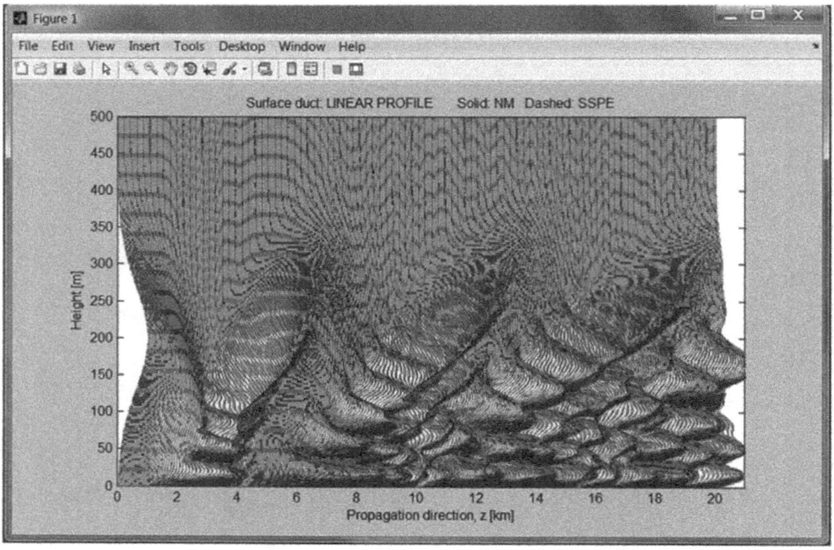

Figure 15.7. Field versus range–height calculated with the mode sum and SSPE models.

propagation tool has three subroutines LIN_EXACT, SSPE, and FEMPE for the numerical calculations of the mathematical exact, the FFT-based PE, and the FEM-based PE solutions, respectively. First, the 2D propagation environment (the maximum range and height) is specified, the frequency is given, and the discretization parameters (Δ_x, Δ_z, the number of range points N_z, height points N_x, etc.) are supplied. Then, the Gauss-shaped antenna pattern (i.e., the antenna height, its beamwidth, and vertical tilt) is given as the initial vertical field profile and then normalized to 1 for all simulation tests. Finally, the slope of the vertically decreasing refractivity profile is supplied.

LINPE propagation tool links to the mode summation subroutine, once all operational parameters are set. The mode subroutine calculates the number of modes and the modal excitation coefficients that are necessary to construct the initial antenna pattern for the given accuracy. These modes are then used for every height and range points to calculate the mathematical solution that will be used as the reference.

LINPE calls all three subroutines one by one, does the numerical computations, and yields the output as the 2D and 3D field strength variations. It yields not only the 3D field strength versus range–height

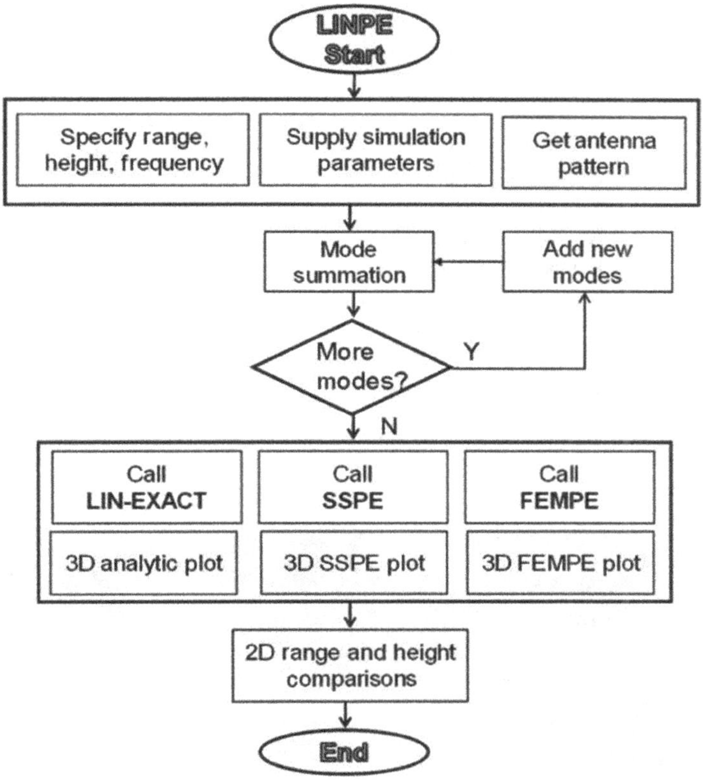

Figure 15.8. Flowchart of the MATLAB-based LINPE code which uses analytical exact representation, SSPE and FEMPE.

color plots for all the three methods but also the two 2D plots for their comparisons: Field strength versus height variations are plotted at three different ranges; field strength versus range variations are given at two different heights.

The first tests and comparisons belong to the scenario specified with the operational parameters listed in Table 15.6. Figure 15.9, Figure 15.10, and Figure 15.11 present the 3D color plots of electric field strength versus range–height where propagation over the flat, nonpenetrable earth is simulated without and with antenna tilt. The antenna has a sectoral vertical radiation beam. The vertical, linearly decreasing refractivity profile causes a surface duct as observed in the figures.

Table 15.6.
The Operational Parameters for Test 1

Operational Parameter	Value
Maximum range (km)	100
Maximum height (m)	750
Number of range pts	1000
Number of height pts	1024
Frequency (MHz)	300
Antenna height (m)	250
Vertical beamwidth (°)	0.5
Antenna tilt (°)	1.0
Refractivity slope (a_0)	1.2×10^{-6}

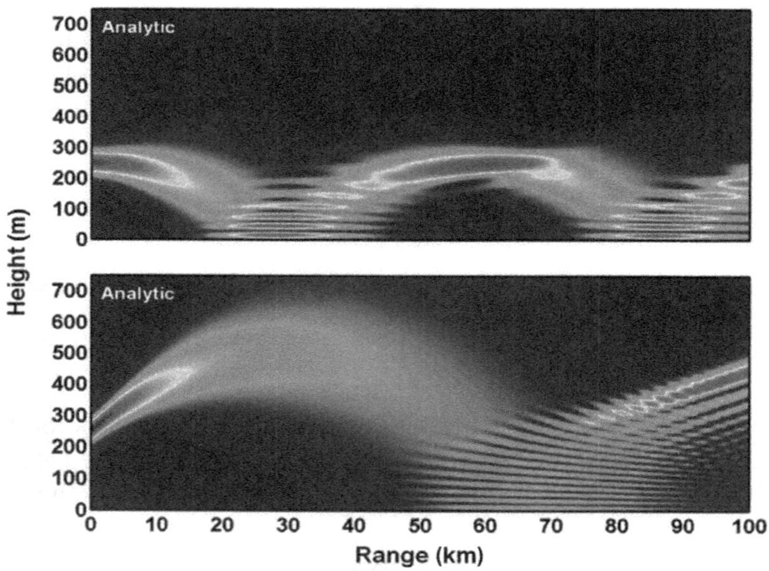

Figure 15.9. Maps of E-field strength versus range–height for a given Gaussian antenna pattern obtained from the analytical representations (top) without a tilt and (bottom) antenna 1° uptilted ($f = 300$ MHz, $a_0 = 1.2 \times 10^{-6}$, $N_x = 256$, $N_z = 1000$).

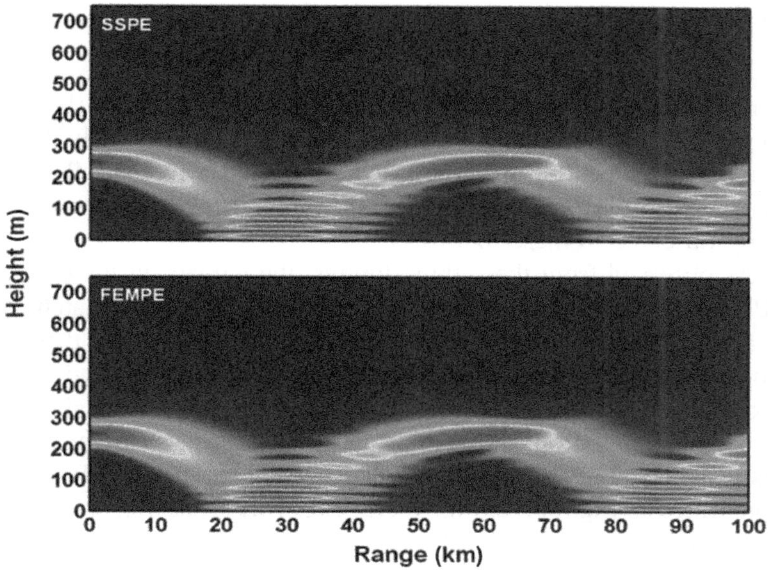

Figure 15.10. Maps of E-field strength versus range–height for a given nontilted Gaussian antenna pattern obtained via (top) SSPE and (bottom) FEMPE (f = 300 MHz, a_0 = 1.2×10^{-6}, $N_x = 256$, $N_z = 1000$).

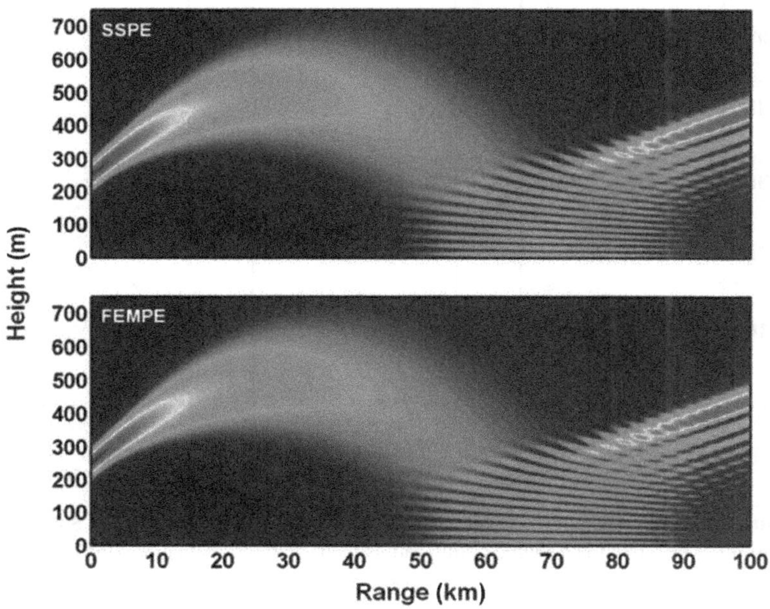

Figure 15.11. Maps of E-field strength versus range–height variations for a given 1° uptilted Gaussian antenna pattern obtained via (top) SSPE and (bottom) FEMPE (f = 300 MHz, a_0 = 1.2×10^{-6}, $N_x = 256$, $N_z = 1000$).

Although the beam is directed horizontally (top) and upward (bottom), EM waves are pushed down as they propagate longitudinally, hit the ground, and reflect upward. The ground-reflected waves and the interference (between the incident and the ground-reflected waves) are clearly observed in these figures.

As observed from these three figures, the agreement between the SSPE and FEMPE numerical propagation tools and their match with the analytical exact solution are impressive. Unfortunately, these 3D color plots are mostly good for the visualization purposes and are not good for the realistic comparisons and the calibration. The calibration of such numerical propagation tools necessitates accurate comparisons. This is achieved via 2D range and/or height illustrations of EM fields as presented in Fig. 15.12, Fig. 15.13, Fig. 15.14, and Fig. 15.15. The frequency is 300 MHz in all these figures. In Fig. 15.12, the vertical

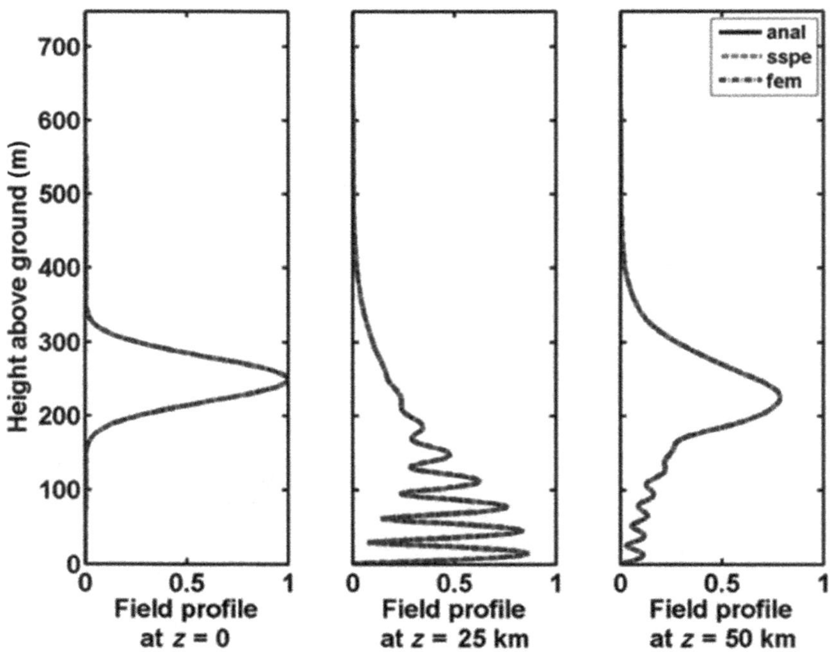

Figure 15.12. E-field versus height at three ranges. Solid: analytic; dashed: SSPE; dash dotted: FEMPE. (Left) The source profile and its mode sum representation, (middle) E-field at 25 km, and (right) E-field at 50 km ($f = 300$ MHz, $a_0 = 1.2 \times 10^{-6}$, $N_x = 256$, $N_z = 1000$, without antenna tilt).

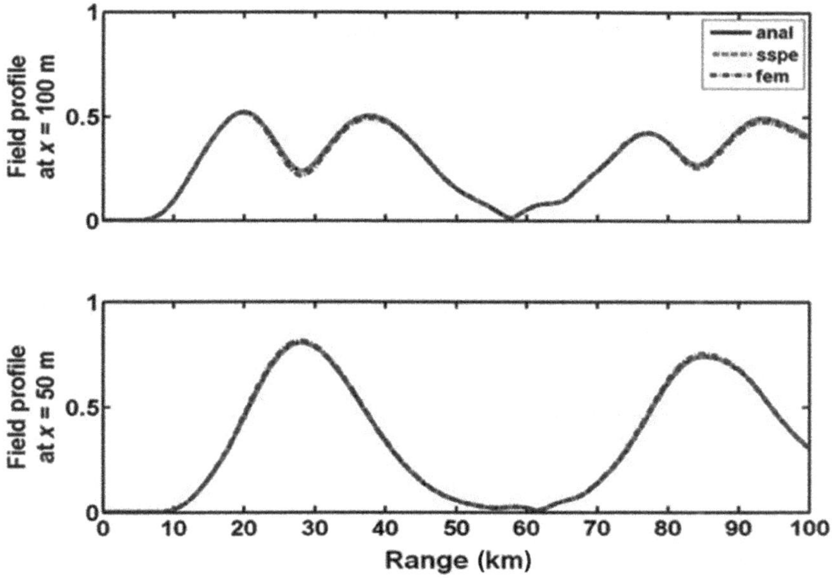

Figure 15.13. E-field versus range at two heights. Solid: analytic; dashed: SSPE; dash dotted: FEMPE. (Top) 100 m and (bottom) 50 m above the ground (f = 300 MHz, $a_0 = 1.2 \times 10^{-6}$, $N_x = 256$, $N_z = 1000$, without antenna tilt).

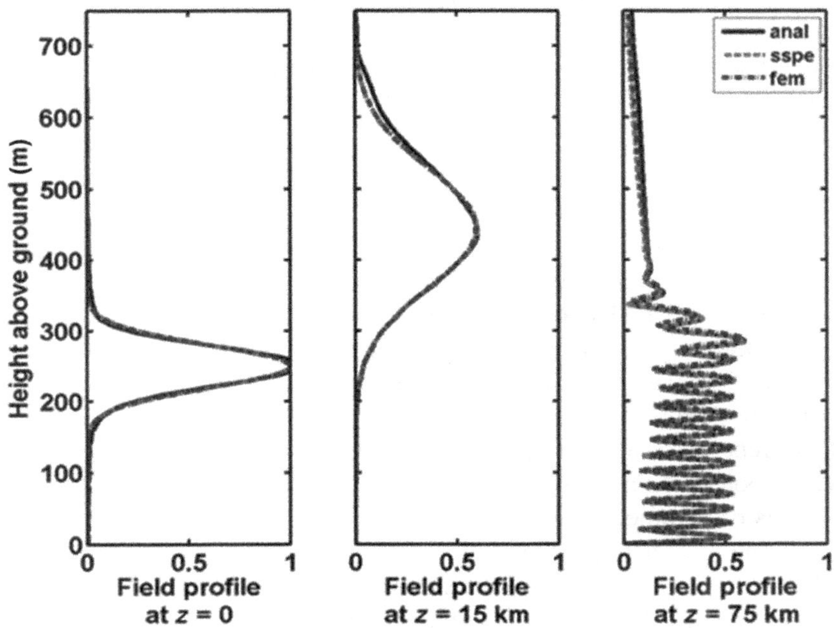

Figure 15.14. E-field versus height at three ranges. Solid: analytic; dashed: SSPE; dash dotted: FEMPE. (Left) The source and its mode sum representation, (middle) E-field at 15 km, and (right) E-field at 75 km (f = 300 MHz, $a_0 = 1.2 \times 10^{-6}$, $N_x = 256$, $N_z = 1000$, with 1° upward antenna tilt).

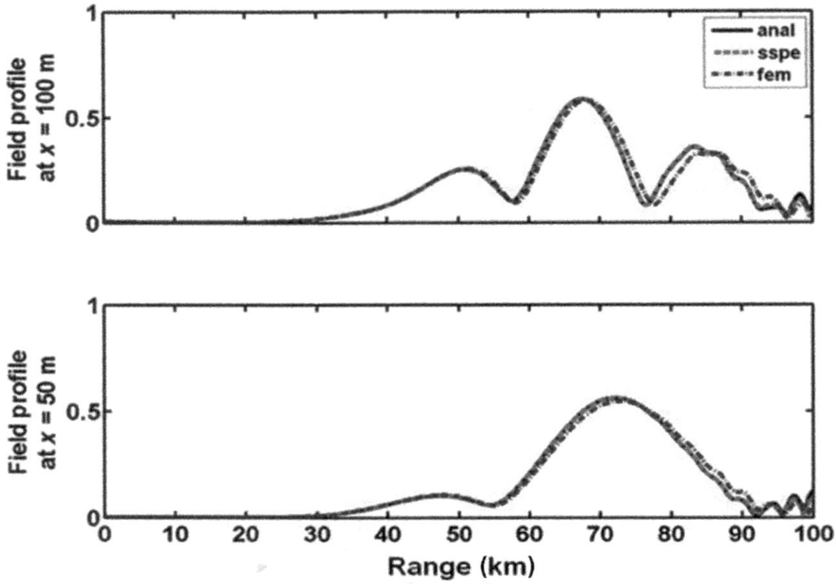

Figure 15.15. E-field versus range at two heights. Solid: analytic; dashed: SSPE; dash dotted: FEMPE. (Top) 100 m and (bottom) 50 m above the ground (f = 300 MHz, $a_0 = 1.2 \times 10^{-6}$, $N_x = 256$, $N_z = 1000$, with 1° upward antenna tilt).

field variations computed via all three methods at three different ranges is given. On the left, the initial field profile (i.e., the antenna radiation pattern) is shown. The other two show the vertical field profiles at 25- and 50-km ranges, respectively. Figure 15.13 shows the same comparison among these three methods as field strengths versus ranges at two different heights, 50 and 100 m, respectively. Figure 15.14 and Figure 15.15 repeat the same comparisons for an upward tilted antenna pattern. Very good agreement in all these figures among all three methods for the sets of parameters listed in Table 15.3 is clearly observed.

Table 15.7 lists the computation times and memories used during all three computations for two different computers with single and dual processors. As observed from the table, FEMPE is the fastest when the same discretization parameters are used in all three simulations, but occupy much more memory. The analytical calculations are much faster if the antenna beamwidth is wide (i.e., the number of modes is less).

Table 15.7.
Computation Times and Memory Requirements for the Same Discretization Parameters

(a) Sony Vaio, Notebook, Single Processor, 1.2 GHz, 1 GB RAM				
	Without Antenna Tilt		With 1° Antenna Tilt	
	55 Modes, $N_x = 256$		90 Modes, $N_x = 256$	
Method	Time (s)	RAM (MB)	Time (s)	RAM (MB)
Analytic	21.2	4.7	49.3	3.9
SSPE	42.5	5.6	42.7	4.3
FEMPE	23.4	24.1	23.4	25.8

(b) HP Compaq, NC6320 Notebook, Dual Processor, 2 GHz, 1.5 GB RAM				
	Without Antenna Tilt		With 1° Antenna Tilt	
	55 Modes, $N_x = 256$		90 Modes, $N_x = 256$	
Method	Time (s)	RAM (MB)	Time (s)	RAM (MB)
Analytic	9.3	4.0	12	5
SSPE	21.1	16	20	16
FEMPE	3.1	22	3.0	14

The narrower the vertical antenna pattern, the higher the number of modes used to reconstruct this pattern; therefore, the longer the computation time. Note that the tilt of the antenna in analytical exact representations is satisfied by introducing the complex modal excitation coefficients. The dual processor uses more dynamic memories, but reduces computation time drastically. Note that doubling the number of samples in range doubles the computation time; there is a linear dependence between the number of samples in range and computation burden. But, doubling the number of samples in height causes four times an increase in the computation time. This is because FEMPE solves a tridiagonal matrix system in height.

The second test is performed with the same operational parameters listed in Table 15.6 under three times stronger ducting refractivity profile (i.e., for $a_0 = 3.6 \times 10^{-6}$). The results are given in Fig. 15.16, Fig. 15.17, Fig. 15.18, and Fig. 15.19. Although the agreement in 3D color plots (see Fig. 15.16) is visually impressive, the 2D plots show

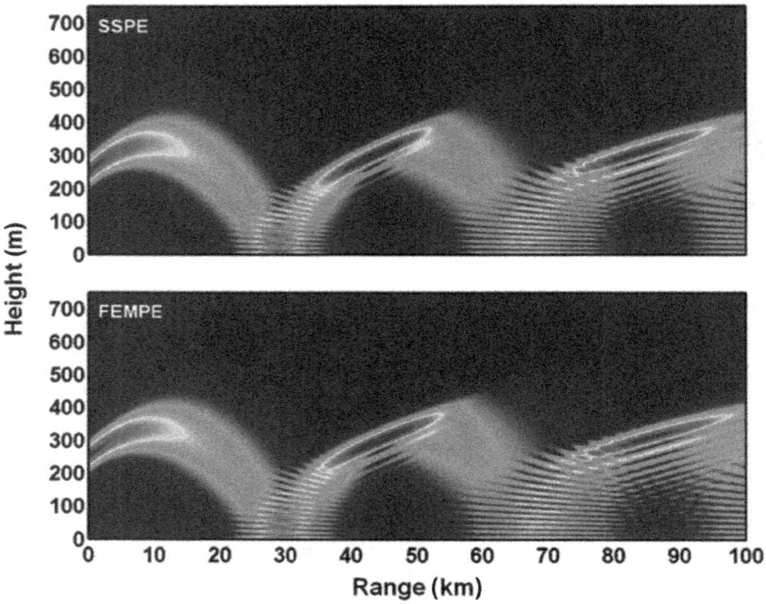

Figure 15.16. The 3D color plots of E-field versus range–height variations for a given 1° upward tilted Gaussian antenna pattern obtained via (top) SSPE and (bottom) FEMPE ($f = 300$ MHz, $a_0 = 3.6 \times 10^{-6}$, $N_x = 256$, $N_z = 1000$).

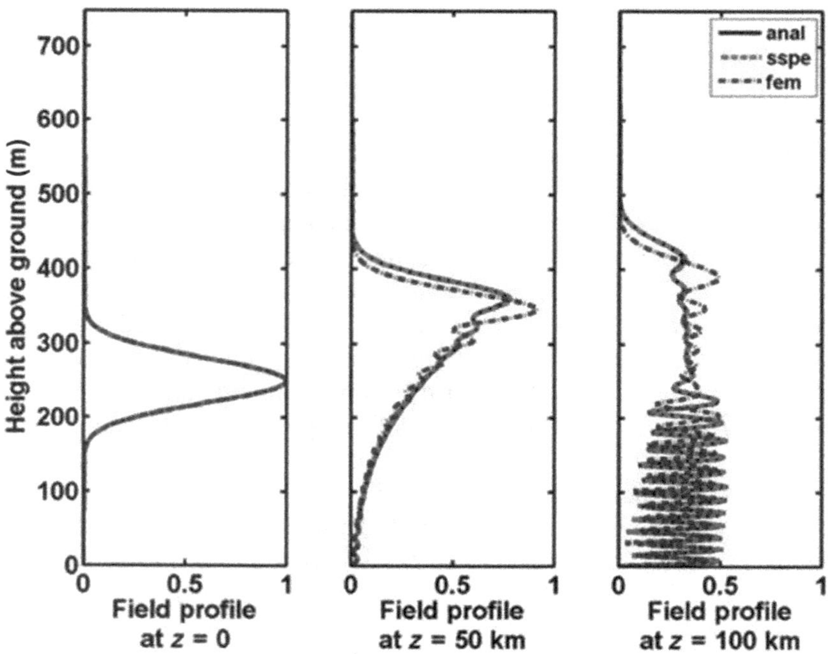

Figure 15.17. E-field versus height at three ranges. Solid: analytic; dashed: SSPE; dash dotted: FEMPE: (Left) The source and its mode sum representation, (middle) E-field at 50 km, and (right) E-field at 100 km ($f = 300$ MHz, $a_0 = 3.6 \times 10^{-6}$, $N_x = 256$, $N_z = 1000$, with 1° upward antenna tilt).

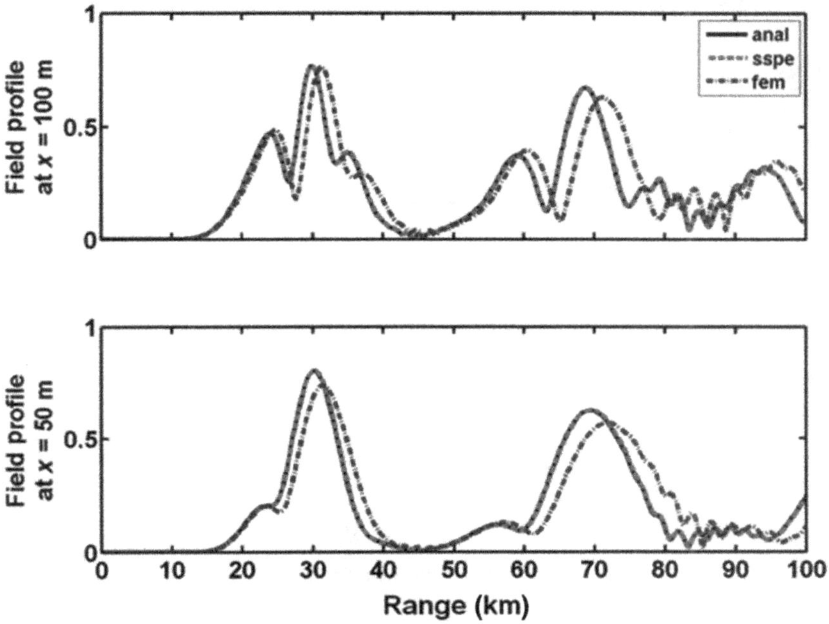

Figure 15.18. E-field versus range at two heights. Solid: analytic; dashed: SSPE; dash dotted: FEMPE. (Top) 100 m and (bottom) 50 m above the ground (f = 300 MHz, $a_0 = 3.6 \times 10^{-6}$, $N_x = 256$, $N_z = 1000$, with 1° upward antenna tilt).

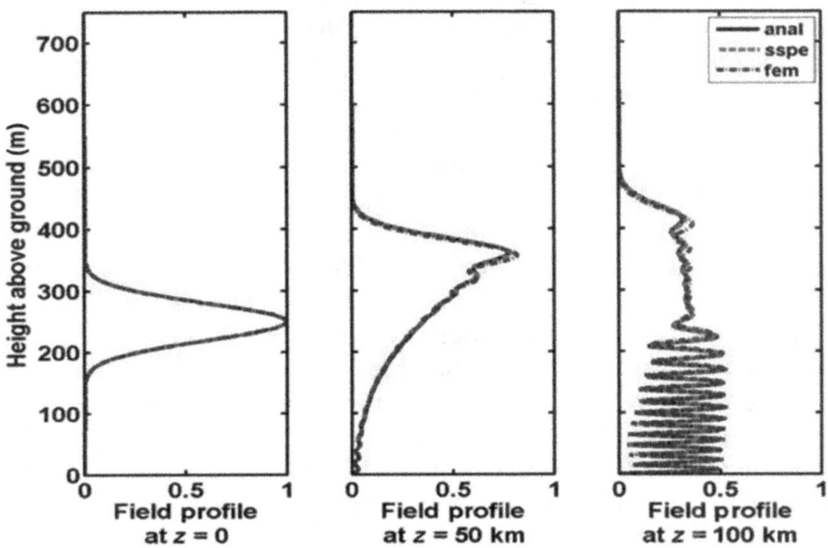

Figure 15.19. The same as in Fig. 15.12 but with SSPE:$N_x = 256$, $N_z = 1000$, FEMPE: $N_x = 512$, $N_z = 2000$.

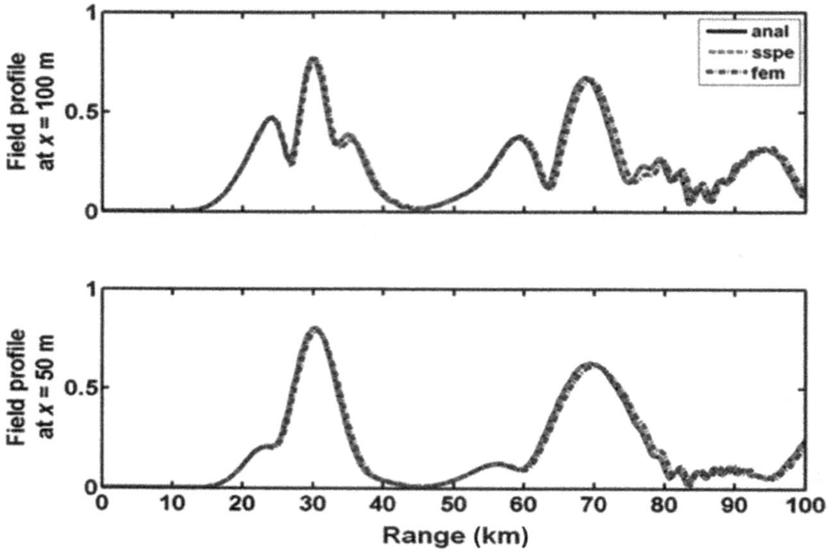

Figure 15.20. The same as in Fig. 15.13 but with SSPE:$N_x = 256$, $N_z = 1000$; FEMPE: $N_x = 512$, $N_z = 2000$.

some discrepancies between FEMPE and the other two, especially at longer distances. The agreement between SSPE and the analytical results is very good, but there are mismatches between FEMPE curves and the other two at certain heights and ranges. Note that everything is kept the same in these calculations.

The accuracy of the FEMPE results may be improved by increasing the number of samples vertically and horizontally. Figure 15.19 and Figure 15.20 belong to the same height and range comparisons (as in Fig. 15.12 and Fig. 15.13) with the number of samples doubled both in height and range in FEMPE (i.e., SSPE:$N_x = 256$, $N_z = 1000$; FEMPE: $N_x = 512$, $N_z = 2000$). As observed, better agreement is obtained when finer discretization is used in the FEMPE calculations; but this increases the computation time.

The tests performed earlier and comparisons against the analytical exact solution which may be accepted as reference calibrate both the SSPE and FEMPE propagation tools. It is only after these tests and calibration when one can perform further simulations to predict the field strength variations along nonflat terrain profiles. Both SSPE and

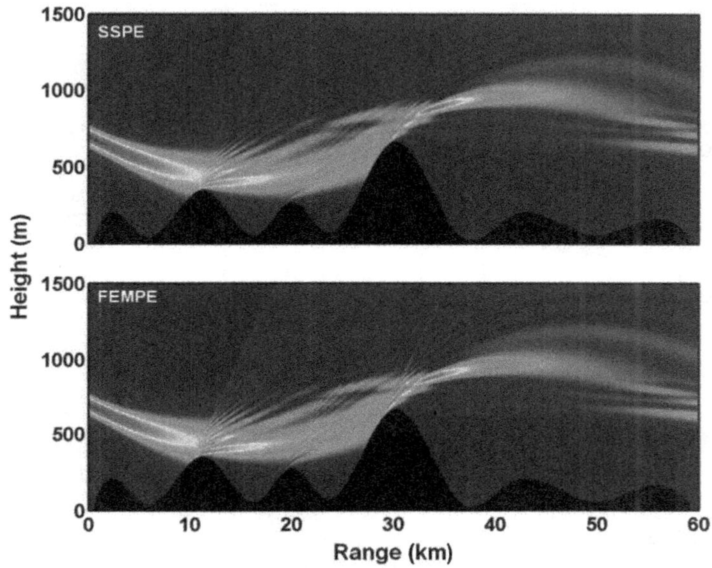

Figure 15.21. Maps of E-field strength versus range–height over a user-specified nonflat terrain profile. (Top) SSPE and (bottom) FEMPE results (the parameters are taken from fig. 13 in Ref. 4, $f = 100$ MHz, $N_x = 256$, $N_z = 1000$).

FEMPE can handle propagation problems above the irregular terrain profiles through variable atmosphere. The examples presented in Fig. 15.21, Fig. 15.22, Fig. 15.23, Fig. 15.24, Fig. 15.25, and Fig. 15.26 belong to the tests and comparisons over two different user-specified irregular terrain profiles and through atmosphere with piecewise linear, bilinear, and trilinear vertical refractivity variations. The parameters of these examples belong to fig. 15.13 and fig. 15.17 in Ref. 4. As observed, very good agreement is obtained in these simulations between the SSPE and FEMPE results.

The last two examples belong to two vertical source cases. In Fig. 15.27, 3D visualization of both analytical and numerical solutions is presented where the transmitter contains two Gaussian patterns (i.e., two antennas) at 200 and 400 m, with $-0.5°$ and $0.5°$ tilts, respectively. In Fig. 15.28, vertical field profiles at two different ranges obtained with all three (analytical, SSPE, and FEMPE) codes are shown.

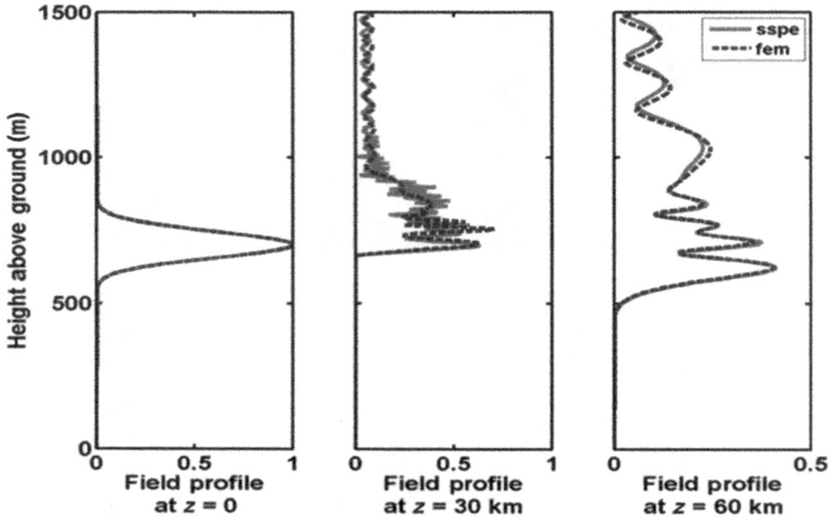

Figure 15.22. E-field versus height at three ranges. Solid: SSPE; dashed: FEMPE. (Left) The source and its mode sum representation, (middle) E-field at 30 km, and (right) E-field at 60 km ($f = 100$ MHz, $N_x = 256$, $N_z = 1000$).

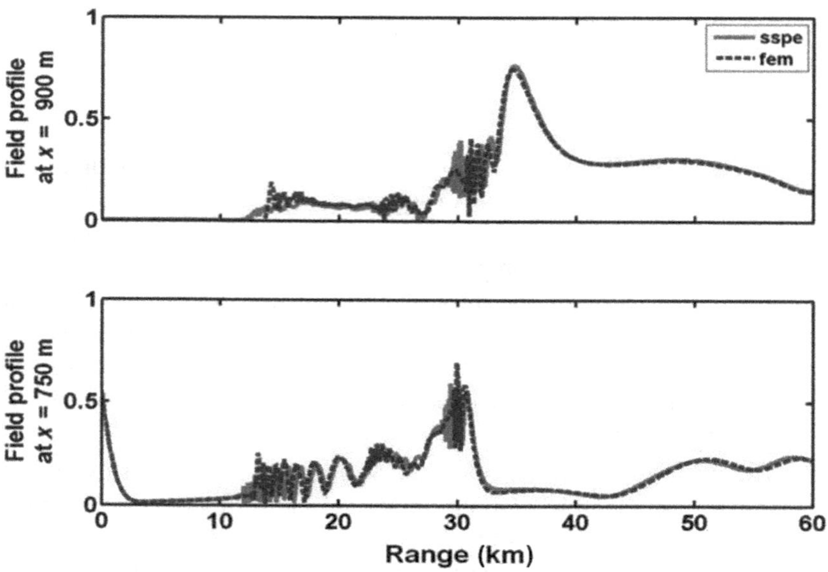

Figure 15.23. E-field versus range at two heights. Solid: SSPE; dashed: FEMPE. (Top) 900 m and (bottom) 750 m above the ground ($f = 100$ MHz, $N_x = 256$, $N_z = 1000$).

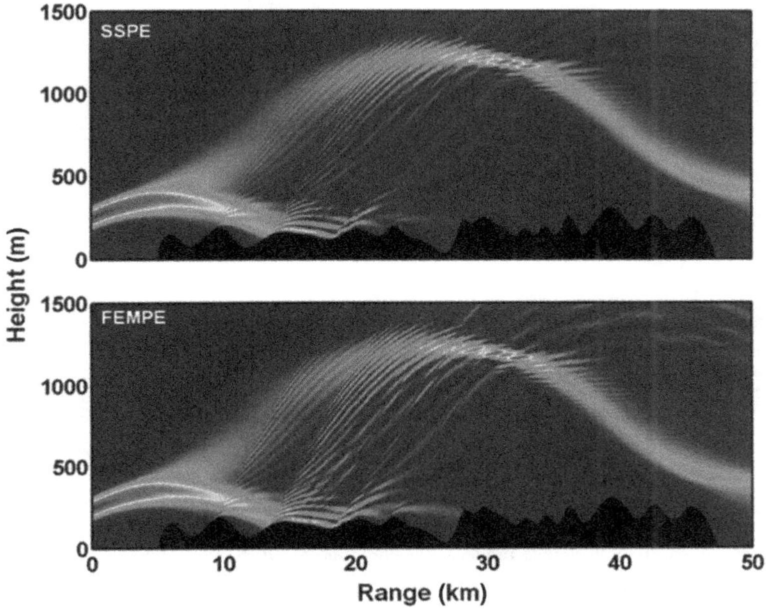

Figure 15.24. Maps of E-field versus range–height variations over a user-specified nonflat terrain profile. (Top) SSPE and (bottom) FEMPE results (the parameters are taken from fig. 17 in Ref. 4, $f = 100$ MHz, $N_x = 256$, $N_z = 1000$).

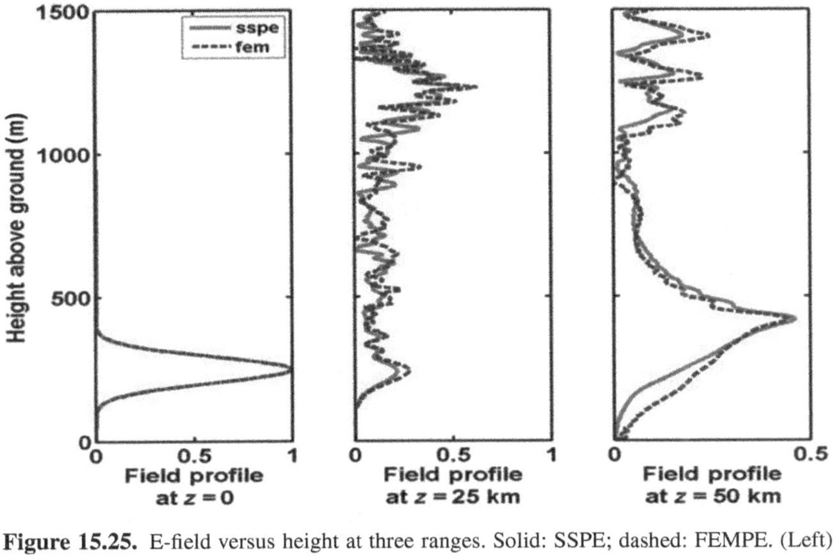

Figure 15.25. E-field versus height at three ranges. Solid: SSPE; dashed: FEMPE. (Left) The source and its mode sum representation, (middle) E-field at 25 km, and (right) E-field at 50 km ($f = 100$ MHz, $N_x = 256$, $N_z = 1000$).

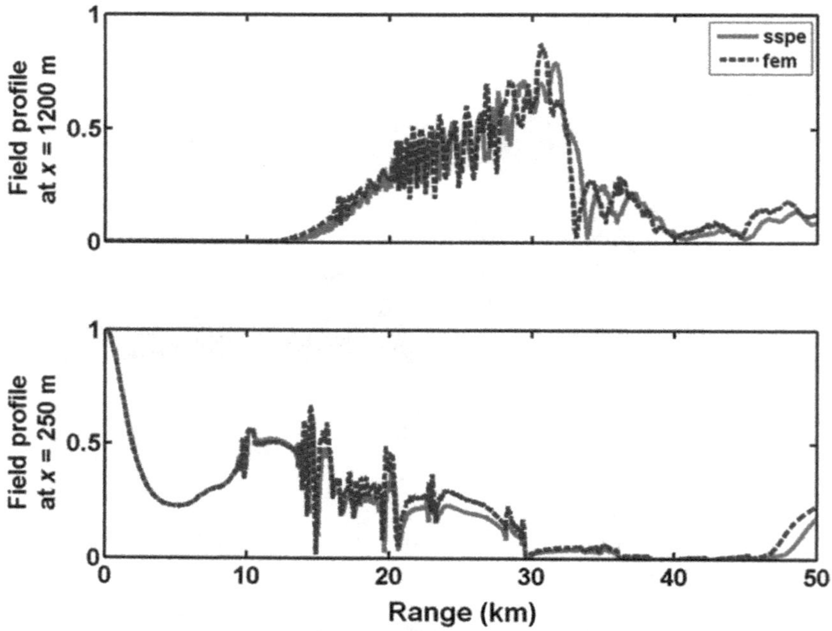

Figure 15.26. E-field versus range at two heights. Solid: SSPE; dashed: FEMPE. (Top) 1200 m and (bottom) 250 m above the ground ($f = 100$ MHz, $N_x = 256$, $N_z = 1000$).

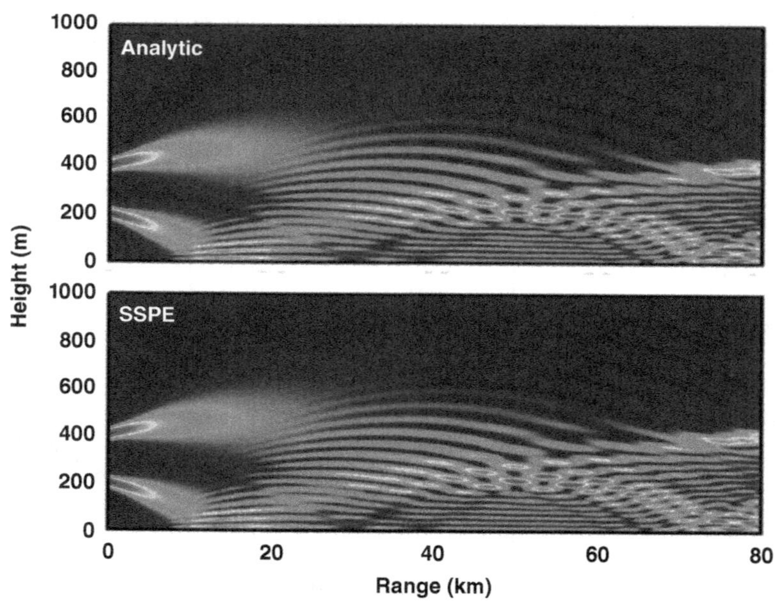

Figure 15.27. SSPE and analytic propagators with tilted waves at 200 and 400 m with $-0.5°$ and $0.5°$ tilts, respectively.

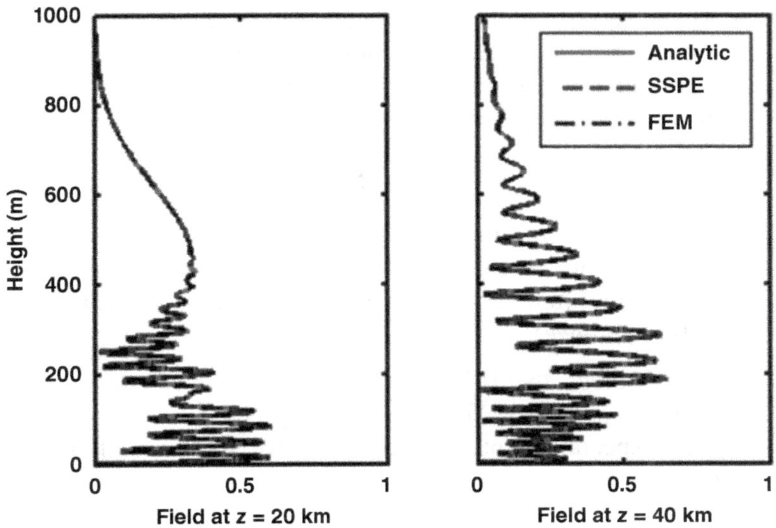

Figure 15.28. Vertical field profiles at two different ranges ($dM/dx = -600\mathrm{M/km}$).

Important aspects of these tests, comparisons, and calibration may be listed as follows.

- The modes of the analytical exact solution are confined between the ground and their (modal) caustics for the linearly decreasing vertical refractivity profile. The height of the caustics increases as the number of modes increases. Therefore, the construction of the initial field profile necessitates the contributions of higher order modes for higher antenna heights. But, as the antenna height increases, the contributions of the lower order modes become negligible, so the best way to reconstruct the antenna pattern in terms of the modal summation is to use modes from M_1 to M_2 which are the lowest and the highest mode numbers to be determined from operational parameters.
- The modal excitation coefficients are found from the given antenna pattern using the orthonormality property. These coefficients are real for the antenna pattern without tilt but becomes complex if a tilt is used. The complex excitation coefficients are obtained from a numerical complex integration.

- An antenna pattern may be reconstructed with a number of modes for a given set of operational parameters. But one should take higher order modes into account (although they might not be necessary to construct antenna pattern) if antenna is tilted upward.
- The analytical solution is exact, and, therefore is *referenc*e when infinite number of modes is taken into account and if accurately computed numerically. On the other hand, the analytical solutions with a finite number of modes should be accurate enough to be reference!
- SSPE solves the PE in the transverse domain so is not as sensitive as FEMPE to the vertical discretization. On the other hand, special attention is required for the vertical boundary treatments.

FEMPE uses discretization on the vertical domain and solves the discrete Crank–Nicholson system horizontally. The computation burden mostly comes from the selection of this tridiagonal system. The computation time increases with N_z in range but N_x^2 in height.

15.8 THE GrSSPE PACKAGE

A MATLAB radio wave propagation package GrSSPE is developed for the visualization of EM propagation over non-flat terrain through non-homogeneous atmosphere, for waves radiated by a horizontally oriented antenna over the ground [4]. Fundamental design principles of the GrSSPE package may be listed as follows:

- The user should be able to investigate radio wave propagation problems through non-homogeneous atmosphere over non-flat terrain.
- The user should be able to design his/her own non-flat terrain profiles graphically, without any difficulty, and the package should perform calculations accordingly.
- Specification of refractivity should also be simple, and the user should have a choice among a variety of different refractivity profiles.

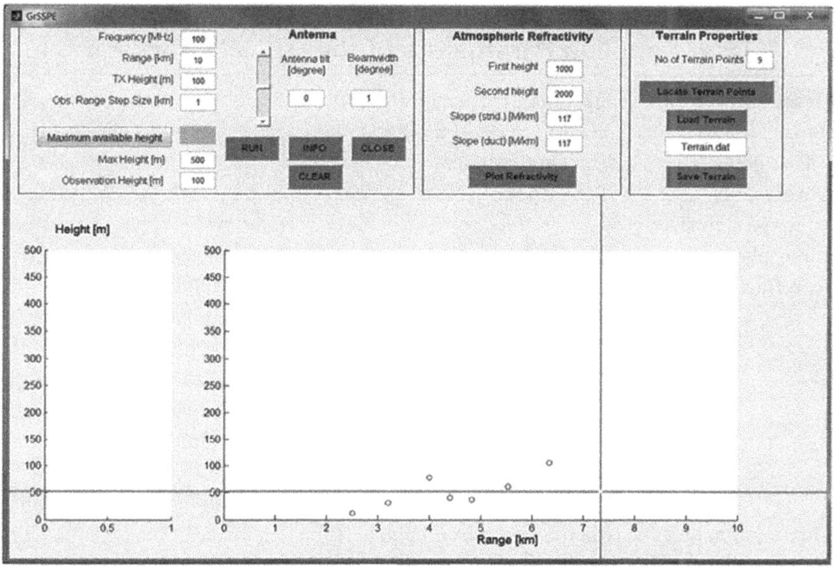

Figure 15.29. Front panel of the GrSSPE MATLAB package. The user locates terrain points on the screen by just left clicking the mouse.

The GrSSPE package designed according to these requirements is given in Fig. 15.29. The user-specified parameters are explained in Table 15.8. The front panel of the package is divided into three sub regions. On top, geometrical and operational parameters are supplied by the user. The command push buttons are also located in this region. The large window at right is the range–height window (where range and height are displayed in kilometers and meters, respectively). In this window, the vertical transmit antenna pattern is supplied at left and propagation is simulated toward right. The user-defined terrain profile is displayed in this window together with the SSPE simulated numerical result, which is plotted in 3D as field strength versus range–height color graphics. Different colors from red to blue correspond to high-to-low level field strength values, respectively.

The window at left is reserved for height variations of both field strength and refractivity. Field strengths versus height at different ranges (with range intervals supplied by the user via "Obs. Range step size" box) are displayed one by one as the simulation progresses from the initial range to the last distance. It should be noted that normalized

Table 15.8.
User-Supplied Parameters of the GrSSPE Package

Frequency	Operating frequency (MHz)	
Range	Maximum desired range in kilometers defined by user	
TX height	Transmitter height	
Observation range step size	The difference between observation ranges for height profile	
Atmospheric refractivity	*First height*	The atmosphere is standard until the first height
	Second height	Elevated duct between first and second height
	Slope 1 (M/km)	Slope of the standard atmosphere from ground to first height
	Slope 2 (M/km)	Slope of duct from first height and second heights
Maximum height	Maximum desired height	
Antenna tilt	Antenna tilt in degrees	
Beamwidth	Beamwidth in degrees for antenna pattern	
Number of terrain points	Number of terrain points that are going to be located for the terrain profile by the user	

refractivity (not to scale) is plotted in this window just to show qualitative height variation of the refractivity.

Operating frequency, maximum range, transmit antenna height, vertical beamwidth of the Gaussian pattern, and the antenna tilt are supplied first. Then, range-independent, vertical atmospheric refractivity profile is specified. The ground terrain profile is built by selecting the number of terrain points and locating them via the mouse between the transmitter and receiver in arbitrary order. When the last terrain point is located, the package branches to a subroutine which predicts a range-dependent function that passes from all marked terrain points. The maximum available height depends on the operating frequency and maximum slope of the chosen terrain; it is calculated accordingly if the user presses the "*Maximum available height*" button. The transmitter and receiver heights may then be supplied by the user.

The ground terrain profile can be supplied via two different ways. The user can choose the number of terrain points and locate them (by

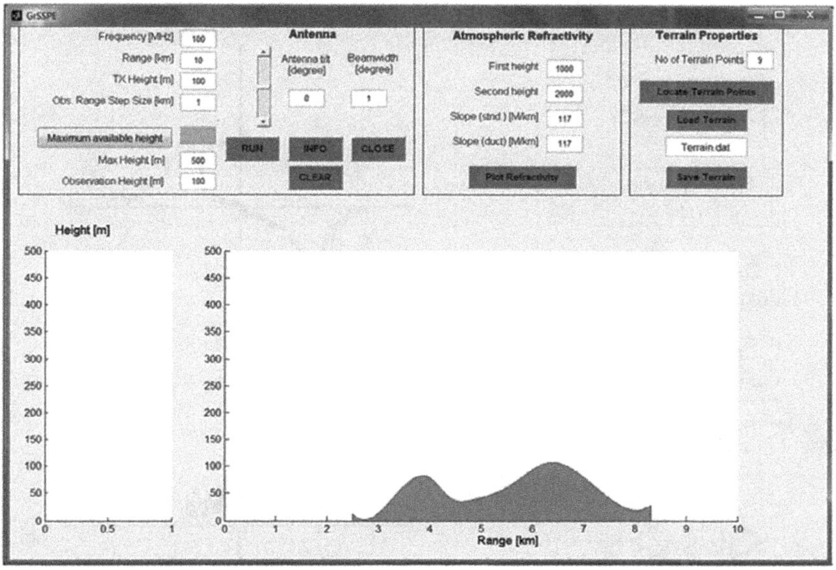

Figure 15.30. Terrain profile is generated from the marked points by using cubic-spline curve fitting technique.

clicking "*Locate Terrain Points*" button) in a way to trace the ground irregularity (see Fig. 15.29).

Figure 15.30 shows the ground terrain profile calculated from the number of points marked in Fig. 15.29. User-generated terrain profile may be stored in a file, in two columns: range versus height, by using "*Save Terrain*" button. The default name of the terrain file is "*terrain.dat*," but it can be any name supplied by the user. Any of pregenerated terrain files may also be chosen by using the "*Load Terrain*" button.

The main MATLAB function of the GrSSPE package is short and simple (see Fig. 15.31 for the flowchart). If N is the number of height points, then the arrays $X(N)$ and $Field(N)$ contain the vertical height points and field values (initial profile), respectively. The SSPE procedure at each range step contains FFT, multiplication by a complex exponent, inverse FFT, and multiplication with another complex exponent. The height array is build when maximum height (*hmax*), number of height points (N), and height increment (*delx*) are supplied by the user. The array of the transverse wave number values $k_x(N)$ are specified according to Nyquist criteria. The Gaussian initial field profile is

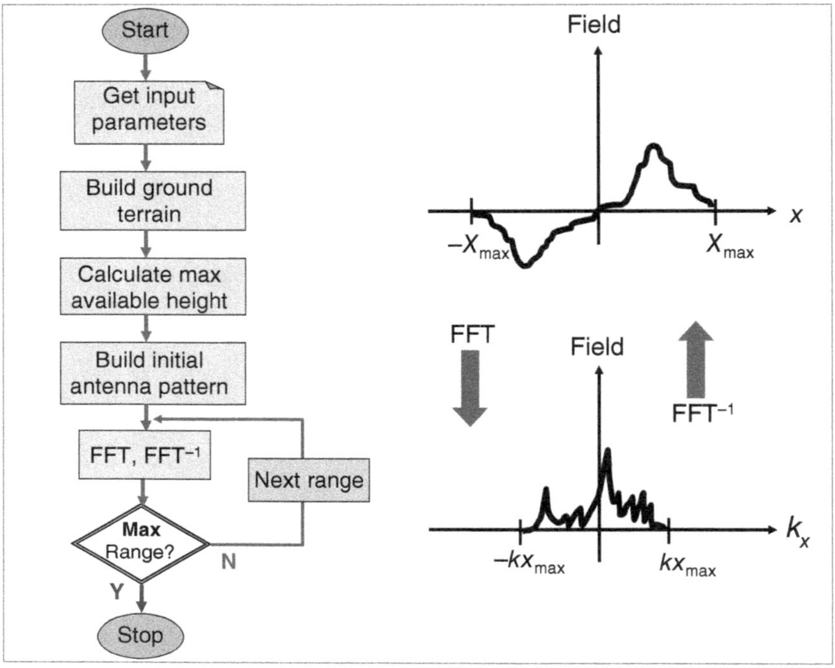

Figure 15.31. (Left) Flowchart of the main SSPE routine and (right) the relation between the transverse spatial and wave number domains.

specified according to the beamwidth ($WW = \sqrt{2\ln 2}/\{k_0 \sin(\theta_{bw}/2)\}$), transmitter height (*hv*), and antenna tilt (*elev*) values supplied by the user.

The user may develop his/her own SSPE MATLAB codes by using these short MATLAB segments and modify according to his/her requirements. Some of the parameters are prespecified and some are restricted to certain ranges in order to keep it simple and efficient. The FFT size is kept $N = 1024$, which means 512 height points are used to construct initial vertical field profile between the surface ($x = 0$) and maximum height ($x = x_{max}$) supplied by the user. This field profile is then extended *odd symmetric* (to satisfy Dirichlet-type BC at earth's surface) from $x = 0$ to $x = -x_{max}$. The 1D arrays reserved for the vertical fields, the refractivity profile, and the transverse wave number values have dimensions of $N = 1024$.

Since height step size (or maximum available height) is related to the maximum angle of propagation considered in the analysis, antenna tilt is hard limited to a range of $\pm 10°$ (i.e., above and below the horizon). The tilt value may be supplied either from the data box or by using the slider. The higher the tilt angle is chosen, the smaller height step size is required, causing a decrease in maximum available height.

Atmospheric refractivity is specified by giving the slopes of piecewise linear segments and two height values in meters. For example, giving first height as 1000 m and first slope as 300 M/km yields a refraction index of $n = 1 - 300 \times 10^{-9} \times 1000 = 0.9997$ at 1000 m height above the surface. Although refractivity becomes less than unity above the surface, this is not important in the SSPE application, since SSPE accounts for the refractivity deviations, not the exact refractivity values. Moreover, this can be prevented by specifying n_0 (refraction index at the surface) in (15.6) greater than unity. The button "Plot refractivity" is reserved to plot complete transverse refractivity variation from $-x_{\max}$ to $+x_{\max}$.

Finally, the open boundary (free space) when $(x \to \infty)$ is simulated by adding artificial lossy (absorbing) layers above (below) $x_{\max}$ $(-x_{\max})$. The vertical region is extended from $x_{\max}$ to $2x_{\max}$ and from $-x_{\max}$ to $-2x_{\max}$, above and below the surface, respectively, and regions between $3x_{\max}/2$ to $2x_{\max}$ and from $-3x_{\max}/2$ to $-2x_{\max}$ are assumed lossy. This is achieved by making the refractive indices in these regions complex.

It should be noted that the GrSSPE package in its present form calculates and stores field strength values at every range and height points in a 2D field array; therefore, the higher the height and longer the range, the larger the dimension of this 2D field array. Because of the memory requirement of this storage, the computation slows down for long range and/or high altitude propagation problems. If only the field strength versus height (range) at a specified range (height) is of interest, then the user may speed up the simulations by just removing the storage command from the GrSSPE.m file.

It should also be noted that the solution of (15.1) in 2D yields waves that are attenuated by the square root of distance as they propagate away from the transmitter (i.e., cylindrical wave spread for an infinite-length line source which allows reducing the 3D wave equation into 2D). Therefore, the user should divide the field strength results by the square root of distance in order to obtain path loss versus range variations in a realistic (3D) environment.

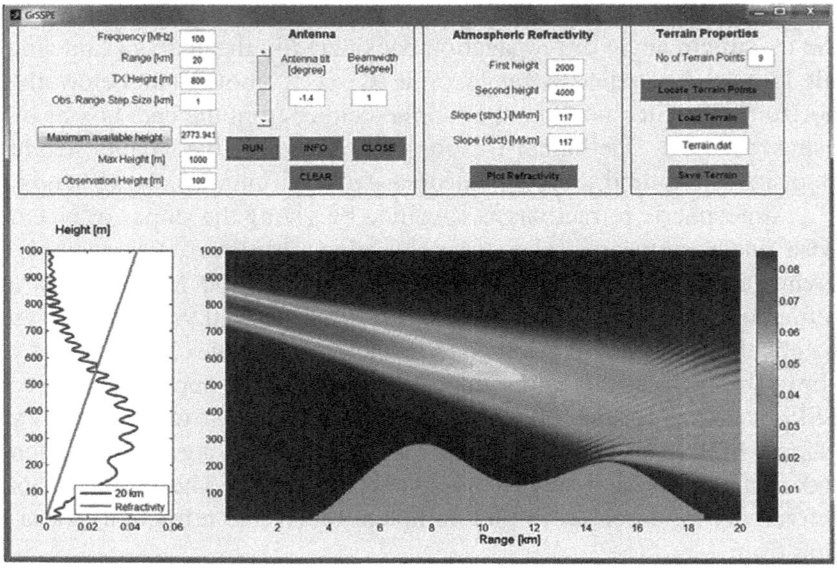

Figure 15.32. Propagation through standard atmosphere over flat earth: frequency of 100 MHz, maximum range of 20 km, maximum height of 1000 m, transmitter height of 800 m, antenna is tilted 1.4° downward, and beamwidth is 1°. Linearly increasing refractivity with a slope of 117 M/km corresponds to the standard atmosphere (including earth's curvature).

To show the power and beauty of the GrSSPE package, some examples and typical results are presented in this section. The first case belongs to propagation over a user-specified terrain profile in standard atmosphere conditions and the results are presented in Fig. 15.32. Here, frequency is 100 MHz, maximum range is 20 km, standard atmosphere includes earth's curvature (i.e., 117 M/km), transmitter height is 800 m, and antenna beamwidth is 1°. Simulations over flat PEC surface (without terrain) can be done by just locating one or two terrain points on the ground by the mouse. The antenna is tilted 1.4° and 4° below the horizon. Note that the beam touches upon the second hill and forward reflections occur. The interference of these two is clearly seen in the figure.

The second case belongs to another terrain profile through bilinear elevated ducting atmosphere. All other parameters are kept the same. The results are given in Fig. 15.33. The user must keep in mind that

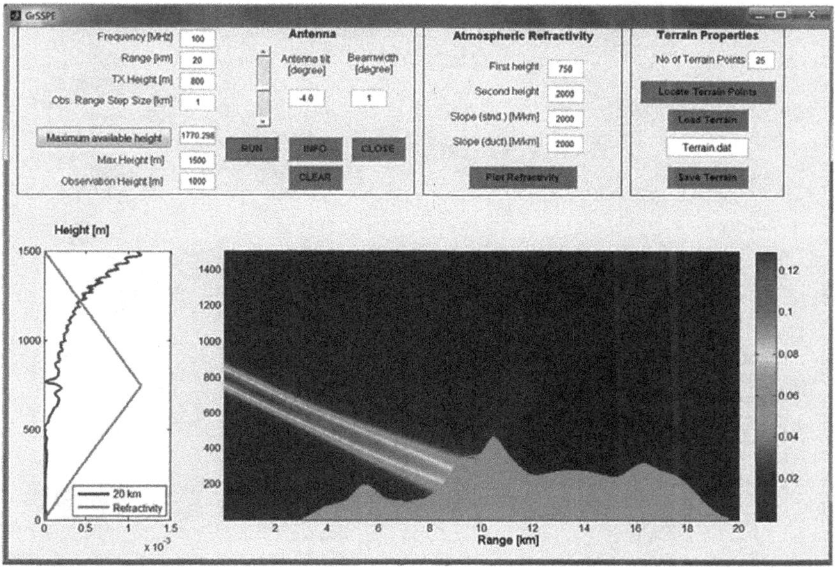

Figure 15.33. Propagation over a user-plotted terrain, through bilinear atmosphere (including earth's curvature).

PE is a one-way propagator; therefore, it can only handle forward propagated waves and cannot take backward reflections into account. Note that the beam hits the terrain; one should see the strong reflections. Unfortunately, backward reflections and/or reflections above 25°–30° cannot be observed.

The last example aims to show the effect of concave–convex transition. The parameters are as follows: $f = 100$ MHz, maximum range is 60 km, antenna tilt is 2° downward, and beamwidth is 1.5°. The transmit antenna is located at 700 m. The results are presented in Fig. 15.34.

In summary, the package is based on the well-known FFT-based, longitudinally marching, step-by-step numerical solution of the 2D PE which accounts for the one-way (forward) propagation problems. Any kind of a range-dependent nonflat terrain profile can be designed by the user by marking a number of points on the terrain surface by just clicking the mouse. Various refractivity profiles may also be supplied with a few height and slope parameters. The package presents a good virtual lab environment that can be used in undergraduate and advanced

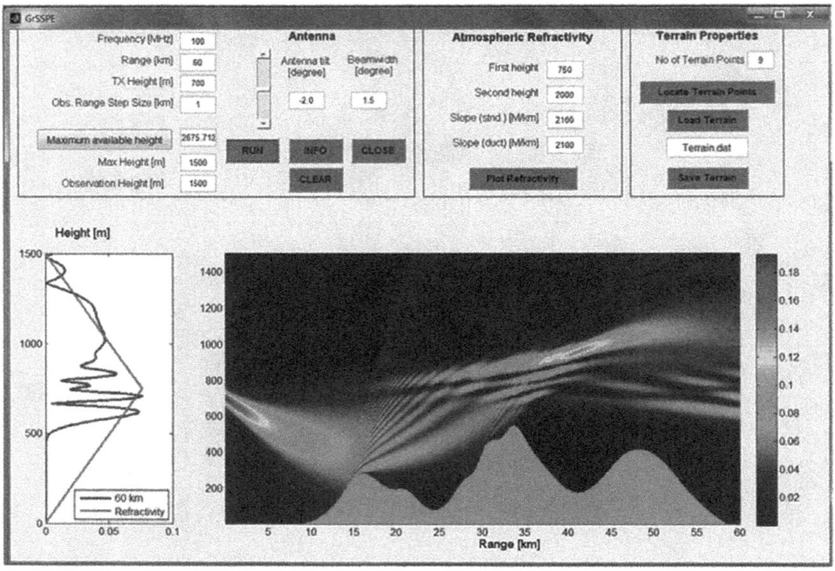

Figure 15.34. Propagation through bilinear atmosphere over irregular terrain: frequency is 100 MHz, maximum range of 60 km, maximum height of 1500 m, transmitter height of 750 m, antenna tilt 2° downward, and beamwidth is 1.5°. Atmospheric refractivity increases up to 700 m with a slope of 2000 M/km, and then decreases with the same slope.

level electromagnetic lectures, as well as an effective engineering tool to train propagation/wireless/communication practitioners.

The user should keep in mind that the MATLAB code in the GrSSPE package is not optimized according to complete professional needs. Although this is achievable, the philosophy accepted here is its simplicity and effectiveness. Therefore, parameter restrictions and optimization are done accordingly.

15.9 THE SINGLE-KNIFE-EDGE PROBLEM

It appears that the surface duct model in Section 15.5 can be used within the paraxial region because the numerical instabilities and insufficiencies meet there during the generation of the reference data [39]. The single-knife-edge problem, the four ray model (4Ray), or Fresnel integral representations can be used as an alternative model from which reliable reference data can be generated. The single-knife-edge problem is discussed in Chapter 12 (see Section 12.6) and a short MATLAB script is given for the 4Ray model (see Table 12.3).

Here 4Ray model is compared with the SSPE model. The MATLAB code which compares SSPE and 4Ray models is listed in Table 15.9. Results are pictured in Fig. 15.35 and Fig. 15.36.

In Fig. 15.35, 3D field maps generated via the SSPE tool and the 4Ray model are presented. The scenario belongs to one-way propagation for horizontal polarization over PEC ground with 75-m-high wall at 15-km range. The line source is at 15-m height at $z = 0$. The FEMPE result is exactly the same with the SSPE map; therefore, it is not included in this figure.

The PFs versus heights in front of and beyond the wedge-type obstacle are plotted in Fig. 15.36. Here, four vertical field profiles obtained with both SSPE and 4Ray model are plotted. The first plot on

Table 15.9.
MATLAB-Based SSPEvs4RAY.m Code

```
function Knife_SSPE4Raynew
%% -----------------------------------------------------
%   Program: Knife_SSPE4Ray.m
%   Purpose: Knife-Edge problem, SSPE vs 4Ray models
%              Plot Field VS. Range,
%% -----------------------------------------------------
clear all; clc; close all;
hw=50;    %input('Knife-Edge height [m] = ? ');
d1=800;   %input('Source-Obstacle distance [m] = ? ');
d=1500;   %input('Max distance [m] = ? ');
pol=1; lam=1; k0=2*pi/lam; elv=-5; bw=15; ht=60;
hr=60;
Nz=1500; Dz=d/Nz; dd=(0:Dz:d).';   Xmax=400; Nx=801;
%% -------------- SSPE ---------------------------------
dx=Xmax/(Nx-1); X=0:dx:2*Xmax; N=length(X);
fr   = [1.5 2];   % filter is applied to [r1 r2]*Xmax
FILFUN = @(x)  (x<=fr(1)*Xmax)+(x>fr(1)*Xmax).*. . .
    (1+cos(pi*(x-fr(1)*Xmax)/((fr(2)-fr(1))*Xmax)))/2;
```

(*Continued*)

Table 15.9.
(Continued)

```
ww = sqrt(2*log(2))/(k0*sind(bw/2)); % beamwidth
cx=1/sqrt(sqrt(pi/2)*ww);
ufs=@(xx) cx*exp(-1i*k0*sind(elv)*xx).*exp(-((xx-ht)/
ww).^2);
u=zeros(N,Nz+1); Field=ufs(X); Field(1)=0;
u(:,1)=Field;    kX=linspace(0,pi/dx,N).';
CC1=exp(1i*Dz*kX.^2./(k0*(1+sqrt(1-kX.^2/k0^2)))).';
for n=1:Nz      % SSPE Range Loop
 if pol==1,
     Field(2:N-1)=idst(CC1(2:N-1).*dst(Field(2:N-1)));
 else
     Field=idct(CC1.*dct(Field));
 end
     Field=Field.*FILFUN(X);
     if n*Dz==d1, Field(1:round(hw/dx)+1)=0; end
     u(:,n+1)=Field;
end
uz=exp(-1i*k0*dd').*u(round(hr/dx)+1,:); % Analyze
Results
%% ------------------ 4Ray Model ---------------------
Wth=@(th) exp(-(sin(th)-sind(elv)).^2*log(2)/
(2*sind(bw/2)^2));
cth=1/sqrt(sum(chebfun(@(t) Wth(t).^2,[-
pi/2,pi/2]+elv/180*pi))); uRay=zeros(size(dd));
for k=1:length(dd),
   d=dd(k); d2=d-d1;
   if d==d1, removeind=k; else
      h0=sqrt((lam*d1*d2)/d); sq2=sqrt(2);
      % Calculate r1, r2, r3, and r4
      r1=sqrt(d^2+(hr-ht)^2); r4=r1;
      r2=sqrt(d^2+(hr+ht)^2); r3=r2;
      % Calculate v1, v2, v3 and v4
      delh=( hr-ht)*d1/d;   v1=sq2/h0*( ht+delh-hw);
      delh=( hr+ht)*d1/d;   v2=sq2/h0*(-ht+delh-hw);
      delh=(-hr-ht)*d1/d;   v3=sq2/h0*( ht+delh-hw);
      delh=(-hr+ht)*d1/d;   v4=sq2/h0*(-ht+delh-hw);
      % Reflection coefficient
      if pol==1, refl=-1; refr=-1; else refl=1; refr=1;
end
      % Calculate field contribution for RAY1
```

Table 15.9.
(Continued)

```
     [ic,is]=determineintegral(v1);  ang1=atan((hr-ht)/d);
     uRay1=cth*Wth(ang1)*(0.5-0.5i)*exp(1i*k0*r1)/sqrt(r1
)*((ic+0.5)+1i*(is+0.5));
     % Calculate field contribution for RAY2
     [ic,is]=determineintegral(v2);
ang2=-atan((hr+ht)./d);
      uRay2=refl*cth*Wth(ang2)*(0.5-0.5i)*exp(1i*k0*r2)/
sqrt(r2)*((ic+0.5)+1i*(is+0.5));
     % Calculate field contribution for RAY3
     [ic,is]=determineintegral(v3);   ang3=ang2;
     uRay3=refr*cth*Wth(ang3)*(0.5-0.5i)*exp(1i*k0*r3)/
sqrt(r3)*((ic+0.5)+1i*(is+0.5));
     % Calculate field contribution for RAY4
     [ic,is]=determineintegral(v4);        ang4=ang1;
     uRay4=refl*refr*cth*Wth(ang4)*(0.5-
0.5i)*exp(1i*k0*r4)/sqrt(r4)*((ic+0.5)+1i*(is+0.5));
     % Total contribution
     uRay(k)=uRay1+uRay2+uRay3+uRay4;
     end
end
ddz=dd; uRay(removeind)=[]; dd(removeind)=[];
%% ------------ to show 2D fields ----------------------
plot(ddz,abs(uz),dd,abs(uRay),'k:');%ylim([0 0.4])
xlabel('Distance [m]','Fontweight','bold');
ylabel('Field','Fontweight','bold');legend('SSPE','4Ray');
title('Knife-Edge Problem: 4Ray vs. PEM','Fontweight','b
old');
% --------------------------------------------------------
function [ic,is]=determineintegral(v)
if v>0,
     ic = sum(chebfun(@(t) cos(pi/2*t.^2),[0,v]));
     is = sum(chebfun(@(t) sin(pi/2*t.^2),[0,v]));
elseif v<0,
     ic=-sum(chebfun(@(t) cos(pi/2*t.^2),[v,0]));
     is=-sum(chebfun(@(t) sin(pi/2*t.^2),[v,0]));
else
     ic=0; is=0;
end
% ------------------------ END -------------------------
```

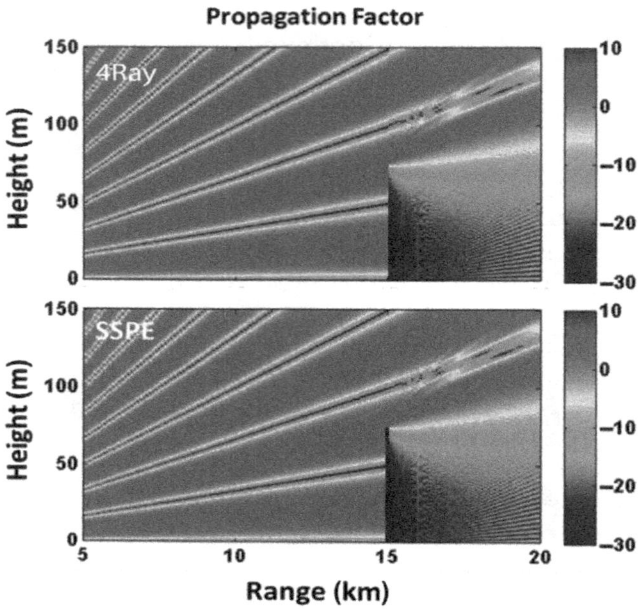

Figure 15.35. PF versus range/height over PEC flat earth 75-m-high knife edge at 15 km (Hor pol, source height = 20 m, $f = 3$ GHz).

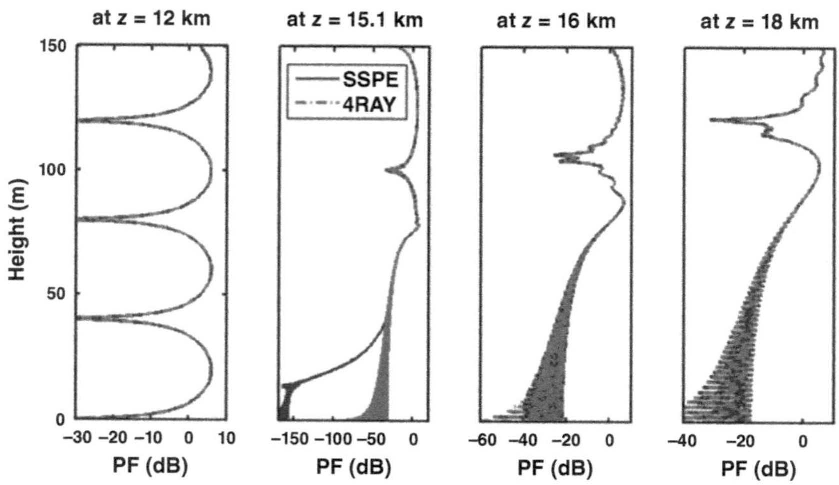

Figure 15.36. PF versus height at four ranges: 12, 15.1, 16, and 18 km ($f = 3$ GHz; solid: 4Ray; dashed: SSPE; $\Delta x = 0.1$ m; $\Delta z = 50$ m).

the left belongs to the interference region (before the obstacle) and the other three are in the diffraction region (beyond the obstacle). As observed, excellent agreement is obtained in all of the plots. Note that the height of the edge of the obstacle is 75 m and the distances of these three profiles from the obstacle are 100 m, 1 km, and 3 km, respectively. The maximum diffraction angles of these three are around 36.9°, 4.3°, and 1.5°, respectively. It is expected that the PE models are not effective and accurate for the angles beyond 30°–35°. As observed, the PE models are insufficient in modeling the diffracted fields in the deep shadow regions.

15.10 ACCURATE SOURCE MODELING

In most of the 2D propagation models, the source is modeled as a transmit antenna pattern. This is achieved in PE propagators by giving a vertical Gauss-shaped input field profile. The vertical beamwidth and beam tilt can also be adjusted. This profile is then propagated above user-specified irregular, lossy terrain through inhomogeneous atmosphere. On the other hand, most of analytical solutions (i.e., mode/ray models) which may serve as reference use line source excitation. In order to compare these two and perform VV&C, excitations must be identical. Hence, source adaptation is crucial.

Widely used sources and their vertical wave number domain contents are pictured in Fig. 15.37. The Green's function based on mode summation uses line source excitation. Geometric optic (GO) ray solutions are also derived under line source assumption. Line source means rays emanating in all directions, as pictured in Fig. 15.37. The other three sources in the figure are used in PE models. As observed from Fig. 15.37b, transverse wave number domain contents are quite different. Nevertheless, identical source effects can be obtained by introducing ray excitation coefficients.

The most common source model which also smoothen the results is the (untilted/tilted) vertical Gaussian profile. This is also suitable for the mode summation-based analytical models. On the other hand, the use of (untilted/tilted) Gaussian profile in ray-based analytical models may be essential in many VV&C studies. Untilted/tilted Gaussian beam can be injected in either domains; x or k_x which are the Fourier transform pairs.

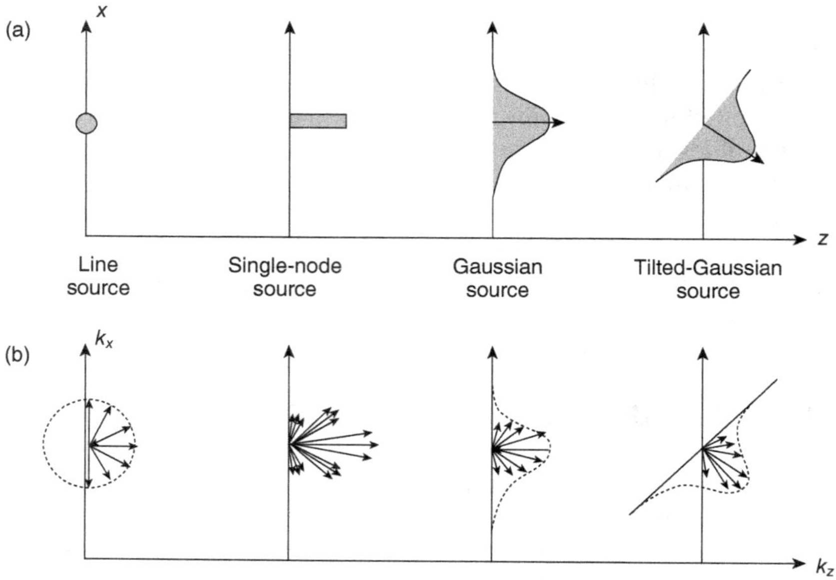

Figure 15.37. (a) Vertical field distributions of some sources and (b) their wave number domain behaviors.

Usually, SSPE is calibrated against mode and/or ray solutions as well as other numerical models such as MoM. Representation of the same source with the mode sum methods is easy. Modal excitation coefficients are extracted from the orthonormality property. Note that modal excitation coefficients are real for untilted Gaussian beams, but become complex if the beam is tilted upward or downward.

On the other hand, the other methods (ray sum, image method, and MoM, all of which are ray-based models) need special care. In these cases, ray excitation coefficients are calculated according to the departure angles of the radiating rays which are extracted from the wave number domain data of the given untilted/tilted Gaussian profile (see (15.5c)).

The ray excitation coefficients are obtained as

$$W(\theta) = \exp\left(-\frac{[\sin\theta - \sin\theta_{\text{tilt}}]^2 \ln 2}{2\sin^2(\theta_{\text{bw}}/2)} \right), \tag{15.41}$$

for every ray using their departure angles (θ) and these coefficients are used in the ray summation model for the antenna pattern. Some ray excitation coefficients for various beamwidth and tilt angles are shown in Fig. 15.38.

Table 15.10 lists a MATLAB code for the modeling of propagation over a PEC flat earth using SSPE, MoM, and 2Ray models. Field versus

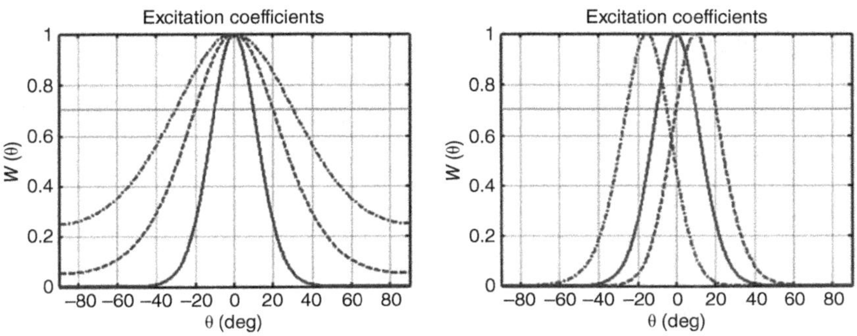

Figure 15.38. Excitation coefficients of various Gaussian beams. (Left) Untilted, (solid) $\theta_{bw} = 20°$, (dashed) $\theta_{bw} = 40°$, (dashed dotted) $\theta_{bw} = 60°$; (right) tilted, (solid) $\theta_{bw} = 20°$, (dashed) untilted, (dashed dotted) $\theta_{tilt} = 10°$.

Table 15.10.
A MATLAB Code for SSPE, MoM, and 4RAY Model Comparisons under Tilted Beam-Type Excitations

```
function SSPEvsMoMvs4Ray
%% ----------------------------------------------------
%  Program: FlatEarth.m
%  Purpose: MoM / SSPE / 2Ray comparisons
%  Tilted Beam excitation, Only TEz mode,
%% ----------------------------------------------------
clear all; clc; Zmax=1000; Nz=1000; Xs=Zmax/5; Dz=Zmax/
Nz;
Xobs=Xs; lambda=10*Dz; freq=3e8/lambda;
k0=2*pi*freq/3e8;
elv=-15; bw=30; % Antenna tilt and beamwidth
Xmax=2000; Nx=801; % between 0-2Xmax
minval=0;  maxval=0;  minpf=0;  maxpf=0;
xmax3d=2*Xs;
%% ----------- MoM Method ---------------------------
```

(*Continued*)

Table 15.10.
(Continued)

```
Zt = (0.5:Nz-0.5)*Dz; Xt = zeros(1,Nz); % for
Flat-Earth
%% STEP 2: Calculate segment voltages from source Xs
ds=sqrt((Xt-Xs).^2+Zt.^2);    % Source-Segment distances
Rayangle=atand((Xt-Xs)./Zt); % Source-to-Segment Ray
angles
WRay=exp(-(sind(Rayangle)-sind(elv)).^2*log(2)/
(2*sind(bw/2)^2));
V_inc=-WRay.*exp(-1i*k0*ds)./sqrt(k0*ds);   % segment
voltages
imp_matrix=-0.25*Dz*120*pi*k0*(1-1i*2/
pi*log(exp(0.577215665). . .
            *k0*Dz/(4*exp(1))))*ones(Nz,1);
for m=1:Nz-1
     dd=Dz*m; val=-0.25*Dz*120*pi*k0*besselh(0,2,k0*dd);
     imp_matrix=[val*ones(Nz,1) imp_matrix
val*ones(Nz,1)];
end
imp_matrix=spdiags(imp_matrix,-(Nz-1):Nz-1,Nz,Nz);
imp_matrix=full(imp_matrix);
ind_current=imp_matrix\transpose(V_inc);
scat_el=zeros(1,Nz);
inc_el=zeros(1,Nz); tot_el=zeros(1,Nz);
PF_el=zeros(1,Nz);
for n=1:Nz
     scat_el(n)=0.0; Zobs=Zt(n);
     for m=1:Nz
          dd=sqrt((Xobs-Xt(m))^2+(Zobs-Zt(m))^2);
          scat_el(n)=scat_el(n)+ind_current(m)*besselh(0,
2,k0*dd);
     end
     scat_el(n)=-0.25*Dz*k0*120*pi*scat_el(n);
     dist=sqrt((Xobs-Xs)^2+Zobs^2);
Rayang=atand((Xobs-Xs)/Zobs);
     WR1=exp(-(sind(Rayang)-sind(elv))^2*log(2)/
(2*sind(bw/2)^2));
     inc_el(n)=WR1.*exp(-1i*k0*dist)./sqrt(k0*dist);
```

Table 15.10.
(Continued)

```
     tot_el(n)=inc_el(n)+scat_el(n);
     PF_el(n)=tot_el(n)/inc_el(n);
end
%% ----------- Two Ray Method ------------------------
[gtotz,GPFz]=RAY2(Zt,Xobs,Xs,elv,bw,k0);
plot3d2ray(Dz,Zmax,Xmax,Nx,elv,bw,k0,Xs,xmax3d,minval,max
val);
norm2ray=mean(abs(gtotz(round(0.8*Nz):end)));
%% ---------- SSPE -----------------------------------
dx=2*Xmax/(Nx-1); X=0:dx:(Nx-1)*dx;
fil_r  = [1.5 2]; % filter is applied to [r1 r2]*Xmax
FILFUN = @(x) (abs(x)<=fil_r(1)*Xmax)+. . . % filter
for Refl.
     (abs(x)>fil_r(1)*Xmax).*(1+cos(pi*(abs(x)-
fil_r(1)*Xmax)/((fil_r(2)-fil_r(1))*Xmax)))/2;
FF=zeros(Nx,Nz);  ww = sqrt(2*log(2))/(k0*sind(bw/2)); %
beamwidth
N=2*Nx-1; unscaledaxis=-ceil((N-1)/2):N-1-ceil((N-1)/2);
% initial
Kxg=2*pi/(dx*N)*unscaledaxis; PP1=Kxg-k0*sind(elv);
Field2=exp(-(PP1*ww/2).∧2).*exp(-1i*PP1*Xs); % from
source at Xs
Field2=fftshift(sqrt(2*pi)*ifft(ifftshift(Field2),N));
Field2=Field2/max(abs(Field2)); Field=Field2(round(N/2):
end);
FF(:,1)=Field; kX=linspace(0,pi/dx,Nx).';
CC1=exp(-1i*Dz*kX.∧2./(k0*(1+sqrt(1-kX.∧2/k0∧2)))).';  %
Wide PE
for n=1:Nz     % SSPE Range Loop
     for k=1:Nx
          if X(k)<=Xt(n);
               Field(k)=complex(0,0);
          end
     end
     Field(2:end-1)=idst(CC1(2:end-1).
*dst(Field(2:end-1)));
     Field=Field.*FILFUN(X); FF(:,n+1)=Field;
end
```

(Continued)

Table 15.10.
(Continued)

```
FFfree=findsspefree(Nx,dx,FILFUN,Field2,k0,Nz,Dz); % SSPE
free space
% Analyze Results
PEPF=FF./FFfree(round(N/2):end,:); PEPFz=PEPF(round(Xobs/
dx)+1,:);
FFz=FF(round(Xobs/dx)+1,:);
normpe=mean(abs(FFz(round(0.8*Nz):end))); FFz=norm2ray/
normpe*FFz;
%% ----------- to show 2D fields ----------------------
figure,
plot(Zt,abs(PF_el),Zt,abs(GPFz),'r-',(0:Dz:Zmax),abs(PEPFz
),'k:');
xlabel('Distance [m]','Fontweight','bold');
ylabel('Propagation Factor','Fontweight','bold');legend('MoM',
'2Ray','PE');
title('Propagation Factor over Irregular Earth: MoM vs.
PEM','Fontweight','bold');
figure,
plot(Zt,abs(tot_el),Zt,abs(gtotz),'r-',(0:Dz:Zmax),abs(FFz
),'k:');
xlabel('Distance [m]','Fontweight','bold');
ylabel('Field []','Fontweight','bold');legend('MoM','2Ray',
'PE');
title('Field over Irregular Earth: MoM vs. PEM','Fontweigh
t','bold');
% --------------------------------
function [gtot,GPF]=RAY2(Z,X,Xs,elv,bw,k0)
R1=sqrt(Z.^2+(Xs-X).^2); Rayang1=atand((X-Xs)./Z);
WRay1=exp(-(sind(Rayang1)-sind(elv)).^2*log(2)/
(2*sind(0.5*bw)^2));
R2=sqrt(Z.^2+(Xs+X).^2); Rayang2=-atand((X+Xs)./Z);
WRay2=exp(-(sind(Rayang2)-sind(elv)).^2*log(2)/
(2*sind(0.5*bw)^2));
gfree=WRay1.*exp(-1i*k0*R1)./sqrt(k0*R1);
gref =WRay2.*exp(-1i*k0*R2)./sqrt(k0*R2);
gtot=gfree-gref; GPF=gtot./gfree;
% --------------------------------
```

Table 15.10.
(Continued)

```
function plot3d2ray(Dz,Zmax,Xmax,Nx,elv,bw,k0,Xs,xmax3d,m
inval,maxval)
z=0:Dz:Zmax; x=linspace(0,2*Xmax,Nx);
[Z,X]=meshgrid(z,x);
[gtot,GPF]=RAY2(Z,X,Xs,elv,bw,k0);
plot3dPF(z,x,gtot,minval,maxval,xmax3d,'2-RAY');
% ----------------------------
function plot3dPF(z,x,PF,miv,mav,xmax,titlename)
f=figure('Position',[10 45 760 280]);
if maxval==0,
axes1=axes('Parent',f,'LineWidth',2,'FontSize',16);
else
axes1=axes('Parent',f,'LineWidth',2,'FontSize',16,'Clim',[miv
mav]);
end
hold on, box on, surf(z,x,abs(PF));
ylim([0 xmax]); shading interp; colormap(jet);colorbar;
view([0 90]),
colorbar('peer',axes1,'FontWeight','bold','FontSize',14);
xlabel('Range [m]'); ylabel('Height [m]');
title(titlename,'FontWeight','bold');
% ----------------------------
function FF=findsspefree(Nx,dx,FILFUN,Field,k0,Nz,Dz)
X=-(Nx-1)*dx:dx:(Nx-1)*dx; kX=linspace(0,pi/
dx,length(X)).';
FF(:,1)=Field; CC1=exp(-1i*Dz*kX.^2./
(k0*(1+sqrt(1-kX.^2/k0^2)))).'; % WPE
for n=1:Nz      % SSPE Range Loop
  Field(2:end-1)=idst(CC1(2:end-1).*dst(Field(2:end-1)));
  Field=Field.*FILFUN(X); FF(:,n+1)=Field;
end
end% ---------------------- END -----------------------
```

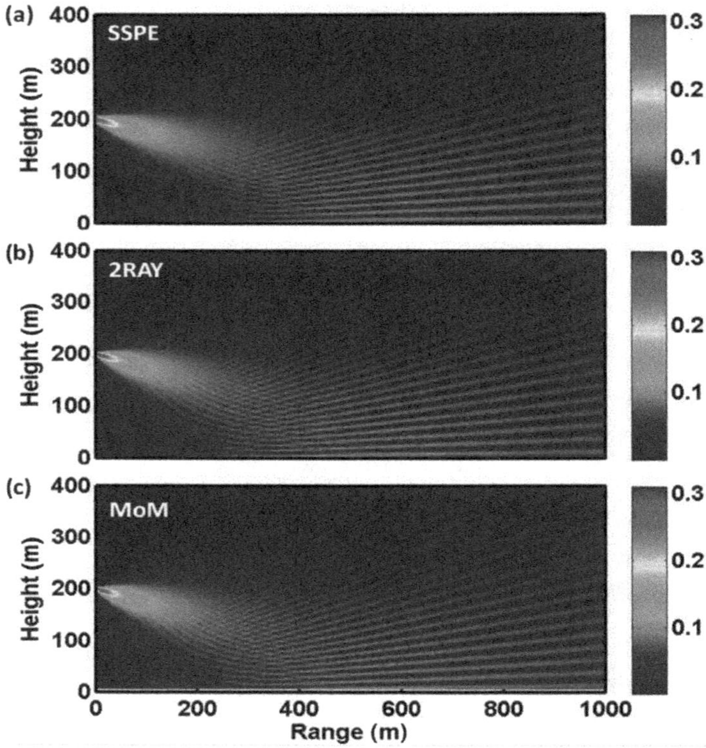

Figure 15.39. Field versus range–height for a given tilted Gaussian source: (a) SSPE, (b) 2Ray, (c) MoM (one-way propagation, DBC, $f = 30$ MHz, source height $= 200$ m, $\theta_{bw} = 30°$, $\theta_{tilt} = -15°$, $\Delta z = 1$ m, $\Delta x = 5$ m).

range–height 3D maps obtained with SSPE, 2Ray, and MoM models (using this code) are plotted in Fig. 15.39. Here, downtilted Gaussian antenna with 30° beamwidth is used. The tilt angle is $\theta_{tilt} = -15°$. As observed, the agreement in the results is impressive. To better visualize the agreement, 2D field variations are presented in Fig. 15.40.

Similar accurate source modeling is tested with the knife-edge problem and the results are presented in Fig. 15.41 and Fig. 15.42.

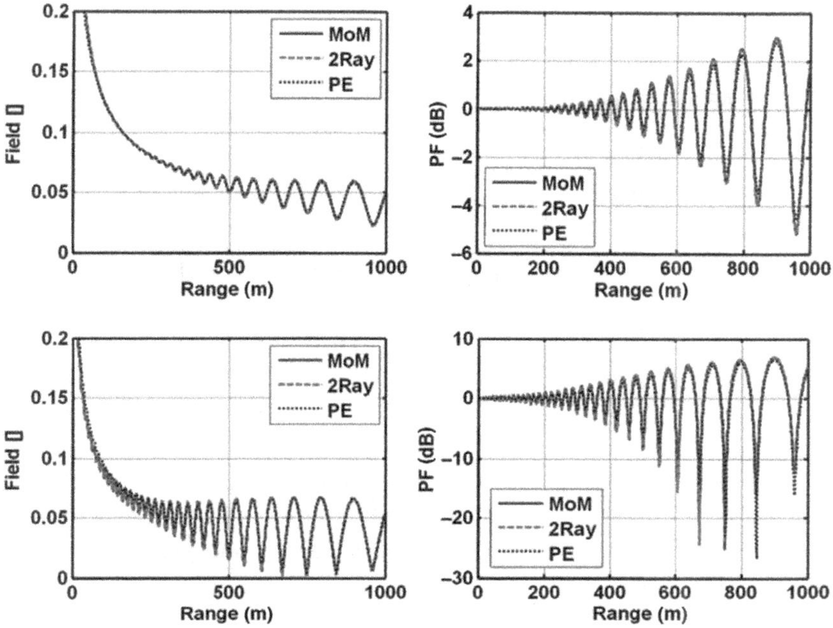

Figure 15.40. Field versus range for a given tilted Gaussian source (DBC): (Top) Untilted source and (bottom) tilted source (at 200 m height, $\theta_{bw} = 30°$, $f = 30$ MHz, $\theta_{tilt} = -15°$).

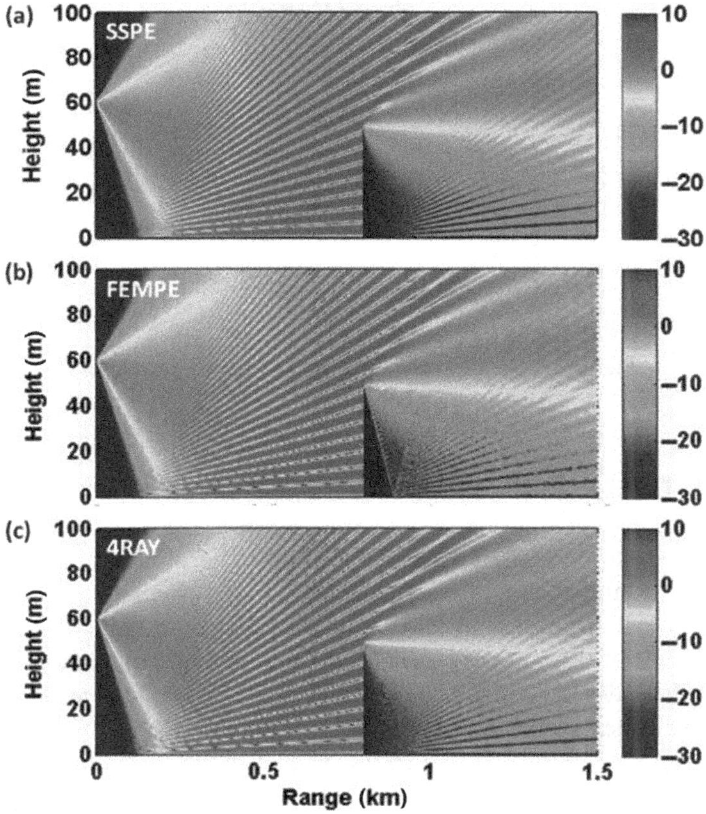

Figure 15.41. PF versus range–height for a given tilted Gaussian source: (a) SSPE, (b) FEMPE, and (c) 4Ray (one-way propagation, DBC, $f = 300$ MHz, with 50-m-high knife edge at 800 m range, source height = 60 m, $\theta_{bw} = 15°$, $\theta_{tilt} = -5°$, $\Delta z = 1$ m, $\Delta x = 0.2$ m).

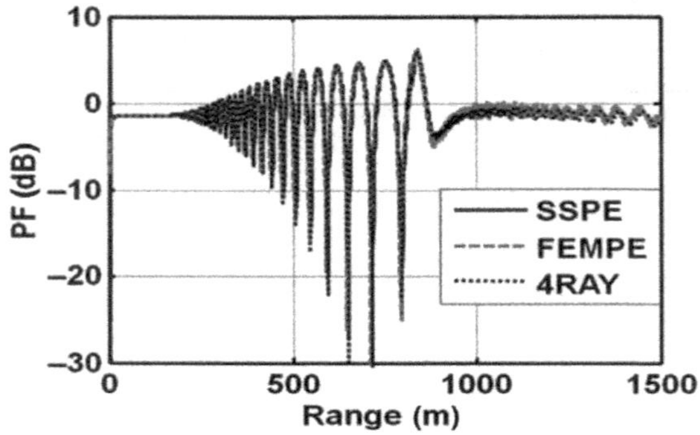

Figure 15.42. PF versus range at 60-m height for the scenario given in Fig. 14.43 ($\Delta z = 1$ m, $\Delta x = 0.2$ m).

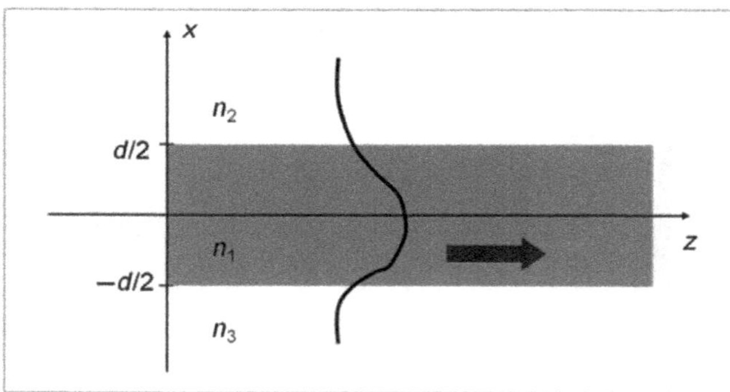

Figure 15.43. The canonical 2D dielectric slab waveguide.

15.11 DIELECTRIC SLAB WAVEGUIDE

EM waves can be guided inside a dielectric layer which has a relative permittivity higher than outside. Dielectric slab waveguides is one of the canonical structures in EM as well as in optics. The structure is pictured in Fig. 15.43. It is a dielectric layer having a thickness d transversely in x-domain, infinitely extends in the y- and z-domains. The refractive index of the layer is n_1. The top layer vertically extends to infinity and has the refractive index of n_2. Similarly, the bottom layer extends vertically to infinity and has the refractive index of n_3. Most of the EM energy excited inside the layer is trapped inside the layer, but leaks out vertically since the boundaries are penetrable. The trapped energy propagates horizontally as pictured in the figure. The structure becomes full symmetric if $n_2 = n_3$, where the solutions can be expressed in terms of even- and odd-symmetric modes.

The structure is canonical, since mathematical solutions of EM fields for this structure can be derived starting with Maxwell equations in the frequency domain or with the 2D wave equation,

$$\left(\frac{\partial^2}{\partial x^2} + \frac{\partial^2}{\partial z^2} + k^2\right) E_{yi}(x, z) = 0, \quad k^2 = k_0^2 n_i^2, i = 1, 2, 3. \tag{15.42}$$

The wave function $E_{yi}(x, z)$ must satisfy radiation BC on both x- and z-directions since the 2D region extends to infinity on the left, right, up, and down. If the refractive index of the film region is greater than the other two, regions waves are trapped inside. It means that the transverse wave function oscillates inside the film but exponentially decays outside. The transverse wave functions will then be different in these three regions, but are related, imposing the continuity conditions (i.e., both functions and their derivatives with respect to transverse variable x must be equal) at the boundaries of the film.

Assuming $\exp(j\omega t)$ time dependency (i.e., $\partial/\partial z \equiv -j\beta$) and $\partial/\partial y \equiv 0$ for TM_z case, the E-field can be written as $E_{yi}(x, z) = E_{yi}(x)\exp(-j\beta_i z)$ and the transverse (x) domain 1D wave equation,

$$\frac{\partial^2}{\partial x^2} E_y + (k^2 - \beta^2) E_y = 0, \quad k^2 = k_0^2 n_i^2, i = 1, 2, 3, \tag{15.43}$$

may be used to derive $E_{yi}(x)$, satisfying radiation BCs when $x \to \pm\infty$.

The dielectric layer may be located anywhere vertically, which does not affect the form of the solutions. Assume, for example, that the layer extends from $x = 0$ to $x = -d$ with refractivity n_1 and symmetric outside with refractivity n_0. In this case, transverse part of the solution can be written separately in three parts:

$$E_y(x) = A\exp(-\gamma_0 x), \quad 0 < x, \tag{15.44a}$$

$$E_y(x) = B\cos(\kappa_1 x) + C\sin(\kappa_1 x), \quad -d < x < 0, \tag{15.44b}$$

$$E_y(x) = D\exp(\gamma_0 (x + d)), \quad x < -d, \tag{15.44c}$$

where $\gamma_0 = \sqrt{\beta^2 - k_0^2 n_0^2}$, $\kappa_1 = \sqrt{k_0^2 n_1^2 - \beta^2}$. The vertical field component $E_y(x)$ and its normal derivative $\partial E_y/\partial x$ are continuous at $x = 0$ and $x = -d$, therefore

$$A\exp(-\gamma_0 x)\big|_{x=0} = B\cos(\kappa_1 x) + C\sin(\kappa_1 x)\big|_{x=0} \Rightarrow A = B, \quad (15.45a)$$

$$-\gamma_0 A\exp(-\gamma_0 x)\big|_{x=0} = -\kappa_1 B\sin(\kappa_1 x) + \kappa_1 C\sin(\kappa_1 x)\big|_{x=0} \Rightarrow C = -A\frac{\gamma_0}{\kappa_1}, \quad (15.45b)$$

$$D = A\left(\cos(\kappa_1 d) + \frac{\gamma_0}{\kappa_1}\sin(\kappa_1 d)\right) \quad (15.45c)$$

are obtained. Vertical E-field can then be written as

$$E_y(x) = A\exp(-\gamma_0 x), \quad 0 < x, \quad (15.46a)$$

$$E_y(x) = A\left[\cos(\kappa_1 x) - \frac{\gamma_0}{\kappa_1}\sin(\kappa_1 x)\right], \quad -d < x < 0, \quad (15.46b)$$

$$E_y(x) = A\left[\cos(\kappa_1 x) + \frac{\gamma_0}{\kappa_1}\sin(\kappa_1 x)\right]\exp(\gamma_0(x+d)), \quad x < -d. \quad (15.46c)$$

The continuity of normal derivative of the vertical field at $x = -d$ is used to obtain the eigenvalue equation

$$\tan(\kappa_1 h) = \frac{2\gamma_0\kappa_1}{\kappa_1^2 - \gamma_0^2}. \quad (15.47)$$

This is a transcendental equation that must be solved numerically or graphically. Together with the eigenvalues derived from (15.45), Equation (15.46) can be used to plot transverse modal shapes. The first four modes of the dielectric slab waveguide are plotted in Fig. 15.44.

15.11.1 Even and Odd Symmetric Solutions

It is possible to solve the symmetric slab waveguide by dividing the slab into two regions: the half under and the other half above $x = 0$. Then, the odd and even symmetric modes can be obtained separately. In Fig. 15.44, mode 1 and mode 3 are even-symmetric modes, while mode 2 and mode 4 are odd symmetric.

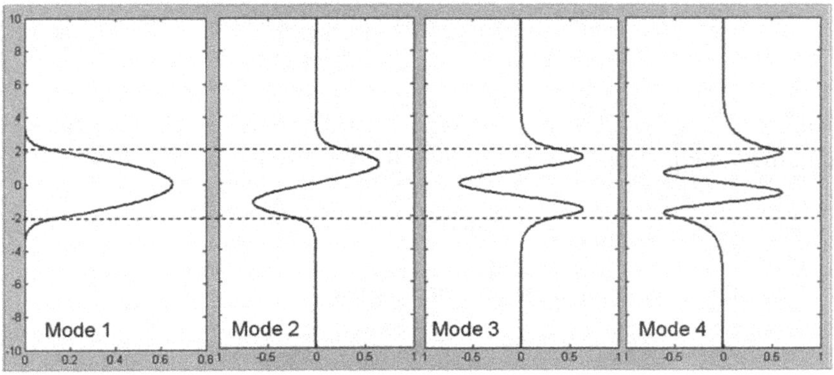

Figure 15.44. The first four modes of the dielectric slab waveguide.

15.11.1.1 Odd Symmetric Slab

Mode 2 and mode 4 (and the rest of the even-indexed modes) are the solutions of odd-symmetric structure. The odd-symmetric solutions are

$$E_y(x) = \begin{cases} A\exp(-\gamma_0(x-d)), & \beta > k_0 n_0, \quad x > d \\ B\sin(\kappa_1 x), & \beta < k_0 n_1, \quad 0 < x < d \end{cases}, \quad (15.48)$$

where $\gamma_0 = \sqrt{\beta^2 - k_0^2 n_0^2}$, $\kappa_1 = \sqrt{k_0^2 n_1^2 - \beta^2}$, and $B = A/\sin(\kappa_1 d)$. The eigenvalue equation for the odd-symmetric modes is

$$\tan(\kappa_1 d) = -\frac{\kappa_1}{\gamma_0}. \quad (15.49)$$

The last coefficient A is obtained from the transverse normalization property (i.e., all modes must carry unit energy) as

$$A = \left[\frac{1}{\gamma_0} + \frac{2\kappa_1 d - \sin(2\kappa_1 d)}{2\kappa_1 \sin^2(\kappa_1 d)}\right]^{-\frac{1}{2}} \quad \text{(odd mode).} \quad (15.50)$$

15.11.1.2 Even Symmetric Slab

Mode 1 and mode 3 (and the rest of the odd-indexed modes) are the solutions of even-symmetric structure. The even-symmetric solutions are

$$E_y(x) = \begin{cases} A\exp(-\gamma_0(x-d)), & \beta > k_0 n_0, \quad x > d \\ B\cos(\kappa_1 x), & \beta < k_0 n_1, \quad 0 < x < d \end{cases}, \tag{15.51}$$

where $\gamma_0 = \sqrt{\beta^2 - k_0^2 n_0^2}$, $\kappa_1 = \sqrt{k_0^2 n_1^2 - \beta^2}$, and $B = A/\cos(\kappa_1 d)$. The eigenvalue equation for the odd-symmetric modes is

$$\tan(\kappa_1 d) = \frac{\kappa_1}{\gamma_0}. \tag{15.52}$$

Similarly, normalization property yields

$$A = \left[\frac{1}{\gamma_0} + \frac{\sin(2\kappa_1 d) + 2\kappa_1 d}{2\kappa_1 \cos^2(\kappa_1 d)}\right]^{-\frac{1}{2}} \quad \text{(even mode).} \tag{15.53}$$

15.11.2 The SSPE Propagator and Eigenvalue Extraction

As also explained in Chapter 8 (Section 8.5), the eigenvalues of the problem can be extracted from the propagation characteristics as an alternative to the link between source-free and source-driven formulations. Source-driven SSPE propagator can be used to obtain eigenvalues for propagation through the slab guide. Any transverse wave function (e.g., a transmitting antenna field profile) accumulates eigencharacteristics of the medium while longitudinally propagating. If the eigenfunction expansion of the initial profile at the initial range $z = z_0$ is

$$E_y(x, z_0) = A\sum_{n=1}^{M} K_n \sin(\kappa x)\exp(-j\beta_n z_0), \tag{15.54}$$

with excitation and normalization coefficients K_n and A, respectively, and M is the maximum number of modes that are required to represent the initial field profile "adequately." A longitudinal correlation function may be built while propagating this field profile through the waveguide,

$$\begin{aligned} P(z) &= \int_{x=0}^{d} E(x, z_0)E(x, z)dx \\ &= \sum_{n=1}^{M} K_n^2 \exp(-j\beta_n z) \int_{x=0}^{d} A\sin(\kappa x)dx = \sum_{n=1}^{M} K_n^2 \exp(-j\beta_n z). \end{aligned} \tag{15.55}$$

Since the integral over transverse domain is equal to "1" (because of the orthonormality condition), the Fourier transformation of this correlation function yields the line spectra

$$P(\beta) = K_n^2 \delta(z - z_0). \tag{15.56}$$

that is, eigenvalues and the excitation coefficients. The Fourier transform may be performed numerically via DFT or FFT; therefore, the longitudinal correlation function should be observed long enough to meet maximum longitudinal wave number ($\beta_{\max}$) and wave number resolution ($\Delta\beta$) via Nyquist sampling criterion,

$$\beta_{\max} = \frac{\pi}{2\Delta z} \quad \text{and} \quad \Delta\beta = \frac{2\pi}{z_{\max}}, \tag{15.57}$$

where Δz and $Z_{\max}$ are range step size and maximum range, respectively.

15.11.3 The Matlab-Based DiSLAB Package

The DiSLAB package is designed to investigate wave propagation through a 2D dielectric waveguide. Both analytical formulations given above and the SSPE propagator explained in Chapter 8 are used. The front panel of the DiSlab is given in Fig. 15.45. The front panel is divided into four regions. Necessary input parameters listed in Table 15.11 that the user should supply are grouped at top left.

The user should choose the operating frequency, slab width, and refractivity indices of both core and cladding regions. Then the user can find the number of propagating modes by pressing the "# of modes" button. The modes that will be included in the computation are defined by choosing the first and last modes. The user can choose one or both of the methods—analytic and/or SSPE—for calculation. The maximum desired range should also be chosen, which is required to obtain the observation ranges for the calculation of $E_y(x)$ variation at the left-hand side figure by pressing the related "Calculate" button. The maximum range, half of it, and $z = 0$ are used as observation ranges during the calculation. The longitudinal propagation constant is calculated using a correlation function and its Fourier transform, which are computed and depicted by pressing the "Calculate" button at the right-hand side.

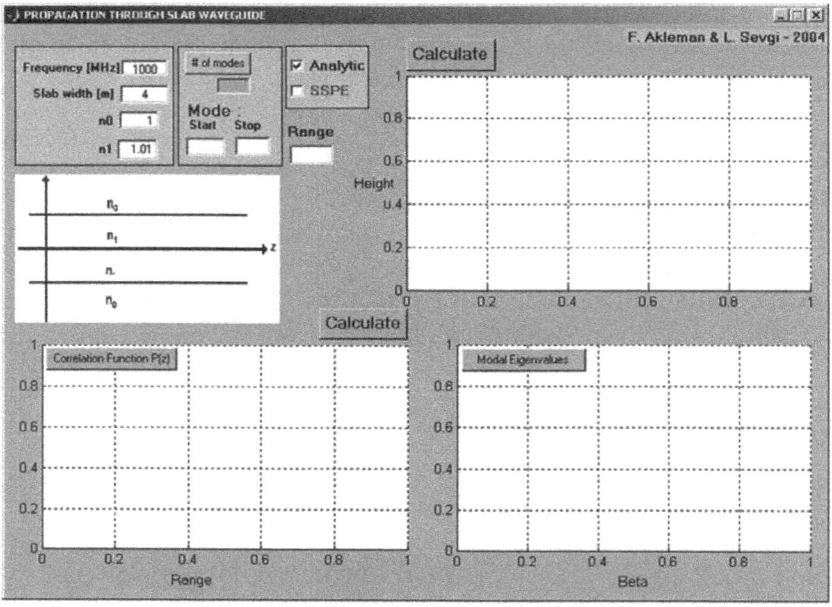

Figure 15.45. Front panel of the DiSLAB package.

Table 15.11.
User-Supplied DiSlab Input Parameters

Frequency	Operating frequency (MHz)
Slab width	Width of slab in meters
$n0$	Refractive index at the cladding region
$n1$	Refractive index at the core region
Mode: Start	First mode that will be included in calculation
Mode: Stop	Last mode that will be included in calculation
Analytic	Choose one or both of analytic and SSPE solutions
SSPE	
Range	Maximum desired range

The graphic window at top right shows the transverse profile of $E_y(x)$ at three different observation ranges (at the beginning, at mid-range, and at the last chosen range). The last range is user supplied in the "range box." The two graphics at the bottom show the correlation function variation with respect to range (left plot) obtained by the integration of the multiplication of transverse field variation at each range

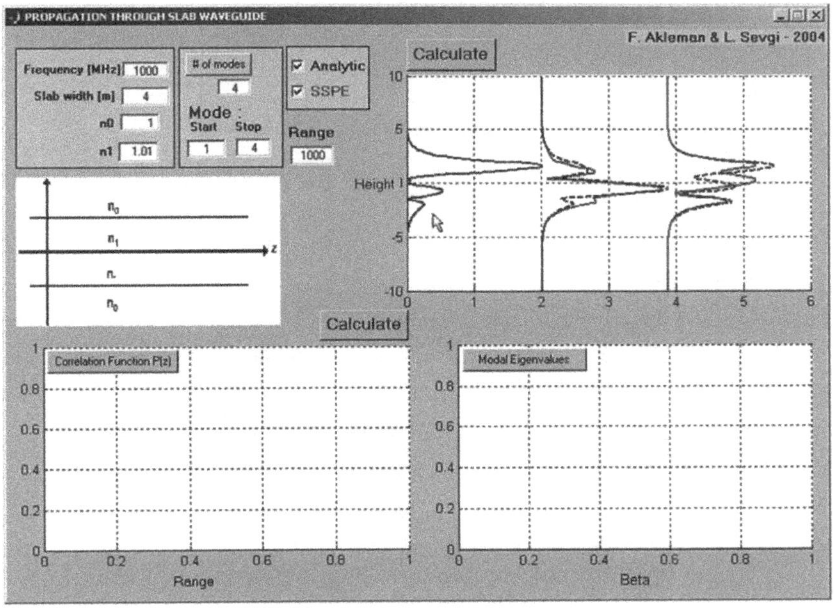

Figure 15.46. DiSLAB screen after run for typical values.

with the transverse field variation at $z = 0$ over x-axis and its Fourier transformation (right plot), where the longitudinal propagation constants β of different modes are easily distinguished from each other.

A typical example is given in Fig. 15.46. Here a 4-m-wide dielectric slab with $n_0 = 1.01$ inside and $n_0 = 1.00$ outside is chosen as an example. The user specified the operating frequency as 1 GHz, which means that only the first four modes have real propagation constant; therefore, there will be only four modes available along the slab waveguide. The user took all four modes into account by specifying start mode as 1 and stop mode as 4. The program builds an initial transverse field $E_y(x)$ profile in terms of the superposition of these four modes with equal amplitudes. It should be noted that any vertical source profile at the initial range can be chosen and many modes are required to build this source function. But, as it propagates through the slab only modes, real propagation constants (i.e., eigenvalues) can propagate and the rest high order modes will be deceased. Therefore, it is appropriate to take only the propagating modes into account when dealing with field variations at ranges far from the source.

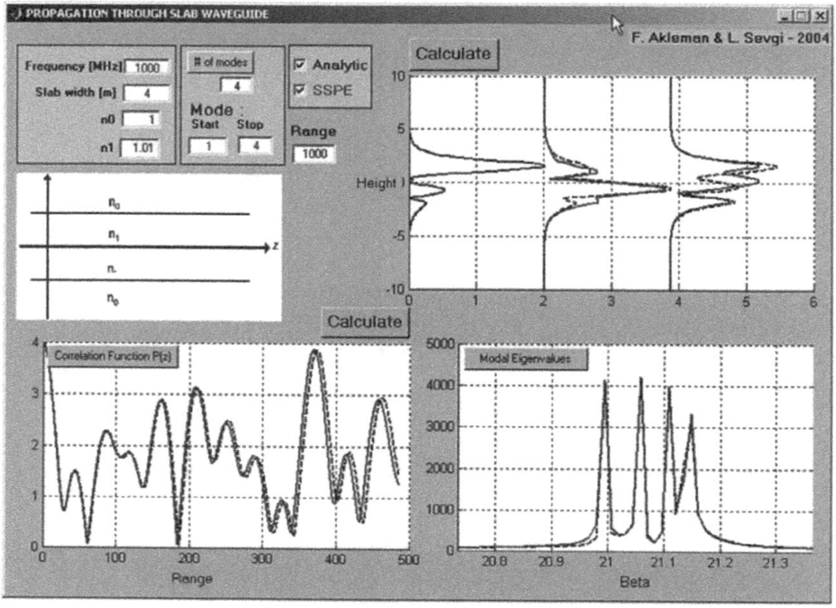

Figure 15.47. Full DiSLAB screen (slab width = 4 m, f = 1 GHz, n_0 = 1.01, n_0 = 1.00).

If the vertical field profiles as shown in Fig. 15.46 are correlated with the initial profile as it propagates according to (15.55), one can obtain modal eigenvalues. The package can apply this procedure from both analytical and numerical (SSPE) propagators. This is illustrated in Fig. 15.47.

The eigenvalues calculated analytically for the first four propagation modes are $\beta_1 = 21.1427, \beta_1 = 21.1111, \beta_1 = 21.0597, \beta_1 = 20.9922$.

A MATLAB module (Slab.m) in DiSLAB, which includes only the analytical calculations and excludes routines required for graphical user interface (GUI) design and root search, is given in Table 15.12. In Slab.m, E_y field is calculated analytically.

Two other examples are given in Fig. 15.48 and Fig. 15.49. In Fig. 15.48, all parameters of the first example are kept same, but refractivity contrast is increased; n_0 = 1.05 inside and n_0 = 1.00 outside. As shown in Fig. 15.47, the number of propagating modes is increased to 9. In order to discriminate eigenvalues of these modes, maximum correlation range should be increased. In this example, it is doubled (changed from

Table 15.12.
A Matlab Script that Plots Even Mode Profiles

```
% ---------------------------------------------------
% Program  : slab.m
% Purpose: Propagation analysis in slab guide (even
  Modes)
% Supply:
%  frequency (freq) in MHz, slab width (2*d)
%  cladding refractivity (n0), core refractivity (n1)
%  maximum range (zmax), first mode, last mode
%  evenroot() : contains roots of the eigenvalue
  equation
% ---------------------------------------------------
% Parameters
ModeNo=input('Even Mode No [1,3, 5, . . .] = ?');
d=input('Slab width [m] = ?');
N=500;  dh=5*d/N;
for n=1:N
  x(n) = -(n-1)*dh;
  ey(n) = 0;
  kappa1 = sqrt((k0*n1)∧2-evenroot(ModeNo)∧2);
  gamma0 = sqrt(evenroot(ModeNo)∧2-(k0*n0)∧2);
  norm = 1/sqrt(1/gamma0+(sin(2*kappa1*d) . . .
         +2*kappa1*d)/(2*kappa1*(cos(kappa1*d))∧2));
  if x(n)<d && x(n)>-d
      ey(n) = cos(kappa1*x(n))/cos(kappa1*d)*norm;
  elseif x(n)>=d
      ey(n) = exp(-gamma0*(x(n)-d))*norm;
  elseif x(n)<=-d
      ey(n) = exp(gamma0*(x(n)+d))*norm;
  end
end
plot(x,ey);
% -------------------- END ----------------------
```

1 to 2 km). Observe that, although can be seen, the highest eigenvalues are not well discriminated.

The last example has the same parameters of the previous one, except the slab width is decreased to 2 m. In this case, the number of propagation modes becomes 5. The results are given in Fig. 15.49.

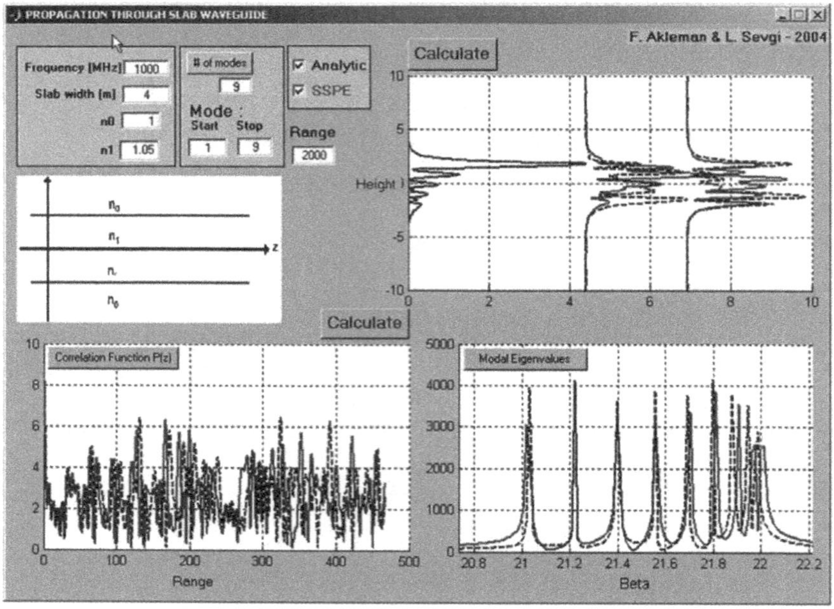

Figure 15.48. Full DiSLAB screen for example 2: Slab width = 4 m, f = 1 GHz, n_0 = 1.05 inside, and n_0 = 1.00 outside. Nine beta values of the propagating modes are 21.9790, 21.9426, 21.8818, 21.7969, 21.6878, 21.5549, 21.3989, 21.2219, and 21.0308.

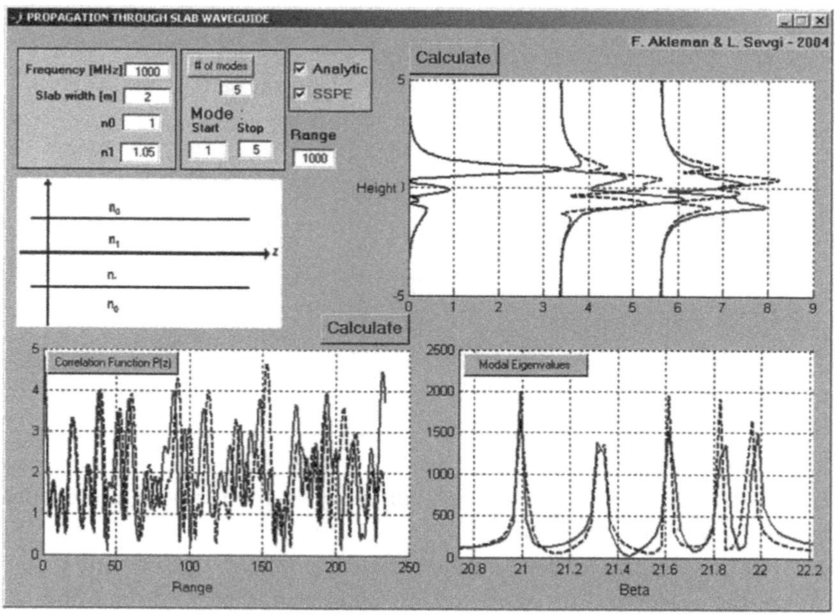

Figure 15.49. Full DiSLAB screen for example 2: Slab width = 4 m, f = 1 GHz, n_0 = 1.05 inside, and n_0 = 1.00 outside. Five beta values of the propagating modes are 21.9487, 21.8219, 21.6128, 21.3286, and 21.0026.

15.12 CONCLUSIONS

The parabolic equation (PE) method is capable of simulating one-way, forward-scatter propagation problems within the paraxial approximation. It takes into account all types of BCs at the ground, the irregular terrain effects, as well as the atmospheric refractivity variations. Widely applied numerical routines are based on either the discrete Fourier split-step solution or the FEM based on the Crank—Nicholson scheme. This chapter discusses two numerical propagation tools based on the Fourier split-step and the FEMs. Simple MATLAB-based propagation tools—SSPE and FEMPE—are introduced. The systematic comparisons on some canonical test scenarios are performed and the propagation tools are calibrated against the mathematical exact solutions. Both SSPE and FEMPE can take any type of refractivity variations as well as user-specified irregular terrain profiles into account. Advanced topics such as impedance boundary implementation and two-way PE approaches are left to the reader [40–47].

REFERENCES

1. L. Sevgi, "Groundwave Modeling and Simulation Strategies and Path Loss Prediction Virtual Tools," (special issue on Electromagnetic Wave Propagation in Complex Environments: A Tribute to L. B. Felsen), *IEEE Transactions on Antennas and Propagation*, **55** (6), 2007, pp. 1591–1598.
2. L. Sevgi, *Complex Electromagnetic Problems and Numerical Simulation Approaches*, IEEE & John Wiley Press, New York, 2003.
3. L. Sevgi, F. Akleman, L. B. Felsen, "Groundwave Propagation Modeling: Problem-Matched Analytical Formulations and Direct Numerical Techniques," *IEEE Antennas and Propagation Magazine*, **44** (1), 2002, pp. 55–75.
4. L. Sevgi, C. Uluisik, F. Akleman, "A MATLAB-Based Two-Dimensional Parabolic Equation Radiowave Propagation Package," *IEEE Antennas and Propagation Magazine*, **47** (4), 2005, pp. 164–175.
5. C. A. Tunc, F. Akleman, V. B. Erturk, A. Altintas, L. Sevgi, "Fast Integral Equation Solutions: Application to Mixed Path Terrain Profiles and Comparisons with Parabolic Equation Method," *Complex Computing Networks*, Springer in Physics Series, **104**, 2006, pp. 55–63.
6. F. Akleman, L. Sevgi, "A Novel MoM- and SSPE-Based Groundwave-Propagation Field-Strength Prediction Simulator," *IEEE Antennas and Propagation Magazine*, **49** (5), 2007, pp. 69–82.
7. F. Akleman, L. Sevgi, "A Novel Finite-Difference-Time-Domain Wave Propagator," *IEEE Transactions on Antennas and Propagation*, **48** (3), 2000, pp. 839–841.

8. M. O. Ozyalcin, F. Akleman, L. Sevgi, "A Novel TLM Based Time Domain Wave Propagator," *IEEE Transactions on Antennas and Propagation*, **51** (7), 2003, pp. 1680–1683.
9. L. Sevgi, "A Mixed-Path Groundwave Field Strength Prediction Virtual Tool for Digital Radio Broadcast Systems in Medium and Short Wave Bands," *IEEE Antennas and Propagation Magazine*, **48** (4), 2006, pp. 19–27.
10. L. Sevgi, L. B. Felsen, "A New Algorithm for Ground Wave Propagation Based on a Hybrid Ray-Mode Approach," *International Journal of Numerical Modelling*, **11** (2), 1998, pp. 87–103.
11. L. Sevgi, "A Ray-Shooting Visualization MATLAB Package for 2D Groundwave Propagation Simulation," *IEEE Antennas and Propagation Magazine*, **46** (4), 2004, pp. 140–145.
12. M. A. Leontovich, V. A. Fock, "Solution of Propagation of Electromagnetic Waves along the Earth's Surface by the Method of Parabolic Equation," *Journal of Physics of the USSR*, **10**, 1946, pp. 13–23.
13. V. A. Fock, *Electromagnetic Diffraction and Propagation Problems*, Pergamon Press, New York, 1965.
14. F. D. Tappert, The Parabolic Approximation Method, in J. B. Keller, J. S. Papadakis, eds., *Wave Propagation and Underwater Acoustics*, Lecture Notes in Physics, vol. 70, Springer, New York, 1977.
15. K. H. Craig, "Propagation Modelling in the Troposphere: Parabolic Equation Method," *IEE Electronics Letters*, **24** (18), 1988, pp. 1136–1139.
16. G. D. Dockery, J. R. Kuttler, "An Improved Impedance Boundary Algorithm for Fourier Split-Step Solutions of the Parabolic Wave Equation," *IEEE Transactions on Antennas and Propagation*, **44** (12), 1996, pp. 1592–1599.
17. M. F. Levy, "Horizontal Parabolic Equation Solution of Radiowave Propagation Problems on Large Domains," *IEEE Transactions on Antennas and Propagation*, **43** (2), 1995, pp. 137–144.
18. G. D. Dockery, "Modeling Electromagnetic Wave Propagation in the Troposphere Using Parabolic Equation," *IEEE Transactions on Antennas and Propagation*, **36** (10), 1988, pp. 1464–1470.
19. A. E. Barrios, "Parabolic Equation Modeling in Horizontally Inhomogeneous Environments," *IEEE Transactions on Antennas and Propagation*, **40** (7), 1992, pp. 791–797.
20. L. Sevgi, "Split Step Parabolic Equation Solutions in Surface Duct-to-Elevated Duct Transition," *Turkish Journal of Physics*, **19** (3), 1995, pp. 541–551.
21. A. E. Barrios, "A Terrain Parabolic Equation Model for Propagation in the Troposphere," *IEEE Transactions on Antennas and Propagation*, **42** (1), 1994, pp. 90–98.
22. D. J. Donohue, J. R. Kuttler, "Propagation Modeling over Terrain Using the Parabolic Wave Equation," *IEEE Transactions on Antennas and Propagation*, **48** (2), 2000, pp. 260–277.

23. M. Levy, *Parabolic Equation Methods for Electromagnetic Wave Propagation*, IEE Institution of Electrical Engineers, London, 2000.
24. I. D. Sirkova, M. Mikhalev, "Parabolic Wave Equation Method Applied to the Tropospheric Ducting Propagation Problem: A Survey," *Electromagnetics*, **26**, 2006, pp. 155–173.
25. D. Huang, "Finite Element Solution to the Parabolic Wave Equation," *Journal of the Acoustical Society of America*, **84** (4), 1988, pp. 1405–1413.
26. V. M. Deshpande, M. D. Deshpande, "Study of Electromagnetic Wave Propagation through Dielectric Slab Doped Randomly with Thin Metallic Wires Using Finite Element Method," *IEEE Microwave and Wireless Components Letters*, **15** (5), 2005, pp. 306–308.
27. K. Arshad, F. A. Katsriku, A. Lasebae, "An Investigation of Tropospheric Wave Propagation Using Finite Elements," *WSEAS Transactions on Communications*, **4** (11), 2005, pp. 1186–1192.
28. K. Arshad, F. A. Katsriku, A. Lasebae, "An Investigation of Tropospheric Wave Propagation over Irregular Terrain and Urban Streets Using Finite Elements," *7th WSEAS Conference on Telecommunications and Informatics*, March 22–24 2007, Dallas, TX, USA.
29. J.-M. Jin, *The Finite Element Method in Electromagnetics*, John Wiley Press, New York, 2002.
30. J. Volakis, A. Chatterjee, L. Kempel, *Finite Element Method for Electromagnetics: Antennas, Microwave Circuits, and Scattering Applications*, IEEE Press and Oxford University Press, Piscataway, NJ, 1998.
31. W. Y. Yang, W. Cao, T. Chung, J. Morris, *Applied Numerical Methods Using MATLAB*, John Wiley Press, Hoboken, NJ, 2005.
32. AREPS (Advanced Refractive Effects Prediction System), visit http://www.public.navy.mil/
33. O. Ozgun, G. Apaydin, M. Kuzuoglu, L. Sevgi, "PETOOL: MATLAB-Based One-Way and Two-Way Split Step Parabolic Equation Tool for Radiowave Propagation over Variable Terrain," *Computer Physics Communications (CPC)*, Elsevier, **182**, 2011, pp. 2638–2654.
34. L. B. Felsen, L. Sevgi, "Adiabatic and Intrinsic Modes for Wave Propagation in Guiding Environments with Longitudinal and Transverse Variations: Formulation and Canonical Test," *IEEE Transactions on Antennas and Propagation*, **39** (8), 1991, pp. 1130–1136.
35. L. B. Felsen, L. Sevgi, "Adiabatic and Intrinsic Modes for Wave Propagation in Guiding Environments with Longitudinal and Transverse Variations: Continuously Refracting Media," *IEEE Transactions on Antennas and Propagation*, **39** (8), 1991, pp. 1137–1142.
36. G. Apaydin, L. Sevgi, "The Calibration of Three-Dimensional Parabolic Equation Models on the Rectangular Waveguide," *IEEE Antennas and Propagation Magazine*, **54** (6), 2012, pp. 102–116.

37. O. Ozgun, L. Sevgi, "Comparative Study of Analytical and Numerical Techniques in Modeling Electromagnetic Scattering from Single and Double Knife-Edge in 2D Ground Wave Propagation Problems," *ACES, International Journal on Applied Computational Electromagnetics*, **27** (5), 2012, pp. 376–388.
38. G. Apaydın, L. Sevgi, "Validation, Verification and Calibration in Applied Computational Electromagnetics," *ACES, International Journal on Applied Computational Electromagnetics*, **25** (12), 2010, pp. 1026–1035.
39. M. L. Meeks, *Radar Propagation at Low Altitudes*, Artech House, Dedham, MA, 1982.
40. G. Apaydın, L. Sevgi, "The Split Step Fourier and Finite Element Based Parabolic Equation Propagation Prediction Tools: Canonical Tests, Systematic Comparisons, and Calibration," *IEEE Antennas and Propagation Magazine*, **52** (3), 2010, pp. 66–79.
41. G. Apaydin, L. Sevgi, "FEM-Based Surface Wave Multi-Mixed-Path Propagator and Path Loss Predictions," *IEEE Antennas and Wireless Propagation Letters*, **8**, 2009, pp. 1010–1013.
42. G. Apaydın, L. Sevgi, "Propagation Modeling in and Path Loss Prediction Tools for High Frequency Surface Wave Radars," (special issue on HF Radars) *TJEECS, Turkish Journal of Electrical Engineering and Computer Sciences*, **18** (3), 2010, pp. 469–484.
43. G. Apaydın, L. Sevgi, "A Novel Split-Step Parabolic Equation Package for Surface Wave Propagation Prediction along Multi-Mixed Irregular Terrain Paths," *IEEE Antennas and Propagation Magazine*, **52** (4), 2010, pp. 90–97.
44. G. Apaydin, L. Sevgi, "Numerical Investigations of and Path Loss Predictions for Surface Wave Propagation over Sea Paths Including Hilly Island Transitions," *IEEE Transactions on Antennas and Propagation*, **58** (4), 2010, pp. 1302–1314.
45. O. Ozgun, "Recursive Two-Way Parabolic Equation Approach for Modeling Terrain Effects in Tropospheric Propagation," *IEEE Transactions on Antennas and Propagation*, **57** (9), 2009, pp. 2706–2714.
46. G. Apaydın, Ö. Özgün, M. Kuzuoğlu, L. Sevgi, "A Novel Two-Way Finite-Element Parabolic Equation Groundwave Propagation Tool: Tests With Canonical Structures and Calibration," *IEEE Transactions on Geoscience and Remote Sensing*, **49** (8), 2011, pp. 2887–2899.
47. G. Apaydin, L. Sevgi, "Two-Way Propagation Modeling in Tunnels with Three Dimensional Finite Element and Split-Step Fourier based PE Approaches," *IEEE Antennas and Wireless Propagation Letters*, 2011, pp. 975–978.

CHAPTER 16

Parallel Plate Waveguide Problem

16.1 INTRODUCTION

Electromagnetic (EM) problems (from nanoscale to kilometer wide systems) are complex in nature, but the theory is well established with Maxwell equations (see Refs. 1–3, for a few classical EM books and Appendix B for many others). The strategy for the solution may be grouped into three: analytical modeling, numerical simulations, and measurements. Measurement in EM is time consuming, expensive, and, in most cases, extremely difficult to do. A limited number of analytical solutions are available only for a few, highly idealized problems; therefore, numerical simulation becomes the only mean for almost all real-life EM engineering problems. This is why EM modeling and simulation (EM-MODSIM) has made a significant progress for the last couple of decades.

EM-MODSIM has been taught for sometime worldwide in many universities. A few books toward EM-MODSIM may be listed as Refs. 4–9. EM-MODSIM approaches may be divided into two: analytical-model-based and numerical-model-based approaches [10–12]. The critical issue in EM-MODSIM is the (model) validation, (data) verification, and (code) calibration (VV&C) [13]. The best VV&C way is to compare numerical models with accurately computed analytical models. In this content, high frequency asymptotics (HFA) plays an essential

Electromagnetic Modeling and Simulation, First Edition. Levent Sevgi.

role in generating reference data [14–20]. HFA models such as geometric optics (GO), geometric theory of diffraction (GTD), physical optics (PO), physical theory of diffraction (PTD), uniform theory of diffraction (UTD), with their advantages/disadvantages and regions of validity not only yield reference data but also give physical insight in understanding EM wave scattering pieces. The two HFA tutorials published in Refs. 21 and 22 review fundamental diffraction models.

In terms of solution approaches, EM problems may be grouped into three:

i. radiation/antenna problems,
ii. scattering and radar cross section (RCS) modeling, and
iii. waveguiding structures.

They all start with Maxwell equations and definition of boundary conditions (BCs).

In (i), it is better to define auxiliary scalar and vector potential functions, and solve the equations which contain these auxiliary functions. This is better because components of the vector potential fit in practical excitations. Scattered field representation is used in RCS modeling because the excitation (i.e., incident field) is elsewhere. Finally, Maxwell equations are decomposed into transverse and longitudinal components in guided wave problems. This not only yields decomposed fields but also scalarizes and solves the problem. Eigenfunctions and eigenvalues are derived from transverse components; excitation coefficients are obtained from longitudinal components. Completeness and orthonormalization relate these two via the well-known Sturm–Liouville equation [23].

Most of EM-MODSIM studies, commercial or educational, are based on one of the listed models:

- Method of moments (MoM) discretizes the object under investigation into N pieces, called segments or patches, and constructs an $N \times N$ system of equations using the Green's function of the problem [24]. It is closed form and stable, but suffers from lack of high memory and high speed computers.

- Parabolic equation (PE) method, based on either finite element (FE) or split step PE (SSPE) discretization, is widely used in complex propagation modeling [25].
- Finite-difference time-domain (FDTD) method discretizes the physical environment into small cells by replacing partial derivatives with their finite-difference equivalents [26–28]. It is open form and iterative, therefore suffers from stability.
- Transmission line matrix (TLM) method uses circuit equivalent of Maxwell equations [29,30]. This is also open form and iterative with similar stability problems.

This last chapter aims to compare widely used numerical models on a simple, canonical problem: *Propagation inside a parallel plate waveguide with perfectly electrical conducting (PEC) boundaries in two-dimension* (2D) [31,32] (see Chapter 8 for details). The same can be applied to wedge waveguide also [33] (see Chapter 9 for details). Propagation inside a parallel plate waveguide is an interesting EM problem where both analytical and numerical models can be tested one against the others.

- First of all, the structure in 2D is nonphysical; acts as a low pass filter (LPF).
- The Green's function solution (i.e., EM response of a line source) is exact but requires infinite number of eigenfunction (mode) summation [31]. This is a numerical challenge especially in the near vicinity of the line source, but is excellent for the analysis of truncation errors.
- Eigenfunctions (modes) are grouped into two: propagating modes with real eigenvalues and evanescent modes with imaginary eigenvalues. The number of propagating modes depends on the frequency and the width of the waveguide.
- A tilted directional antenna can also be located inside and can be modeled in terms of modes, but modal excitation coefficients are now complex. This is another numerical challenge, especially at high frequencies when the number of propagating modes is high.

- The modes (eigenfunctions) are global, therefore do not suffer from local problems, but extraction of modal excitation coefficients is crucial when generating reference solutions.
- Analytical exact solution can also be constructed in terms of rays which are local wave pieces; again infinite number of ray summation is required for the line source excitation. This may be achieved via either eigenray specification or ray shooting. The solution of eigenray equations is difficult especially for rays with high number of reflections from top and bottom boundaries [31]. Ray shooting is very time consuming and is highly sensitive to source/observer locations. Their computer implementations are both numerical challenges.
- Another ray-based analytical solution is the image method (IM). The IM overcomes numerical difficulties of both eigenray extraction and ray shooting approaches but very high number of images are required for accurate field computations.
- One-way, FE and/or fast Fourier transformation (FFT)-based PE models have been used in propagation modeling in waveguiding environments [13,34,35]. The PE models can handle tilted and directed excitation antennas easily; but one needs to use wide-angle PE models since narrow-angle models suffer paraxial propagation restriction (i.e., it is valid up to 10°–15° vertical propagation angles).
- The two powerful time-domain (FDTD and TLM) methods may also be used in modeling propagation inside a parallel plate waveguide. This is possible even for very long ranges by using sliding window approaches [36,37]. Both line source and short dipoles can be used in wave excitation in these models, but a special care is essential for the tilted and directional transmit antenna modeling.
- Finally, MoM can be used in propagation modeling inside a parallel plate waveguide. This also requires a special treatment because MoM has rarely been used in resonating structures (a novel approach called Mi-MoM is introduced in Ref. 38). MoM is based on the solution of an $N \times N$ system of equations which yields currents induced on the segments. The procedure is first applied and currents induced by the source are obtained. The

consecutive applications yield segment currents induced by other segment currents. This has to be continued until their contributions diminish.

- A hybrid, MoM plus image method (MoM-IM) increases both accuracy and computation time. In this case, $N \times N$ system of equations is solved once for segment currents induced by the source. The rest is handled via the IM.

In summary, a one-semester CEM lecture may be built on analytical and numerical methods mentioned earlier through a single structure, parallel plate waveguide.

16.2 PROBLEM POSTULATION AND ANALYTICAL SOLUTIONS: REVISITED

To make this chapter compact, the 2D parallel plate waveguide is mathematically revisited first. Details are given in Chapter 8. The 2D parallel plate waveguide is pictured in Fig. 16.1. Here, x and z are the transverse and longitudinal coordinates, respectively. The structure is infinite along y-direction. The height of the waveguide is a. The PEC boundaries are assumed for Dirichlet-type for the TE_z (transverse electric with respect to z) problem and Neumann-type for the TM_z (transverse magnetic with respect to z) problem (see Ref. 39 for TE/TM discussions).

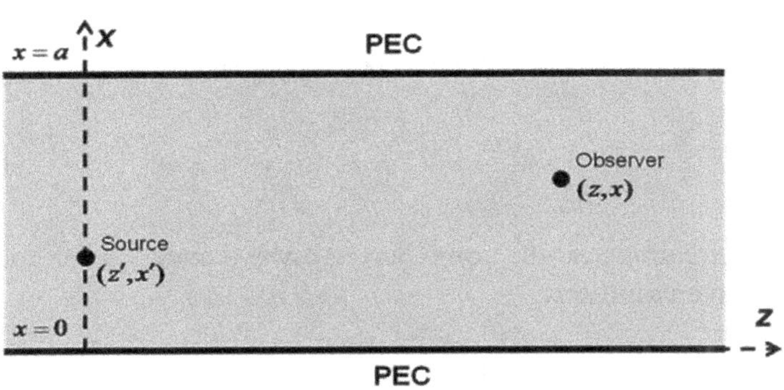

Figure 16.1. Parallel plate waveguide and source–observer locations.

Assuming that the medium between the plates is free space and isotropic, the Maxwell curl equations, under $\exp(j\omega t)$ time dependence, are given as

$$\nabla \times \mathrm{E} = -j\omega\mu_0 \mathrm{H} - \mathrm{M}, \tag{16.1a}$$

$$\nabla \times \mathrm{H} = j\omega\varepsilon_0 \mathrm{E} + \mathrm{J}, \tag{16.1b}$$

where boldface symbols denote vector quantities, the electric and magnetic field intensities (**E** and **H**) and the nonphased electric and magnetic source currents (**J** and **M**), ω is the angular frequency, and ε_0 and μ_0 are the free space permittivity and permeability. Consider there is no variation of field with respect to y ($\partial/\partial y \equiv 0$), the vector equations in (16.1) are reduced to given scalar equations for TE_z and TM_z, respectively as follows:

TE_z equations:

$$\frac{\partial E_y}{\partial z} = M_x + j\omega\mu_0 H_x, \tag{16.2a}$$

$$\frac{\partial H_x}{\partial z} - \frac{\partial H_z}{\partial x} = J_y + j\omega\varepsilon_0 E_y, \tag{16.2b}$$

$$\frac{\partial E_y}{\partial x} = -M_z - j\omega\mu_0 H_z \tag{16.2c}$$

TM_z equations:

$$\frac{\partial H_y}{\partial z} = -J_x - j\omega\varepsilon_0 E_x, \tag{16.3a}$$

$$\frac{\partial E_x}{\partial z} - \frac{\partial E_z}{\partial x} = -M_y - j\omega\mu_0 H_y, \tag{16.3b}$$

$$\frac{\partial H_y}{\partial x} = J_z + j\omega\varepsilon_0 E_z. \tag{16.3c}$$

These equations can be combined into wave equations as follows:

TE_z wave equation:

$$\left\{\frac{\partial^2}{\partial x^2} + \frac{\partial^2}{\partial z^2} + k_0^2\right\} E_y = j\omega\mu_0 J_y - \frac{\partial M_z}{\partial x} + \frac{\partial M_x}{\partial z}, \tag{16.4}$$

with BCs

$$E_y = 0 \quad \text{at } x = 0, x = a, \tag{16.5a}$$

$$E_y = 0 \quad \text{as } z \to \pm\infty. \tag{16.5b}$$

The remaining two TE_z field components can be derived from

$$H_x = \frac{1}{j\omega\mu_0}\left\{\frac{\partial E_y}{\partial z} - M_x\right\}, \tag{16.6a}$$

$$H_z = -\frac{1}{j\omega\mu_0}\left\{\frac{\partial E_y}{\partial x} + M_z\right\}. \tag{16.6b}$$

TM_z wave equation:

$$\left\{\frac{\partial^2}{\partial x^2} + \frac{\partial^2}{\partial z^2} + k_0^2\right\} H_y = j\omega\varepsilon_0 M_y + \frac{\partial J_z}{\partial x} - \frac{\partial J_x}{\partial z}, \tag{16.7}$$

with BCs

$$\frac{\partial H_y}{\partial x} = 0 \quad \text{at } x = 0, x = a, \tag{16.8a}$$

$$H_y = 0 \quad \text{as } z \to \pm\infty. \tag{16.8b}$$

The remaining two TM_z field components can be derived from

$$E_x = -\frac{1}{j\omega\varepsilon_0}\left\{\frac{\partial H_y}{\partial z} + J_x\right\}, \tag{16.9a}$$

$$E_z = \frac{1}{j\omega\varepsilon_0}\left\{\frac{\partial H_y}{\partial x} - J_z\right\}. \tag{16.9b}$$

Here, $k_0 = 2\pi/\lambda = \omega\sqrt{\varepsilon_0\mu_0}$ is the free-space wave number and λ is the free-space wavelength. Since the TE_z and TM_z sets are decoupled, each can be excited independently of the other by appropriate selection of the sources, **J** and **M**. The line sources M_x, M_z, J_y excite the TE_z set, whereas the line sources M_y, J_x, J_z excite the TM_z set. Further simplification can be obtained by setting the source components $M_x = 0$, $M_z = 0$ for the TE_z set, and $J_x = 0$, $J_z = 0$ for the TM_z set. These simplifications will be assumed in what follows.

16.2.1 Green's Function in Terms of Mode Summation

The Green's function problem associated with both the TE_z set (when $M_x = M_z = 0$) and the TM_z set (when $J_x = J_z = 0$) is defined by the equation

$$\left\{\frac{\partial^2}{\partial x^2}+\frac{\partial^2}{\partial z^2}+k_0^2\right\}g(x,z;x',z')=\delta(x-x')\delta(z-z'), \quad (16.10)$$

with BCs

$$g(x,z;x',z')=0 \quad \text{at } x=0,a \quad (\mathrm{TE}_z), \quad (16.11\text{a})$$

$$\frac{\partial}{\partial x}g(x,z;x',z')=0 \quad \text{at } x=0,a \quad (\mathrm{TM}_z), \quad (16.11\text{b})$$

$$g(x,z;x',z')=0 \quad \text{as } z\to\pm\infty. \quad (16.11\text{c})$$

Here, (x', z') and (x, z) specify source and observation points, respectively, and $\delta(x)$ and $\delta(z)$ are the Dirac delta functions. The Green's function $g(x, z; x', z')$ can be obtained as

$$g(x,z;x',z')=\tilde{g}(z;z')+\frac{2}{a}\sum_{m=1}^{\infty}\frac{e^{-jk_{zm}|z-z'|}}{2jk_{zm}}B(k_{xm}x)B(k_{xm}x'), \quad (16.12)$$

$$\tilde{g}(z;z')=0, \quad B(x)=\sin(x) \quad (\mathrm{TE}_z), \quad (16.13\text{a})$$

$$\tilde{g}(z;z')=\frac{1}{a}\frac{e^{-jk_0|z-z'|}}{2jk_0}, \quad B(x)=\cos(x) \quad (\mathrm{TM}_z), \quad (16.13\text{b})$$

where $k_{xm} = m\pi/a$, $k_{zm} = \sqrt{k_0^2 - k_{xm}^2}$. The line-source-excited fields are then given by either $E_y = j\omega\mu_0 g$ or $H_y = j\omega\varepsilon_0 g$ for TE_z and TM_z cases, respectively.

A short MATLAB code is prepared for the calculation of field distribution inside the parallel plate waveguide in terms of mode summation for both polarizations. Table 16.1 lists this code. An example is shown in Fig. 16.2. Here, field versus range inside a 1-m-wide plate at 0.7 m is pictured. Source is at 0.3 m height. The number of propagating modes for the sets of parameters listed in the figure is 15. The two curves belong to 15 mode and 100 mode summations. As observed, at a distance beyond 0.5 m (i.e., after 3–4(λ) distance), only propagating

Table 16.1.
A MATLAB Mode Sum Code for Field versus Range/Height Calculations

```
function output=modesummation(E0,fr,a,zmax,xs,zo,xo,NX,NZ
,mm,pol)
% -----------------------------------------------------------
%% Mode Summation inside a 2D-PEC parallel plate
waveguide
%   for the calculation of
%       Field vs Range at a given height (xo)
%       Field vs Height at a given range (zo)
% E0: Field amplitude
% fr: frequency [MHz], a: plate height [m], zmax:
maximum range [m]
% xs: source height [m], (zo,xo): observer range [m],
height [m]
% NX: # of height points, NZ: number of range points,
mm: mode number
% pol: polarization type, 0 for TEz case, 1 for TMz
case
% output.zz:     range points
% output.xx:     height points
% output.fx0:    vertical field at z=0
% output.fx:     vertical field at z=zo
% output.fz:     horizontal field at x=xo
% -----------------------------------------------------------
k0=2*pi*fr/300; % wavenumber in free space
kz=conj(sqrt(k0^2-((1:mm)*pi/a).^2));
xx=(0:a/(NX-1):a).';
zz=(0:zmax/(NZ-1):zmax).';
zz=0.5*(zz(1:end-1)+zz(2:end));

if pol==0 % TEz case
     funTE=@(MM,XX,KZ,ZZ)2/a*sum(sind(180*MM*xs/a) . . .
          .*sind(180*MM.*XX/a).*exp(-1i*KZ.*ZZ)./
(2*1i*KZ));
     % Field vs Height
     [~,MM]=meshgrid(xx,1:mm); [XX,KZ]=meshgrid(xx,kz);
     Fv0=funTE(MM,XX,KZ,0);          % at z=0
     Fv1=funTE(MM,XX,KZ,zo);         % at z=zo
```

(Continued)

Table 16.1.
(Continued)

```
     % Field vs Range
     [~,MM]=meshgrid(zz,1:mm); [ZZ,KZ]=meshgrid(zz,kz);
     Fh1=funTE(MM,xo,KZ,ZZ).';    % at x=xo

     Fv0=E0*Fv0;       Fv1=E0*Fv1;       Fh1=E0*Fh1;
elseif pol==1 % TMz case
     funTM=@(MM,XX,KZ,ZZ) 2/a*sum(cosd(180*MM*xs/a)
. . .
          .*cosd(180*MM.*XX/a).*exp(-1i*KZ.*ZZ)./
(2*1i*KZ));
     % Field vs Height
     [~,MM]=meshgrid(xx,1:mm); [XX,KZ]=meshgrid(xx,kz);
     Fv0=funTM(MM,XX,KZ,0)+1/a*exp(-1i*k0*0)/(2*1i*k0);
% at z=0
     Fv1=funTM(MM,XX,KZ,zo)+1/a*exp(-1i*k0*zo)/(2*1i*k0);
% at z=zo
     % Field vs Range
     [~,MM]=meshgrid(zz,1:mm); [ZZ,KZ]=meshgrid(zz,kz);
     Fh1=funTM(MM,xo,KZ,ZZ).'+1/a*exp(-1i*k0*zz)/
(2*1i*k0); %  at x=xo
     Fv0=E0/(120*pi)*Fv0;     Fv1=E0/(120*pi)*Fv1;
Fh1=E0/(120*pi)*Fh1;
end
% Results
output.xx=xx;    output.zz=zz;
output.Fv0=Fv0.'; output.Fv1=Fv1.'; output.Fh1=Fh1.';
% ----------------------- END -----------------------
```

modes contribute. Figure 16.3 displays field versus height at two different distances inside the same plate. As observed, the contribution of only propagating modes at two wavelength distance is not enough to build the correct field.

16.2.2 Mode Summation for a Tilted/Directive Antenna

A directive antenna is used in many propagation applications. This is modeled by using a vertical, tilted Gaussian function in analytical and

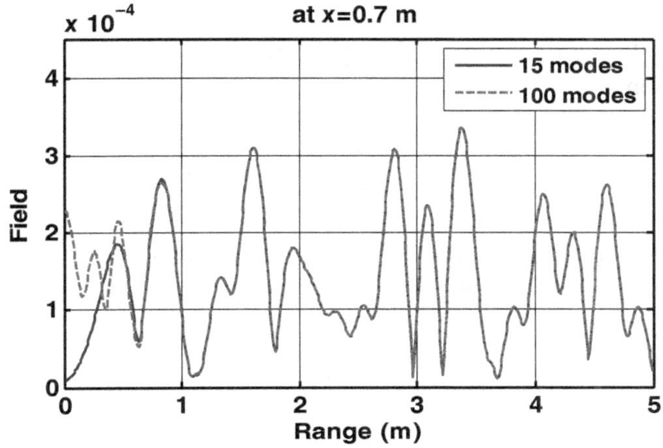

Figure 16.2. Magnetic field versus range with Green's function: TM_z case, $a = 1$ m, $z' = 0$, $x' = 0.3$ m, $x = 0.7$ m, $k_0a = 50$. (Solid) Only propagating modes (15 modes) and (dashed) 100 modes.

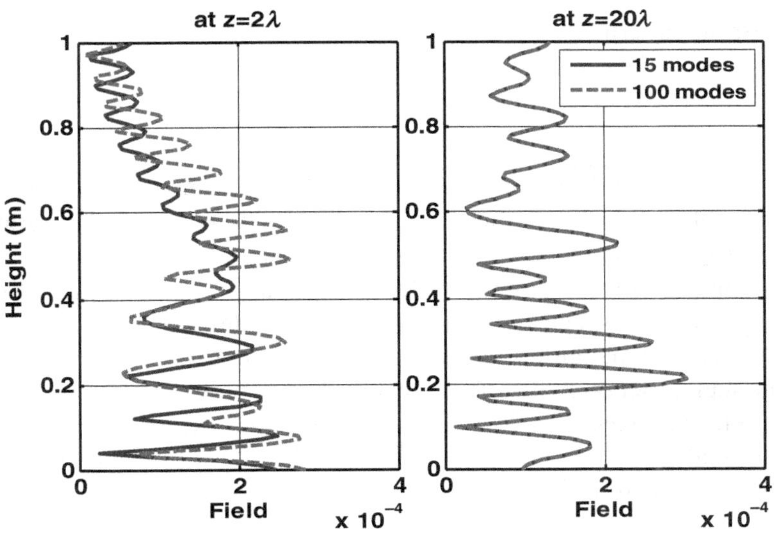

Figure 16.3. Magnetic field versus height with Green's function: TM_z case, $a = 1$ m, $z' = 0$, $x' = 0.3$ m, $k_0a = 50$, (a) at $z = 2\lambda$ and (b) at $z = 20\lambda$. (Solid) Only propagating modes (15 modes) and (dashed) 100 modes.

numerical simulations. This tilted Gaussian source inside a PEC parallel plate waveguide at $z = z'$ may be represented in terms of modal summation as

$$f(x, z') = \sum_{m=m_0}^{M} c_m(z') v_m B(k_{xm}x), \tag{16.14}$$

where M is the highest mode that should be included for the specified excitation, v_m is the normalization constant

$$v_m = \left(\int_{x=0}^{a} B^2(k_{xm}x)\,dx \right)^{-1/2}, \tag{16.15}$$

and $c_m(z')$ is the modal excitation coefficient, numerically derived from the vertical orthonormality condition as

$$c_m = v_m \int_0^a f(x, z) B(k_{xm}x)\,dx. \tag{16.16}$$

The initial field profile $f(x, z')$ at $z = z'$ is generated from a tilted Gaussian pattern,

$$f(x, 0) = \exp\left[-jk_0 x \sin\theta_{\text{elv}} - \frac{(x - x')^2}{w^2} \right], \tag{16.17}$$

where $w = \sqrt{2\ln 2}/(k_0 \sin(\theta_{\text{bw}} / 2))$. The tilted antenna pattern is specified by its vertical position (x'), beamwidth (θ_{bw}), and tilt (elevation) angle (θ_{elv}). Note that B again shows either sine or cosine function starting from either $m_0 = 1$ or $m_0 = 0$ for horizontal (TE_z) and vertical (TM_z) polarizations, respectively. The number of modes would be finite for numerical computation. It is common to choose a vertically extending Gaussian function with arbitrary location having vertical elevation angle in the range of $\pm 90°$. Note that the modal excitation coefficient c_m is real for a real source function without any tilt and becomes complex if there is a tilted source.

16.2.3 Eigenray Representation

Modes are global wave pieces and can be used to calculate fields at any point. They do not contain signatures of the local paths. On the other hand, rays which are local wave objects trace all the paths between source and observer. Fields inside a parallel plate waveguide can also be constructed in terms of rays [31,32]. The integral representation of the eigenfunctions can be used for this purpose.

The integral representation for eigenfunctions can be restructured as

$$g(x,z;x',z')=\frac{1}{2\pi}\int_{-\infty}^{+\infty}\frac{\sin(k_x x_<)\sin[k_x(a-x_>)]}{k_x\sin(k_x a)}e^{-jk_z(z-z')}dk_z \qquad (16.18)$$

in a manner that expresses the x-domain Green's function behavior in terms of multiply reflected traveling plane waves between the boundaries at $x=0$ and $x=a$, rather than the sinusoidal standing waves. Here, $x_>(x_<)$ represent max (min) of (x', x). This is achieved by expanding the resonant denominator in the integrand (16.18) into an infinite power series,

$$(1-\Omega)^{-1}=\sum_{n=0}^{\infty}\Omega^n, \Omega=\exp(-j2k_x a), \qquad (16.19)$$

obtained by writing $\sin(k_x a) = [(0.5j)\,(1-\Omega)\exp(jk_x a)]$. This series can be ordered into four (i)-indexed infinite periodic "image arrays" in the x-domain. Interchanging the orders of summation and integration, each of the integrals represents an exact 2D "image" line source Green's function $\hat{g}_n^{(i)}$ in *free space*, which can be identified asymptotically with a ray-optical field reaching the observer along a direct or multiply reflected ray path, and is frequently referred to as an "eigenray" from a specified source point to a specified observation point. Following this procedure, one obtains the Green's function fields as

$$g(x,z;x',z')=\sum_{i=1}^{4}\sum_{n=0}^{\infty}\hat{g}_n^{(i)}(x,z;x_n^{(i)},z'), \qquad (16.20)$$

where $\hat{g}_n^{(i)}(x,z;x_n^{(i)},z')$ is the image line source Green's function in free space. Note that the contribution of $\hat{g}_n^{(2)}$ and $\hat{g}_n^{(3)}$ must be negative for TE_z case. In the high frequency range, (16.20) can be approximated as

$$g(x,z;x',z')\approx\frac{j}{4\pi}\sum_{i=1}^{4}\sum_{n=0}^{\infty}\left[\frac{2j\pi}{k_0 q_n''^{(i)}(W_n^{(i)})}\right]^{1/2}e^{-jk_0 q_n^{(i)}(W_n^{(i)})}, \qquad (16.21)$$

where a double prime in the denominator denotes the second derivative with respect to the argument. The four different ray species are defined by the following phase accumulations [31,32]:

$$q_n^{(1)} = (x_> - x_<)\cos W + z\sin W + 2an\cos W, \tag{16.22a}$$

$$q_n^{(2)} = (x_> + x_<)\cos W + z\sin W + 2an\cos W, \tag{16.22b}$$

$$q_n^{(3)} = -(x_> + x_<)\cos W + z\sin W + 2a(n+1)\cos W, \tag{16.22c}$$

$$q_n^{(4)} = -(x_> - x_<)\cos W + z\sin W + 2a(n+1)\cos W. \tag{16.22d}$$

Here, the *angular* spectrum variable W denotes the angle between the ray and the normal to the boundaries, whereas n is the number of reflections from both boundaries. The eigenrays emanating from the source and reaching the observer are determined from the eigenray (stationary phase) condition [31,32]

$$\frac{d}{dW} q_n^{(i)}(W) = 0 \Rightarrow W = W_n^{(i)}, \quad i = 1, 2, 3, 4. \tag{16.23}$$

The four eigenray species and their trajectories are organized according to their departure directions (up and down) at the source x_s and their arrival directions (up and down) at the observer x_o.

- $i = 1$: n reflections at both boundaries (down/down for $x_> = x'$, $x_< = x$ and up/up for $x_> = x$, $x_< = x'$).
- $i = 2$: n reflections at upper boundary and $n + 1$ reflections at lower boundary (down, up).
- $i = 3$: $n + 1$ reflections at upper boundary and n reflections at lower boundary (up, down).
- $i = 4$: $n + 1$ reflections at both boundaries (up/up for $x_> = x'$, $x_< = x$ and down/down for $x_> = x$, $x_< = x'$).

Table 16.2 lists a short MATLAB script for the calculation of the field inside the parallel plate waveguide with the eigenray summation model for the line source. Eigenrays for $n = 0$ (i.e., the first four eigenrays) and $n = 1$ (i.e., the first eight eigenrays) are plotted in Fig. 16.4.

Table 16.2.
A MATLAB Eigenray Code for the Calculation of Field versus Range/Height

```
function output=eigenraysum(E0,fr,a,zmax,xs,zo,xo,NX,NZ,n
ray,pol)
% -----------------------------------------------------
% EigenRay Summation inside a 2D-PEC parallel plate
waveguide
%    for the calculation of
%    Field vs Range at a given height (xo)
%    Field vs Height at a given range (zo)
% E0: Field amplitude, nray: number of reflections
% fr: frequency [MHz], a: plate height [m], zmax: max
range [m]
% xs: source height [m], (zo, zo): observer range,
height [m]
% NX: number of height points, NZ: number of range
points
% nray: number of rays
% pol: polarization type, 0 for TEz case, 1 for TMz case
% output.zz:     range points
% output.xx:     height points
% output.fx0:    vertical field at z=0
% output.fx:     vertical field at z=zo
% output.fz:     horizontal field at x=xo
% -----------------------------------------------------
k0=2*pi*fr/300; % wavenumber in free space
xx=(0:a/(NX-1):a).';
zz=(0:zmax/(NZ-1):zmax).';
zz=0.5*(zz(1:end-1)+zz(2:end));
Fv0=zeros(NX,1); Fv1=zeros(NX,1);
Fh1=zeros(length(zz),1);
for k=1:NX     % Field vs Height at z=0
     g1=0; g2=0; g3=0; g4=0;
     for n=0:nray
        qq1= abs(xx(k)-xs)+2*a*n; qq2=abs(xx(k)+xs)+2*a*n;
        qq3=-abs(xx(k)+xs)+2*a*(n+1);
qq4=-abs(xx(k)-xs)+2*a*(n+1);
        % Calculate four rays
        g1=g1+1i/(4*pi)*sqrt((2*pi*1i)/
(k0*-qq1))*exp(-1i*k0*qq1);
```

(Continued)

Table 16.2.
(Continued)

```
        g2=g2+1i/(4*pi)*sqrt((2*pi*1i)/
(k0*-qq2))*exp(-1i*k0*qq2);
        g3=g3+1i/(4*pi)*sqrt((2*pi*1i)/
(k0*-qq3))*exp(-1i*k0*qq3);
        g4=g4+1i/(4*pi)*sqrt((2*pi*1i)/
(k0*-qq4))*exp(-1i*k0*qq4);
      end
      if pol==0,
          Fv0(k)=Fv0(k)+g1-g2-g3+g4;
      elseif pol==1
          Fv0(k)=Fv0(k)+g1+g2+g3+g4;
      end
end
for k=1:NX    % Field vs Height z=zo
      g1=0; g2=0; g3=0; g4=0;
      for n=0:nray
        qq1=@(t)
abs(xx(k)-xs)*cos(t)+zo*sin(t)+2*a*n*cos(t);
        qq2=@(t)  abs(xx(k)+xs)*cos(t)+zo*sin(t)+2*a*n*c
os(t);
        qq3=@(t) -abs(xx(k)+xs)*cos(t)+zo*sin(t)+2*a*(n+
1)*cos(t);
        qq4=@(t) -abs(xx(k)-xs)*cos(t)+zo*sin(t)+2*a*(n+1
)*cos(t);
        % Calculate eigenray angles
        wn1=atan(zo/(abs(xx(k)-xs)+2*a*n));
        wn2=atan(zo/(abs(xx(k)+xs)+2*a*n));
        wn3=atan(zo/(-abs(xx(k)+xs)+2*a*(n+1)));
        wn4=atan(zo/(-abs(xx(k)-xs)+2*a*(n+1)));
        % Calculate four rays
        g1=g1+1i/(4*pi)*sqrt((2*pi*1i)/
(k0*(-qq1(wn1)))). . .
            *exp(-1i*k0*qq1(wn1));
        g2=g2+1i/(4*pi)*sqrt((2*pi*1i)/
(k0*(-qq2(wn2)))). . .
            *exp(-1i*k0*q2(wn2));
        g3=g3+1i/(4*pi)*sqrt((2*pi*1i)/
(k0*(-qq3(wn3)))). . .
```

Table 16.2.
(Continued)

```
            *exp(-1i*k0*qq3(wn3));
        g4=g4+1i/(4*pi)*sqrt((2*pi*1i)/
(k0*(-qq4(wn4)))). . .
            *exp(-1i*k0*qq4(wn4));
    end
    if pol==0,
        Fv1(k)=Fv1(k)+g1-g2-g3+g4;
    elseif pol==1
        Fv1(k)=Fv1(k)+g1+g2+g3+g4;
    end
end
for k=1:length(zz)  % Field vs Range at x=xo
    g1=0; g2=0; g3=0; g4=0;
    for n=0:nray
        qq1=@(t)  abs(xo-xs)*cos(t)+zz(k)*sin(t)+2*a*n*co
s(t);
        qq2=@(t)  abs(xo+xs)*cos(t)+zz(k)*sin(t)+2*a*n*c
os(t);
        qq3=@(t) -abs(xo+xs)*cos(t)+zz(k)*sin(t)+2*a*(n+
1)*cos(t);
        qq4=@(t) -abs(xo-xs)*cos(t)+zz(k)*sin(t)+2*a*(n+1
)*cos(t);
        % Calculate eigenray angles
        wn1=atan(zz(k)/(abs(xo-xs)+2*a*n));
        wn2=atan(zz(k)/(abs(xo+xs)+2*a*n));
        wn3=atan(zz(k)/(-abs(xo+xs)+2*a*(n+1)));
        wn4=atan(zz(k)/(-abs(xo-xs)+2*a*(n+1)));
        % Calculate four rays
        g1=g1+1i/(4*pi)*sqrt((2*pi*1i)/
(k0*(-qq1(wn1)))). . .
            *exp(-1i*k0*qq1(wn1));
        g2=g2+1i/(4*pi)*sqrt((2*pi*1i)/
(k0*(-qq2(wn2)))). . .
            *exp(-1i*k0*qq2(wn2));
        g3=g3+1i/(4*pi)*sqrt((2*pi*1i)/
(k0*(-qq3(wn3)))). . .
            *exp(-1i*k0*qq3(wn3));
        g4=g4+1i/(4*pi)*sqrt((2*pi*1i)/
(k0*(-qq4(wn4)))). . .
```

(Continued)

Table 16.2.
(Continued)

```
          *exp(-1i*k0*qq4(wn4));
     end
     if pol==0,
          Fh1(k)=Fh1(k)+g1-g2-g3+g4;
     elseif pol==1
          Fh1(k)=Fh1(k)+g1+g2+g3+g4;
     end
end
% Results
output.xx=xx;    output.zz=zz;
if pol==0,
     output.Fv0=E0*Fv0; output.Fv1=E0*Fv1; output.
Fh1=E0*Fh1;
elseif pol==1,
     output.Fv0=E0/(120*pi)*Fv0; output.Fv1=E0/
(120*pi)*Fv1; output.Fh1=E0/(120*pi)*Fh1;
end
% ------------------------ END ----------------------
```

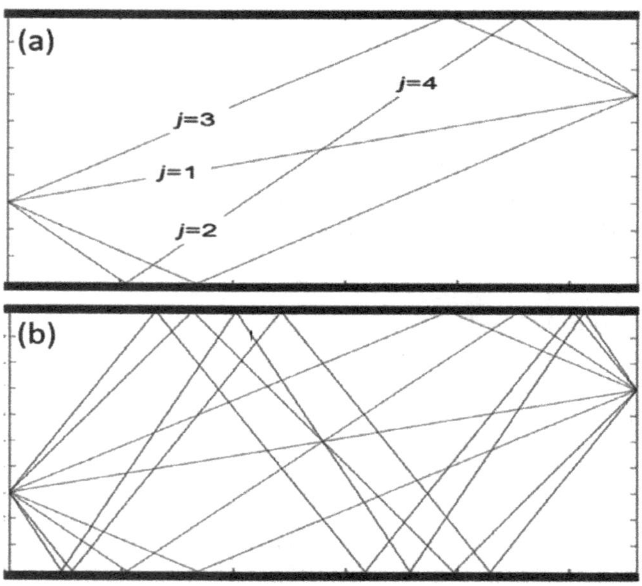

Figure 16.4. Eigenrays with (a) $n = 0$ and (b) $n = 1$ inside a 1-m-wide 300-m-long parallel plate waveguide; source height = 0.3 m, receiver height = 0.7 m.

16.2.4 Hybrid Ray + Image Method

The most difficult task of the eigenray representation is the numerical solution of the stationary phase equation (16.23). This becomes a real challenge as the number of top/bottom boundary reflections increases. Another approach in ray model which smoothen numerical computations is to use the image method (IM).

The IM can be used to satisfy infinite number of reflections with infinite number of images between two plates. The images are placed with respect to the position of the plates. The unit field (E_y for TE_z case; H_y for TM_z case) can be created by an incident point source at (z', x') as

$$F_i = \frac{e^{-jk_0 r}}{\sqrt{k_0 r}}, \tag{16.24}$$

where $r = \sqrt{(x-x')^2 + (z-z')^2}$. First image sources appear at $(z', -x')$ with respect to $x = 0$ plate and at $(z', 2a - x')$ with respect to $x = a$ plate.

Table 16.3 lists a short MATLAB script for the calculation of the field inside the parallel plate waveguide in terms of IM.

A triple comparison among mode summation, eigenray summation, and IM for the line source excitation is done and results are given in Fig. 16.5. The figure shows field versus range for specified source and observer heights. Here, 100 modes are used in mode summation, eigenray summation is carried out until 50 reflections from both boundaries, and 20 images are used in the IM. As observed, the agreement among the results is very impressive.

16.3 NUMERICAL MODELS

It was shown in the previous section that analytical reference data can be generated using any of the three models: mode/eigenray summations or IM. Three well-known numerical models are discussed in this section.

16.3.1 Split Step Parabolic Equation Model

The PE wave propagator for a parallel plate waveguide filled with air uses the PE

Table 16.3.
A MATLAB Ray + Image Code for Field versus Range/Height Calculations

```
function output=imageraysummation(E0,fr,a,zmax,xs,zo,xo,N
X,NZ,pol)
% ---------------------------------------------------------
% Ray Sum + Image method inside a 2D-PEC parallel plate
waveguide
%    for the calculation of
%    Field vs Range at a given height (xo)
%    Field vs Height at a given range (zo)
% E0: Field amplitude
% fr: frequency [MHz], a: plate height [m], zmax:
maximum range [m]
% xs: source height [m], (zo,zo): observer range, height
 m]
% NX: number of height points, NZ: number of range
points
% pol: polarization type, 0 for TEz case, 1 for TMz
case
% output.zz:     range points
% output.xx:     height points
% output.fx0:    vertical field at z=0
% output.fx:     vertical field at z=zo
% output.fz:     horizontal field at x=xo
% ---------------------------------------------------------
k0=2*pi*fr/300; % wavenumber in free space
xx=(0:a/(NX-1):a).';
zz=(0:zmax/(NZ-1):zmax).';
zz=0.5*(zz(1:end-1)+zz(2:end));
Fv0=zeros(NX,1); Fv1=zeros(NX,1);
Fh1=zeros(length(zz),1);
% Field vs Height at z=0 and z=zo
for ii=0:zo:zo
     Fvfree=rayfld(k0,ii,xx,xs);
     Fvp1=rayfld(k0,ii,xx,-xs);   Fvp2=rayfld(k0,ii,xx,2*a
-xs);
     if pol==0,
          Fvtot=Fvfree-Fvp1-Fvp2;
     elseif pol==1,
```

Table 16.3.
(Continued)

```
          Fvtot=Fvfree+Fvp1+Fvp2;
     end
     ind=1; ep=sum(abs(Fvp1+Fvp2))/1e3;
     while sum(abs(Fvp1+Fvp2))>ep, ind=ind+1;
         Fvp1=rayfld(k0,ii,xx,(-1)^ind*xs+2*ceil(ind/2)*a);
         Fvp2=rayfld(k0,ii,xx,(-1)^ind*xs-2*floor(ind/2)*a);
         if pol==0,
             Fvtot=Fvtot+(-1)^ind*(Fvp1+Fvp2);
         elseif pol==1,
             Fvtot=Fvtot+Fvp1+Fvp2;
         end
         if ind==1000, display('Max iter.(1000)!');Fvp1=0;
Fvp2=0;end
     end
     if pol==0, Fv=E0*Fvtot; elseif pol==1, Fv=E0/
(120*pi)*Fvtot;  end
     if ii==0, Fv0=Fv; else Fv1=Fv; end
end
% Field vs Range at x=xo
Fhfree=rayfld(k0,zz,xo,xs);
Fhp1=rayfld(k0,zz,xo,-xs);   Fhp2=rayfld(k0,zz,xo,2*a-xs);
if pol==0,
     Fhtot=Fhfree-Fhp1-Fhp2;
elseif pol==1,
     Fhtot=Fhfree+Fhp1+Fhp2;
end
ind=1; ep=sum(abs(Fhp1+Fhp2))/1e3;
while sum(abs(Fhp1+Fhp2))>ep, ind=ind+1;
     Fhp1=rayfld(k0,zz,xo,(-1)^ind*xs+2*ceil(ind/2)*a);
     Fhp2=rayfld(k0,zz,xo,(-1)^ind*xs-2*floor(ind/2)*a);
     if pol==0,
        Fhtot=Fhtot+(-1)^ind*(Fhp1+Fhp2);
     elseif pol==1,
        Fhtot=Fhtot+Fhp1+Fhp2;
     end
     if ind==1000, display('Max iter.(1000)
reached!');Fhp1=0;Fhp2=0;end
end
```

(Continued)

Table 16.3.
(Continued)

```
if pol==0, Fh1=E0*Fhtot; elseif pol==1, Fh1=E0/
(120*pi)*Fhtot;   end
% Results
output.xx=xx;    output.zz=zz;
output.Fv0=Fv0; output.Fv1=Fv1; output.Fh1=Fh1;
function  F=rayfld(k0,zo,xo,xs)
R=sqrt(zo.^2+(xo-xs).^2);
F=exp(-1i*k0*R)./sqrt(k0*R);
% ---------------------- END ----------------------
```

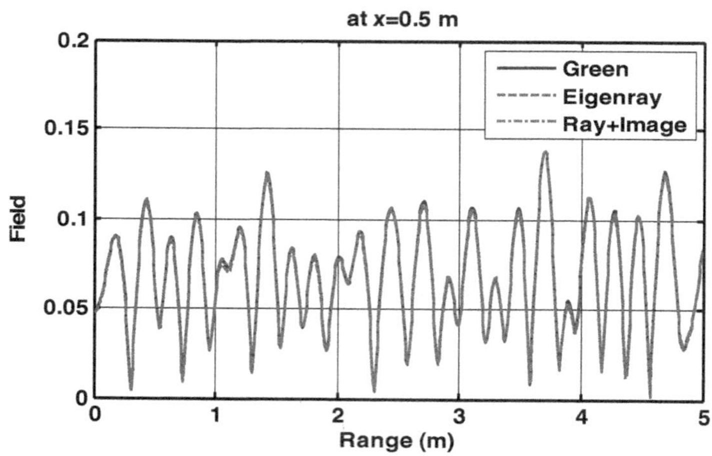

Figure 16.5. E-field versus range with (solid) Green's function (100 modes), (dashed) eigenray representation ($n = 50$), and (dashed dotted) ray + image method (20 images): TE_z case, $a = 1$ m, $z' = 0$, $x' = 0.3$ m, $x = 0.5$ m, $k_0 a = 50$.

$$\left(\frac{\partial^2}{\partial x^2} + \frac{\partial^2}{\partial z^2} - 2jk_0\frac{\partial}{\partial z}\right)u(z,x) = 0 \tag{16.25}$$

under $\exp(j\omega t)$ time dependence and represents one-way, forward propagation under the paraxial approximation (i.e., near-axial propagation). Since the direction of wave propagation is predominantly along z-axis, the reduced function $u(z, x)$ is used by separating the rapidly varying phase term from $\psi(z, x) = \exp(-jk_0z)u(z, x)$, which corresponds to either electric or magnetic fields for horizontal (TE_z) and vertical (TM_z) polarizations, respectively.

The SSPE uses longitudinally marching procedure with the use of discrete Fourier transform (DFT). First, an antenna pattern is injected as the initial vertical field profile. Then, this initial field is propagated longitudinally and the transverse field profile at the next range is obtained from

$$u(z+\Delta z, x) = F^{-1}\{C(k_x)F[u(z, x)]\}, \tag{16.26}$$

where $C(k) = \exp\left(jk^2\Delta z \Big/ \left(k_0 + \sqrt{k_0^2 - \hat{c}k^2}\right)\right)$. Here, F refers to the discrete sine/cosine transform (DST/DCT) while considering (TE_z/TM_z) BCs and k_x is the spectral variable. The parameter $\hat{c}$ is 0 (1) for the narrow (wide) angle model.

The new profile $u(z + \Delta z, x)$ is then used in (16.26) as the initial profile and the procedure is repeated; the profile $u(z + 2\Delta z, x)$at the next range is obtained. The procedure is repeated continuously until the propagator reaches the desired range. Table 16.4 lists a short MATLAB script for the calculation of SSPE fields. An example produced with this code is given in Fig. 16.6.

Field versus range/height simulated via narrow- and wide-angle SSPE models are shown in Fig. 16.6a,b, respectively. Downward tilted Gaussian beam hits the bottom plate first and then bounced back and hits the upper plate. As observed, there is a significant difference between these two plots. In order to see which is accurate, the same field versus range/height plot is produced and reference data are generated via the mode summation model. This is shown in Fig. 16.6c. As observed, the agreement between wide-angle SSPE (Fig. 16.6b) and the reference data (Fig. 16.6c) is impressive. Two vertical cuts are taken from these plots and are shown separately in Fig. 16.7 as field versus height at a constant range. These plots show that the agreement between analytical reference solution and wide-angle SSPE, and disagreement with narrow-angle SSPE is better.

16.3.2 Finite-Difference Time-Domain Model

The 2D parallel plate waveguide can also be modeled with the FDTD method (see Refs. 26–28 for details of FDTD modeling). A sample

Table 16.4.
A MATLAB SSPE Code for Field versus Range/Height Calculations

```
function output=sspeanalysis(fr,a,zmax,zo,xo,NX,NZ,fieldx1
d,pol)
% -----------------------------------------------------
% SSPE analysis inside a 2D-PEC parallel plate waveguide
% for the calculation of
%   Field vs Range at a given height (xo)
%   Field vs Height at a given range (zo)
% fieldx1d: Initial Field at z=0
% fr: frequency [MHz], a: plate height [m], zmax:
maximum range [m]
% zo: observer range [m], xo: observer height [m]
% NX: number of height points, NZ: number of range
points,
% pol: polarization type, 0 for TEz case, 1 for TMz
case
%
% output.zz:     range points
% output.xx:     height points
% output.fx0:    vertical field at z=0
% output.fx:     vertical field at z=zo
% output.fz:     horizontal field at x=xo
% -----------------------------------------------------
k0=2*pi*fr/300; % wavenumber in free space
dx=a/(NX-1);  dz=zmax/(NZ-1);
xx=(0:dx:a).';  zz=(0:dz:zmax).';
FF=zeros(NX,NZ);
FF(:,1)=fieldx1d;    % initial field profile
%% start SSPE Analysis
KX=linspace(0,pi/dx,NX).';
CC1=exp(1i*dz*KX.^2./(conj(sqrt(k0^2-KX.^2))+k0));% WA
Propagator
ind=1; Nind=zmax/dz;
while(ind<Nind+1), ind=ind+1;
  if pol==0 % TEz case
    Field=CC1(2:end-1).*dst(fieldx1d(2:end-1));
    Field2=zeros(size(CC1));
Field2(2:end-1)=idst(Field);
```

Table 16.4.
(Continued)

```
      fieldx1d=Field2;
   elseif pol==1 % TMz case
      Field=CC1.*dct(fieldx1d); fieldx1d=idct(Field);
   end
   FF(:,ind)=fieldx1d*exp(-1i*k0*(ind-1)*dz);
end
%% Results
output.xx=xx;     output.zz=zz;
output.Fv0=FF(:,1);
output.Fv1=FF(:,round(zo/dz)+1);
output.Fh1=FF(round(xo/dx)+1,:);
end% ---------------------- END -----------------------
```

virtual tool presented in Ref. 40 can be used for this purpose. The time-domain versions of 2D Maxwell equations in (16.2) and (16.3) are given as follows:

TE$_z$ case:

$$\frac{\partial E_y}{\partial t} = \frac{1}{\varepsilon_0}\left(\frac{\partial H_x}{\partial z} - \frac{\partial H_z}{\partial x}\right), \tag{16.27a}$$

$$\frac{\partial H_z}{\partial t} = -\frac{1}{\mu_0}\frac{\partial E_y}{\partial x}, \tag{16.27b}$$

$$\frac{\partial H_x}{\partial t} = \frac{1}{\mu_0}\frac{\partial E_y}{\partial z}. \tag{16.27c}$$

TM$_z$ case:

$$\frac{\partial H_y}{\partial t} = -\frac{1}{\mu_0}\left(\frac{\partial E_x}{\partial z} - \frac{\partial E_z}{\partial x}\right), \tag{16.28a}$$

$$\frac{\partial E_z}{\partial t} = \frac{1}{\varepsilon_0}\frac{\partial H_y}{\partial x}, \tag{16.28b}$$

$$\frac{\partial E_x}{\partial t} = -\frac{1}{\varepsilon_0}\frac{\partial H_y}{\partial z}. \tag{16.28c}$$

The FDTD method is based on the discretization of these equations directly in the time domain where the physical geometry is divided into

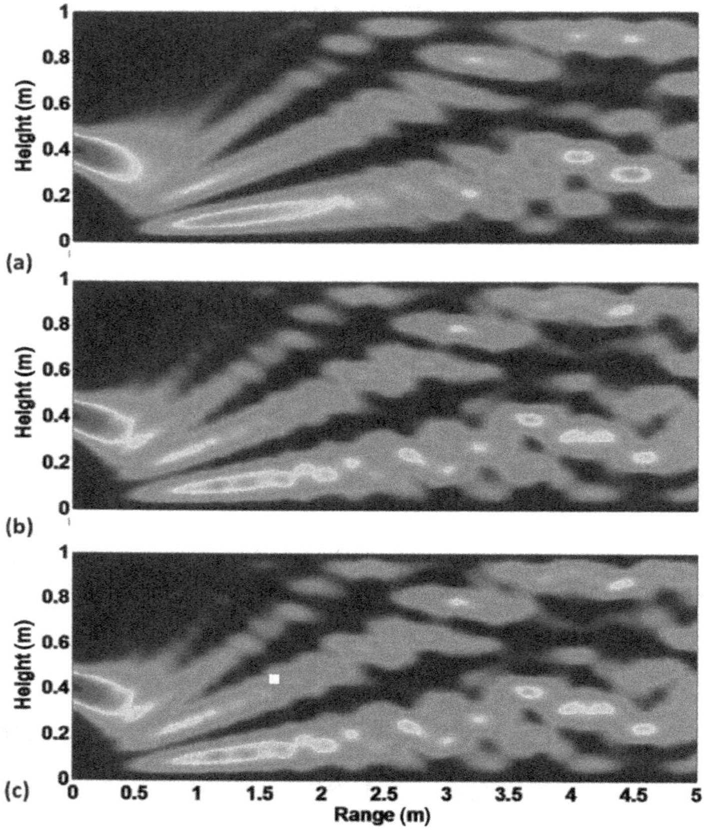

Figure 16.6. E-field versus range/height with a tilted Gaussian beam. (a) Narrow SSPE, (b) wide SSPE, and (c) Green's function (31 modes): TE_z case, $a = 1$ m, $z' = 0$, $x' = 0.4$ m, $k_0 a = 50$, $\theta_{bw} = 30°$, $\theta_{elv} = -10°$.

small cells. Both time and spatial partial derivatives are handled with the finite central difference approximation and the solution is obtained with a marching scheme in iterative form as follows:

TE_z case:

$$\begin{aligned}(E_y)^n_{i,k} = (E_y)^{n-1}_{i,k} &+ \frac{\Delta t}{\varepsilon_0 \Delta z}\left[(H_x)^{n-1/2}_{i,k+1/2} - (H_x)^{n-1/2}_{i,k-1/2}\right] \\ &- \frac{\Delta t}{\varepsilon_0 \Delta x}\left[(H_z)^{n-1/2}_{i+1/2,k} - (H_z)^{n-1/2}_{i-1/2,k}\right],\end{aligned} \tag{16.29a}$$

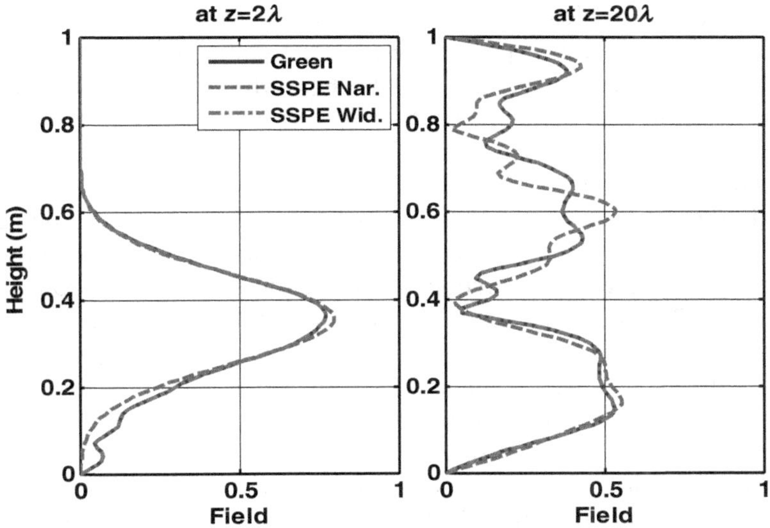

Figure 16.7. E-field versus height for a tilted Gaussian beam: TE_z case, $a = 1$ m, $z' = 0$, $x' = 0.4$ m, $k_0a = 50$, $\theta_{\text{bw}} = 30°$, $\theta_{\text{elv}} = -10°$ (a) at $z = 2\lambda$ and (b) $z = 20\lambda$. (Solid) Green's function (31 modes), (dashed) narrow SSPE, and (dashed dotted) wide SSPE.

$$(H_z)^{n+1/2}_{i+1/2,k} = (H_z)^{n-1/2}_{i+1/2,k} - \frac{\Delta t}{\mu_0 \Delta x}\left[(E_y)^n_{i+1,k} - (E_y)^n_{i,k}\right], \quad (16.29\text{b})$$

$$(H_x)^{n+1/2}_{i,k+1/2} = (H_x)^{n-1/2}_{i,k+1/2} + \frac{\Delta t}{\mu_0 \Delta z}\left[(E_y)^n_{i,k+1} - (E_y)^n_{i,k}\right]. \quad (16.29\text{c})$$

TM_z case:

$$\begin{aligned}(H_y)^n_{i,k} = (H_y)^{n-1}_{i,k} &- \frac{\Delta t}{\mu_0 \Delta z}\left[(E_x)^{n-1/2}_{i,k+1/2} - (E_x)^{n-1/2}_{i,k-1/2}\right] \\ &+ \frac{\Delta t}{\mu_0 \Delta x}\left[(E_z)^{n-1/2}_{i+1/2,k} - (E_z)^{n-1/2}_{i-1/2,k}\right],\end{aligned} \quad (16.30\text{a})$$

$$(E_z)^{n+1/2}_{i+1/2,k} = (E_z)^{n-1/2}_{i+1/2,k} + \frac{\Delta t}{\varepsilon_0 \Delta x}\left[(H_y)^n_{i+1,k} - (H_y)^n_{i,k}\right], \quad (16.30\text{b})$$

$$(E_x)^{n+1/2}_{i,k+1/2} = (E_x)^{n-1/2}_{i,k+1/2} - \frac{\Delta t}{\varepsilon_0 \Delta z}\left[(H_y)^n_{i,k+1} - (H_y)^n_{i,k}\right]. \quad (16.30\text{c})$$

The characteristics of the medium are defined in terms of permittivity, conductivity, and permeability. The electric and magnetic field components are calculated at different locations of each cell (see fig. 1 in Ref. 39 for the cell structures). Beside the spatial differences in field components, there is also a half-time-step difference between electric and magnetic field components. In order to provide good accuracy, the time step should satisfy Courant limit and the sampling space should be smaller than the wavelength (such as $\Delta x < \lambda/10$, $\Delta z < \lambda/10$). Using the FDTD, although it is possible to analyze EM fields in time domain for a broad range of frequencies with one simulation, high memory size and high speed computer requirements should be taken into consideration. FDTD has been extensively used in complex EM simulations for the last few decades.

Field versus range/height at three different time instants for a tilted Gaussian beam computed with the FDTD for the TE_z polarization is shown in Fig. 16.8. As observed, downtilted Gaussian beam first hits the bottom plate and interference patterns are formed from the interaction of direct and bottom-reflected beams. Then beams that hit the top plate reflects. Interaction of the direct beam with both bottom- and top-reflected beams is clearly observed from the FDTD simulation.

Figure 16.9 shows electric field versus height for a tilted Gaussian beam for the TE_z polarization computed with the mode summation, SSPE, and FDTD. Note that the mode summation and SSPE are frequency domain methods and yield the results directly. On the other hand, time data are accumulated at the specified observation points during the FDTD simulations and frequency domain result is then obtained via offline FFT.

16.3.3 Method of Moments (MoM)

Method of moments (MoM) [24] is one of the oldest numerical EM methods. In this method, first a discrete model of the object under investigation is created from small (compared with wavelength) pieces called segments or patches, and then an $N \times N$ system of equations is built with N unknown segment/patch currents, N known segment voltages calculated from the Green's function of the problem, and known (analytically, but requires numerical computation) $N \times N$ segment/patch impedances (it requires N^3 operations). The model is closed form and stable, but necessitates high memory and high speed computers especially for high frequency applications. Also, it requires the derivation of the Green's function of the problem.

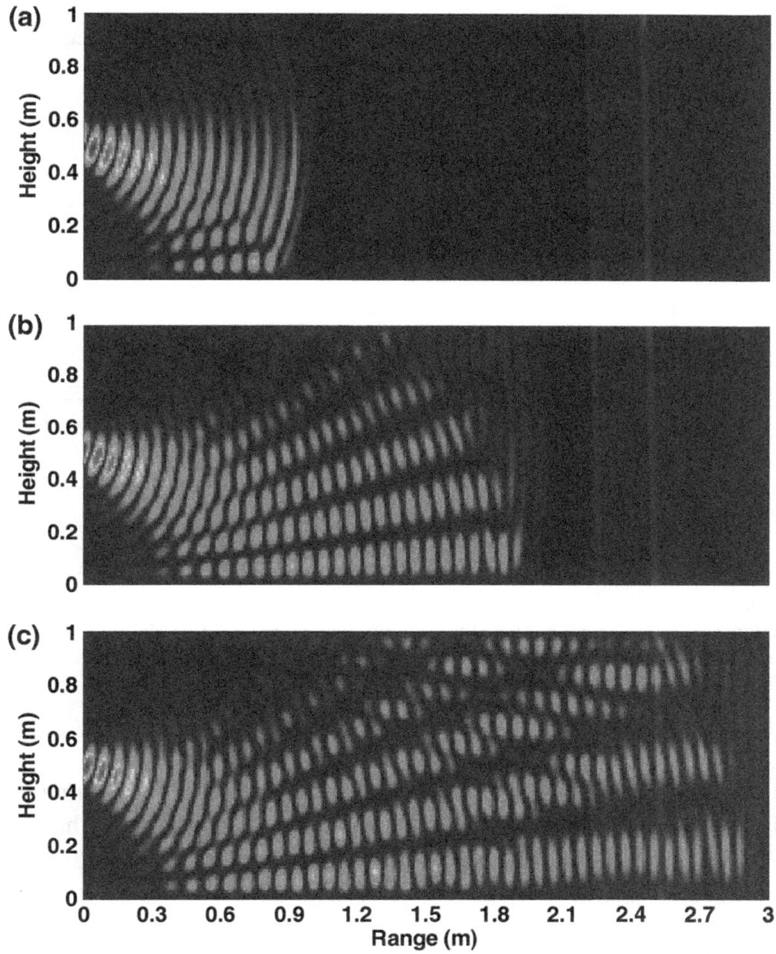

Figure 16.8. E-field versus range/height at three different time instants for a tilted Gaussian beam computed with FDTD: TE_z case, $a = 1$ m, $z' = 0$, $x' = 0.5$ m, $k_0a = 50$, $\theta_{bw} = 40°$, $\theta_{elv} = -20°$, $t_a = 3.42 \times 10^{-9}$ s, $t_b = 6.84 \times 10^{-9}$ s, $t_c = 1.03 \times 10^{-9}$ s.

Propagation modeling inside waveguides with irregular and lossy boundaries has become very important because of signaling requirements through railway tunnels, communication in mines, screening in printed circuit boards (PCBs), and so on. MoM can be used to find the propagation of horizontally (TE_z case) and vertically (TM_z case) polarized waves by using the electric field integral equation (EFIE) and the

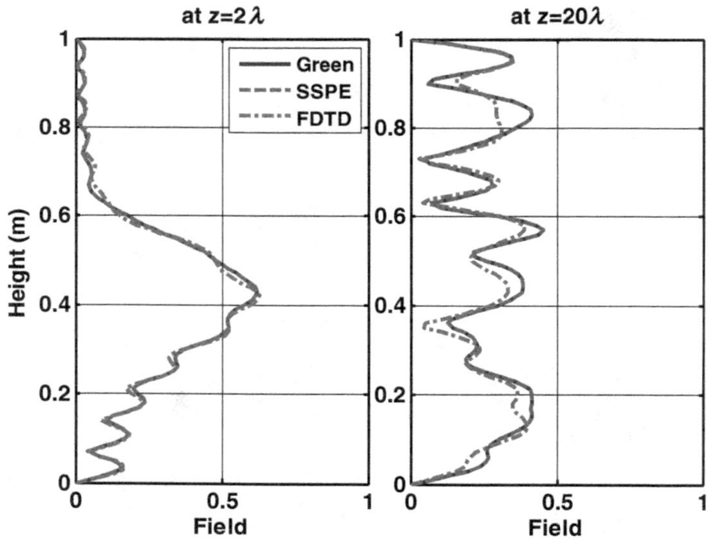

Figure 16.9. E-field versus height for a tilted Gaussian beam: TE_z case, $a = 1$ m, $z' = 0$, $x' = 0.5$ m, $k_0a = 50$, $\theta_{bw} = 40°$, $\theta_{elv} = -20°$ (a) at $z = 2\lambda$ and (b) $z = 20\lambda$. (Solid) Green's function (42 modes), (dashed) SSPE, and (dashed dotted) FDTD $dx = dz = 0.5$ m, $dt = 1.14 \times 10^{-11}$ s.

magnetic field integral equation (MFIE), respectively. Unfortunately, it suffers from resonances in these waveguiding structures [41]; therefore, its direct application is a challenge. Recently, a novel multiple MoM model (Mi-MoM) is introduced for this purpose [38]. In this method, the integral equation is converted to the corresponding matrix equation via the discretization of the plates at $x = 0$ and $x = a$ (see Fig. 16.10). The lengths of line segments on plates are chosen to be less than one-tenth of the wavelength to satisfy the surface current constant at each line segment. Then, an $N \times N$ system of equations is constructed with the unknowns of N segment currents and the system is solved numerically.

Point-matching MoM technique [24,42] with the closed form matrix equation $[V] = [Z]\ [I]$ was used, where $[I]$ contains the unknown segment currents, $[V]$ corresponds to excitation voltage matrix using the incident field evaluated at the midpoints of each line segments, and $[Z]$ is the $N \times N$ impedance matrix for the PEC plate. Using segment

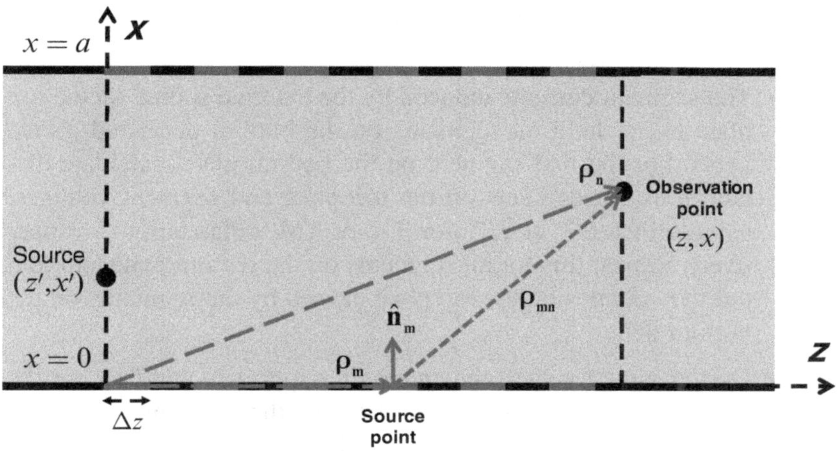

Figure 16.10. Parallel plate waveguide and source/observer for the MoM (ρ_m: source segment; ρ_n: observer segment).

currents, the scattered fields are obtained. Finally, fields at a specified observation point are calculated in terms of incident field and scattered fields caused by the source-induced segment currents on plates (see also Refs. 43 and 44).

The necessary equations for the elements of excitation voltage matrix and impedance matrix and scattered fields are summarized in Chapter 13 (see (13.32)–(13.41)).

The Mi-MoM procedure for propagation modeling inside resonating structures is based on taking into account multiple-induced segment currents and may be outlined as follows [38].

- First, discretize top and bottom boundaries. Use $N/2$ segments for the lower and $N/2$ segments for the upper boundaries. Label all segments from 1 to N in a way that segment 1 and segment $N+1$ have the same horizontal (i.e., z) coordinate (i.e., parallel to each other).
- Calculate segment currents $[I]$ from $[I] = [Z]^{-1}[V]$ and scattered/total fields using either E_y^{inc} or H_y^{inc} for TE_z or TM_z polarizations, respectively.
- For a given source point, calculate distances to all segments and segment voltages, using either E_y^{inc} or H_y^{inc} for TE_z or TM_z cases, respectively. This will yield $[V]$.

- Calculate the impedance matrix Z_{mn} for TE_z or TM_z polarizations.
- The segment currents induced by the external source on the top plate excite field on segments on the bottom plate and a vice versa. For the first segment on the bottom plate, calculate distances to all segments on the top plate and segment voltages, using either E_y^{inc} or H_y^{inc} for TE_z or TM_z polarizations, respectively. Repeat this for all segments on the bottom plate and find out the voltages on the top plate caused by the segments on the bottom plate.
- Do the same for the segments on the top plate and find out the voltages on the bottom plate caused by the segments on the top plate. This will yield second round $[V]$.
- Use the same impedance matrix Z_{mn} and calculate second round segment currents $[I]$ from $[I] = [Z]^{-1}[V]$ and scattered/total fields for TE_z or TM_z polarizations, respectively.
- Repeat the procedure and find out third round segment currents and scattered/total fields caused by these current.
- Repeat the whole procedure until a desired accuracy is reached.

An alternative way is to find out first round segment currents and then use the image method. First, all segment currents of upper and lower plates are obtained. Then, boundaries are removed and image segments are added with respect to the upper and lower plates. Finally, field contributions from the currents of segments and image segments are superposed at the receiver. A short MATLAB script for the Mi-MoM calculation of field versus range/height at a given height/range point is listed in Table 16.5. The classical MoM can also be used together with IM in hybrid form. This is achieved by finding segment currents once and then using images of all segments on both sides of the boundaries. A short MATLAB script for the hybrid MoM + IM calculation of field versus range/height at a given height/range is listed in Table 16.6.

Two examples for the Mi-MoM procedure are given in Fig. 16.11 and Fig. 16.12. Figure 16.11 shows propagation factor (PF) versus range at a fixed height inside the parallel plate waveguide calculated with the mode summation and Mi-MoM methods. As shown, very good agreement is obtained. As expected, Mi-MoM suffers from end-point

Table 16.5.
A MATLAB MoM Code for Field versus Range/Height Calculations

```
function output=momanalysis(E0,fr,a,zmax,xs,xo,NX,NZ,numi
tr,pol)
% ----------------------------------------------------
% MOM analysis inside a 2D-PEC parallel plate waveguide
% for the calculation of
%    Field vs Range at a given height (xo)
% E0: Field amplitude
% fr: frequency [MHz], a: plate height [m], zmax:
maximum range [m]
% xs: source height [m], xo: observer height [m]
% NX: number of height points, NZ: number of range
points,
% numitr: number of iteration
% pol: polarization type, 0 for TEz case, 1 for TMz
case
%
% output.zz:     range points
% output.xx:     height points
% output.Fh1:     horizontal field at x=xo
% ----------------------------------------------------
k0=2*pi*fr/300; % wavenumber in free space
dx=a/(NX-1);  dz=zmax/(NZ-1); xx=(0:dx:a).';
zz=(0:dz:zmax);
xter=zeros(size(zz)); zz=0.5*(zz(1:end-1)+zz(2:end));
Nz=length(zz);
FF=zeros(NX,NZ-1);
%% Calculate the impedance matrix
if pol==1 % TMz (vert. case)
  imp_matrix1=0.5*eye(Nz);     imp_matrix2=0.5*eye(Nz);
elseif pol==0 % TEz (hor. case)
  imp_matrix1=zeros(2*Nz-1,1);
  for m=1:Nz-1
    dd=dz*m;
    imp_matrix1(Nz+m,1)=-0.25*dz*120*pi*k0*besselh(0,2,k
0*dd);
  end
  imp_matrix1(1:Nz-1)=imp_matrix1(2*Nz-1:-1:Nz+1);
```

(Continued)

Table 16.5.
(Continued)

```
  imp_matrix1(Nz,1)=-0.25*dz*120*pi*k0*. . .
    (1-1i*2/pi*log(exp(0.577215665)*k0*dz/(4*exp(1))));%
if m==n
  imp_matrix1=meshgrid(imp_matrix1,ones(Nz,1));
  imp_matrix1=spdiags(imp_matrix1,-(Nz-1):Nz-1,Nz,Nz);

imp_matrix1=full(imp_matrix1);  imp_matrix2=imp_matrix1;
end
% Calculate initial voltages on the surface coming from
source
% lower part at x=0
ds=sqrt((xter-xs).^2+zz.^2); inc_magx0=exp(-1i*k0*ds)./
sqrt(k0*ds);
% upper part at x=a
ds=sqrt((a-xs)^2+zz.^2);  inc_magxa=exp(-1i*k0*ds)./
sqrt(k0*ds);
V_inc1=-inc_magx0;     V_inc2=-inc_magxa;
%% Calculate currents on each segment
ind_current1=imp_matrix1\V_inc1.';
ind_current2=imp_matrix2\V_inc2.';
%% Calculate scattering fields
zobs = (0.5:Nz-0.5)*dz;
scat_mag=zeros(1,Nz);   sx0=zeros(1,Nz);
sxa=zeros(1,Nz);
for m=1:Nz
    % lower part contribution for observer range

scat_mag=scat_mag+findscatfld(ind_current1(m),k0,zobs,. . .
      xo-xter(m),zz(m),pol);
    % upper part contribution for observer range

scat_mag=scat_mag+findscatfld(ind_current2(m),k0,zobs,. . .
      a-xo,zz(m),pol);
    % field on each segment

sx0=sx0+findscatfld(ind_current2(m),k0,zobs,a-
xter,zz(m),pol);

sxa=sxa+findscatfld(ind_current1(m),k0,zobs,a-
xter(m),zz(m),pol);
```

Table 16.5.
(Continued)

```
end
if pol==1,
  scat_mag =0.25*dz*1i*k0*scat_mag;  % scattering fields
on x=xobs
  sx0      =0.25*dz*1i*k0*sx0;       % scattering
fields on x=0
  sxa      =0.25*dz*1i*k0*sxa;       % scattering
fields on x=b
else
  scat_mag =-0.25*dz*120*pi*k0*scat_mag;  % scat fields
on x=xobs
  sx0      =-0.25*dz*120*pi*k0*sx0;       % scattering
fields on x=0
  sxa      =-0.25*dz*120*pi*k0*sxa;       % scattering
fields on x=b
end
for m=1:numitr
  [newscat_mag,sx0,sxa]=findmomscatfld(sx0,sxa,
imp_matrix1,. . .
  imp_matrix2,Nz,k0,zobs,a,xo,zz,dz,pol,xter);
  scat_mag=scat_mag+newscat_mag; % sum all contributions
end
dist=sqrt((xo-xs)^2+zobs.^2);
inc_mag=exp(-1i*k0*dist)./
sqrt(k0*dist); tot_hor=inc_mag+scat_mag;
output.xx=xx;   output.zz=zz;
if pol==1, % TMz (ver)
   tot_hor=(E0/120*pi)*tot_hor;
elseif pol==0, % TEz (hor)
   tot_hor=E0*tot_hor;
end
output.Fh1=tot_hor;
function scat_mag=findscatfld(ind_current,k0,zobs,xobs,zs,
pol)
dd=sqrt(xobs.^2+(zobs-zs).^2);
if pol==1, % TMz (ver)
  prinner=abs(xobs)./dd;
  scat_mag=ind_current.*besselh(1,2,k0*dd).*prinner;
elseif pol==0 % TEz (hor)
  scat_mag=ind_current.*besselh(0,2,k0*dd);
end
```

(Continued)

Table 16.5.
(Continued)

```
function [scat_mag,sx0,sxa]=findmomscatfld(sx0,sxa,. . .
    imp_matrix1,imp_matrix2,Nz,k0,zobs,a,xobs,zs,delz,pol,
xter)
% Calculate new voltages on each segment
V_inc1=-sx0;     V_inc2=-sxa;
% Calculate new currents on each segment
ind_current1=imp_matrix1\V_inc1.';% transpose(V_inc)
ind_current2=imp_matrix2\V_inc2.';% transpose(V_inc)
% Calculate new scattering fields
scat_mag=zeros(1,Nz); sxa=zeros(1,Nz); sx0=zeros(1,Nz);
for m=1:Nz
   % lower part contribution for observer range

scat_mag=scat_mag+findscatfld(ind_current1(m),k0,zobs,. . .
     xobs-xter(m),zs(m),pol);
   % upper part contribution for observer range

scat_mag=scat_mag+findscatfld(ind_current2(m),k0,zobs,. . .
     a-xobs,zs(m),pol);
   % field on each segment

sxa=sxa+findscatfld(ind_current1(m),k0,zobs,a-
xter(m),zs(m),pol);

sx0=sx0+findscatfld(ind_current2(m),k0,zobs,a-
xter,zs(m),pol);
end
if pol==1
   scat_mag=0.25*delz*1i*k0*scat_mag;   % scattering fields
on x=xobs
   sx0=0.25*delz*1i*k0*sx0;               % scattering
fields on x=0
   sxa=0.25*delz*1i*k0*sxa;               % scattering
fields on x=b
elseif pol==0
   scat_mag=-0.25*delz*120*pi*k0*scat_mag;   % scat fields
on x=xobs
   sx0=-0.25*delz*120*pi*k0*sx0;               % scat
fields on x=0
   sxa=-0.25*delz*120*pi*k0*sxa;               % scat
fields on x=b
end
% ----------------------- END ----------------------
```

Table 16.6.
A MATLAB MoM + IM Code for Field versus Range/Height Calculations

```
function output=momimganalysis(E0,fr,a,zmax,xs,zo,xo,NX,N
Z,numimg,pol)
% -----------------------------------------------------
% MOM analysis with IM inside a 2D-PEC parallel plate
waveguide
% for the calculation of
%   Field vs Range at a given height (xo)
%   Field vs Height at a given range (zo)
% E0: Field amplitude, fr: frequency [MHz], a: plate
height [m], zmax: maximum range [m]
% xs: source height [m], xo: observer height [m], zo:
observer range [m]
% NX: number of height points, NZ: number of range
points,
% numimg: number of image
% pol: polarization type, 0 for TEz case, 1 for TMz
case
% output.zz:    range points,  output.xx:    height
points
% output.Fh1:    horizontal field at x=xo
% output.Fv0:    vertical field at z=0
% output.Fv1:    vertical field at z=zo
% -----------------------------------------------------
k0=2*pi*fr/300; % wavenumber in free space
dx=a/(NX-1);  dz=zmax/(NZ-1); xx=(0:dx:a).';
zz=(0:dz:zmax);   zz=0.5*(zz(1:end-1)+zz(2:end));
Nz=length(zz);
% Calculate initial voltages on the surface coming from
source at (0,xs)
% lower part at x=0
ds1=sqrt(xs^2+zz.^2); inc_fld1=exp(-1i*k0*ds1)./
sqrt(k0*ds1);
% upper part at x=a
ds2=sqrt((a-xs)^2+zz.^2); inc_fld2=exp(-1i*k0*ds2)./
sqrt(k0*ds2);
V_inc1=-inc_fld1; V_inc2=-inc_fld2;
% Calculate the impedance matrix
```

(*Continued*)

Table 16.6.
(Continued)

```
if pol==1 % TMz (vert. case)
   imp_matrix1=0.5*eye(Nz);      imp_matrix2=0.5*eye(Nz);
elseif pol==0 % TEz (hor. case)
   imp_matrix1=zeros(2*Nz-1,1);
   for m=1:Nz-1
       dd=dz*m;
       imp_matrix1(Nz+m,1)=-0.25*dz*120*pi*k0*besselh(0,2,
k0*dd);
   end
   imp_matrix1(1:Nz-1)=imp_matrix1(2*Nz-1:-1:Nz+1);
   imp_matrix1(Nz,1)=-0.25*dz*120*pi*k0*. . .
    (1-1i*2/pi*log(exp(0.577215665)*k0*dz/(4*exp(1))));%
if m==n
   imp_matrix1=meshgrid(imp_matrix1,ones(Nz,1));
   imp_matrix1=spdiags(imp_matrix1,-(Nz-1):Nz-1,Nz,Nz);
   imp_matrix1=full(imp_matrix1);
   imp_matrix2=imp_matrix1;
end
ind_current1=imp_matrix1\V_inc1.';
ind_current2=imp_matrix2\V_inc2.';
% -------------- RANGE PROFILE --------------
scat_ver0=zeros(NX,1); scat_ver1=zeros(NX,1); %
scattering fields
for m=1:Nz
  % lower part at x=0
  for ii=0:numimg,
    if mod(ii,2)==0, kkk=xx+ii*a; else kkk=-
xx+(ii+1)*a; end
     s1=findscatfld(ind_current1(m),k0,0,kkk,zz(m),pol);
    if pol==0
       scat_ver0=scat_ver0+(-1)^ii*s1;
    elseif pol==1
       scat_ver0=scat_ver0+s1;
    end
    s1=findscatfld(ind_current1(m),k0,zo,kkk,zz(m),pol);
    if pol==0
      scat_ver1=scat_ver1+(-1)^ii*s1;
    elseif pol==1
      scat_ver1=scat_ver1+s1;
    end
  end
```

Table 16.6.
(Continued)

```
  % upper part at x=a
  for ii=0:numimg,
    if mod(ii,2)==0, kkk=-xx+(ii+1)*a; else
kkk=xx+ii*a; end
     s2=findscatfld(ind_current2(m),k0,0,kkk,zz(m),pol);
     if pol==0
        scat_ver0=scat_ver0+(-1)^ii*s2;
     elseif pol==1
        scat_ver0=scat_ver0+s2;
     end
     s2=findscatfld(ind_current2(m),k0,zo,kkk,zz(m),pol);
     if pol==0
        scat_ver1=scat_ver1+(-1)^ii*s2;
     elseif pol==1
        scat_ver1=scat_ver1+s2;
     end
  end
end
if pol==1, % TMz (ver)
  dist=sqrt((xx-xs).^2+0.^2);
  inc_ver0=exp(-1i*k0*dist)./sqrt(k0*dist);
  tot_ver0=inc_ver0+0.25*dz*1i*k0*scat_ver0;
  dist=sqrt((xx-xs).^2+zo.^2);
  inc_ver1=exp(-1i*k0*dist)./sqrt(k0*dist);
  tot_ver1=inc_ver1+0.25*dz*1i*k0*scat_ver1;
elseif pol==0 % TEz (hor)
  dist=sqrt((xx-xs).^2+0.^2);
  inc_ver0=exp(-1i*k0*dist)./sqrt(k0*dist);
  tot_ver0=inc_ver0-0.25*dz*120*pi*k0*scat_ver0;
  dist=sqrt((xx-xs).^2+zo.^2);
  inc_ver1=exp(-1i*k0*dist)./sqrt(k0*dist);
  tot_ver1=inc_ver1-0.25*dz*120*pi*k0*scat_ver1;
end
scat_hor=zeros(1,Nz); % scattering fields
for m=1:Nz
  % lower part at x=0
  for ii=0:numimg,
    if mod(ii,2)==0, kkk=xo+ii*a; else kkk=-
xo+(ii+1)*a; end
     s1=findscatfld(ind_current1(m),k0,zz,kkk,zz(m),pol);
```

(Continued)

Table 16.6.
(Continued)

```
    if pol==0
      scat_hor=scat_hor+(-1)^ii*s1;
    elseif pol==1
      scat_hor=scat_hor+s1;
    end
  end
  % upper part at x=a
  for ii=0:numimg,
    if mod(ii,2)==0, kkk=-xo+(ii+1)*a; else
kkk=xo+ii*a; end
    s2=findscatfld(ind_current2(m),k0,zz,kkk,zz(m),pol);
    if pol==0
      scat_hor=scat_hor+(-1)^ii*s2;
    elseif pol==1
      scat_hor=scat_hor+s2;
    end
  end
end
dist=sqrt((xo-xs)^2+zz.^2); inc_hor=exp(-1i*k0*dist)./
sqrt(k0*dist);
if pol==1,
  tot_hor=inc_hor+0.25*dz*1i*k0*scat_hor;
elseif pol==0
  tot_hor=inc_hor-0.25*dz*120*pi*k0*scat_hor;
end
if pol==1, % TMz (ver)
  tot_ver0=E0/(120*pi)*tot_ver0;    tot_ver1=E0/
(120*pi)*tot_ver1;
  tot_hor=(E0/120*pi)*tot_hor;
elseif pol==0, % TEz (hor)

tot_ver0=E0*tot_ver0;    tot_ver1=E0*tot_ver1;    tot_
hor=E0*tot_hor;
end
output.xx=xx;    output.zz=zz;
output.Fv0=tot_ver0;    output.Fv1=tot_ver1;    output.
Fh1=tot_hor;
function scat_mag=findscatfld(ind_current,k0,zobs,xobs,zs,
pol)
```

Table 16.6.
(Continued)

```
dd=sqrt(xobs.^2+(zobs-zs).^2);
if pol==1, % TMz (ver)
   prinner=abs(xobs)./dd;
   scat_mag=ind_current.*besselh(1,2,k0*dd).*prinner;
elseif pol==0 % TEz (hor)
   scat_mag=ind_current.*besselh(0,2,k0*dd);
end
% --------------------- END ----------------------
```

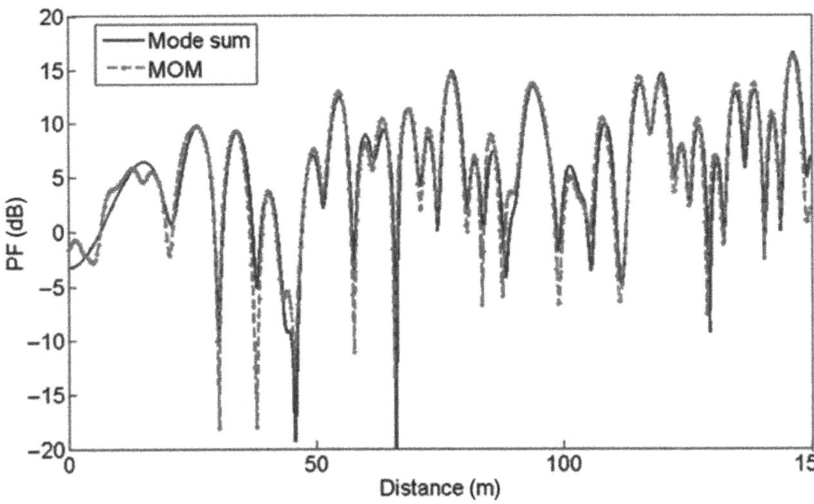

Figure 16.11. Propagation factor (PF) versus range (TM_z case): (solid) mode sum and (dashed) Mi-MoM, $a = 100$ m, $z' = 0$, $x' = 50$ m, $x = 5$ m, $k_0a = 209.5$.

effects (as MoM does), since segments before the first one and after the last one are neglected [38].

Figure 16.12 shows field versus height at two different ranges inside the parallel plate waveguide calculated with the mode summation and Mi-MoM methods. Again, the agreement is very good. Note that the agreement in Fig. 16.11 is better than the agreement in Fig. 16.12;

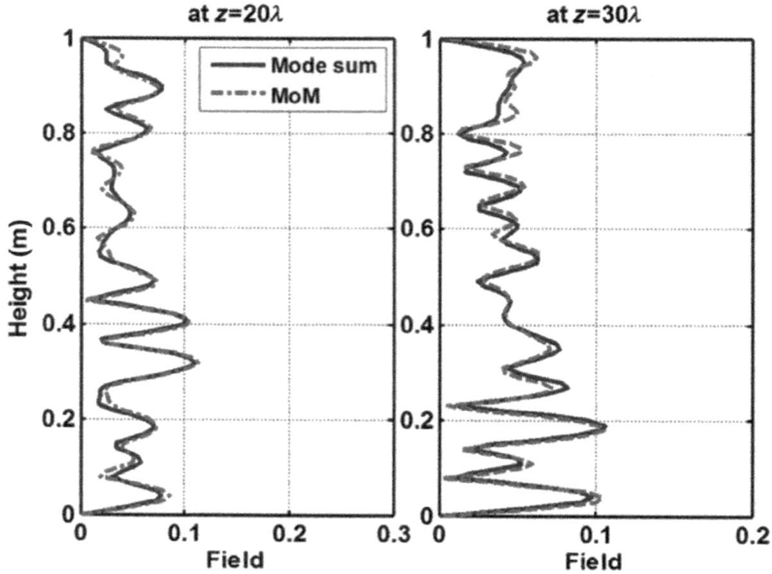

Figure 16.12. Field versus height (TE_z case): (solid) mode sum and (dashed) multi-MoM, $a = 1$ m, $z' = 0$, $x' = 0.4$ m, $k_0a = 50$.

this is merely because of the frequency used in these examples ($k_0a = 209.5$ in Fig. 16.11, but $k_0a = 50$ in Fig. 16.12). The accuracy of Mi-MoM solution increases with frequency (i.e., with ka).

Figure 16.13, Figure 16.14, and Figure 16.15 belong to comparisons for directive antennas. As observed, the agreement between the Mi-MoM result and the reference data is impressive even for these highly resonating/oscillatory variations.

The final example belongs to a more realistic case. Figure 16.16 presents Mi-MoM versus SSPE comparisons for propagation inside a PEC parallel plate waveguide having irregularities on the bottom plate. The top plot shows three-dimensional (3D) field map inside the plate with two Gaussian-shaped irregularities on the bottom plate. The bottom plot includes field versus range variations at $x = 0.4$ m for the TE_z polarization showing the comparison between Mi-MoM and SSPE models.

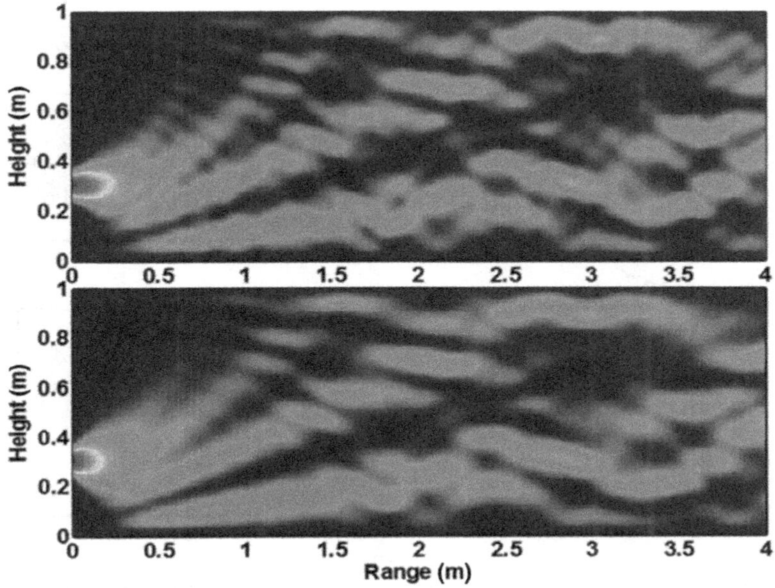

Figure 16.13. Field versus range/height (TE_z case): (top) mode sum with 42 modes and (bottom) Mi-MoM with 40 iterations, $a = 1$ m, $z' = 0$, $x' = 0.3$ m, $k_0a = 50$, $dz = dx = 0.01$ m, $\theta_{bw} = 45°$, no tilt.

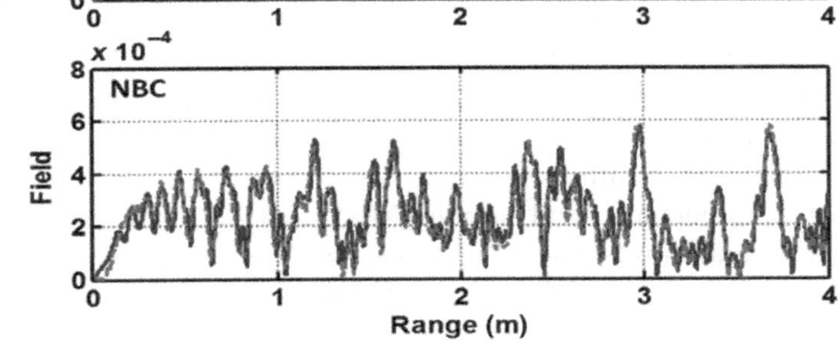

Figure 16.14. Field versus height at $x = 0.2$ m: (top) TE_z case, (bottom) TM_z case, (solid) mode sum with 282 modes, and (dashed) Mi-MoM with 50 iterations, $a = 1$ m, $z' = 0$, $x' = 0.4$ m, $k_0a = 200$, $dz = dx = 0.0025$ m, $\theta_{bw} = 80°$, no tilt.

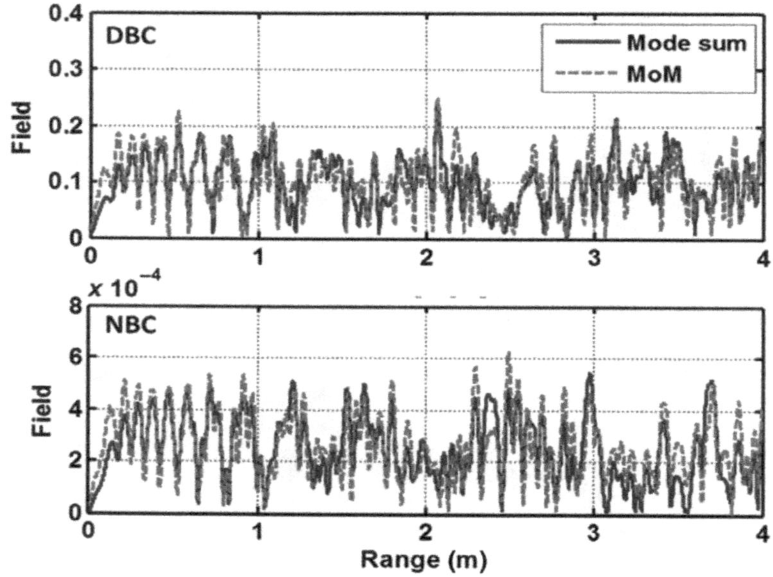

Figure 16.15. Field versus height at $x = 0.2$ m: (top) TE_z case, (bottom) TM_z case, (solid) mode sum with 298 modes, and (dashed) multi-MoM with 50 iterations, $a = 1$ m, $z' = 0$, $x' = 0.4$ m, $k_0a = 200$, $dz = dx = 0.0025$ m, $\theta_{bw} = 45°$, $\theta_{tilt} = -20°$.

16.4 CONCLUSIONS

A simple, canonical structure is investigated within computational EM (CEM) context. Wave propagation inside a parallel plate waveguide with nonpenetrable boundaries is chosen for this purpose. Details of this problem are given in Chapter 8, but analytical solutions in terms of mode and ray summation as well as IM are discussed here for the sake of completeness and simple MATLAB scripts are presented. Various excitation models are also introduced. Fundamental numerical models such as finite-difference time-domain (FDTD) and parabolic equation (PE) methods, and method of moments (MoM) are also taken into account and codes developed using these models are calibrated against analytical reference solutions.

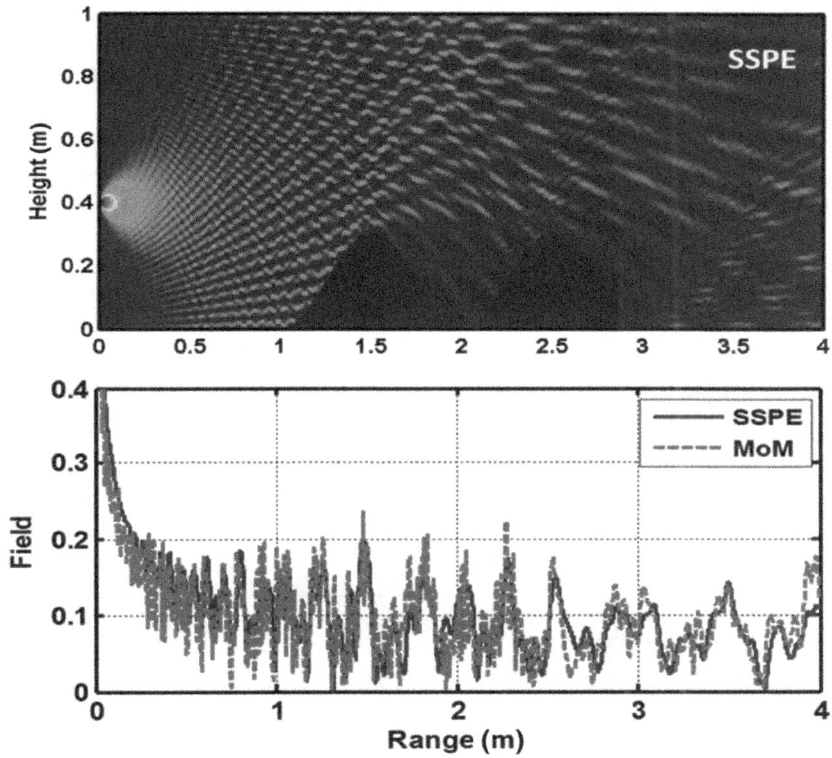

Figure 16.16. (Top) field versus range/height produced with SSPE, (bottom) field versus range at $x = 0.4$ m, both for TE_z case, (solid) SSPE, and (dashed) Mi-MoM with 50 iterations, $a = 1$ m, $z' = 0$, $x' = 0.4$ m, $k_0a = 200$, $dz = dx = 0.0025$ m, $\theta_{bw} = 80°$, no tilt.

REFERENCES

1. C. A. Balanis, *Advanced Engineering Electromagnetics*, John Wiley, New York, 1989.
2. L. B. Felsen, N. Marcuvitz, *Radiation and Scattering of Waves*, Prentice-Hall, Englewood Cliffs, NJ, 1973, Classic reissue IEEE Press, 1994.
3. D. G. Dudley, *Mathematical Foundations for Electromagnetic Theory*, IEEE Press (Series on Electromagnetic Waves), New York, 1994.
4. M. N. O. Sadiku, *Numerical Techniques in Electromagnetics with MATLAB*, 3rd ed., CRC Press, Boca Raton, FL, 2009.

5. A. Bondeson, T. Rylander, P. Ingelström, *Computational Electromagnetics*, Springer, New York, 2005.
6. A. F. Peterson, S. L. Ray, R. Mittra, *Computational Methods for Electromagnetics*, IEEE Press, New York, 1998.
7. J. M. Jin, *Theory and Computation of Electromagnetic Fields*, John Wiley, Hoboken, NJ, 2010.
8. R. Garg, *Analytical and Computational Methods in Electromagnetics*, Artech House, Boston, 2008.
9. L. Sevgi, *Complex Electromagnetic Problems and Numerical Simulation Approaches*, IEEE Press/Wiley, Piscataway, NJ, 2003.
10. L. Sevgi, "Modeling and Simulation in Electromagnetic Engineering: Validation, Verification and Calibration," Invited Tutorial, *10th Mediterranean Microwave Symposium MMS'2010*, August 25–27 2010, Middle East Technical University—Northern Cyprus Campus, (METU-NCC).
11. L. Sevgi, "Modeling and Simulation Approaches in EM Engineering: Strategies, Problems, and Challenges," 1-Hour Keynote Speech, *IBCAST 2011*, The 8th International Bhurban Conference on Applied Sciences & Technology, Jan 13–16, 2010, Islamabad, Pakistan.
12. L. Sevgi, "Modeling and Simulation Approaches in Electromagnetic Engineering: Perspectives, Problems, and Challenges," Institute for InfoComm (I2R) Research, Agency for Science, Technology and Research (A*STAR), Singapore, May 23, 2012.
13. G. Apaydin, L. Sevgi, "Validation, Verification and Calibration in Applied Computational Electromagnetics," *Applied Computational Electromagnetics Society Journal*, **25** (12), 2010, pp. 1026–1035.
14. V. A. Fock, *Electromagnetic Diffraction and Propagation Problems*, Pergamon Press, London, 1965.
15. J. B. Keller, "Geometrical Theory of Diffraction," *Journal of the Optical Society of America*, **52** (2), 1962, pp. 116–130.
16. C. Hansen, ed., *Geometrical Theory of Diffraction*, IEEE Press, New York, 1981.
17. R. G. Kouyoumjian, P. H. Pathak, "A Uniform Geometrical Theory of Diffraction for an Edge in a Perfectly Conducting Surface," *Proceedings of the IEEE*, **62**, 1974, pp. 1448–1461.
18. P. Ya. Ufimtsev, "Transverse Diffusion for Diffraction by a Wedge," *Radio Engineering and Electronic Physics*, **10** (6), 1965, pp. 866–875.
19. P. Ya. Ufimtsev, *Fundamentals of the Physical Theory of Diffraction*, Wiley & Sons, Inc., Hoboken, NJ, 2007.
20. P. Ya. Ufimtsev, *Theory of Edge Diffraction in Electromagnetics: Origination & Validation of PTD*, SciTech Publishing, Inc., Raleigh, NC, 2009.
21. F. Hacivelioglu, L. Sevgi, P. Ya. Ufimtsev, "Electromagnetic Wave Scattering from a Wedge with Perfectly Reflecting Boundaries: Analysis of Asymptotic Techniques," *IEEE Antennas and Propagation Magazine*, **53** (3), 2011, pp. 232–253.

22. F. Hacivelioglu, M. A. Uslu, L. Sevgi, "A MATLAB-Based Simulator for the Electromagnetic Wave Scattering from a Wedge with Perfectly Reflecting Boundaries," *IEEE Antennas and Propagation Magazine*, **53** (6), 2011, pp. 234–243.

23. L. Sevgi, "Sturm-Liouville Equation: The Bridge between Eigenvalue and Green's Function Problems," *ELEKTRIK, Turkish Journal of Electrical Engineering and Computer Sciences*, **14** (2), 2006, pp. 293–311.

24. R. F. Harrington, *Field Computation by Moment Method*, 1st ed. 1968, IEEE Press, New York, 1993.

25. M. Levy, *Parabolic Equation Methods for Electromagnetic Wave Propagation*, IEE Institution of Electrical Engineers, London, 2000.

26. K. S. Kunz, R. J. Luebbers, *Finite Difference Time Domain Method for Electromagnetics*, CRC Press LLC, Boca Raton, FL, 1993.

27. D. M. Sullivan, *Electromagnetic Simulation Using the Finite-Difference Time-Domain Method*, IEEE Press, New York, 2000.

28. A. Taflove, *Computational Electrodynamics: The Finite-Difference Time-Domain Method*, Artech House, Boston, First Edition, 1995, Second Edition, 2000.

29. C. Christopoulos, *The Transmission Line Modeling Method: TLM*, IEEE Press, Piscataway, NJ, 1995.

30. P. Russer, *Electromagnetics, Microwave Circuit and Antenna Design for Communications Engineering*, 2nd ed., Artech House, Boston, 2006.

31. L. B. Felsen, A. H. Kamel, "Hybrid Ray-Mode Formulation of Parallel Plate Waveguide Green's Functions," *IEEE Transactions on Antennas and Propagation*, **29** (4), 1981, pp. 637–649.

32. L. B. Felsen, F. Akleman, L. Sevgi, "Wave Propagation Inside a Two-Dimensional Perfectly Conducting Parallel Plate Waveguide: Hybrid Ray-Mode Techniques and Their Visualizations," *IEEE Antennas and Propagation Magazine*, **46** (6), 2004, pp. 69–89.

33. L. Sevgi, F. Akleman, L. B. Felsen, "Visualizations of Wave Dynamics in a Wedge-Waveguide with Non-Penetrable Boundaries: Normal, Adiabatic, and Intrinsic Mode Representations," *IEEE Antennas and Propagation Magazine*, **49** (3), 2007, pp. 76–94.

34. G. Apaydin, L. Sevgi, "Two-Way Propagation Modeling in Waveguides with Three Dimensional Finite Element and Split-Step Fourier Based PE Approaches," *IEEE Antennas and Wireless Propagation Letters*, **10**, 2011, pp. 975–978.

35. G. Apaydin, L. Sevgi, "The Split Step Fourier and Finite Element Based Parabolic Equation Propagation Prediction Tools: Canonical Tests, Systematic Comparisons, and Calibration," *IEEE Antennas and Propagation Magazine*, **52** (3), 2010, pp. 66–79.

36. F. Akleman, L. Sevgi, "A Novel Finite Difference Time Domain Wave Propagator," *IEEE Transactions on Antennas and Propagation*, **48** (5), 2000, pp. 839–841.

37. M. O. Ozyalcın, F. Akleman, L. Sevgi, "A Novel TLM Based Time Domain Wave Propagator," *IEEE Transactions on Antennas and Propagation*, **51** (7), 2003, pp. 1679–1683.

38. G. Apaydin, L. Sevgi, "Method of Moments (MoM) Modeling for Resonating Structures: Propagation Inside a Parallel Plate Waveguide," *ACES, Applied Computational Electromagnetics Society Journal*, **27** (10), 2012, pp. 842–849.

39. L. Sevgi, "Guided Waves and Transverse Fields: Transverse to What?" *IEEE Antennas and Propagation Magazine*, **50** (6), 2008, pp. 221–225.

40. G. Çakır, M. Çakır, L. Sevgi, "A Multipurpose FDTD-Based Two Dimensional Electromagnetic Virtual Tool," *IEEE Antennas and Propagation Magazine*, **48** (4), 2006, pp. 142–151.

41. R. Ding, L. Tsang, H. Braunisch, "Wave Propagation in a Randomly Rough Parallel-Plate Waveguide," *IEEE Transactions on MTT*, **57** (5), 2009, pp. 1216–1223.

42. E. Arvas, L. Sevgi, "A Tutorial on the Method of Moments," *IEEE Antennas and Propagation Magazine*, **54** (3), 2012, pp. 260–275.

43. L. L. Tsai, C. E. Smith, "Moment Methods in Electromagnetics for Undergraduates," *IEEE Transactions on Education*, **E-31**, 1978, pp. 14–22.

44. J. Moore, R. Pizer, eds., *Moment Methods in Electromagnetics: Techniques and Applications*, Research Studies Press, New York, 1984.

APPENDIX A

Introduction to MATLAB

VECTORS, ARRAYS, AND MATRICES

Assign a value to a variable (observe the effect of the semicolon):

`>> x = 3;`
`>>`
	`>> x = 3`
	`>> x =`
	`>> 3`

Create a row vector of size (1 × 4): Create a column vector of size (1 × 4):

`>> x = [1 2 3 5]`
`>> x =`
`1 2 3`
`5`
	`>> x = [1; 2; 3; 5]`
	`>> x =`
	`>> 1`
	`2`
	`3`
	`5`

Creating an array of size (1 × 5) using the semicolon sign (;):

`>> x = 0.5:0.5:2.5;`
`>> x =`
`0.5000 1.0000`
`1.5000 2.0000 2.5000`
	`>> `***size*** `(x)`
	`>> ans =`
	`1 5`

Electromagnetic Modeling and Simulation, First Edition. Levent Sevgi.

Creating an array of size (1 × 5) using FOR loop:

```
>>   for k = 1: 5
>>   x = k*0.5;
>>   end
```

Creating a matrix of size (3 × 3):

```
>>   A = [2 3 5; 6 3 8; 7 1 9]
>>   A =
           2    3    5
           6    3    8
           7    1    9
```

BASIC MATRIX OPERATIONS

```
>> A = [2 3 5; 6 3 8; 7 1 9];
>> B = [0 5 1; 4 2 6; 2 7 0];
>> det (A)
>> ans =
            -31.0000
>> inv (A)
>> ans =
  -0.6129  0.7097 -0.2903
  -0.0645  0.5484 -0.4516
  0.4839  -0.6129 0.3871
>> A*B
>> ans =
           22     51     20
           28     92     24
           22     100    13
```

```
>>   size (A)
ans =
       3     3
>>   A + B
>>ans =
  2     8    6
  10    15   14
  9     8    9
>>   A.*B
>>   ans =
  0    15    0
  24    6   48
  14    7    0
```

EIGENVECTORS AND EIGENVALUES OF A MATRIX

```
>> A = [2 3 5; 6 3 8; 7 1 9];
>> eig (A)
>> ans =
            14.5411
            -1.7555
            1.2144
>>
>>   [Eigenvector, Eigenvalues] = eig (A)
```

```
Eigenvectors =
     -0.4099    -0.8336    -0.4618
     -0.6539    0.1649     -0.7267
     -0.6359    0.5272     0.5086
Eigenvalues =
     14.5411    0           0
     0          -1.7555     0
     0          0           1.2144
```

COMPLEX NUMBERS

Defining a complex number:

`>> x = 3 + 4i`
`>> ans =`
`3.0000 + 4.0000i`
`>> `***abs*** `(x)`
`>> ans =`
`5`
	`>> y = x' % Conjugate`
	`>> y =`
	
	`3.0000 - 4.0000i`
	`>> `***angle*** `(x)`
	`>> ans =`
	`0.9273`

Complex arithmetic:

`>> y = 2 - 5i;`
`>> x*y`
`>> ans =`
`26.0000 - 7.0000i`
	`>> x/y`
	`>> ans =`
	`-0.4828 + 0.7931i`
	`>> x+y`
	`>> ans =`
	`5.0000 - 1.0000i`

DATA LOADING AND SAVING

`>> A = `***load*** `('myresult.`
`dat')`
`A =`
`0 0.0000`
`1 0.2500`
`2 0.7500`
`3 1.0000`
`4 1.2500`
`5 1.5000`
	`>> B(:,2) = A(:,1) ;`
	`>> B(:,1) = 2*A(:,2)`
	`B =`
	`0.0000 0`
	`0.2500 2`
	`0.7500 4`
	`1.0000 6`
	`1.2500 8`
	`1.5000 10`

Number of digits:

`>> format short`
`>> x = 1/3`
`x =`
`0.3333`
	`>> format long`
	`>> x = 1/3`
	`x =`
	`0.33333333333333`

Defining a polynomial: $P(x) = x^3 + 2x^2 + 3x + 4$:

`>> coef = [1 2 3 4]`
`coef =`
`1 2 3 4`
	`>> roots (coef)`
	`ans =`
	`-1.6506`
	`-0.1747 + 1.5469i`
	`-0.1747 - 1.5469i`

Symbolic integration:	Numeric integration:

`>>syms x`
`>>y = int(1/(x-3))`
`y =`
`log(x-3)`
	`>>x = 0:0.1:1;`
	`>>y = x.*x;`
	`>>Result = trapz (x,y)`
	`Result =`
	`0.3350`

Approximating a number:

```
>>    round (6.7)
          ans =
                7
>>    floor (6.7)
          ans =
                6
>>    ceil (6.2)
          ans =
                7
```

STATISTICS

Mean value, variance, and standard deviation:

```
>>a = [1 3 -4 5 6 7 2 1 0 -14 2 9 11 7 8];
>>mean(a)
      ans =
            2.9333
>>var(a)
```

```
ans =
        37.6381
>>std(a)
        ans =
               6.1350
```

PRESENTING OUTPUT

```
>>fprintf ('This will appear on the screen');
        This will appear on the screen.
>>fprintf ('%4.2f', pi);
        3.14
>>fprintf ('%4.5f ', pi);
        3.14159
>>fprintf ('%4.2f\n%6.3f%',pi,1000.456798);
        3.14   12
        100.457
```

GRAPHICS UTILITIES

```
>>   ezplot ('cos(x)');
>>   grid on;
```

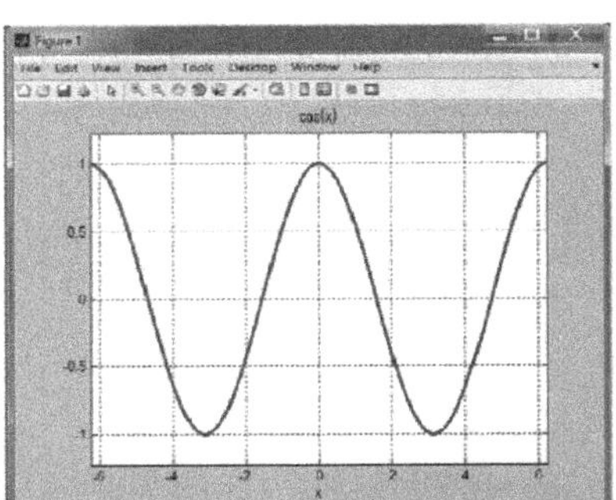


```
>>   ezplot ('(x-2)*(x +
   2)*x');
>>   grid on;
```

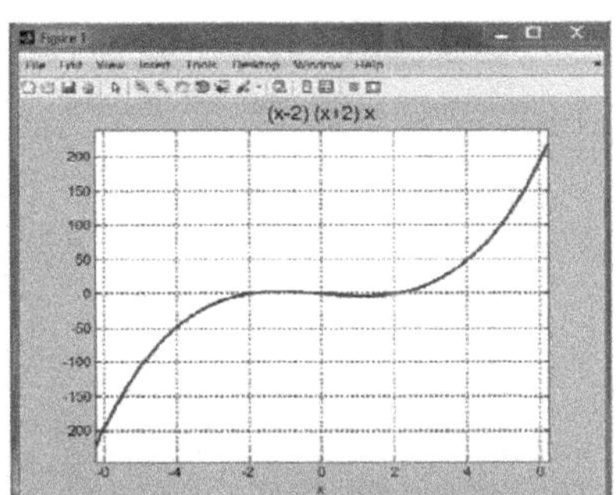

```
>>x = 0.0:0.1:2*pi;y =
   sin(x);
>>area(x,y,-1.5,'Facecolor',
    0.1 1.0 0.0]);
>>ylim([-1.5 2]);
```

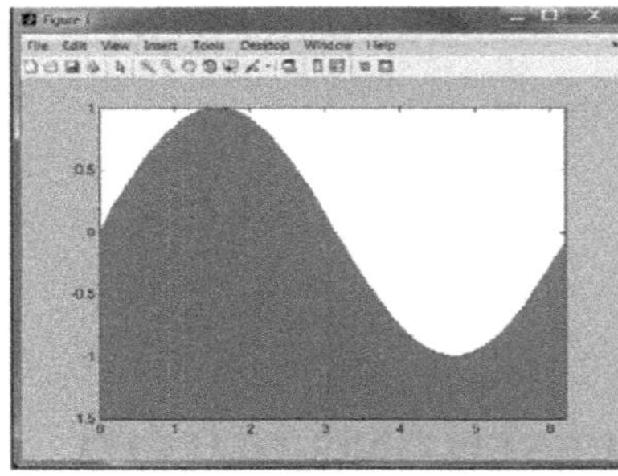


```
>>x1 = -3.5:0.01:3.5;
>>y1 =
   1/sqrt(2*pi)*exp(-x1.*x1/2);
>>plot (x1,y1); hold on;
>>x2 = -1.96:0.01:1.96;
>>y2 =
   1/sqrt(2*pi)*exp(-x2.*x2/2);
>>area (x2,y2,'Facecolor', [0.1
   1.0 0.0])
```

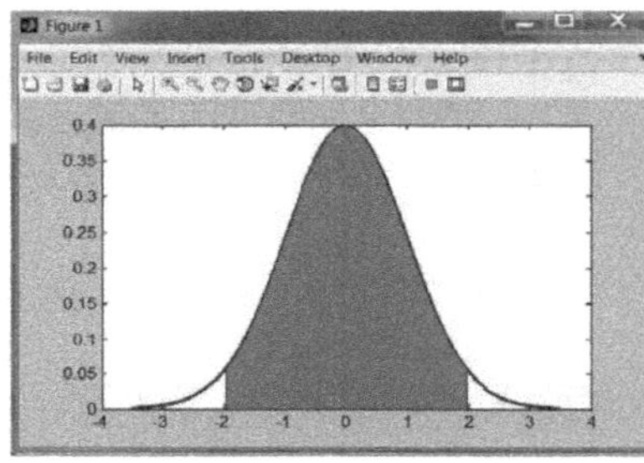


```
>>[X,Y] = meshgrid(-8:.5:8);
>>R = sqrt(X.^2+Y.^2) + eps;
>>Z = sin(R)./R;
>>Z = surf(X,Y,Z);
>>colormap jet;
>>colorbar;
```

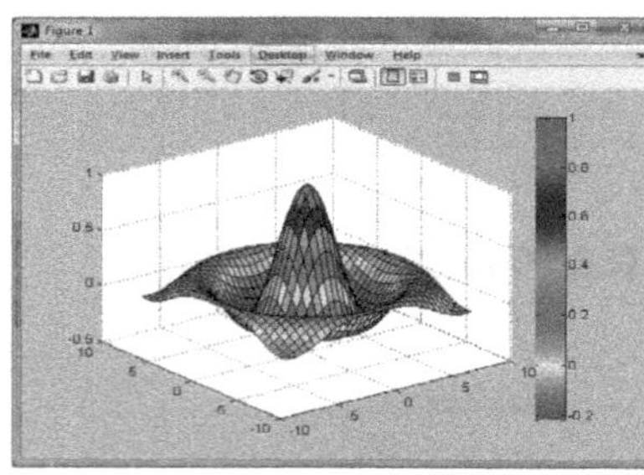

Sample CODE 1:

Sum of the first n natural numbers:

```
clear all, clc
n=input('Give n: ');
% a)
Total1=n*(n+1)/2
% b)
Total2=0;
for i=1:n
  Total2=Total2+i;
end
Total2
% c)
k=1:n;
Total3=sum(k)
```

Sample CODE 2:

Calling a function

```
function cagatay
clear all;
clc;
a=5.345;
b=3.123;
c=7.547;
average(a,b,c)

function res=average(a,b,c)
res=(a+b+c)/3;
```

SAMPLE CODE 3

Load a data file and plot a curve:

```
>> load out.dat;
>> plot(out(:,1),out(:,2))
>> xlabel ('Time [ns]');ylabel
('Voltage [V]'); title ('Voltage vs Time')
>> axis([0 50 0 1])
```

SAMPLE CODE 4

Plot f(x) = sin(pi*x) from x = 0 to x = 5 at 101 points:

```
>> dx = 5/100;
>> x = 0:dx:5;
>> fx = sin(pi*x);
>> plot(x,fx)
```

SAMPLE CODE 5

Accept a function from the keyboard:

```
>> s = input ('Enter f(x) function :');
>>Enter f(x) function :'x*x-5'
>> f = inline(s,'x');
>> f
>> f =
>> Inline function:
     f(x) = x*x-5
```

SAMPLE CODE 6

To calculate e^x function using the first n terms in its Taylor expansion (a) via the calculation of the error when the user supplies the number of terms, (b) via the calculation of the number of terms when the error is given, and (c) via vector summation.

```
clear all;
clc;
x=input('Give x');
n=input('Number of terms = ?');

exp1=exp(x);
exp2=1;
for k=1:n
    exp2=exp2+x^k./factorial(k);
end
error=exp1-exp2
```

```
clear all
clc
x=input('Give x:    ');
n=input('Number of terms =   ');

exp1=exp(x);
k=0:n-1;
terms=x.^k./factorial(k);
exp2=sum(terms);
error=exp1-exp2
```

```
clear all;
clc;
x=input('Give x');
eps=input('Give the error');

exp1=exp(x);
exp2=1;

error=1+eps;
n=1;
while error >= eps
     exp2=exp2+x^n/factorial(n);
     error=exp1-exp2;
     n=n+1;
end
n
```

SAMPLE CODE 7

Generate 20 random numbers with uniform distribution:

```
>> yy=rand(1,20);
```

Generate 35 random numbers with normal distribution:

```
>> yy=randn(1,35);
```

Generate 100 random bits:

```
>> txdata=round(rand(1,100));
```

SAMPLE CODE 8

Finding roots of the function f(x)=4x^2 + 5x−4:

```
>> f=[4 5 -4];
>> r=roots(f)
            ans = -1.8042  0.5542
```

SAMPLE CODE 9

FFT of cosine function:

```
>> n=0:50;
>> x=cos(2*pi*n/20);
>> N=2048;
>> X1=abs(fft(x,N));
>> F1=(0:N-1)/N;
>> plot(F1,X1,'r');
```

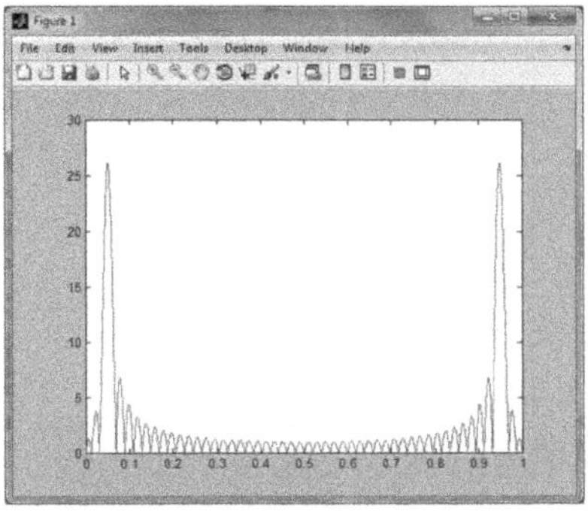

APPENDIX B

Suggested References

EM BOOKS

C. A. Balanis, *Antenna Theory*, John Wiley & Sons, New York, 1982, 2nd ed., 1997.
C. A. Balanis, *Advanced Engineering Electromagnetics*, 2nd ed., John Wiley & Sons, Inc., Hoboken, NJ, 2012.
R. E. Colin, *Antennas and Radiowave Propagation*, McGraw Hill Book Co., New York, 1985.
D. G. Dudley, *Mathematical Foundations for Electromagnetic Theory*, IEEE Press (Series on Electromagnetic Waves), New York, 1994.
L. B. Felsen, N. Marcuvitz, *Radiation and Scattering of Waves*, Prentice-Hall, Englewood Cliffs, NJ, 1973.
V. A. Fock, *Electromagnetic Diffraction and Propagation Problems*, Ch. 10, p. 191 and Ch. 12, p. 235, Pergamon, Oxford, 1965.
R. C. Hansen, ed., *Geometric Theory of Diffraction*, IEEE Press, New York, 1981.
R. F. Harrington, *Field Computation by Moment Method*, 1st ed. 1968, IEEE Press, New York, 1993.
E. Jordan, K. Balmain, *Electromagnetic Waves and Radiating Systems*, Prentice Hall, Englewood Cliffs, NJ, 1950, 2nd ed., 1968.
D. E. Kerr, ed., *The Propagation of Short Radio Waves*, vol. 1.S, *Radiation Lab. Series*, McGraw-Hill, New York, 1951.
S. Kingsley, S. Quegan, *Understanding Radar Systems*, McGraw Hill Co., London, 1992.
J. D. Kraus, *Antennas*, 2nd ed., McGraw Hill Book Co., New York, 1988.
C. R. Paul, *Introduction to Electromagnetic Compatibility*, Wiley Interscience, New York, 1992.
W. L. Shutzman, G. A. Thiele, *Antenna Theory and Design*, 2nd ed., John Wiley & Sons, New York, 1998.
J. A. Stratton, *Electromagnetic Theory* (Series on EM Theory), Characteristics (ROC), IEEE Press, Piscataway, NJ, 2007.

Electromagnetic Modeling and Simulation, First Edition. Levent Sevgi.

P. Ya. Ufimtsev, *Fundamentals of the Physical Theory of Diffraction*, John Wiley & Sons, Inc., Hoboken, NJ, 2007.

CEM BOOKS

C. S. Cesari, J. F. Abel, *Introduction to the Finite Element Method: A Numerical Approach for Engineering Analysis*, Van Nostrand Reinhold, New York, 1972.
C. Christopoulos, *The Transmission-Line Modeling Method*, Oxford University Press, Oxford, 1995.
K. S. Kunz, R. J. Luebbers, *The Finite-Difference Time-Domain Method for Electromagnetics*, CRC Press, Boca Raton, FL, 1993.
M. Levy, *Parabolic Equation Methods for Electromagnetic Wave Propagation*, IEE Institution of Electrical Engineers, London, 2000.
M. N. O. Sadiku, *Numerical Techniques in Electromagnetics*, CRC Press, Boca Raton, FL, 1992.
L. Sevgi, *Complex Electromagnetic Problems and Numerical Simulation Approaches*, IEEE Press/Wiley, Piscataway, NJ, 2003.
J. F. Shaeffer, M. T. Tuley, E. F. Knot, *Radar Cross Section*, Artech House, Norwood, MA, 1985.
D. M. Sullivan, *Electromagnetic Simulation Using the Finite-Difference Time-Domain Method*, IEEE Press, New York, 2000.
A. Taflove, *Computational Electrodynamics: The Finite-Difference Time-Domain Method*, Artech House, Boston, 1st ed., 1995, 2nd ed., 2000.
A. Taflove, *Advances in Computational Electrodynamics: The Finite-Difference Time Domain Method*, Artech House, Boston, 1998.
A. Taflove, S. C. Hagless, *Computational Electrodynamics: The Finite-Difference Time Domain Method*, 2nd ed., Artech House, Boston, 2000.

APPENDIX C

Suggested Tutorials and Feature Articles

IEEE AP Magazine

Testing Ourselves Column Tutorials

G. Apaydin, L. Sevgi, "A Canonical Test Problem for Computational Electromagnetics (CEM): Propagation in a Parallel Plate Waveguide," *IEEE Antennas and Propagation Magazine*, **54** (4), 2012, pp. 290–315.

G. Apaydin, L. Sevgi, "Groundwave Propagation at Short Ranges and Accurate Source Modeling," *IEEE Antennas and Propagation Magazine*, **55** (3), 2013, pp. 244–262.

E. Arvas, L. Sevgi, "A Tutorial on the Method of Moments," *IEEE Antennas and Propagation Magazine*, **54** (3), 2012, pp. 260–275.

C. Goknar, "Models and Modeling: Be Careful and Use Your Imagination," *IEEE Antennas and Propagation Magazine*, **50** (5), 2008, pp. 215–221.

C. Goknar, "Reciprocity and Anti-Reciprocity Revisited," *IEEE Antennas and Propagation Magazine*, **52** (2), 2010, pp. 211–220.

F. Hacıvelioğlu, L. Sevgi, P. Ya. Ufimtsev, "Electromagnetic Wave Scattering from a Wedge with Perfectly Reflecting Boundaries: Analysis of Asymptotic Techniques," *IEEE Antennas and Propagation Magazine*, **53** (3), 2011, pp. 232–253.

L. Sevgi, "Innumeracy: The Meaning of the Numbers We Use," *IEEE Antennas and Propagation Magazine*, **49** (2), 2007, pp. 185–190.

L. Sevgi, "Numerical Fourier Transforms DFT and FFT," *IEEE Antennas and Propagation Magazine*, **49** (3), 2007, pp. 238–243.

L. Sevgi, "Digital Multimeters and Basic Measurements," *IEEE Antennas and Propagation Magazine*, **49** (4), 2007, pp. 232–237.
L. Sevgi, "Synthetic Radar Signal Environment: Computer Generation of Signal, Noise, and Clutter," *IEEE Antennas and Propagation Magazine*, **49** (5), 2007, pp. 192–198.
L. Sevgi, "The Antenna as a Transducer: Simple Circuit and Electromagnetic Models," *IEEE Antennas and Propagation Magazine*, **49** (6), 2007, pp. 211–218.
L. Sevgi, "Review of Discrete Solutions of Poisson, Laplace, and Wave Equations," *IEEE Antennas and Propagation Magazine*, **50** (1), 2008, pp. 246–254.
L. Sevgi, "Guided Waves and Eigenvalue Extraction from Propagation Characteristics," *IEEE Antennas and Propagation Magazine*, **50** (2), 2008, pp. 222–234.
L. Sevgi, "Guided Waves and Transverse Fields: Transverse to What?" *IEEE Antennas and Propagation Magazine*, **50** (6), 2008, pp. 221–225.
L. Sevgi, "Receiver Operating Characteristics (ROC), Hypothesis Operation Characteristics (HOC)," *IEEE Antennas and Propagation Magazine*, **51** (3), 2009, pp. 218–224.
L. Sevgi, "Biostatistics and Epidemiology: Hypothetical Tests on Cell Phone Users," *IEEE Antennas and Propagation Magazine*, **52** (1), 2010, pp. 267–273.
L. Sevgi, "Reciprocity: Some Remarks from Field Point of View," *IEEE Antennas and Propagation Magazine*, **52** (2), 2010, pp. 205–210.
L. Sevgi, "Electromagnetic Modeling and Simulation: Teaching as a Graduate Level Course," *IEEE Antennas and Propagation Magazine*, **54** (5), 2012, pp. 261–269.

Electromagnetic Virtual Tool Tutorials

F. Akleman, L. Sevgi, "A Novel MoM- and SSPE-Based Groundwave Propagation Field Strength Prediction Simulator," *IEEE Antennas and Propagation Magazine*, **49** (5), 2007, pp. 69–82.
C. Balanis, L. Sevgi, P. Ya. Ufimtsev, "Fifty Years of High Frequency Asymptotics," *RFMiCAE, International Journal of RF and Microwave Computer-Aided Engineering*, **23** (4), 2013, pp. 394–402.
G. Çakır, M. Çakır, L. Sevgi, "A Multipurpose FDTD-Based Two Dimensional Electromagnetic Virtual Tool," *IEEE Antennas and Propagation Magazine*, **48** (4), 2006, pp. 142–151.
G. Çakır, M. Çakır, L. Sevgi, "A Novel Virtual FDTD-Based Microstrip Circuit Design and Analysis Tool," *IEEE Antennas and Propagation Magazine*, **48** (6), 2006, pp. 161–173.
G. Çakır, M. Çakır, L. Sevgi, "Radar Cross Section (RCS) Modeling and Simulation: Part II—A Novel FDTD-Based RCS Prediction Virtual Tool," *IEEE Antennas and Propagation Magazine*, **50** (2), 2008, pp. 81–94.
M. Çakır, G. Çakır, L. Sevgi, "A Two-Dimensional FDTD-Based Virtual Metamaterial—Wave Interaction Visualization Tool," *IEEE Antennas and Propagation Magazine*, **50** (3), 2008, pp. 166–175.

L. B. Felsen, F. Akleman, L. Sevgi, "Wave Propagation inside a Two-Dimensional Perfectly Conducting Parallel Plate Waveguide: Hybrid Ray-Mode Techniques and Their Visualisations," *IEEE Antennas and Propagation Magazine*, **46** (6), 2004, pp. 69–89.

F. Hacıvelioğlu, M. A. Uslu, L. Sevgi, "A Matlab-Based Simulator for the Electromagnetic Wave Scattering from a Wedge with Perfectly Reflecting Boundaries," *IEEE Antennas and Propagation Magazine*, **53** (6), 2011, pp. 234–243.

O. Ozgun, G. Apaydin, M. Kuzuoglu, L. Sevgi, "PETOOL: MATLAB-Based One-Way and Two-Way Split Step Parabolic Equation Tool for Radiowave Propagation over Variable Terrain," *Computer Physics Communications (CPC)*, Elsevier, **182**, 2011, pp. 2638–2654.

L. Sevgi, "A Ray Shooting Visualisation Matlab Package for 2D Groundwave Propagation Simulations," *IEEE Antennas and Propagation Magazine*, **46** (4), 2004, pp. 140–145.

L. Sevgi, "A Mixed-Path Groundwave Field Strength Prediction Virtual Tool for Digital Radio Broadcast Systems in Medium and Short Wave Bands," *IEEE Antennas and Propagation Magazine*, **48** (4), 2006, pp. 19–27.

L. Sevgi, Ç. Uluışık, "A Matlab-Based Visualization Package for Planar Arrays of Isotropic Radiators," *IEEE Antennas and Propagation Magazine*, **47** (1), 2005, pp. 156–163.

L. Sevgi, Ç. Uluışık, "A Matlab-Based Transmission Line Virtual Tool: Finite-Difference Time-Domain Reflectometer," *IEEE Antennas and Propagation Magazine*, **48** (1), 2006, pp. 141–145.

L. Sevgi, Ç. Uluışık, F. Akleman, "A Matlab-Based Two-Dimensional Parabolic Equation Radiowave Propagation Package," *IEEE Antennas and Propagation Magazine*, **47** (4), 2005, pp. 184–195.

L. Sevgi, F. Akleman, L. B. Felsen, "Visualizations of Wave Dynamics in a Wedge-Waveguide with Non-Penetrable Boundaries: Normal, Adiabatic, and Intrinsic Mode Representations," *IEEE Antennas and Propagation Magazine*, **49** (3), 2007, pp. 76–94.

G. Toroglu, A. Uslu, L. Sevgi, "RCS2D: A 2D MoM and FDTD Based Simulator," *RFMiCAE, International Journal of RF and Microwave Computer-Aided Engineering*, **23** (4), 2013, pp. 437–443.

Ç. Uluışık, L. Sevgi, "Bessel Functions and Numerical Evaluations of Bessel Integrals," *IEEE Antennas and Propagation Magazine*, **51** (6), 2009, pp. 222–233.

Ç. Uluışık, L. Sevgi, "Web-Based Virtual Engineering Tool Design," *IEEE Antennas and Propagation Magazine*, **53** (4), 2011, pp. 252–260.

Ç. Uluışık, L. Sevgi, "A MATLAB-Based Visualization Package for Complex Functions, and Their Mappings and Integrals," *IEEE Antennas and Propagation Magazine*, **54** (1), 2012, pp. 243–253.

Ç. Uluışık, G. Çakır, M. Çakır, L. Sevgi, "Radar Cross Section (RCS) Modeling and Simulation: Part I—Definitions, Strategies, and Canonical Examples," *IEEE Antennas and Propagation Magazine*, **50** (1), 2008, pp. 115–126.

A. Uslu, L. Sevgi, "Matlab-Based Filter Design Program: From Lumped Elements to Microstriplines," *IEEE Antennas and Propagation Magazine*, **53** (1), 2011, pp. 213–224.

M. A. Uslu, L. Sevgi, "Matlab-Based Virtual Wedge Scattering Tool for the Comparison of High Frequency Asymptotics and FDTD Method," *ACES, International Journal on Applied Computational Electromagnetics*, **27** (9), 2012, pp. 697–705.

Index

Electromagnetic Modeling and Simulation, First Edition. Levent Sevgi.

IEEE PRESS SERIES ON ELECTROMAGNETIC WAVE THEORY

Andreas C. Cangellaris, *Series Editor*
University of Illinois, Urbana-Champaign, Illinois

Field Theory of Guided Waves, Second Edition
Robert E. Collin

Field Computation by Moment Methods
Roger F. Harrington

Radiation and Scattering of Waves
Leopold B. Felsen, Nathan Marcuvitz

Methods in Electromagnetic Wave Propagation, Second Edition
D. S. J. Jones

Mathematical Foundations for Electromagnetic Theory
Donald G. Dudley

The Transmission-Line Modeling Method: TLM
Christos Christopoulos

The Plane Wave Spectrum Representation of Electromagnetic Fields, Revised Edition
P. C. Clemmow

General Vector and Dyadic Analysis: Applied Mathematics in Field Theory
Chen-To Tai

Computational Methods for Electromagnetics
Andrew F. Peterson

Plane-Wave Theory of Time-Domain Fields: Near-Field Scanning Applications
Thorkild B. Hansen

Foundations for Microwave Engineering, Second Edition
Robert Collin

Time-Harmonic Electromagnetic Fields
Robert F. Harrington

Antenna Theory & Design, Revised Edition
Robert S. Elliott

Differential Forms in Electromagnetics
Ismo V. Lindell

Conformal Array Antenna Theory and Design
Lars Josefsson, Patrik Persson

Multigrid Finite Element Methods for Electromagnetic Field Modeling
Yu Zhu, Andreas C. Cangellaris

Electromagnetic Theory
Julius Adams Stratton

Electromagnetic Fields, Second Edition
Jean G. Van Bladel

Electromagnetic Fields in Cavities: Deterministic and Statistical Theories
David A. Hill

Discontinuities in the Electromagnetic Field
M. Mithat Idemen

Understanding Geometric Algebra for Electromagnetic Theory
John W. Arthur

The Power and Beauty of Electromagnetic Theory
Frederic R. Morgenthaler

Modern Lens Antennas for Communications Engineering
John Thornton, Kao-Cheng

Electromagnetic Modeling and Simulation
Levent Sevgi